Heavy Metal Toxicity: Environmental Concerns, Remediation and Opportunities

Rajeev Pratap Singh • Pooja Singh •
Amrita Srivastava
Editors

Heavy Metal Toxicity: Environmental Concerns, Remediation and Opportunities

Springer

Editors
Rajeev Pratap Singh
Institute of Environment and Sustainable Development
Banaras Hindu University
Varanasi, Uttar Pradesh, India

Pooja Singh
Department of Science, Institute of Computer Science and Technology, SHEPA
Varanasi, Uttar Pradesh, India

Amrita Srivastava
Department of Life Science
Central University of South Bihar
Gaya, Bihar, India

ISBN 978-981-99-0396-2 ISBN 978-981-99-0397-9 (eBook)
https://doi.org/10.1007/978-981-99-0397-9

This Springer imprint is published by the registered company Springer Nature Singapore Pte Ltd.
The registered company address is: 152 Beach Road, #21-01/04 Gateway East, Singapore 189721, Singapore

Preface

The book *Heavy Metal Toxicity: Environmental Concerns, Remediation and Opportunities* is a comprehensive account of the sources, toxic biological as well as environmental impacts and possible remediation strategies for contamination by heavy metals that include cadmium (Cd), lead (Pb), arsenic (As), mercury (Hg), and chromium (Cr). Latter is a major threat in the current scenario that has inflicted critical damage to all life forms. Sources of these metallic elements include industrialization, urbanization, domestic effluents, agricultural route, and technological advancement. Being non-biodegradable in nature, they enter the food chain and get bioaccumulated causing abiotic stress. Thus, toxicity, bioaccumulation, and persistence of heavy metals universally affect hydro ecosystem, agriculture, air quality, and ultimately human health.

In biological systems, toxic metals affect integrity of cellular organelles, cell membrane, DNA damage repair system, cell cycle checkpoints, and metabolic enzymes. They act as carcinogens causing chromosomal aberrations or as systemic toxicants leading to cardiovascular, neurobehavioral, and immunological disorders. In plants, they interfere with photosynthesis, fertility, and metabolite and chlorophyll synthesis. Toxicity induced by heavy metals involves mechanistic approaches that need to be understood properly.

They cannot be degraded by biological or chemical means, thus can only be converted to less harmful forms. The conventional detection methods include biosensors, voltammetry, atomic absorption spectrometry, and inductively coupled plasma with atomic emission spectrometry. Strategies for metal detoxification include biosorption, bioaccumulation, biotransformation, biomineralization, phytoremediation, immobilization of metal-resistant microbes on carrier, use of chelators, metal precipitation or metal detoxification. These can be used as in situ or ex situ remedial process. Techniques for metal removal include precipitation, oxido-reduction, adsorption, ion exchange and electrochemical technique.

This book is an attempt to identify and evaluate all these aspects in detail. It shall incorporate the classical views along with modern scientific approaches to develop an understanding of the subject matter suitable for academicians, researchers, planners, policymakers, NGOs, environmental consultancies and raise awareness on this concern. Topics representing diverse sections namely environmental impacts, biological effects and methods used for detection and remediation have been included to address all possible contemporary issues on the topic in one concise volume.

Varanasi, Uttar Pradesh, India	Rajeev Pratap Singh
Varanasi, Uttar Pradesh, India	Pooja Singh
Bihar, India	Amrita Srivastava

Introduction

Chapter 1 deals with cadmium toxicity in plants and its remediation strategies. Being toxic and unwanted metal, it imposes several health-related issues in organisms around the globe when consumed through the food chain. Plants get exposed to Cd ions through various means and exert a negative impact in their growth and development. The oxidative stress generated through metal stress and its defense strategy is a prerequisite which is present in this chapter. There are several strategies evolved by plants for its removal or toxicity minimization such as phytoextraction, phytofiltration, and phytostimulation. Current information regarding its uptake and transport has been included which will be useful for the researchers.

Chapter 2 illustrates that heavy metals are considered as the metallic elements having density greater than water or atomic density above 5 gm cm^{-3}. These metals or metalloids are present in the environment either through natural or anthropogenic sources like industries and automobiles. The effects of some of the metals are lethal and are carcinogenic like arsenic, cadmium, lead, and mercury. Being non-biodegradable in nature they bioaccumulate in the ecosystem. The present book chapter highlights the nature, properties, and effect of heavy metals particularly As, Cd, Pb, and Hg. The adverse effects of these heavy metals in food, water, and soil give a clear picture of the current situation around the globe.

Chapter 3 emphasizes the impact of industrialization. Due to rapid increase in industrialization and urbanization, the extent of heavy metals has increased dynamically and imposes severe health threats to humans. The presence of heavy metals like arsenic, lead, cadmium, and mercury can significantly impact several primary metabolic systems including respiratory, neurological, hematological, reproductive, dermal, and skeletal functions. The major diseases associated are encephalopathy, peripheral vascular disease, bone marrow depression, hepatomegaly, diarrhea, nephropathy, long QT syndrome, cancer, diabetes, cardiovascular diseases, kidney failure, memory loss, DNA breakage and mutation in the worst situation. The mitigation strategies from water through oxidation, precipitation, electrokinetic, ion exchange, membrane filtration and adsorption process are dealt with in this chapter. Heavy metal removal strategies from soil and food crops like source reduction, bioremediation, phytoremediation, chemical or physicochemical remediation techniques, and nanoparticle methods are also illustrated in this book chapter.

Chapter 4 deals with atmospheric pollution which is a major cause of concern around the world, and presence of heavy metals led to greater toxicity as they possess abilities like non-biodegradability, bioaccumulation, environmental stability, persistence, and biotoxicity. The present chapter aims to deliver a summary of heavy metal pollution in the atmosphere and its concerns on air quality along with the implications for human health. The authors have explored environmentally relevant HMs and their sinks and remobilization processes between air, soil, and water compartments by analyzing their transports, exposure pathways together with fates and behaviors in air-water-soil-biota. Metals in particulate matter, characteristics, and its health implications are further elaborated. Long-Range Transboundary Air Pollution (LRTAP) is an important aspect which requires special attention and has also been included by the authors.

Chapter 5 explores the case study of Bhopal related to heavy metal contamination due to construction and demolition. The identification of contaminated zones, drainage area of Betwa basin and lower Chambal basin, soil classification, rainfall pattern, and ecological status of lakes has been carried out by researchers in the chapter. Manganese and boron overload was observed in the upper lake region of Bhopal which requires immediate attention.

Chapter 6 illustrates soil contamination occurring due to heavy metals requiring a solution to work upon. Thorough discussions on heavy metal contamination due to anthropogenic activities are covered in this book chapter. The bottom-up approach that causes food chain contamination is a cause of concern. The anthropogenic sources like mining, sewage irrigation, and application of pesticides, traffic emission, and waste dumping are well documented in this chapter. Current status of heavy metals contamination in different land-use patterns is summarized here. Approaches involved in major risk assessment analysis like carcinogenic risk are calculated through calculation of chronic daily intake which is highlighted in this book chapter along with ecological risk assessment.

Chapter 7 discusses over-pollution of groundwater due to heavy metal, a serious cause of concern for all life forms. Disease caused due to the presence of arsenic has made humans helpless. Thus there is a need to monitor the level of toxic heavy metals inside the ground. The traces of route of these heavy metals may give a clue about its remedial process. The book chapter is devoted to providing routes and factors affecting metal contamination. The environmental concern due to groundwater pollution and how it is changing the properties of groundwater and agro economic systems is also reported. The physico-chemical and biological treatment technologies for groundwater body are also elaborated in this chapter.

Chapter 8 precisely covers the adverse results of heavy metal pollution activated by the increased number of vehicles on the vegetation along the roadside. The heavy metals change soil pH thus affecting vegetable crops, its anatomical, physiological, and fertility characteristics. The impact of copper, zinc, lead, cadmium, and nickel in plants is demonstrated. The chapter highlights the lethal effects of heavy metal pollution from car exhaust on vegetation crops.

The transportation sector has deteriorated air quality by emitting organic and inorganic pollutants. Thus the need to create an effective policy aiming at sustainable development like biofuels etc. along with a control plan for metropolitan cities by introducing advanced technologies is a prerequisite. The aim of chap. 9 is to present elaborate information regarding emission levels of heavy metals from vehicles and types of fuel used in order to evaluate the health risk assessment of human beings. Factors affecting heavy metal pollution due to vehicle emission like tires type, climate and environmental factors, traffic density factor, etc. are well documented in this chapter.

Chapter 10 deals with a comprehensive analysis of heavy metal toxicity and environmental contamination. The chapter outlines the properties and effects of heavy metals. Important chemical properties and different forms of heavy metals are summarized. Environmental contamination through heavy metals occurrence through metal corrosion, atmospheric deposition, soil erosion of metal ions, and leaching of heavy metals has been summarized.

Chapter 11 explains industrialization, agriculture, and other events that released huge amounts of heavy metals into the environment with deleterious effects on agricultural fields, water, and air. Heavy metals stress conditions and plant growth-promoting bacteria (PGPB) developed several strategies to fight toxicity created by heavy metals. This chapter also assembles various reports on microbe-mediated bioremediation and recent development in this field.

Chapter 12 includes detailed studies about the phytoremediation of heavy metals or green technology used to remove pollutants from environmental components and their reaction mechanism. It illustrates pollution pathways of heavy metals. It explains in detail mechanisms used for the phytoremediation of heavy metals in the environment. The chapter summarizes phytoremediation techniques like phytoaccumulation, phytofiltration, phytostabilization, phytovolatilization, phytodegradation, rhizodegradation, and phytoextraction.

Chapter 13 summarizes industrial wastewater treatment strategies and includes essential details regarding different strategies in industrial wastewater treatment. Explanation of wastewater treatment in an industrial facility has been provided and it covers a description of different strategies like denitrification, reuse or recovery approach, energy conversion approach, bioaugmentation application, waste reduction/zero waste approach, and integrated approach.

Chapter 14 is dedicated to potential phytoremediation by the important oilseed crop *Brassica juncea* L. of various heavy metals. Also, concise importance of toxic effects of heavy metals on human health along with details for bioremediation and phytoremediation has been illustrated. The chapter includes details of morphology and growth of *Brassica juncea* L., phytovolatilization, phytoextraction, phytostabilization, rhizofiltration, rhizoremediation, the role of *Brassica juncea* L. in phytoremediation, studies on *Brassica juncea* L. as a phytoremediator, and enhancement of phytoremediation process.

Chapter 15 covers phytoremediation prospects for heavy metal removal by means of traditional strategies for removing heavy metals and their advantages and challenges. It also covers brief knowledge regarding future prospects of medicinal and aromatic plants for phytoremediation and biotechnology-based strategies to enhance the phytoremediation potential of plants.

Chapter 16 illustrates bioremediation of mining sites and a sustainable approach to restoring the healthy ecosystem. It mainly focuses on methods of bioremediation and its factors, especially in industrial and mining activities. It also gives brief knowledge on microbial bioremediation techniques like biosparging, bioventing, and bioaugmentation.

Chapter 17 explains industrial pollution management approach and gives concise knowledge about different types of industrial effluents, waste characterization, risk assessment involved in waste management, methods of industrial wastewater treatment, and its treatment levels. It summarizes methods of industrial effluent treatment. The chapter reports a case study about the status of Gorakhpur, Uttar Pradesh, in effluent treatment. It also includes a brief account of control of industrial pollution.

Industrialization has resulted in extensive mining and processing leading to deterioration and depletion of natural resources. Thus there is a need to look for alternatives that should aim at sustainable technology. Bio-electrochemical systems have the potential of metal recovery, wastewater treatment, and bioelectricity generation. Chapter 18 deals with types and configuration that include bio-electrochemical systems, microbial fuel cell, microbial electrolysis cell, and microbial desalination cell. The approach and mechanism involved to recover precious metal as well as factors affecting this process is extensively studied. How transfer of electrons through microbial extracellular electron transfer and electroactive bacteria affect BES and bioelectricity generation is also documented in this chapter.

Siderophores are metal chelators with various roles and bioremediation through chelation is one of the major aspects. Cadmium ion is known to adversely affect wheat at a very large scale. Chapter 19 encompasses the role of hydroxamate, a kind of siderophore isolated from *Aspergillus nidulans* in minimizing cadmium toxicity. The positive impact of these non-ribosomal peptides is evident as the seedlings managed to recover growth and development which were assessed through morphological, physiological, and biochemical assays.

Heavy metal presence in wastewater is obnoxious and leads to several diseases at different trophic levels. Chapter 20 focusses on various methods of removal of heavy metals from water bodies so as to make them suitable for other downstream processes.

Varanasi, Uttar Pradesh, India — Rajeev Pratap Singh
Varanasi, Uttar Pradesh, India — Pooja Singh
Bihar, India — Amrita Srivastava

Contents

About the Editors

Rajeev Pratap Singh works in the area of waste management and has worked on various kinds of wastes, i.e., fly ash, sewage sludge, tannery sludge, palm oil mill waste, contaminated water irrigation, etc. He has many publications in reputed journals on waste management and similar topics. He has been honored with several international awards including "Green Talent" Award from the Federal Ministry of Education and Research (BMBF), Germany; Prosper.Net Scopus Young Scientist Award; DST Young Scientist Award; and Water Advanced Research and Innovation (WARI) Fellowship, a fellowship supported by the Department of Science and Technology (DST), Government of India, the Indo-US Science and Technology Forum (IUSSTF), University of Nebraska-Lincoln (UNL), and the Robert Daugherty Water for Food Institute (DWFI). He has coauthored 4 books and more than 40 highly cited research and review articles on solid waste management.

Pooja Singh holds a PhD from the School of Industrial Technology, University Sains Malaysia, Penang, Malaysia; an MSc (Biotechnology) from Devi Ahilya University, Indore; and an MTech in Biotechnology from Rajiv Gandhi Proudyogiki Vishwavidyalaya, Bhopal, MP. She is currently a faculty member at ICST, SHEPA, Varanasi. Her current research interests include enzymes, biopulping of oil palm biomass, and vermicomposting using various types of wastes. She has published 18 papers in various reputed international journals, 5 book chapters, and 1 edited book.

Amrita Srivastava holds a PhD from the Department of Botany at Banaras Hindu University and works at the Central University of South Bihar, Gaya. She works in the area of microbial secondary metabolite applications. Her major interest is in the characterization of microbial secondary metabolites called siderophores. Recently, she has started working on heavy metal stress, specifically arsenic and cadmium stress. The focus of her work lies in the extraction and purification of microbial secondary metabolite siderophores and their application for heavy metal mitigation in crops, bioremediation, and various symbiotic/antibiotic interactions mediated by them. This includes antifungal and anticancer properties of siderophore and physiological effects on plants and other microbes. She has published 23 research and review articles in journals of international repute, edited 1 book, and authored over 13 book chapters.

Cadmium Toxicity in Plants: Uptake, Translocation and Phytoremediation Strategy

1

Anita, Suman Parihar, and Gyan Singh Shekhawat

Abstract

Cadmium (Cd) is a highly noxious, unnecessary and heavy metal element with no known biotic purpose. Cd harms not only plants but also humans by entering the food chain/web. Subsequently entering the root system, it can be transported from vascular organs into leaves and fruits, therefore becoming a major global ecological and health threat problem. Furthermost normal Cd toxicity symptom is inductive oxidative stress in plant cells. Nevertheless, plants have evolved several strategies to evade Cd toxicity, with the primary out-turn of reactive oxygen or nitrogen species for signalling purposes. Understanding the plant under Cd stress would enable high Cd uptake for the phytoremediation and accumulation potential of Cd in crop production. Hyperaccumulator plants' ability to phytoremediation Cd-polluted soil has shown to be an effective option for dealing with pollutants.

Keywords

Cadmium · Phytoremediation · Nitric oxide · Heme oxygenase · Carbon monoxide · Signalling

Abbreviations

As	Arsenic
Ca	Calcium
Cd	Cadmium

Anita · S. Parihar · G. S. Shekhawat (✉)
Plant Biotechnology and Molecular Biology Laboratory, Department of Botany (UGC-CAS), Jai Narain Vyas University, Jodhpur, Rajasthan, India

R. P. Singh et al. (eds.), *Heavy Metal Toxicity: Environmental Concerns, Remediation and Opportunities*, https://doi.org/10.1007/978-981-99-0397-9_1

Cu Copper
Fe Ferrous
Mn Manganese
Pb Lead
Zn Zinc

1.1 Introduction

Nowadays, soil contamination is a primary environmental concern due to the spread of industrial, urban emissions and anthropogenic activities. Numerous anthropogenic activities include emissions of metal mines, coal burning, widespread use of herbicides or fertilizers and direct deposits specially phosphate and numerous forms of heavy metal elements (Clemens and Ma 2016; Clemens et al. 2013). One of the most common types of soil contaminants is heavy metals, which include cadmium (Cd), copper (Cu), lead (Pb), arsenic (As) and zinc (Zn) (Mathur and Shekhawat 2013; Pantola and Shekhawat 2012; Shekhawat et al. 2009). Heavy metal elements have become a crucial issue for plant scientists because of their detrimental effects on soil plants as well as other living biota in the environment (Chen et al. 2006).

Cadmium (Cd) is an unnecessary metal element for human beings and plants but is extant in excessive amounts in the soil. However, Cd comes to the food chain, and its stances threaten living organisms. The control of Cd accumulation is difficult in plants because most essential nutrient transporters, for example, Cu, Zn, Mn, Fe and Ca, promote Cd accumulation. Cd impairs or toxicates plant development at morphological and physiological levels (Shanying et al. 2017). Its toxicity involves leaf chlorosis, growth rate delay, prevention of respiration processes and photosynthesis machinery, reduced yield, amplified oxidative damage and reduced mineral uptake capacity in the entire plant (Mohamed et al. 2012). In addition, Cd toxicity harms human physiology condition through food, water and air. For instance, Cd contact affects the lungs and reproductive system by impairing gametogenesis, semen quality and hormonal synthesis/release, causing cancer (Genchi et al. 2020; Kumar and Sharma 2019).

Well-organized and economical treatment of polluted agricultural land is important for sustainable agrarian progress. The agricultural land replacement method uses clean soil to wholly and partially replace polluted soil. Bringing in new soil reduces the polluted soil, which is only helpful for small-scale, harshly polluted soils. Due to mechanical restrictions, logistical complications, time, cost and some face limitations. Numerous biochemical and physical methods such as electrodynamics and phytoremediation remove heavy metal element contaminants from soil (Paz-Ferreiro et al. 2018; Yao et al. 2012). Physical remediation contains both high soil replacement and thermal absorption methods. Phytoremediation is a profitable and green technique for soil treatment. Particularly, phytoremediation plants only uptake and deposit cadmium through their roots, shoots, leaves and fruits (Paz-Ferrciro et al. 2018).

The purpose of this chapter is to describe cadmium as heavy metal element and the most recent data about the biochemistry, phytoremediation, transportation, toxicity in plants and plant defence mechanisms.

1.2 Biochemistry of Cd

Cadmium ($_{48}Cd^{112.41}$) is a transition metal element [Kr] $4d^{10}5s^2$, placed in the d-block, group 12 (IIB), with an atomic number of 48 and an atomic mass of 112.41 g of the periodic table. Although Cd usually has an oxidation state of +2, it is also extant in the +1 state. The average concentration of Cd in the Earth's crust is among 0.1 to 0.5 ppm. Cd has the third most hazardous environmental toxic metal element after mercury (Hg) and lead (Pb) according to the US Environmental Protection Agency (Jamla et al. 2021; Jaishankar et al. 2014). The majority of Cd is found in the ionic form (Cd^{2+}), which is most soluble in water and has a long half-life (~30 years). Because of these characteristics, Cd is easily adsorbed to water, soil and the atmosphere (Clemens and Ma 2016). Although cadmium is non-essential for plant development and metabolism or quickly bio-accumulated in upper levels of the food chain by plant uptake to the root system (Shahid et al. 2016), Cd has been analysed in almost food sources as nuts, tubers, cereals and vegetables. In plants, Cd uptakes harshly limit plant productivity and have been also considered a significant health threat for humans (Clemens and Ma 2016). Low concentrations of Cd are very hazardous to plants and can have various negative impacts on their mobility and solubility (Pinto et al. 2004). High concentrations of cadmium toxicity have related to the affinities of necessary elements such as calcium, ferrous, manganese, copper and zinc, which permit Cd to enter plant cell membranes and transform these metal elements into proteins (Verbruggen et al. 2009). A low acidity typically increases the uptake of cadmium by plants in the zinc or phosphate. Supplements like silicon and organic substances can reduce Cd consumption by the plant (Romero-Puertas et al. 2012). As exposed in several reports, oxidative stress has a mechanism in plants that causes Cd toxicity (Cuypers et al. 2016). This results in various effects on reactive oxygen species (ROS), important for the excess of these molecules that damage nucleic acids, proteins and lipids (Sandalio et al. 2012; Foyer and Noctor 2005).

1.3 Phytoremediation of Cd

The phytoremediation process discusses the biotic cleaning of plants' biosphere, i.e. soil, water and air. Plants form symbiotic associations with microbes that benefit soil remediation from heavy metal elements and other organic contaminants. Phytoremediation is commonly considered an eco-friendly technique as a result of the excellent refining capacity of heavy metal elements with minimal effluent to the environment (Shah and Daverey 2020). Instead, phytoremediation is widely accepted among people because of its comfort of use, low cost and being eco-friendly. Nevertheless, inhibited growth process like decreased biomass and improved Cd sensitivity have been detected in plants involved in phytoremediation processes (Shah and Daverey 2020). Phytoremediation includes numerous processes, for instance, phytoextraction, phytovolatilization, phytoaccumulation, phytotransformation and phytostabilization. Phytoextraction and phytoaccumulation mechanisms act in tandem; for example, throughout phytoextraction, plants take up heavy metal elements, like Cd, Zn, Ni, Cr, As and other elements, from the soil. These metal elements are deposited in the root, shoots, leaves and fruits by phytoaccumulation processes. Several species of plants have been formerly described for their extremely growth potential, which are likely entrants for phytoremediation (Shah and Daverey 2020). In phytoremediation of Cd, plants are frequently absorbed and transfer Cd to aerial parts of the plant. Plants have evolved some various adaptations, including detoxification mechanisms, to sustain average growth level under high Cd-contaminated soils. Cd's concentration in plant parts demonstrates a subsequent movement: fruits < leaves < stem < root (Ahmadpour and Soleimani 2015). There are many techniques used to enhance the efficiency of Cd phytoremediation.

1.3.1 Phytoextraction

Its procedure involved organic or inorganic pollutants via stem and root systems. By this time, plants growing in the environment are selected for this procedure. Hyperaccumulator families, for example, Asteraceae, Scrophulariaceae, Euphorbiaceae, Lamiaceae and Brassicaceae, are important for this procedure. In addition, certain plant species of these families, for example, *Cassia alata* (Silva et al. 2018), *Celosia argentea*, *Solanum melonaena*, *Salix mucronata* (El-Mahrouk et al. 2019), *Vigna unguiculata*, *Nicotiana tabacum* (Liu et al. 2011), *Momordica charantia* (Ali et al. 2016), *Swietenia macrophylla* and *Kummerowia striata* (Liu et al. 2011), are potential plants for enhancing the process of Cd phytoextraction. Phytoextraction supports the reduction of metalloid toxicity through the development of substrate geochemistry for the upcoming establishment of native plants. It is a reasonable, eco-friendly and potentially cost-effective technology to recover soil (Ranieri et al. 2020). Despite the mostly accepted benefits of phytoextraction, there are difficulties, like the time mandatory to treat highly polluted soils, which can be decades and restrictions for mine waste submissions. Most hyperaccumulator plants have established an ability to store one metal which is likely to be sensitive to the occurrence of other metals (Ernst 2005).

1.3.2 Phytostabilization

There has been a progressive change from phytoextraction to phytostabilization. Phytostabilization is the capacity of plants to accumulate and stabilize heavy metal, which inhibits metal passage and changes it to fewer noxious elements. Most of the increasing plants in polluted soil act not only as hyperaccumulators but also as exfoliators. Exclusion converts metal elements and compounds into a somewhat harmful transportable substrate form depriving them removal from the soil. It absorbs and accumulates these substances inside the rhizosphere or else stores them in the roots. Lately, the favourable consequences of *Virola surinamensis* (white ucuuba) for cadmium phytostabilization have been recognized (Dalvi and Bhalerao 2013). Similarly, *Miscanthus giganteus* (Zgorelec et al. 2020), oats or white mustard (Boros-Lajszner et al. 2020) also has phytostabilization of Cd. Phytostabilization is an emerging eco-friendly phytotechnology that stabilizes toxins. Roots led to some degree of phytostabilization, reducing metal access to plants, thus reducing contact with other tropical levels in the biosphere. Furthermore, the significant difficulty is that the toxic residues in the soil and root system are usually in the rhizosphere (Ghosh and Singh 2005).

1.3.3 Phytofiltration

It is classified as rhizofiltration, including blastofiltration and caulofilteration. Rhizofiltration is a water treatment process where roots successfully absorb toxins. In rhizofiltration, the contaminant sticks to the roots or assimilates and can be carried to the plants (Galal et al. 2018). This process is mainly used to disinfect wastes and contaminated water. Furthermost radioactive metals or materials are separated through this process. Phytofiltration techniques increased cadmium uptake from aquatic system using yellow velvetleaf (*Limnocharis flava*) as a probe flowering plant (Abhilash et al. 2009). In an additional experiment, Islam et al. (2015) described the phytofiltration ability of *Micranthemum umbrosum* (shade mud flower) to eliminate cadmium from the aquatic system. It is an economical technique, and the plants play as a solar-powered system to remove toxins from the atmosphere. However, none of the toxic substances is extracted below the depth of the root system. It is time-taking and will not be sufficient to remove organic substrates and toxic metal element (Islam et al. 2015).

1.3.4 Phytostimulation

It is a method used to promote the phytoremediation process through exciting root-derived substances to increase microbial life. These transducers increase microbial life through meeting their mineral and nutrient demands. Its procedure is utilized in rhizoremediation procedures. It's an affordable technology for removing Cd and other organic substrates (Jia et al. 2016). Another approach is adding resistant microbial inoculants to the soil, leading to Cd metals' growth (Yanai et al. 2006). It is an additional operative procedure to convert toxic substrates to non-toxic substrates. Nevertheless, it is further a time-consuming process, and making use of unstable and decomposable substrates is not an uncomplicated procedure. The procedure quickly responds to the toxicity level in the soil, and in few cases, a partial failure of organic substrates is detected. Furthermore, this technique requires a well-controlled monitoring (Jia et al. 2016).

1.3.5 Consequence of Phytoremediation on Cd Exclusion from Soil

Phytoremediation of soils polluted with Cd has been a considerable international problem. Hyperaccumulator plants are important in phytoremediation, as it is commonly included in the uptake of Cd from the soil. Diverse hyperaccumulator plants differ in their ability to remove Cd from the soil. It has low affinity due to the mobile nature of Cd (Mahajan and Kaushal 2018). Although Cd is transported from a plant's root system to its aerial portions, plants can easily transfer Cd into the soil (Mahajan and Kaushal 2018). Some features of soil like pH, temperature and other heavy metal elements facilitate to the transfer of Cd by various plants, For example, *Conyza sumatrensis*, *Gynura pseudochina*, *Nicotiana tabacum* and *Chromolaena odorata* have established under in-ground conditions (Khaokaew and Landrot 2015).

1.4 Cadmium Uptake and Transporter in Plants

Recently, several families of Cd transporters have been exposed to understand the function better. The possible recognized transporters are bridged underneath the following captions with a diagrammatic representation (Fig. 1.1).

1.4.1 Cd Enter the Root Cell Membrane

Dihydrogen carbonate [H_2CO_3 ($H_2O + CO_2$)] severs into HCO_3^- and H^+ at the root cell membrane through respiration, and subsequently, absorbed H^+ quickly interchanges with Cd. Formerly, Cd is absorbed via the surface of root epidermis cells, in which exchange in the layers of root epidermis cells occurs by apoplastic and symplastic pathways (Yamaguchi et al. 2011). Root hairs provide a greater surface area for Cd to assimilate from the soil via diffusion (Seregin and Ivanov 1997). Roots of plant also exclude some organic substances, for example, chelates, to form complex Cd ion ligands, allowing particulars to be entered into the epidermis cell of root (Sidhu et al. 2019). In addition, Cd is also uptaken through cation transporters/channels including Zn/Fe-regulated transporters (Tan et al. 2020) and MTPI (Yuan et al. 2012). Additionally, specific protein transporters, for instance, NRAMP (natural resistance-associated macrophage protein) (Song et al. 2014),

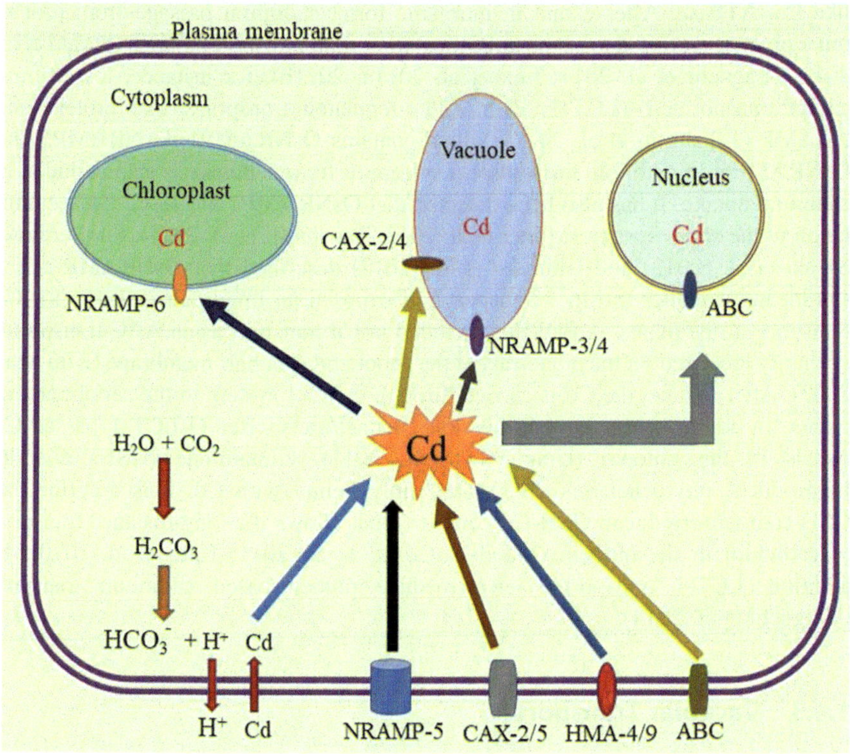

Fig. 1.1 Cd uptake through membrane transporters of plant cell and its accumulation in various cell organelles

AtHMA4 and AtHMA9 forms of P-type ATPase (Bækgaard et al. 2010), ABC transporter (Moons 2003) and AtCAX2 and AtCAX5 forms of the CAX family (Shigaki et al. 2003; Hirschi et al. 2000), play significant acts in Cd transport across the cell membrane of the root. Generally, after plant roots uptake the Cd, the extreme part of the Cd comes to root regions, and some amount comes to other higher regions of plants.

1.4.2 Cd Transporters for Plant Aerial Parts

Cd's accumulation is regulated via numerous methods, as well as vacuolar appropriation, xylem uploading, crossways of cell membrane of aerial parts, energy-dependent transport and cadmium apoplastic or symplastic routes into the aerial plant system (Loix et al. 2017). Future imperatives for the bioremediation of Cd are appropriately carried out in the aerial parts of plants. Upward transport significantly gives the shoot system a low Cd concentration (Verret et al. 2004). These procedures are intermediated via various families associated with metal and metalloid channels, like P_{1B}-ATPase. After extant in inorganic form, cadmium passage from root to other organs is referred to as the three most important transport systems (Sasaki et al. 2012; Uraguchi et al. 2011; Lux et al. 2011a, 2011b), for instance, low-affinity cation transporter-1 (LCT-1) ZIP, Zn/Fe-regulated transporter-like protein and NRAMP (Takahashi et al. 2011), which contains OsNRAMP1, OsNRAMP5 and OsNRAMP6. In addition, the transport system of ferrous uptake is also included in cadmium uptake. It has also been detected that OsNRAMP1 increased Cd accumulation in the shoot system (Milner et al. 2013; Takahashi et al. 2011). Furthermore, Sasaki et al. (2012) and Ishimaru et al. (2012) described that OsNRAMP acts a significant character in Mn^{2+} transport and exposed an important pathway for Cd transport in rice plants. AtPDR8 is included in Cd transport as an ABC transporter primarily localized on the epidermis of the shoot and root hair membrane (Kim et al. 2007). After uptake, the Cd is carried through the root system to the aerial parts of plants. In addition, the low-affinity ion channel/transporter OsLCT-1 passes Cd metals in the phloem (Uraguchi et al. 2011). Glutathione (GSH) and its by-products, phytochelatins (PCs), are firmly bound with Cd. It is described as GSH-cadmium (reduced GSH-Cd complex) that allows the long-distance transport of cadmium in the phloem (Mendoza-Cózatl et al. 2011; Kato et al. 2010). In addition, LCT-1 transporters also mediate phloem-based cadmium transport (Uraguchi et al. 2014).

1.4.3 Vacuolar Transporter

Numerous families of channels/transporters like ABCs, CAXs, NRAMP and HMAs have been considered in the vacuolar appropriation of cadmium (Zhang et al. 2016). Park et al. (2012) described that ABC1 and ABC2 are necessary vacuolar channels/transporters that consult to transports of Cd. Primarily, AtABC3 shows an act in phytochelatin-mediated Cd transporter (Brunetti et al. 2015). NRAMPs are important transporter, transporting numerous divalent metal elements, like Zn, Mn, Fe and Cd. NRAMP-3 and NRAMP-4 are transporters of vacuolar membrane, i.e. tonoplast, that play a significant role in remobilizing important metal elements from the vacuole to cytosol (Lanquar et al. 2005). GSH and PCs both are proteins that bind to Cd and allow the transport of Cd in the vacuole. CAXs are transporters of tonoplast that have precise transport of Ca^{2+}. Nevertheless, Korenkov et al. (2009) defined that the AtCAX-2 and AtCAX-4 channels/transporters transport Ca^{2+} ions also with Cd.

1.5 Cd Toxicity to Plants

Cd is a non-essential and toxic metal element for plant development that prevents plant growth and metabolism process. The initial symptom of Cd toxicity includes stress causing an imbalance of macro and micro nutrients, especially ferrous and calcium (Loix et al. 2017; Gupta et al. 2017). A redox status non-equivalence causes oxidative stress primarily as a conclusive result of an evident imbalance in antioxidant defence, respiratory process (Cuypers et al. 2016) and enzyme deactivation due to binding of Cd to Cys residues and direct movement of important metal elements through cadmium at particular structural or functional sites (Sharma and Dietz 2009; Valko et al. 2005). For example, whenever cadmium is used to replace calcium (Ca) in calmodulin, this prevents calmodulin-dependent phosphodiesterase activity in radish and PS II reaction centre (photosystem II), which in turn inhibits PS II photoactivation (Clemens 2006; Faller et al. 2005). Even though disparities in mineral uptake and distribution are defined in an extensive variety of species, for example, *Brassica*, *Vigna*, wheat, pea and sunflower, responses to cadmium toxicity diverge between genotypes and in the context under the cadmium stress (Khator et al. 2021; Khator and Shekhawat 2020; Mahawar et al. 2018; Rizwan et al. 2016; Shekhawat et al. 2010; Verma et al. 2008). Growth can also be repressed by reducing chlorophyll content, rate of transpiration and efficiency of water use commonly detected in plants under Cd stress (Qian et al. 2009; Faller et al. 2005). All signs of cadmium toxicity eventually decrease plant development or produce and lead to periodic genotoxicity, thus, after plant death. However, several impacts of toxicity depend on the plant's developmental stage, metal concentration, time duration of treatment and plant's coping mechanism. The best-graded plant response to cadmium entering the cell is the production of phytochelatins, a group of Cys-rich peptides formed enzymatically for Cd chelation and sequestration. Nevertheless, phytochelatin synthesis, sulphate acclimatization and Cys and GSH synthesis are essential to be regulated (Mendoza-Cózatl et al. 2010; Clemens 2006). Additionally, reliant on the species, phytochelatin synthase activity is controlled at both transcriptional and post-translational stages in response to the high affinity of Cd (Vatamaniuk et al. 1999). Cadmium-phytochelatin complexes are dragged into the vacuole via ABC transporters, Cd/proton antiporters, V-ATPase and V-Pase activity (Sharma et al. 2016; Berezin et al. 2008; Korenkov et al. 2007). Even the shape and volume of the vacuole play an essential role in the Cd detoxification process (Sharma et al. 2016). An additional feature defined for Cd is the upregulation of plant response stress proteins, like heat-shock and antioxidant proteins.

Cd is included in cellular redox reactions and directly produces ROS. However, oxidative stress is most important to Cd toxicity in plant cells (Chmielowska-Bąk et al. 2018). ROS formed through aerobic metabolism indicates species resulting from oxygen reduction composed of free radicals and non-radical substrates (Halliwell and Gutteridge 2015). The most significant free radicals are hydroxyl, alkoxyl, peroxyl and hydroperoxyl superoxide, and the major non-radical substrates are singlet oxygen, hydrogen peroxide (H_2O_2), ozone (O_3), hypochlorous acid ($HClO^-$) or peroxynitrite. Interactions between ROS-producing and ROS-scavenging processes observe ROS steady-state levels. The ROS-scavenging mechanism contains enzymatic antioxidant-free oxygen radicals like SOD that remove catalase, peroxidase and H_2O_2 (Halliwell and Gutteridge 2015; Gallego et al. 2012). ROS excess is mainly hazardous as a result of reactions with lipids, proteins and nucleic acids, which seriously damage the plant cell (Chmielowska-Bąk et al. 2018).

1.6 Heme Oxygenase as an Antioxidant Defence Against Oxidative Stress

Heme oxygenases (HOs) are universal, rate-limited and highly active family of antioxidant enzymes that catalyses the oxidative cleavage of heme to biliverdin, ferrous ion (Fe II) and carbon monoxide (CO) (Shekhawat et al. 2019; Shekhawat and Verma 2010). HO has been initially acknowledged in animals by Tenhunen et al. (1969) but has been now identified in all living biota. HO executes numerous functions in diverse living biota, but the process of heme deprivation is common to all living biota (Shekhawat et al. 2019). HO and its by-products are important constituents of the antioxidant defence system and protected against oxidative stress. Many reports have presented on the antioxidant properties of HO and its by-product in ROS scavenging in the plant system. The act of HO provided antioxidant defence counter to ROS caused by Cd stress in glycine (Yannarelli et al. 2006; Balestrasse et al. 2005). *Vigna* (Mahawar et al. 2018; Shekhawat et al. 2011) and *Medicago* (Cui et al. 2011) plants are well described. Here, an increase in HO-1 expression was found to involve in the defence of cells against ROS leading to oxidative stress (Mahawar and Shekhawat 2022; Mahawar et al. 2021; Dixit et al. 2014).

Carbon monoxide (CO) is a ubiquitous, poisonous, diatomic gaseous molecule identified as a silent killer. Nevertheless, CO has recently been recognized as one of the most important cellular mechanisms regulating various biological procedures in the living organisms like animals and plants (Xie et al. 2008). It has been well reported as gaseous signalling molecule in animals, but its experimental evidence has limited information on plants. Wilks (1959) first reported CO biosynthesis in plants and claimed HO as its main productive pathway. CO plays as a compound through hormonal effects, seed germination, root development and stomatal closure (Cui et al. 2015; Cao et al. 2007; Dekker and Hargrove 2002). CO not only plays as a signalling molecule throughout plant growth and development but also interacts with other signalling molecules in plant stress conditions. CO is also produced against oxidative damage under Cd stress in alfalfa (Han et al. 2008). Furthermore, Cd-induced oxidative damage has been reduced through CO pre-treatment by controlling glutathione metabolism in alfalfa and led to the conversion of oxidized glutathione disulphide (GSSG) to glutathione (GSH) to accumulate the GSH/GSSG ratio (Fig. 1.2; Han et al. 2008).

1.7 Crosstalk Between CO and Other Signalling Molecules

The carbon monoxide signal transduction pathway does not continuously operate autonomously but is nearly associated with nitric oxide. The two cellularly formed gas molecules share various regular downstream signalling routes and have few similar characteristics. For instance, in animals, carbon monoxide and nitric oxide are associated with ferrous ions of the heme protein of soluble guanylate cyclase (sCG) to stimulate the enzyme and enhance intracellular production of secondary messenger cGMP (cyclic guanosine monophosphate), therefore eliminating

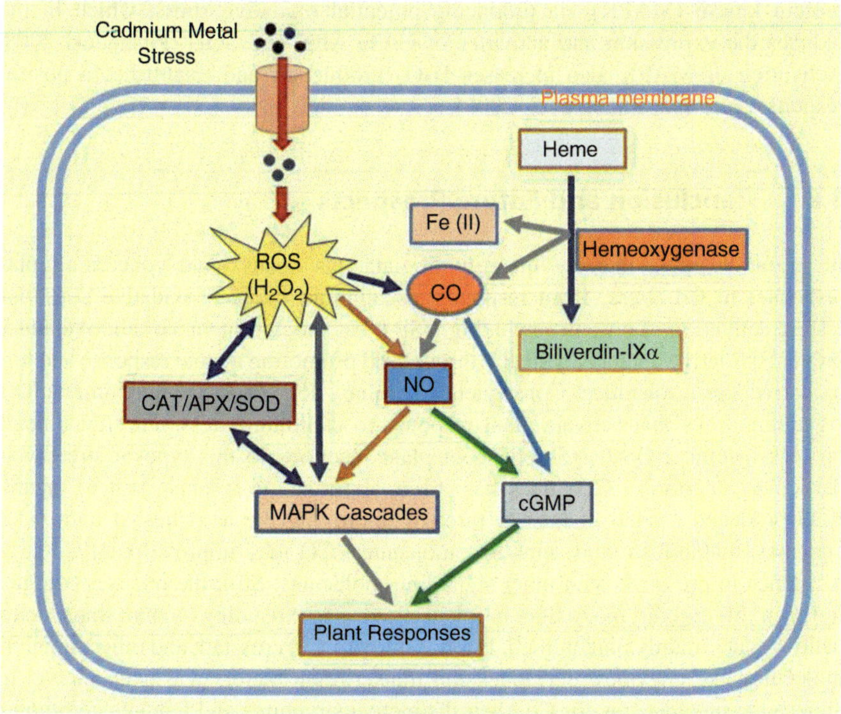

Fig. 1.2 Carbon monoxide-mediated signalling in plant responses to Cd stress

numerous biological functions like regulating vascular tone, preventing platelet accumulation and reducing blood pressure (Snyder et al. 1998). Nevertheless, this phenomenon has been well considered in animals, but in plants, CO research is now recognized. CO could somewhat dose-dependently mimic the effect of NO in stomatal closure and the K/Na ratio (Song et al. 2008; Xie et al. 2008). Furthermore, it has become increasingly more assertive that CO can potentiate nitric oxide synthase (NOS) activity in plant cells. Song et al. (2008) contained that *V. faba* guard cells might have present HO-1 enzymes and NOS-like enzyme activity. CO was included in the dark-induced NO synthesis by an enzyme such as NOS (Song et al. 2008). Nitric oxide donor SNP reduces cadmium-induced chlorophyll damage through expression of HO-1 at transcriptional levels (Noriega et al. 2007). Wheat (*Triticum aestivum*) seedling roots treated with exogenous carbon monoxide can synthesize nitric oxide, which indicates that nitric oxide may be a share of a signal molecule downstream of carbon monoxide action (Xie et al. 2008). The balance between NO and ROS is important for triggering the antioxidant response against oxidative stress (Santa-Cruz et al. 2010). CO secures endothelial cells from several stimuli-induced programmed cell death through inhibiting ROS formation, which can affect the activity of several transcription factors and kinases, like NF-kB and p38 (Brouard et al. 2002; Wang et al. 2007). CO stimulates p38 mitogen-activated protein kinase (MAPK) via producing potential oxidative stress, which in turn induces the expression and activities of CAT, APX and SOD (Piantadosi 2008). Activation of MAPK also increases H_2O_2 production and establishes a positive response (Fig. 1.2; Zhang et al. 2006).

1.8 Conclusion and Future Prospects

In recent years, a massive quantity of data has been found concerning plant responses to Cd stress. Plant responds to cadmium induced oxidative stress and restore activation of proteins including HSP (heat shock proteins) or antioxidants to prevent oxidative damage, typically through Cd promoting a rapid response and high oxidative stress tolerance. Consequently, primary ROS/RNS production results in signalling ways that activate plant response to cadmium stress. The fundamental process-specific ROS-/RNS-dependent plant reactions to this type of stress want particular attention. CO that has been exposed to plants protect against ROS-produced oxidative damage through improving the activities of antioxidant enzymes, metabolism and signalling molecules. CO may improve plant Cd stress resistance to crosstalk with other signalling molecules. Still, the precise biological function of carbon monoxide in plants and its expanding signal transduction pathways are mainly unidentified. Further study of CO crosstalk and other signalling molecules like enzymes, hormones and transcription factors will help unravel the underlying mechanisms for Cd, their distinctive structures and features common to different species.

Acknowledgement The authors wish to thank the University Grants Commission, New Delhi, for providing financial assistance through the Center for Advanced Study programme.

References

Abhilash PC, Pandey VC, Srivastava P, Rakesh PS, Chandran S, Singh N, Thomas AP (2009) Phytofiltration of cadmium from water by *Limnocharis flava* (L.) Buchenau grown in free-floating culture system. Int J Phytoremediation 170(2–3):791–797

Ahmadpour P, Soleimani M (2015) Cadmium accumulation and translocation in *Jatropha curcas* grown in contaminated soils. J Sci Technol Agric Natur Resour Water Soil Sci 19(73):179–190

Ali SY, Banerjee SN, Chaudhury S (2016) Phytoextraction of cadmium and lead by three vegetable-crop plants. Plant Sci Today 3(3):298–303

Bækgaard L, Mikkelsen MD, Sørensen DM, Hegelund JN, Persson DP, Mills RF, Yang Z, Husted S, Andersen JP, Palmgren MG (2010) A combined zinc/cadmium sensor and zinc/cadmium export regulator in a heavy metal pump. J Biol Chem 285(41):31243–31252

Balestrasse KB, Noriega GO, Batlle A, Tomaro ML (2005) Involvement of haemoxygenase as antioxidant defense in soybean nodules. Free Radic Res 39:145 –151

Berezin I, Mizrachy-Dagry T, Brook E, Mizrahi K, Elazar M, Zhuo S, Shaul O (2008) Overexpression of AtMHX in tobacco causes increased sensitivity to Mg^{2+}, Zn^{2+}, and Cd^{2+} ions, induction of V-ATPase expression, and a reduction in plant size. Plant Cell Rep 27(5): 939–949

Boros-Lajszner E, Wyszkowska J, Kucharski J (2020) Application of white mustard and oats in the phytostabilisation of soil contaminated with cadmium with the addition of cellulose and urea. J Soils Sediments 20(2):931–942

Brouard S, Berberat PO, Tobiasch E, Seldon MP, Bach FH, Soares MP (2002) Heme oxygenase-1-derived carbon monoxide requires the activation of transcription factor NF-κB to protect endothelial cells from tumor necrosis factor-α-mediated apoptosis. J Biol Chem 277(20): 17950–17961

Brunetti P, Zanella L, De Paolis A, Di Litta D, Cecchetti V, Falasca G, Barbieri M, Altamura MM, Costantino P, Cardarelli M (2015) Cadmium-inducible expression of the ABC-type transporter AtABCC3 increases phytochelatin-mediated cadmium tolerance in Arabidopsis. J Exp Bot 66(13):3815–3829

Cao Z, Huang B, Wang Q, Xuan W, Ling T, Zhang B, Shen W (2007) Involvement of carbon monoxide produced by heme oxygenase in ABA-induced stomatal closure in *Vicia faba* and its proposed signal transduction pathway. Chin Sci Bull 52(17):2365–2373

Chen A, Komives EA, Schroeder JI (2006) An improved grafting technique for mature Arabidopsis plants demonstrates long-distance shoot-to-root transport of phytochelatins in Arabidopsis. Plant physiol 141(1):108–120

Chmielowska-Bąk J, Izbiańska K, Ekner-Grzyb A, Bayar M, Deckert J (2018) Cadmium stress leads to rapid increase in RNA oxidative modifications in soybean seedlings. Front Plant Sci 8: 2219

Clemens S (2006) Toxic metal accumulation, responses to exposure and mechanisms of tolerance in plants. Biochimie 88(11):1707–1719

Clemens S, Ma JF (2016) Toxic heavy metal and metalloid accumulation in crop plants and foods. Annu Rev Plant Biol 67:489–512

Clemens S, Aarts MG, Thomine S, Verbruggen N (2013) Plant science: the key to preventing slow cadmium poisoning. Trends Plant Sci 18(2):92–99

Cui W, Fu G, Wu H, Shen W (2011) Cadmium-induced heme oxygenase-1 gene expression is associated with the depletion of glutathione in the roots of *Medicago sativa*. Biometals 24(1): 93–103

Cui W, Qi F, Zhang Y, Cao H, Zhang J, Wang R, Shen W (2015) Methane-rich water induces cucumber adventitious rooting through heme oxygenase1/carbon monoxide and Ca^{2+} pathways. Plant Cell Rep 34(3):435–445

Cuypers A, Hendrix S, Amaral dos Reis R, De Smet S, Deckers J, Gielen H, Keunen E (2016) Hydrogen peroxide, signaling in disguise during metal phytotoxicity. Front Plant Sci 7:470

Dalvi AA, Bhalerao SA (2013) Response of plants towards heavy metal toxicity: an overview of avoidance, tolerance and uptake mechanism. Ann Plant Sci 2(9):362–368

Dekker J, Hargrove M (2002) Weedy adaptation in *Setaria spp.V.* Effects of gaseous environment on giant foxtail (*Setaria faberii*) (Poaceae) seed germination. Am J Bot 89(3):410–416

Dixit S, Shekhawat GS, Alam A (2014) Hemeoxygenase-1(Bjho-1) functions in modulating antioxidant defence responses against cadmium induced oxidative stress: an in vitro and in vivo comparative analysis. Int Multidiscip Res J 2(8):27–44

El-Mahrouk ESM, Eisa EAH, Hegazi MA, Abdel-Gayed MES, Dewir YH, El-Mahrouk ME, Naidoo Y (2019) Phytoremediation of cadmium-, copper-, and lead-contaminated soil by *Salix mucronata* (Synonym *Salix safsaf*). HortScience 54(7):1249–1257

Ernst WH (2005) Phytoextraction of mine wastes–options and impossibilities. Geochemistry 65: 29–42

Faller P, Kienzler K, Krieger-Liszkay A (2005) Mechanism of Cd^{2+} toxicity: Cd^{2+} inhibits photoactivation of photosystem II by competitive binding to the essential Ca^{2+} site. Biochim Biophys Acta 1706(1–2):158–164

Foyer CH, Noctor G (2005) Redox homeostasis and antioxidant signaling: a metabolic interface between stress perception and physiological responses. Plant Cell 17(7):1866–1875

Galal TM, Eid EM, Dakhil MA, Hassan LM (2018) Bioaccumulation and rhizofiltration potential of *Pistia stratiotes* L. for mitigating water pollution in the Egyptian wetlands. Int J Phytoremediation 20(5):440–447

Gallego SM, Pena LB, Barcia RA, Azpilicueta CE, Iannone MF, Rosales EP, Benavides MP (2012) Unravelling cadmium toxicity and tolerance in plants: insight into regulatory mechanisms. Environ Exp Bot 83:33 –46

Genchi G, Sinicropi MS, Lauria G, Carocci A, Catalano A (2020) The effects of cadmium toxicity. Int J Environ Res Public Health 17(11):3782

Ghosh M, Singh SP (2005) A review on phytoremediation of heavy metals and utilization of it's by products. AJEE 6(4):214–231

Gupta DK, Pena LB, Romero-Puertas MC, Hernández A, Inouhe M, Sandalio LM (2017) NADPH oxidases differentially regulate ROS metabolism and nutrient uptake under cadmium toxicity. Plant Cell Environ 40(4):509–526

Halliwell B, Gutteridge JM (2015) Free radicals in biology and medicine. Oxford University Press, Cary, NC

Han Y, Zhang J, Chen XY, Gao ZZ, Xuan W, Xu S (2008) Carbon monoxide alleviates cadmium-induced oxidative damage by modulating glutathione metabolism in the roots of *medicago sativa*. New Phytol 177:155 –166

Hirschi KD, Korenkov VD, Wilganowski NL, Wagner GJ (2000) Expression of Arabidopsis CAX2 in tobacco. Altered metal accumulation and increased manganese tolerance. Plant Physiol 124(1):125–134

Ishimaru Y, Takahashi R, Bashir K, Shimo H, Senoura T, Sugimoto K, Nishizawa NK (2012) Characterizing the role of rice NRAMP5 in manganese, iron and cadmium transport. Sci Rep 2(1):1–8

Islam MS, Saito T, Kurasaki M (2015) Phytofiltration of arsenic and cadmium by using an aquatic plant, *Micranthemum umbrosum*: phytotoxicity, uptake kinetics, and mechanism. Ecotoxicology 112:193 –200

Jaishankar M, Tseten T, Anbalagan N, Mathew BB (2014) Toxicity, mechanism and health effects of some heavy metals. Interdiscip Toxicol 7(2):60–72

Jamla M, Khare T, Joshi S, Patil S, Penna S, Kumar V (2021) Omics approaches for understanding heavy metal responses and tolerance in plants. Curr Plant Biol 27:100213

Jia H, Wang H, Lu H, Jiang S, Dai M, Liu J, Yan C (2016) Rhizodegradation potential and tolerance of *Avicennia marina* (Forsk.) Vierh in phenanthrene and pyrene contaminated sediments. Mar Pollut Bull 110(1):112–118

Kato M, Ishikawa S, Inagaki K, Chiba K, Hayashi H, Yanagisawa S, Yoneyama T (2010) Possible chemical forms of cadmium and varietal differences in cadmium concentrations in the phloem sap of rice plants (*Oryza sativa* L.). Soil Sci Plant Nutr 56(6):839–847

Khaokaew S, Landrot G (2015) A field-scale study of cadmium phytoremediation in a contaminated agricultural soil at Mae Sot District, Tak Province, Thailand:(1) determination of Cd-hyperaccumulating plants. Chemosphere 138:883 –887

Khator K, Shekhawat GS (2020) Cd-and Cu-induced phytotoxicity on 2–3 leaf stage of *Cyamopsis tetragonoloba* and its regulation by nitrate reductase and ROS quenching enzyme. Acta Physiol Plant 42(7):1–14

Khator K, Saxena I, Shekhawat GS (2021) Nitric oxide induced Cd tolerance and phytoremediation potential of *B. juncea* by the modulation of antioxidant defense system and ROS detoxification. Biometals 34(1):15–32

Kim DY, Bovet L, Maeshima M, Martinoia E, Lee Y (2007) The ABC transporter AtPDR8 is a cadmium extrusion pump conferring heavy metal resistance. Plant J 50(2):207–218

Korenkov V, Hirschi K, Crutchfield JD, Wagner GJ (2007) Enhancing tonoplast Cd/H antiport activity increases Cd, Zn, and Mn tolerance, and impacts root/shoot Cd partitioning in *Nicotiana tabacum* L. Planta 226(6):1379–1387

Korenkov V, King B, Hirschi K, Wagner GJ (2009) Root-selective expression of AtCAX4 and AtCAX2 results in reduced lamina cadmium in field-grown *Nicotiana tabacum* L. Plant Biotechnol J 7(3):219–226

Kumar S, Sharma A (2019) Cadmium toxicity: effects on human reproduction and fertility. Rev Environ Health 34(4):327–338

Lanquar V, Lelièvre F, Bolte S, Hamès C, Alcon C, Neumann D, Vansuyt G, Curie C, Schröder A, Krämer U, Thomine S (2005) Mobilization of vacuolar iron by AtNRAMP3 and AtNRAMP4 is essential for seed germination on low iron. EMBO J 24(23):4041–4051

Liu L, Li Y, Tang J, Hu L, Chen X (2011) Plant coexistence can enhance phytoextraction of cadmium by tobacco (*Nicotiana tabacum* L.) in contaminated soil. J Environ Sci 23(3):453–460

Loix C, Huybrechts M, Vangronsveld J, Gielen M, Keunen E, Cuypers A (2017) Reciprocal interactions between cadmium-induced cell wall responses and oxidative stress in plants. Front Plant Sci 8:1867

Lux A, Martinka M, Vaculík M, White PJ (2011a) Root responses to cadmium in the rhizosphere: a review. J Exp Bot 62(1):21–37

Lux A, Vaculík M, Martinka M, Lišková D, Kulkarni MG, Stirk WA, Van Staden J (2011b) Cadmium induces hypodermal periderm formation in the roots of the monocotyledonous medicinal plant *Merwilla plumbea*. Ann Bot 107(2):285–292

Mahajan P, Kaushal J (2018) Role of phytoremediation in reducing cadmium toxicity in soil and water. J Toxicol 2018:1 –16

Mahawar L, Kumar R, Shekhawat GS (2018) Evaluation of heme oxygenase 1 (HO 1) in Cd and Ni induced cytotoxicity and crosstalk with ROS quenching enzymes in two to four leaf stage seedlings of *Vigna radiata*. Protoplasma 255(2):527–545

Mahawar L, Popek R, Shekhawat GS, Alyemeni MN, Ahmad P (2021) Exogenous hemin improves Cd^{2+} tolerance and remediation potential in *Vigna radiata* by intensifying the HO-1 mediatedantioxidant defence system. Sci Rep 11:2811 –2822

Mahawar L, Shekhawat GS (2022) Understanding the physiological mechanism of heme oxygenase for enhanced tolerance and phytoremediation of cd2+ in eruca sativa: co-ordinated function of antioxidant defense system. J Plant Growth Regul:1–12

Mathur S, Shekhawat GS (2013) Copper induced phytotoxicity and differential tissue specific response of antioxidant system as plant defense *Vigna radiata* L. Wilkzek. In: Khanna DR, Chopra AK, Singh V, Bhutiani R, Matta G (eds) Climate change effects on agriculture and economy. Biotech Books, New Delhi, pp 173–186

Mendoza-Cózatl DG, Jobe TO, Hauser F, Schroeder JI (2011) Long-distance transport, vacuolar sequestration, tolerance, and transcriptional responses induced by cadmium and arsenic. Curr Opin Plant Biol 14(5):554–562

Mendoza-Cózatl DG, Zhai Z, Jobe TO, Akmakjian GZ, Song WY, Limbo O, Schroeder JI (2010) Tonoplast-localized Abc2 transporter mediates phytochelatin accumulation in vacuoles and confers cadmium tolerance. J Biol Chem 285(52):40416–40426

Milner MJ, Seamon J, Craft E, Kochian LV (2013) Transport properties of members of the ZIP family in plants and their role in Zn and Mn homeostasis. J Exp Bot 64(1):369–381

Mohamed AA, Castagna A, Ranieri A, di Toppi LS (2012) Cadmium tolerance in *Brassica juncea* roots and shoots is affected by antioxidant status and phytochelatin biosynthesis. Plant Physiol Biochem 57:15 –22

Moons A (2003) Ospdr9, which encodes a PDR-type ABC transporter, is induced by heavy metals, hypoxic stress and redox perturbations in rice roots. FEBS Lett 553(3):370–376

Noriega GO, Yannarelli GG, Balestrasse KB, Batlle A, Tomaro ML (2007) The effect of nitric oxide on heme oxygenase gene expression in soybean leaves. Planta 226(5):1155–1163

Pantola R, Shekhawat GS (2012) Copper induced antioxidative enzyme indices in leaves of *Brassica juncea* seedlings. J Pharm Biomed Anal 15:1 –6

Park J, Song WY, Ko D, Eom Y, Hansen TH, Schiller M, Lee TG, Martinoia E, Lee Y (2012) The phytochelatin transporters AtABCC1 and AtABCC2 mediate tolerance to cadmium and mercury. Plant J 69(2):278–288

Paz-Ferreiro J, Gascó G, Méndez A, Reichman SM (2018) Soil pollution and remediation. Int J Environ Res Public Health 15(8):1657

Piantadosi CA, Carraway MS, Babiker A, Suliman HB (2008) Heme oxygenase-1 regulates cardiac mitochondrial biogenesis via Nrf2-mediated transcriptional control of nuclear respiratory factor-1. Circ Res 103(11):1232–1240

Pinto AP, Mota AD, De Varennes A, Pinto FC (2004) Influence of organic matter on the uptake of cadmium, zinc, copper and iron by sorghum plants. Sci Total Environ 326(1–3):239–247

Qian H, Li J, Sun L, Chen W, Sheng GD, Liu W, Fu Z (2009) Combined effect of copper and cadmium on *Chlorella vulgaris* growth and photosynthesis-related gene transcription. Aquat Toxicol 94:56 –61

Ranieri E, Moustakas K, Barbafieri M, Ranieri AC, Herrera-Melián JA, Petrella A, Tommasi F (2020) Phytoextraction technologies for mercury-and chromium-contaminated soil: a review. J Chem Technol Biotechnol 95(2):317–327

Rizwan M, Ali S, Abbas T, Zia-ur-Rehman M, Hannan F, Keller C, Al-Wabel MI, Ok YS (2016) Cadmium minimization in wheat: a critical review. Ecotoxicology 130:43 –53

Romero-Puertas MC, Ortega-Galisteo AP, Rodríguez-Serrano M, Sandalio LM (2012) Insights into cadmium toxicity: reactive oxygen and nitrogen species function. In: Metal toxicity in plants: perception, signaling and remediation. Springer, Berlin, Heidelberg, pp 91–117

Sandalio LM, Rodríguez-Serrano M, Gupta DK, Archilla A, Romero-Puertas MC, Río LAD (2012) Reactive oxygen species and nitric oxide in plants under cadmium stress: from toxicity to signaling. In: Environmental adaptations and stress tolerance of plants in the era of climate change. Springer, New York, NY, pp 199–215

Santa-Cruz DM, Pacienza NA, Polizio AH, Balestrasse KB, Tomaro ML, Yannarelli GG (2010) Nitric oxide synthase-like dependent NO production enhances heme oxygenase up-regulation in ultraviolet-B-irradiated soybean plants. Phytochemistry 71(14–15):1700–1707

Sasaki A, Yamaji N, Yokosho K, Ma JF (2012) Nramp5 is a major transporter responsible for manganese and cadmium uptake in rice. Plant Cell 24(5):2155–2167

Seregin IV, Ivanov VB (1997) Histochemical investigation of cadmium and lead distribution in plants. Russ J Plant Physiol 44(6):791–796

Shah V, Daverey A (2020) Phytoremediation: a multidisciplinary approach to clean up heavy metal contaminated soil. Environ Technol Innov 18:100774

Shahid M, Dumat C, Khalid S, Niazi NK, Antunes P (2016) Cadmium bioavailability, uptake, toxicity and detoxification in soil-plant system. Rev Environ Contam Toxicol 241:73 –137

Shanying HE, Xiaoe YANG, Zhenli HE, Baligar VC (2017) Morphological and physiological responses of plants to cadmium toxicity: a review. Pedosphere 27(3):421–438

Sharma SS, Dietz KJ (2009) The relationship between metal toxicity and cellular redox imbalance. Trends Plant Sci 14(1):43–50

Sharma SS, Dietz KJ, Mimura T (2016) Vacuolar compartmentalization as indispensable component of heavy metal detoxification in plants. Plant Cell Environ 39:1112 –1126

Shekhawat GS, Verma K (2010) Haem oxygenase (HO): an overlooked enzyme of plant metabolism and defence. J Exp Bot 61(9):2255–2270

Shekhawat GS, Dixit S, Verma K, Nasybullina EI, Kosmachevskaya OV, Topunov AF (2011) Heme oxygenase: enzyme with functional diversity. J Stress Physiol Biochem 7(1):88–94

Shekhawat GS, Kumar D, Verma K, Singh K, Jana S, Teotia P, Yadav S (2009) Copper as a biometal: an insight on transport, toxicity and tolerance in photosynthetic organisms. In: Blanc G, Moreau D (eds) Biometals: molecular structures, binding properties and applications. Nova Science Publishers, New York, NY, pp 91–118

Shekhawat GS, Parihar S, Mahawar L, Khator K, Bulchandani N (2019) Bilin metabolism in plants: structure, function and haem oxygenase regulation of bilin biosynthesis. eLS:1–13

Shekhawat GS, Verma K, Jana S, Singh K, Teotia P, Prasad A (2010) In vitro biochemical evaluation of cadmium tolerance mechanism in callus and seedlings of *Brassica juncea*. Protoplasma 239(1):31–38

Shigaki T, Pittman JK, Hirschi KD (2003) Manganese specificity determinants in the Arabidopsis metal/H+ antiporter CAX2. J Biol Chem 278(8):6610–6617

Sidhu GPS, Bali AS, Bhardwaj R (2019) Use of fungi in mitigating cadmium toxicity in plants. In: Cadmium toxicity and tolerance in plants. Academic Press, London, pp 397–426

Silva JRR, Fernandes AR, Silva Junior ML, Santos CRC, Lobato AKS (2018) Tolerance mechanisms in *Cassia alata* exposed to cadmium toxicity–potential use for phytoremediation. Photosynthetica 56(2):495–504

Snyder SH, Jaffrey SR, Zakhary R (1998) Nitric oxide and carbon monoxide: parallel roles as neural messengers. Brain Res Rev 26(2–3):167–175

Song XG, She XP, Zhang B (2008) Carbon monoxide-induced stomatal closure in *Vicia faba* is dependent on nitric oxide synthesis. Physiol Plant 132(4):514–525

Song Y, Hudek L, Freestone D, Puhui J, Michalczyk AA, Senlin Z, Ackland ML (2014) Comparative analyses of cadmium and zinc uptake correlated with changes in natural resistance-associated macrophage protein (NRAMP) expression in *Solanum nigrum* L. and *Brassica rapa*. Environ Chem Lett 11(6):653–660

Takahashi R, Ishimaru Y, Senoura T, Shimo H, Ishikawa S, Arao T, Nakanishi H, Nishizawa NK (2011) The OsNRAMP1 iron transporter is involved in Cd accumulation in rice. J Exp Bot 62(14):4843–4850

Tan L, Qu M, Zhu Y, Peng C, Wang J, Gao D, Chen C (2020) ZINC TRANSPORTER5 and ZINC TRANSPORTER9 function synergistically in zinc/cadmium uptake. Plant Physiol 183(3): 1235–1249

Tenhunen R, Marver HS, Schmid R (1969) Microsomal haem oxygenase characterization of the enzyme. J Biol Chem 244:6388 –6394

Uraguchi S, Kamiya T, Clemens S, Fujiwara T (2014) Characterization of OsLCT1, a cadmium transporter from indica rice (*Oryza sativa*). Physiol Plant 151(3):339–347

Uraguchi S, Kamiya T, Sakamoto T, Kasai K, Sato Y, Nagamura Y, Yoshida A, Kyozuka J, Ishikawa S, Fujiwara T (2011) Low-affinity cation transporter (OsLCT1) regulates cadmium transport into rice grains. Proc Natl Acad Sci U S A 108(52):20959–20964

Valko MMHCM, Morris H, Cronin MTD (2005) Metals, toxicity and oxidative stress. Curr Med Chem 12(10):1161–1208

Vatamaniuk OK, Mari S, Lu YP, Rea PA (1999) AtPCS1, a phytochelatin synthase from Arabidopsis: isolation and in vitro reconstitution. Proc Natl Acad Sci U S A 96(12):7110–7115

Verbruggen N, Hermans C, Schat H (2009) Mechanisms to cope with arsenic or cadmium excess in plants. Curr Opin Plant Biol 12(3):364–372

Verma K, Shekhawat GS, Sharma A, Mehta SK, Sharma V (2008) Cadmium induced oxidative stress and changes in soluble and ionically bound cell wall peroxidase activities in seedling and 3-4 leaf stage plants of *Brassica juncea* (L.) czern. Plant Cell Rep 27(7):1261–1269

Verret F, Gravot A, Auroy P, Leonhardt N, David P, Nussaume L, Vavasseur A, Richaud P (2004) Over expression of AtHMA4 enhances root-to-shoot translocation of zinc and cadmium and plant metal tolerance. FEBS Lett 576(3):306–312

Wang X, Wang Y, Kim HP, Nakahira K, Ryter SW, Choi AM (2007) Carbon monoxide protects against hyperoxia-induced endothelial cell apoptosis by inhibiting reactive oxygen species formation. J Biol Chem 282(3):1718–1726

Wilks SS (1959) Carbon monoxide in green plants. Science 129(3354):964–966

Xie Y, Ling T, Han YI, Liu K, Zheng Q, Huang L, Shen W (2008) Carbon monoxide enhances salt tolerance by nitric oxide-mediated maintenance of ion homeostasis and up-regulation of antioxidant defence in wheat seedling roots. Plant Cell Environ 31(12):1864–1881

Yamaguchi N, Mori S, Baba K, Kaburagi-Yada S, Arao T, Kitajima N, Hokura A, Terada Y (2011) Cadmium distribution in the root tissues of solanaceous plants with contrasting root-to-shoot Cd translocation efficiencies. Environment 71(2):198–206

Yanai J, Zhao FJ, McGrath SP, Kosaki T (2006) Effect of soil characteristics on Cd uptake by the hyperaccumulator *Thlaspi caerulescens*. Environ Pollut 139(1):167–175

Yannarelli GG, Noriega GO, Batlle A, Tomaro ML (2006) Heme oxygenase up-regulation in ultraviolet-B irradiated soybean plants involves reactive oxygen species. Planta 224(5): 1154–1162

Yao Z, Li J, Xie H, Yu C (2012) Review on remediation technologies of soil contaminated by heavy metals. Procedia Environ Sci 16:722 –729

Yuan L, Yang S, Liu B, Zhang M, Wu K (2012) Molecular characterization of a rice metal tolerance protein, OsMTP1. Plant Cell Rep 31(1):67–79

Zgorelec Z, Bilandzija N, Knez K, Galic M, Zuzul S (2020) Cadmium and mercury phytostabilization from soil using Miscanthus × giganteus. Sci Rep 10(1):1–10

Zhang A, Jiang M, Zhang J, Tan M, Hu X (2006) Mitogen-activated protein kinase is involved in abscisic acid-induced antioxidant defense and acts downstream of reactive oxygen species production in leaves of maize plants. Plant Physiol 141(2):475–487

Zhang M, Zhang J, Lu LL, Zhu ZQ, Yang XE (2016) Functional analysis of CAX2-like transporters isolated from two ecotypes of *Sedum alfredii*. Biol Plant 60(1):37–47

Heavy Metal/Metalloid Contamination: Their Sources in Environment and Accumulation in Food Chain

2

Seema Mishra, Geeta Singh, Apoorv Gupta, and Ravi Kumar Tiwari

Abstract

Heavy metals/metalloids are generally defined as having density more than 5 g cm^{-3}. Many of them, like Cu, Co, Zn, Fe, Mn, Mo, etc., are biologically essential in trace amounts, however, toxic at high concentrations. On the other hand, several toxic heavy metals/metalloids, such as arsenic (As), lead (Pb), mercury (Hg), and cadmium (Cd), are nonessential and deleterious to environment and human health. Although traces of these elements are naturally present in rocks and aquifers, however, their high concentrations have been found at several places, both through natural and anthropogenic sources. Mining, industrial production and uses, metal containing pesticides and fertilizers, etc. are major anthropogenic sources of most of the heavy metals in water and soil, while contamination of metalloid As in most parts of the world is from natural geogenic processes, e.g., through dissolution of As-rich sediments under reducing environment of groundwater aquifers and volcanic hot springs. High concentration of these toxicants in water adversely affects not only flora and fauna of aquatic system but also human health through bioaccumulation and biomagnification in fishes. Further, the use of heavy metal-/metalloid-contaminated water for irrigation leads to their high concentration in soil. Crops and vegetables grown in contaminated areas accumulate significant amount of heavy metals/metalloid from soil in different plant parts including in cereal grains. Thus, humans get

S. Mishra (✉)
Department of Chemistry, University of Lucknow, Lucknow, India

G. Singh · A. Gupta
Department of Chemistry, Deen Dayal Upadhyay Gorakhpur University, Gorakhpur, India

R. K. Tiwari
Plant Ecology and Environmental Science Division, CSIR-National Botanical Research Institute, Lucknow, India

R. P. Singh et al. (eds.), *Heavy Metal Toxicity: Environmental Concerns, Remediation and Opportunities*, https://doi.org/10.1007/978-981-99-0397-9_2

exposed to these toxicants through drinking water and contaminated food resulting in many folds higher dietary intake of these elements than their respective maximum tolerable daily/weekly intake values (MTDI/MTWI), set by various regulatory bodies. For instance, it has been reported that consumption of contaminated vegetables alone may contribute to over 50% of provisional tolerable daily intake of individual element. Similarly, rice is the main source of dietary As and Cd for population on rice-based diet, and fish is the main source of Hg. Chronic exposure to these toxicants through water and food results in severe health hazards to humans which is discussed in the subsequent chapter in sequence. In the current chapter, the major sources of different metal/metalloid in environment, the accumulation in food chain and the global status of contamination in water, soil, and food have been discussed.

Keywords

Carcinogenic elements · Contamination in water · Heavy metal and metalloid · MTDI · Permissible limits

2.1 Introduction

Rapid urbanization, industrialization, and land-use changes have led to elevated level of toxic heavy metal and metalloid throughout the globe to an extent that severely perturbed environment and pose serious health hazards to humans. The problem is grave in developing countries with high populations such as India and China. Heavy metals are generally those elements which have density above 5 g cm^{-3}, i.e., five times more than water, such as Pb, Cd, Hg, Cu, Fe, Ni, and Zn. Metalloids, such as arsenic (As) and antimony (Sb), are also included in this category considering their weight, density, and environmental impact, hence, hereafter, cumulatively being termed as heavy metal(loid)s. Heavy metal(loid)s are naturally present in Earth's crust; however, different developmental, industrial, and research-related activities have radically changed the natural geochemical cycle of heavy metal(loid)s leading to their high concentrations in various environmental compartments. For instance, these contaminants primarily release in water, soil, and air, and from there, these heavy metal(loid)s get accumulated in different plant parts and subsequently transferred to animals and humans. Also, in water, these heavy metal(loid)s may present naturally in a colloidal, particulate, or dissolved form (such as eroded minerals within sediments, ore deposit leaching, and extruded products through volcano) (Adepoju-Bello et al. 2009) or from anthropogenic sources (such as industrial or domestic effluents, harbor channel dredging, solid waste disposal) (Marcovecchio et al. 2007). Some of the heavy metals are essential micronutrients required for different physiological and biochemical functions in plants and animals, and their deficiency may cause various diseases (WHO 1996). For example, Co, Cu, Fe, Mn, Mo, and Zn are needed at trace levels to catalyze biological reactions as cofactors in enzymes and proteins (Adepoju-Bello et al.

Table 2.1 Classification of heavy metal(loid) carcinogenicity by the International Agency for Research on Cancer (IARC)

Element	Group	Carcinogenicity	Contaminant	Evidence	References
Arsenic	Group 1	Carcinogenic	Arsenic and inorganic compounds	Sufficient evidence in humans	IARC (2012)
Cadmium	Group 1	Carcinogenic	Cadmium and cadmium compounds	Sufficient evidence in humans	IARC (2012)
Lead	Group 2A	Probably carcinogenic	Lead compounds inorganic	Limited evidence in humans, enough evidence in animals	IARC (2006)
Mercury	Group 2B	Possibly carcinogenic	Methylmercury compounds	Limited evidence in humans, not enough evidence in animals	IARC (1993)
	Group 3	Carcinogenicity not classifiable	Mercury and inorganic mercury compounds	Carcinogenicity not classifiable	IARC (1993)

2009). However, they cause toxicity at high concentrations. Whereas several heavy metals and metalloids have no known biological functions, i.e., they are nonessential and deleterious to plants, animals, and humans. Many of them are known carcinogen and/or are extremely toxic even at low concentrations and can be referred to as high-priority elements with respect to their toxicity to humans, such as As, Pb, Cd, and Hg (Table 2.1). These metal(loid)s are released naturally to environment through weathering process. They can also be naturally high in some environmental conditions, such as wind-borne soil particles, rock weathering, volcanic eruptions, forest fires, and biogenic sources. They are generally found in the form of hydroxides, oxides, sulfides, sulfates, silicates, phosphates, and organic compounds (Ali et al. 2019; Sankhla and Kumar 2019). However, most of the environmental concern arises due to high contamination through anthropogenic activities like industries, agriculture, sewage, mining, metallurgical process, and thermal power plants (Sankhla and Kumar 2019). These metal(loid)s are generally emitted in the form of dusts from industries and thermal power plants and effluents from industries and sewage. Thus, heavy metal contamination is a serious problem worldwide, and their prolong exposure causes severe health issues in humans depending on their nature and quantity (Adepoju-Bello and Alabi 2005). Heavy metal(loid)-contaminated drinking water and food constitute the main dietary exposure route to humans. Further, uptake of these heavy metal(loid) by crops and vegetables hampers plant growth and productivity and obstructs mineral nutrient acquisition, thus compromising quantity and nutritional quality of food. In the current chapter, the natural background level of priority heavy metal(loid)s; their major sources of contamination in environment; the accumulation in food, including agricultural produce and fishes; and the global contamination status with respect to permissible limit in water and maximum tolerable daily/weekly intake limits have been discussed.

2.2 Sources of Heavy Metal(loid) Contamination in Environment

The Earth's crust is the prime source of various element including heavy metals and metalloids in the environment. Ninety-five percent of the Earth's crust is generally made up of igneous rocks and 5% by sedimentary rocks (Sarwar et al. 2017). Thus, most of the heavy metal(loid)s is found naturally in Earth's crust, and their traces are found in air, water, and soil, but in certain environmental compartments, their concentration reaches to the level which is detrimental for human and ecosystem. Both natural geochemical processes and anthropogenic activities contribute to heavy metal(loid) contamination. The natural sources include weathering of rocks, dissolution of ions from sedimentary rocks, volcanic emissions, and geothermal waters. Anthropogenic sources include industrial, agricultural, pharmaceutical, and domestic effluent (Tchounwou et al. 2012; Zamora-Ledezma et al. 2021). The level of contamination is more prominent in point source areas like metal-based industries, mining, foundries and smelters, and volcanic and geothermal areas, while application of agrochemicals and automobile exhaust are diffused sources of contamination contaminating wide areas. Natural chemical forms, background levels, and major sources of contamination of As, Cd, Pb, and Hg have been discussed below. The common and specific point and nonpoint sources of each of these elements have been summarized in Table 2.2.

2.2.1 Arsenic

The As content in Earth's crust ranges 1.5–2 mg/kg in upper crust and 1–1.8 mg/kg in bulk crust (Matschullat 2000). Sulfides are the main As-bearing ores where As occurs in the form of arsenopyrite (FeAsS), orpiment (As_2S_3), realgar (As_4S_4), pyrite (FeS_2), and sphalerite (ZnS) (Leermakers et al. 2006). Arsenic, a redox element, exhibits variable oxidation states (+5, +3, 0, −3) and occurs in several chemical forms. In aquatic and terrestrial environment, inorganic As that is oxyanions of As^{5+} (arsenate, $HAsO_4^{2-}$) and As^{3+} (arsenite, $HAsO_3^{2-}$) and organic forms like dimethylarsinic acid (DMA), monomethylarsonic acid (MMA), and trimethylarsine oxide are the most common As species. Methylated As and several other compounds like arsenobetaines, arsenocholines, and arsenosugars are found in marine environment. Arsenic has numerous industrial and agricultural applications, such as in glass and semiconductor industries; As compounds are used as pesticides, herbicides, wood preservatives, feed additive in poultry, and growth promotor for swine (Leermakers et al. 2006). Arsenicals were widely produced and stockpiled as chemical arsenal far a long time from the First World War, which were later disposed

Table 2.2 Main source of various heavy metal(loid)s in environment

S. no.	Element	Sources of elements in environment	References
1.	Arsenic	Groundwater, wood preservatives, smelting, semiconductor industries, thermal power plants residue, burning of fossil fuels (coal), particulate matter deposition, arsenic-based pesticides (lead arsenate, copper arsenate, and calcium arsenate), arsenic-containing medicines, feed additive in poultry, growth promotor for swine, and arsenic minerals	Nickson et al. (2000), Riethmiller (2005), Leermakers et al. (2006), Rai et al. (2019)
2.	Cadmium	Zinc ores, cadmium alloys, Ni-Cd batteries, sewage sludge, dyes, stabilizers, electroplating industries, power stations, and phosphate fertilizers	Rai et al. (2019)
3.	Lead	Gasoline, mining, paint, smelting, thermal power plant residues, mini blinds, ceramics, pipes and plumbing materials, solders, batteries, ammunition, and cosmetics	Ab Latif Wani and Usmani (2015), Rai et al. (2019)
4.	Mercury	Nonsurgical tools, dental amalgams, chemical/chlor-alkali industries, electrical industries, wood processing, production of caustic soda, thermal power plants residues, lead-acid batteries	Tchounwou et al. (2012), Rai et al. (2019)

of on land and sea. Arsenic compounds have also been used as medicine since ancient time (Riethmiller 2005). Arsenic has also been used as chemotherapeutic agent to treat breast cancer and leukemia (Rust and Soignet 2001). These historical and present-day applications have led to significant contamination of As. However, in recent times, arsenic exposure through water, soil, and food due to naturally arsenic-enriched groundwater aquifers is the main concern in many countries of the world. The sources of As in groundwater aquifers are As-containing sediments from which As is mobilized through reductive dissolution in the anoxic groundwater conditions (Nickson et al. 2000).

2.2.2 Cadmium

In Earth's crust, the average concentration of Cd is about 0.1 mg/kg. Sedimentary rocks and marine phosphates contain most parts of environmental Cd with an average of about 15 mg/kg (Tchounwou et al. 2012). Cadmium is present as cadmium sphalerite (CdS) as an impurity in Zn sphalerite (ZnS), an important Zn ore. Environmental contamination of Cd is mostly anthropogenic through various industrial and mining activities. Manufacturing of Ni-Cd batteries, pigments, alloys, stabilizers, electroplating industries, power stations, and phosphate fertilizers are the major sources of Cd in the environment (ATSDR 2008). Phosphate fertilizers and pesticides are the major source of Cd in agricultural soil (Fatima et al. 2019). Worldwide production of cadmium is about 24,000 metric tons per year (US Geological survey 2022). Although Cd is not mined, it is obtained as a side product of refining zinc, copper, and lead and is enough to meet the industrial demand of Cd.

2.2.3 Lead

The average concentration of lead (Pb) in the Earth's crust is about 13 mg/kg, where it is found as a bluish gray metal. Traces of Pb are present in all environmental compartments including indoors and in homes and offices. Lead has a wide industrial application because of its unique physical and chemical properties like softness, malleability, ductility, poor conductibility, and resistance to corrosion. For a long time, Pb has been used in several agricultural and industrial applications and for domestic purposes. Industrial sources of Pb are mining and smelting, burning of fossil fuels, and various manufacturing industries like lead-acid batteries, synthesis of Pb oxides for paints, and pigments (USEPA 2022, Gabby 2006). According to an estimate, in 2004, about 1.52 million metric tons of lead was used in different industries in the United State, out of which 83% was used for the production of lead-acid batteries. Pb also comes in environment through domestic uses of lead containing products like paint, ceramics, pipes and plumbing materials, solders, gasoline, batteries, ammunition, and cosmetics. Paints are the major source of indoor Pb exposure (Jacobs et al. 2002). In the past, leaded gasoline was one of the most prominent routes of lead in the environment and human exposure; however, since the 1970s, stringent measures were taken to eliminate lead from gasoline and from other domestic uses (ATSDR 1992, 1999). Oral ingestion through water and food and inhalation of Pb-contaminated aerosols or dust particles are the main routes of Pb exposure (ATSDR 1992, 1999), in which 35–50% of Pb absorption in the body occurs through drinking contaminated water with even higher rates of absorption for children.

2.2.4 Mercury

Mercury is a ubiquitous trace element in the Earth's crust with a worldwide mean concentration of 0.03 mg/kg in soil (García-Sánchez et al. 2008). It is derived in the atmosphere by degassing from the Earth's crust, volcanic emissions and evaporation from the seas as well as through anthropogenic sources like industrial applications and mining activities (Langford and Ferner 1999). The industrial application includes application in nuclear reactors, production of caustic soda, as a solvent for reactive and precious metals, as antifungal agents for wood processing, and as a preservative of pharmaceutical products. Mercury is also utilized in electrical industry and for making dental amalgams (Tchounwou et al. 2012). Like As, Hg also

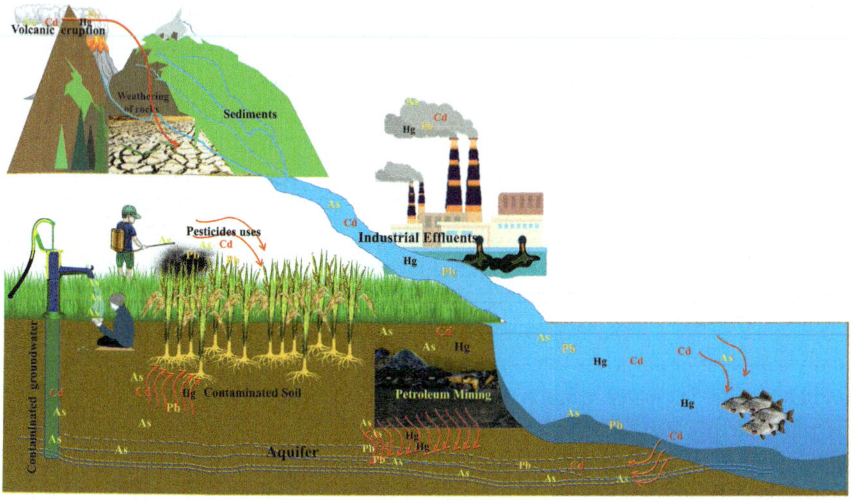

Fig. 2.1 Geologic cycle of heavy metal(loid) including anthropogenic interventions

exhibits variable oxidation states (0, +1, and + 2); and it is also present in environment in various chemical forms, i.e., elemental, inorganic, and organic forms. At room temperature, elemental mercury exists in a liquid form, and it gets vaporize from soil to atmosphere as Hg vapors (Guzzi and La Porta 2008). The elemental Hg can stay in atmosphere for 1–2 years, while its lifetime in soil is over 1000 years (García-Sánchez et al. 2008). Gold mines are one of the main sources of Hg pollution in environment because Hg is used for the extraction of gold by amalgamation process (Olivero and Solano 1998). The organic forms of Hg, i.e., methylated Hg, are formed by the microbial action in soil and water (Dopp et al. 2004). The toxicity of Hg depends on its chemical form (Clarkson et al. 2003). Once Hg enters to water or soil through natural or industrial pollution, it is methylated by the microorganism. The methyl Hg being highly lipophilic enters into the food chain through fish and shellfish and eventually into humans (García-Sánchez et al. 2008). Common geochemical cycle of these heavy metal(loid)s including anthropogenic intervention has been presented in Fig. 2.1.

2.3 Level of Heavy Metal(loid) Contamination in Water

The biogeochemical reactions regulate the cycle of elements in environment and the level of elements in hydrosphere and pedosphere. In the last century, anthropogenic interventions have disturbed the geochemical cycle and have resulted to the high level of many toxic elements in these compartments. Water is one of the most essential components for the living being, and access to clean water is the most challenging task in the twenty-first century. Surface water (rivers, lakes, etc.) and groundwater (borehole and well water) are two major sources of water for drinking

Table 2.3 Permissible limit (μgl^{-1}) of heavy metals in water set by different regulatory bodies

S. no.	Regulatory bodies	Arsenic	Cadmium	Lead	Mercury	References
1	WHO	10	3	10	6 (*i*Hg)	WHO (2011)
2	USEPA	10	5	15	2 (*i*Hg)	USEPA (2009)
3	EU	10	5	10	1	EC (1998), WHO (1993)
4	BIS	50	3	10	1	BIS (2012)

WHO, World Health Organization; *USEPA*, US Environmental Protection Agency; *EU*, European Union; *BIS*, Bureau of Indian Standards

and various other purposes like domestic, industrial, and agricultural uses (VanLoon and Duffy 2005). Water resources have been affected by wide diversity of pollutants, in which heavy metals have gained high concern due to toxicity at very low level (Vodela et al. 1997; Marcovecchio et al. 2007). Water is the predominant source of heavy metal(loid) exposure to human either through drinking water or through food chain contamination. Despite technological developments in modern era, the level of heavy metals in drinking water has been still higher than the recommended limits set by regulatory bodies in various counties (Table 2.3). The untreated discharge of industrial and domestic effluent and runoff from agricultural fields as well as mining areas severely contaminates surface water. There have been several poisoning episodes in the past like Cd poisoning in Japan and China and methyl mercury poisoning in Japan (Cai 1989; Hachiya 2006). Today, the use of groundwater has increased along with the use of surface water to meet the demand of the increasing population for both drinking and agricultural purposes especially in areas of high population density. The use of groundwater was also promoted as a safe alternative for drinking water during the 1980s to avoid waterborne diseases. However, groundwater has been also contaminated by heavy metals and metalloids and is the most important environmental issue and cleaning of groundwater is difficult as well as expensive (Belkhiri et al. 2017; Marcovecchio et al. 2007). The quality of groundwater sources is affected by the characteristics of the media through which the water passes on its way to the groundwater zone of saturation. Thus, heavy metals discharged from municipal wastes, traffic, industries, and hazardous waste sites as well as from the agrochemicals used in agricultural fields and accidental oil spillages result in a steady rise in contaminants in groundwater (Vodela et al. 1997; Igwilo et al. 2006).

In water, As is generally present as inorganic As. Generally, inorganic compounds of As, i.e., pentavalent arsenate ions [As^{5+}] like H_3AsO_4, $H_2AsO_4^-$, $HAsO_4^{2-}$, and AsO_4^{3-}, are predominant in surface water, and trivalent arsenite ions [As^{3+}] like H_3AsO_3, $H_2AsO_3^-$, and $HAsO_3^{2-}$ are predominant in groundwater (Siddiqui and Chaudhry 2018; Siddiqui et al. 2020). The guideline value of As in drinking water is 10 μgl^{-1}; however, several orders of magnitude higher level of As have been reported in many parts of the world (Smedley and Kinniburgh 2002). High level of As in groundwater has been reported India, Bangladesh, Taiwan, China, Mongolia, Myanmar, Nepal, Mexico, Argentina, Chile, and many parts of the United States (Smedley and Kinniburgh 2002; Mishra et al. 2016). More than 200 million people all over the world are in risk of high As exposure (Podgorski and Berg 2020). Southeast Asian countries are more contaminated. In the United States, up to 1000 μgl^{-1} As has been reported in drinking water through natural origin (Steinmaus et al. 2003). In Bengal delta (West Bengal, India, and Bangladesh) and in red river delta, the level of As >50 μgl^{-1} is common with as high as >3000 μgl^{-1} has been reported in many groundwater samples which were used for drinking and irrigation (Meharg et al. 2009; Mishra et al. 2016). The level of As in Pakistan has been reported to be up to 1889 μgl^{-1} in Kalalanwala, Punjab (Rehman et al. 2018), while concentrations of up to 7250 μgl^{-1} in India, 18,600 μgl^{-1} in China, and 9000 μgl^{-1} in Bangladesh are reported from industrial contamination. The historic gold mine area in Obuasi, Ghana, is heavily contaminated with 2250 μgl^{-1} As in dam water and 1400 μgl^{-1} in drinking water (Amasa 1975). Later studies in the same region showed a mean As concentration of 5190 μgl^{-1} (range 2800–10,400 μgl^{-1}) (Amonoo-Neizer and Amekor 1993) and 175 μgl^{-1} in stream water used for drinking (Smedley 1996). The As concentration in water from some contaminated areas of the world is summarized in Table 2.4. Recently, high As concentration has been reported from the hot springs of the Rehai in China. The maximum As concentration is 1350 μgl^{-1} (Guo et al. 2017). These hot springs contain thioarsenates in addition to arsenate and arsenite, probably originating from magmatic fluids due to geothermal activities.

Cadmium compounds are highly water-soluble and thus easily contaminate water and pose high toxicity to living being due to their high bioavailability through water. Environmental contamination of Cd is mostly anthropogenic through various industrial and mining activities. The main Cd compounds are oxide, sulfide, and inorganic salts of Cd, viz., CdO, CdS, $CdSO_4$, $CdCl_2$, and $Cd(NO_3)_2$. The permissible limit of Cd in drinking water is 3 μgl^{-1} (WHO 2011); however, the concentrations of Cd may reach more than three orders of magnitude higher in contaminated areas of the world (Table 2.4). The irrigation water in China contains about 50 μgl^{-1} Cd that resulted in high Cd accumulation in rice and other agricultural produces leading to severe health consequences in the population (Cai 1989). Groundwater samples from Balochistan area of Pakistan contain 24–30 μgl^{-1} Cd, while it ranged from 1 to 210 μgl^{-1} in different provinces of Pakistan (Rehman et al. 2018). The Cd concentration in groundwater samples from West Uttar Pradesh in India lies between 40 and 70 μgl^{-1} with the average concentration 60 μgl^{-1} (Idrees et al. 2018). Back in the 1950s, high Cd concentration was observed in the groundwater of America, ranging between 10 and 1950 μgl^{-1} (Lieber and Welsch 1954).

Pb exists in the form of nitrates, chlorates, and chlorides because other inorganic salts of Pb^{2+} have poor solubility in water (WHO 2001). The permissible limit of lead in drinking water is 10 μgl^{-1}. However, the water from contaminated areas may contain several folds higher Pb in surface water (Table 2.4). In Assam State of India, the level of lead in 25 drinking water samples collected around the tea gardens of Darrang District has been found to contain Pb higher than permissible limit. The level of Pb ranged from 40 to 350 μgl^{-1} with a mean level of 127 μgl^{-1} (Borah et al. 2010), while the level of Pb analyzed in drinking water in Western Bangladesh

Table 2.4 Level of heavy metal(loid)s in water, soil, and food in some contaminated areas of the world

Element	Country/ region	In water ($\mu g l^{-1}$)	In soil (mg kg^{-1})	Food commodity (mg kg^{-1})	References
Arsenic	United States	1000	*72*	**0.753** **0.2–0.46 (rice)**	Steinmaus et al. (2003); *Davis et al. (2009)*; **Brammer and Ravenscroft** (2009)
	Chile	800–900 (Toconce River)	*86–448*	**0.317**	Ferreccio et al. (2000); *Díaz et al. (2004)*; **Carey et al.** (2020)
	Egypt	59	*147.46*	**0.580**	Otero et al. (2016); *Salman et al. (2019)*; **Meharg et al.** (2009)
	Ghana (gold mine area)	2250 dam water 1400 (DW) 175	*157*	**0.277**	**Carey et al.** (2020); Asante et al. (2007); *Amonoo-Niezer and Busari* (1979)
	Pakistan	672–2400	*46.2*	**0.106**	Farooqi et al. (2007); *Arain et al. (2009)*; **Abbas et al.** (2010)
	Bangladesh	4730	*28* 57.5	**1 (rice)** **0.058–1.835 (rice)** **0.79 (reddish leaves)**	Herath et al. (2016); *Mandal and Suzuki (2002)*; **Brammer and Ravenscroft** (2009); **Rahman et al.** (2007)
	China	2000	*626*	**0.460**	Xia and Liu (2004); *Mandal and Suzuki (2002)*; **Meharg et al.** (2009)
	India	7350 (industrial area)	*28*	**0.961 (rice)** **0.526 (potato skin)**	Mishra et al. (2016); *Roychowdhury et al. (2002)*; **Brammer and Ravenscroft** (2009); **Halder et al.** (2014)

(continued)

Table 2.4 (continued)

Element	Country/ region	In water ($\mu g l^{-1}$)	In soil ($mg\ kg^{-1}$)	Food commodity ($mg\ kg^{-1}$)	References
	Japan	293	*70*	**0.511**	Herath et al. (2016); *Mandal and Suzuki (2002)*; **Naito et al.** (2015)
	Taiwan	1820 (black shales) (gw)	*67–438*	**0.063–0.2**	Herath et al. (2016); *Lin et al. (2015)*; **Brammer and Ravenscroft** (2009)
	Malaysia	1400	*57.05*	**0.055–0.103**	Ahmed et al. (2021); *Ong et al. (2013)*; **Carey et al.** (2020)
	Myanmar	57.05		**0.080–0.140**	**Mwale et al.** (2018); Wai et al. (2019)
	South Korea	48.4		**0.145**	Park et al. (2016); **Carey et al.** (2020)
	Sri Lanka	498		**0.171**	**Carey et al.** (2020); Perera et al. (2016)
	Iran	689	21.63	**0.347**	Keshavarz et al. (2012); **Sharafi et al.** (2019)
	Thailand	5000		**0.390**	Herath et al. (2016); **Meharg et al.** (2009)
	Vietnam	3050		**0.440**	Herath et al. (2016); **Carey et al.** (2020)
	Australia	7000		**0.385**	Australia New Zealand Food Standards (2017); **Carey et al.** (2020)
	France	100	5	**0.560**	Marchant et al. (2017); **Carey et al.** (2020)
	Hungary	4000		**0.112–0.145**	Mukherjee et al. (2006); **Mihucz et al.** (2007)
	Ireland	234	0.744	**0.020–0.095**	Carr et al. (2008); **Pogoson et al.** (2021)

(continued)

Table 2.4 (continued)

Element	Country/ region	In water (μgl^{-1})	In soil (mg kg^{-1})	Food commodity (mg kg^{-1})	References
	Italy	6940	*60*	**0.425**	Herath et al. (2016); *Mandal and Suzuki (2002)*; **Carey et al.** (2020)
	Spain	100		**0.820** **0.186 (rice)**	Herath et al. (2016); **Meharg et al.** (2009); **Brammer and Ravenscroft** (2009)
	Germany	150	*2.5–4.6*		Herath et al. (2016); *Hackethal et al. (2021)*
	United Kingdom	80	*61*	**0.396**	Herath et al. (2016); *Signes-Pastor et al. (2016)*; **Menon et al.** (2021)
	Mexico	620	*40*	**0.385**	Herath et al. (2016); *Cave et al. (2013)*; **Carey et al.** (2020)
	Canada	738	*150*	**4.830**	Herath et al. (2016); *Wang and Mulligan (2006)*; **Dabeka et al.** (1993)
	Argentina	5300	22	**0.316**	Smedley et al. (2005); **Carey et al.** (2020)
	Brazil	350	*40*	**0.781**	*de Menezes et al. (2020)*; Herath et al. 2016; **Carey et al.** (2020)
	Uruguay	24.19		**0.244**	Mañay et al. 2014; **Carey et al.** (2020)
Cadmium	China	474	*531 (atmospheric deposition, wastewater, Pb-Zn mining/ refinery)*	**3.71 (rice)**	Li and Chen (2016); *Wen et al. (2015)*; **Wu et al.** (2014)

(continued)

Table 2.4 (continued)

Element	Country/region	In water ($\mu g l^{-1}$)	In soil ($mg\ kg^{-1}$)	Food commodity ($mg\ kg^{-1}$)	References
	Pakistan	210	*10*	*0.055–0.090 (vegetables)*	Lone et al. (2003); *Khan et al.* (2011)
	Nigeria	1.1	*54.8*	**54.8 (fish)**	Asubiojo et al. 1997; *Jimoh and Mohammed* (2016); **Edogbo et al.** (2020)
	Australia	20	*3.62*	**0.870 (vegetables)**	Hart and Lake (1987); *Mann et al.* (2002); **Rahman et al.** (2014)
	France	3600	*0.63*	**0.200 (vegetables)**	Chiffoleau et al. (2001); *Hernandez et al.* (2003); **Malmauret et al.** (2002)
	Hong Kong	150	*4.11*	**1.8 (oyster)**	Ngoc et al. (2020); *Chung et al.* (2020); **Chen et al.** (2014)
	Indonesia	6300	1.8	0.4	Elfidasari et al. (2020)
	Sweden	5			Friberg et al. (1986)
	Saudi Arabia	26			Mustafa et al. (1988)
	Netherlands	0.2			Ros et al. (1987)
	United States	77	32	**0.090–24 (lettuce)**	Page et al. (1987); **Feeney et al.** (1984)
	India	280	*83.6 (textile industry)*	**3.90 (lady finger)**	Deepali and Gangwar (2010); *Panwar and Ahmed* (2018); **Sharma et al.** (2008)
Lead	India	40–350 (dw)	*52.30 (roadside soil)*	**1.133 (fish)**	Borah et al. (2010); *Sharma and Prasad* (2010); **Agarwal et al.** (2007)

(continued)

Table 2.4 (continued)

Element	Country/ region	In water (μgl^{-1})	In soil (mg kg^{-1})	Food commodity (mg kg^{-1})	References
	Pakistan	30–4700	*49.5–121 (sediments) 10,300 (calc-silicate rocks, Pb-Zn sulfide mineralization and mining waste)*	**44 (vegetables)**	Javaid et al. (2008); Rehman et al. (2018); *Siddique et al. (2009); Mashiatullah et al. (2013); Muhammad et al. (2011)*; **Khan et al.** (2011)
	Bangladesh	0.2–17 (dw)			Frisbie et al. (2009)
	Mexico	50–120			Wyatt et al. (1998)
	United States		*135*	**0.032**	*Holmgren et al. (1993)*; **TatahMentan et al.** (2020)
	Australia		*58.6 (sediments)*		*Birch et al. (2001)*
	Ghana	43			Asante et al. (2007)
Mercury	China	0.11 to 27.7	*2920 (Hg mining)*	0.**042 to 0.636 (vegetable and wheat)**	Jiang et al. (2006); *Liu et al. (2019)*; **Zhang et al.** (2022)
	India	0.032–0.21	0.382–2.730 (surface sediments)	0.030–2.85 (fish)	Ramasamy et al. (2017)
	Pakistan	0.154	*144 (surface soil)*	**0.020 (vegetables)**	Khan and Abbas (2021); *Khan et al. (2011)*; **Abbas et al.** (2010)
	Ghana	2.3			Asante et al. (2007)
	Venezuela (gold mine area)	4.6, 0.77		0.32 to 1.92 (fish)	García-Sánchez et al. (2008)
	Brazil	0.2–19.8		**0.011–5.96**	Pfeiffer et al. (1991); **Nriagu et al.** (1992)
	Ecuador (gold mine area)	0.02–0.9			Appleton et al. (2001)

(continued)

Table 2.4 (continued)

Element	Country/ region	In water (μgl^{-1})	In soil (mg kg^{-1})	Food commodity (mg kg^{-1})	References
	Philippines (gold mine area)	0.1–2900			Appleton et al. (1999)
	Spain	<0.11–20.3 (surface water			Berzas et al. (2003); Gray et al. (2004)
	North America	1.4–19			Rytuba (2003)
	Indonesia			**0.057–0.07**	**Kambey et al.** (2001)
	United States (Hg mines)	0.001–0.18		**0.02–0.75**	Ikem and Egiebor (2005); **Hurley et al.** (1998)

The level of heavy metal(loid)s in food is the mean of various food commodities, unless specified. The normal, italic, and bold texts indicate corresponding studies in water, soil, and food, respectively

showed low level of contamination with 1% of samples exceeding the WHO permissible limit (Frisbie et al. 2009). Pb contamination in tap water has been reported in Saudi Arabia lying between 0.13 and 10.58 μgl^{-1}. The historic gold mine area of Tarkwa in Ghana has been found to be heavily contaminated with heavy metals with Pb and Hg concentrations reaching up to 43 and 2.3 μgl^{-1}, respectively, in drinking water (Asante et al. 2007).

The worldwide normal distribution of Hg in water ranges 0.003–0.005 μg/l (García-Sánchez et al. 2008). The drinking water guideline value for total Hg (tHg) is 1 μgl^{-1}, while for inorganic Hg (iHg) it is set to 6 μgl^{-1} by the WHO (1993, 2011). In developing countries, industrial contamination of Hg is a big concern. It has been reported that two-third of the total anthropogenic Hg is emitted by Asian countries with India among the top three (Ramasamy et al. 2017). The level of total Hg (tHg) and methyl Hg has been reported in up to 31.8 and 0.21 ngl^{-1}, respectively, in a backwater lake in southwest coast of India receiving the effluent from Kerala industrial zone. The sediment contained 0.38–2.85 μgl^{-1} tHg which was lower compared to earlier reports (5.5–11.5 μg/g) from the same region (Ramasamy et al. 2017). In Nepal, from 31 sampling sites, 23 samples contained Hg at concentrations exceeding the WHO guideline of 1 μg/L; the highest concentration reported was 300 μgl^{-1} in water from a dug well from an urban area (Khatiwada et al. 2002). In New Jersey Coastal Plain, USA, about 700 wells are known to have high concentration of tHg with concentrations as high as 80 μgl^{-1} (Dr. Judith Louis, New Jersey Department of Environmental Protection, 2010, oral commun.) The gold mine areas all over the world have been reported to cause significant mercury contamination in the water bodies and rivers around, such as in the Philippines, Ecuador, and Venezuela (Appleton et al. 1999, 2001; García-Sánchez et al. 2008). High Hg concentrations up to 4.60 μgl^{-1} were recorded in surface water samples (rivers, springs, mining ponds, streams) of Cuyuni River Basin, Venezuela, which was associated with the amalgam processes used in artisan gold mining (García-Sánchez et al. 2008). Mercury concentration in edible fish tissue in the area reached up to 1.92 $mgkg^{-1}$ (García-Sánchez et al. 2008). Similarly, the water samples from the historic gold mine area of the Philippines showed up to 2900 μgl^{-1} Hg (Appleton et al. 1999). The level of Hg in water from contaminated areas around the world has been also given in Table 2.4.

2.4 Level of Heavy Metal(loid) Contamination in Soil

Traces of most of heavy metal(loid)s are present in all soils originating from the parent rocks; however, industrial waste disposal, landfills, and irrigation with contaminated water have caused severe soil contamination. Most of the toxic heavy metals are cationic (metallic elements whose forms in soil are positively charged cations) like Pb, Cd, and Hg, while As commonly occurs as anionic compounds.

The natural content of As in soil ranges from 0.01 to over 600 $mgkg^{-1}$ globally, with an average range of 2–20 $mgkg^{-1}$ (Huang 1994). Uncontaminated soils typically contain <10 $mgkg^{-1}$ of total As, but the concentration can reach up to hundreds or thousands $mgkg^{-1}$ in contaminated environments (Kabata-Pendias and Pendias 2001). In the United States, a nationwide survey conducted in areas having no anthropogenic sources of As reported that the natural background concentrations of As in soil ranged from less than 1 to 97 $mgkg^{-1}$ (Shacklette and Boerngen 1984). In the United States, the level of As in the soil having used arsenical pesticides in the past was 23 $mgkg^{-1}$ compared to 5 $mgkg^{-1}$ in surrounding soil (Cox et al. 1996). In soil, inorganic As predominates; however, intentional uses of organic As, such as arsenical pesticides, may result in higher organic compounds of As. Further, pentavalent As predominates in soil due to oxidation of trivalent arsenicals in toxic conditions of soil. In Bangladesh, the soils irrigated with As-contaminated groundwater contain >20 $mgkg^{-1}$ As with up to 81 $mgkg^{-1}$ As. Panaullah et al. (2009) reported high gradient in As soil concentration in rice field irrigated with As-contaminated groundwater from 78.6 $mgkg^{-1}$ near the well to 12.8 $mgkg^{-1}$ at far side.

Cadmium naturally occurs in all agricultural soils in small amounts. The global concentration of Cd in uncontaminated soils ranges from 0.06 to 1.1 $mgkg^{-1}$, with a minimum of 0.01 and a maximum of 2.7 $mgkg^{-1}$ (Kabata-Pendias and Pendias 2001). The mean background concentration of Cd in the United States is 0.27 $mgkg^{-1}$, in the United Kingdom 0.6–0.7 $mgkg^{-1}$, in the Netherlands 0.5 $mgkg^{-1}$ (708 samples), and in Germany <0.3–1.0 $mgkg^{-1}$ (Kabata-Pendias and Pendias 2001). Natural process like volcanic activity, sea spray, weathering of serpentine rocks and minerals may cause high soil Cd concentration. The use of soil amendments (e.g., biosolids, fly ash, sewage sludge) and irrigation with industrial or mining wastewater significantly contribute to soil Cd contamination. Phosphate fertilizers are the main source of Cd in agriculture soil. In rock phosphates, Cd is naturally found as an impurity at concentrations 1–200 mg Cd (kg $P_2O_5)^{-1}$, from where it comes in commercial phosphate fertilizers (Roberts 2014). The application of P fertilizers is the main source of annual addition of Cd in the agricultural soils of Europe and the United States and in many other countries. Sewage sludge can contain Cd ranging from >1 to 3650 mgkg^{-1}, and the range of Cd in the land amended with sludge may contain from 30 to 40 gha^{-1} yr.$^{-1}$ (Alloway et al. 1990; Alloway and Steinnes 1999). The range of Cd content in sedimentary and igneous phosphate rocks from different countries varies from <1 to 150 mgkg^{-1} with maximum mean level in United States, i.e., 92 mgkg^{-1} (Roberts 2014). The Cd present in these rocks gets transferred to the superphosphate and triple superphosphate during production of phosphate fertilizer. Depending on the phosphate rock source, superphosphate can contain 2 to >40 mgkg^{-1} Cd and triple superphosphate <10 to >100 mgkg^{-1} Cd, while ammonium phosphates (monoammonium phosphate (MAP) and diammonium phosphate (DAP)) may contain <1 to>100 mgkg^{-1} Cd (Roberts 2014). Robarge et al. (2004) reported the average Cd content in DAP/MAP, 16 to 22 mgkg^{-1}, and in triple superphosphate, 19 mgkg^{-1}. The known cases of Cd toxicity mainly arise due to contamination from industrial/mining waste which contaminated soil and agricultural produces in Japan after World War II and more recently in China. In China, the level of Cd in soil was from 0.89 to 1.49 mgkg^{-1} due to irrigation with contaminated water containing about 50 μgl^{-1} Cd.

The mean background level of Pb in soil is 14 mgkg^{-1} in the United States; however, mining, smelting, and refining activities have resulted in substantial increases in lead levels in the environment, especially near mining and smelting sites with a normal range between 50 and 400 mgkg^{-1} (USEPA). Also, the strict measures taken during the 1980s on the use of Pb in gasoline and some other industrial goods have checked the level of Pb emission. Holmgren et al. (1993) reported the global geometric mean of Cd and Pb in soil, 0.62 and 33.7 mgkg^{-1}, respectively. They analyzed more than 3000 agricultural soil samples from different parts of the United States and reported the mean level of Cd and Pb as 0.265 and 12.3 mgkg^{-1}, respectively. The level of Pb ranged between 3 and 135 mgkg^{-1} while Cd 1.01 and 2 mgkg^{-1}in the soil of the United States. Birch et al. (2001) reported lead up to 58.6 mg/kg in fluvial marine sediments found naturally in Australia. Sharma and Prasad (2010) reported the lead content in the roadside soil up to 52.30 mgkg^{-1}in Uttar Pradesh, India. The level of Pb was higher in the agricultural fields within 5 m to the highway, while it decreased with the distance, i.e., 26 mgkg^{-1} at a distance of 10 m and 18.70 mgkg^{-1} beyond 10 m. They concluded that the source of Pb in the roadside soil is from gasoline. Although the use of leaded gasoline has decreased, the increasing number of automobile is probably compensating its effect. In addition, careless disposal of tires, etc. can also introduce Pb to the roadside soil. The mineralized mining site in Northern Pakistan has been reported to be highly contaminated with heavy metal (Muhammad et al. 2011). The level of Pb in this area ranged between 5 and 10300 mgkg^{-1}, while it ranged 12–1337 mgkg^{-1} in non-contaminated soil in the same region.

Mercury is a ubiquitous trace element in the Earth's crust with a global average content of 0.03 mgkg^{-1} in soil (García-Sánchez et al. 2008). It is derived in the atmosphere by degassing from the Earth's crust, volcanic emissions, and evaporation from the seas as well as through anthropogenic like industrial applications and mining activities (Langford and Ferner 1999). Qu et al. (2019) analyzed 230 soil samples from different land-use areas of Thailand and reported that the level of Hg in top soil (0–30 cm) was higher in the woodland (15.89 mgkg^{-1}) and in the agricultural land (13.48 mgkg^{-1}) in comparison to abandoned agricultural land (4.08 mgkg^{-1}). Liu et al. (2019) reported urban mercury pollution in Shanghai, China, by measuring the level of Hg in surface soils, road dust, foliar dust, and camphor tree leaves and found that the average tHg in surface soils was 0.36 mgkg^{-1} (range 0.078–1362 mgkg^{-1}) in the test area, while the background value of 0.1 soil was mgkg^{-1}. Further, the tHg level in soil was doubled in the last 10 years. The average mercury in road dust in Shanghai was 0.596 mgkg^{-1} (range 0.2–2.2). The level of Hg in soils in other parts of China ranged from 0.012 to 9.4 mgkg^{-1}, while in road dust it ranged from 0.2 to 2.32 mgkg^{-1} (Liu et al. 2019). The mean level of soil mercury in Slovenia, Mexico, the United States, and Greece has been reported as 0.106, 0.45, 0.024, and 0.166 mgkg^{-1}, respectively, with maximum level (5.293 mgkg^{-1}) being in Slovenia (Liu et al. 2019). Li et al. (2009) reviewed the Hg contamination in Asian countries and reported that the annual anthropogenic emission of Hg was maximum in China (605 t), followed by India (150 t) and Japan (144 t), and least in Kazakhstan (44 t), Indonesia, and Israel. The mean soil Hg concentration in Japan ranged from 0.007 to 0.229 mgkg^{-1} near Sakurajima Volcano. In Japan, tHg concentrations in paddy rice fields near Niigata Prefecture ranged from 0.019 to 0.62 μgg^{-1} (mean 0.155 μgg^{-1}) in 1989 and from 0.015 to 0.34 μgg^{-1} (mean 0.146 μgg^{-1}) in 1997. The paddy fields of the Philippines contaminated by gold mine silt contained a mean of tHg 24 mgkg^{-1} (Li et al. 2009).

2.5 Contamination of Heavy Metal(loid)s in Food

From the contaminated water and soil, heavy metal(loid)s get accumulated in algae and plants, thus enters into food chains, and eventually transfers to the human body via fishes and other aquatic animals and cereal grain and vegetables. The comprehensive global status of As, Cd, Pb, and Hg contamination in water, in soil, and in foods has been given in Table 2.4. Aquatic macrophytes and benthic animals accumulate high level of heavy metal(loid)s even at very low water concentration and thus often used as a sign of pollution. Aquatic animals mainly the crustacean species are the bioindicators of toxic metal(loid)s (Ahmed et al. 2017). Subsequently, these heavy metal(loid)s biomagnify in fishes and makes a significant dietary route of exposure. Plants take up heavy metal(loid)s inadvertently through the nutrient uptake pathways. For instance, Cd, Pb, and Hg are taken up by transporters and channels of divalent cations like Ca, Mg, Fe, Zn, Mn, and Cu, while inorganic As is mainly taken up in the form of oxyanions or neutral molecule via phosphate and silicon (as silicic acid) channels, respectively. Methylated forms of As are also taken up via silicon channels (Mishra et al. 2017). In non-hyper accumulator plants, like common agricultural plants and crops, most of the heavy metal(loid)s taken up are confined to the roots; relatively small amounts are transferred to the shoot followed by accumulation in grain. However, it is sufficiently high to exceed the recommended daily/weekly intake values. For example, the permissible limit of As for irrigation water is 100 μgl^{-1}, by FAO, and the maximum acceptable limit for agricultural soil has been recommended up to 20 $mgkg^{-1}$ by the European Union, but an As level of 10 $mgkg^{-1}$ in soil already results in significant accumulation in edible part of plants (Bhattacharya et al. 2010). Arsenic level in different agricultural food produce ranged between 0.0003 and 1.02 $mgkg^{-1}$ from the field containing 5.70–9.71 $mgkg^{-1}$ As in soil. The maximum mean level of As was found in potato (0.654), followed by Boro rice grain (0.451), arum (0.407), amaranth (0.372), radish (0.344), Aman rice grain (0.334), ladyfinger (0.301), cauliflower (0.293), and Brinjal (0.279) (Bhattacharya et al. 2010). Williams et al. (2006) analyzed concentration of As in common agricultural produce of Bangladesh, and they reported that As concentration in common vegetables ranged from 0.69 to 1.93 $mgkg^{-1}$ with max level in arum stolons. The mean level of root and tuber vegetables was max (0.745 $mgkg^{-1}$) followed by fruit vegetables (0.56 $mgkg^{-1}$) and leafy vegetables (0.39 $mgkg^{-1}$). Pulses and spices accumulated comparatively lower As. They further reported eating these vegetables would contribute with about 0.7 to 13.4% of the MTDI. Depending on the level of As and the nature of food, rice has been found to be the greatest contributor of As consumption for the population with rice as a staple food, such as in India, Bangladesh, and elsewhere. The global normal range of As in rice has been suggested to be from 0.08 to 0.2 $mgkg^{-1}$ (Zavala and Duxbury 2008), while for common agricultural plants including rice the range of As has been reported to be from 7 μgkg^{-1} to 7500 μgkg^{-1} around the world (Mandal and Suzuki 2002; Roychowdhury et al. 2002; Dahal et al. 2008). Daily consumption of rice containing 0.08 $mgkg^{-1}$ As would contribute to 25% of MTDI and would be equivalent to drinking water arsenic level of 10 μgl^{-1}, while As level more than 20 times of this value has been reported around the word. The average level of As in rice grain varies from 0.03 to 2.05 $mgkg^{-1}$ in Bangladesh, 0.11 to 0.46 $mgkg^{-1}$ in the United States, and from 0.04 to 1.6 $mgkg^{-1}$ in India. Wheat also accumulates significant level of As when grown in contaminated soil. The level of As in wheat ranged from 0.49 to 1.15 $mgkg^{-1}$ in Bangladesh, 0.01–0.234 $mgkg^{-1}$ in India, 0.02 to 0.365 $mgkg^{-1}$ in China, and 0.01–0.50 $mgkg^{-1}$ in England (Mishra et al. 2021). Apart from As, rice may accumulate Cd and Pb as well in the grains. In a market basket survey in Iran, Sharafi et al. (2019) analyzed Cd and Pb in rice from different origins and reported the mean levels of Cd to be 0.083, 0.07, and 0.146 $mgkg^{-1}$ and the mean levels of Pb to be 0.047, 0.149, and 0.736 $mgkg^{-1}$, respectively, in Iranian, Pakistani, and Indian rice. The mean levels of Cd and Pb in Brazilian rice was 0.025 $mgkg^{-1}$ and 0.111 $mgkg^{-1}$, respectively (Pedron et al. 2019). Rai et al. (2019) reported high level of Pb (from 0.02 to 3.96 $mgkg^{-1}$) and Cd (0.003 to 0.370 $mgkg^{-1}$) in different vegetables and cereal grains grown in regions receiving effluents from sewage, industrial, mining, etc. in different parts of world. Khan et al. (2010) reported high contamination of heavy metals in various vegetables grown in contaminated soil in northern Pakistan. The area is naturally geologically contaminated, with soil Cd ranging from 0.30 to 2.30 $mgkg^{-1}$ and Pb from 29 to 138 $mgkg^{-1}$. The mean concentrations of Cd in different vegetables ranged from 0.24 to 2.10 $mgkg^{-1}$, with the highest accumulation in spinach. The mean concentrations of Pb ranged from 9 to 44 $mgkg^{-1}$ in different vegetables, with the highest mean concentration of Pb being in *Malus sylvestris* (Khan et al. 2010).

Fish is the most important dietary source of protein for the population, particularly residing in the river basins. However, it is also a significant source of heavy metal(loid) exposure to human. It is the main dietary route of Hg for human. Depending on the feeding habits and life span, fish may accumulate significantly high level of heavy metals in edible tissues. For instance, carnivorous fish species accumulated higher concentration of Hg (0.32–1.92 mg/kg) than herbivorous (0.12–0.46 mg/kg) collected from Hg-contaminated gold mine area of Cuyuni River Basin, Venezuela (García-Sánchez et al. 2008). In India, the level of heavy metal in different fish species of the river Ganga ranged from 1.04 to 4.55 $mgkg^{-1}$ for As, 0.2 to 15.90 mg kg^{-1} for Pb, 0.003 to 9.50 $mgkg^{-1}$ for Hg, and 0.003 to 11.9 $mgkg^{-1}$ for Cd (Dwivedi et al. 2018; Djedjibegovic et al. 2020). Agarwal et al. (2007) reported up to 1.133 $mgkg^{-1}$ Pb in *Clarias batrachus*, the walking catfish, and 2.85 $mgkg^{-1}$ tHg in *Cynoglossus semifasciatus* (Bengal tonguesole) found in Gomati River, India. The methyl Hg accounts for 16.68 to 90.25% of THg in various fish species. The level of Hg in Kodai Lake, India, contaminated with thermometer factory effluent, has been reported to 0.356–0.465 μgl^{-1} tHg and 0.050 μgl^{-1} Me-Hg, while fishes in the lake accumulated 120 to 290 μgkg^{-1} tHg (Karunasagar et al. 2006).

2.6 Conclusion

The current review of literature demonstrated that heavy metal(loid)s are naturally present in environment as a part of natural geochemical cycle and cause background level of exposure through drinking water and food. However, intentional (mining and industrial uses) or unintentional (excessive extraction of groundwater) human intervention has disturbed the natural cycle of metal(loid)s. In the last half-century, industrial revolution and rapid urbanization resulted in excessive exploitation of natural resources, and careless disposal of urban and industrial waste has detrimentally affected the environment. Although groundwater As contamination is generally natural geogenically originated, however, excessive extraction of water has played a role in aggravating the problem. In recent decades, several rules and regulations have been made to check toxic waste release; however, their imposition is in question, particularly in developing and highly populated countries. Further, the GDP-based economy has led to the excessive production and consumption of industry-based goods and logistics. Thus, current analysis of data showed that despite several regulatory measures, the level of As, Cd, Pb, and Hg is still radically increasing.

Water is the main route of human exposure, however, with the development of mitigation strategies (for domestic and community level, discussed in the subsequent chapter) and people's awareness about clean drinking water, nowadays, food has become the major human exposure source of these heavy metals and metalloids in many areas. The data summarized in current study show that concentration of each heavy metal(loid) is several folds higher in water and soil in most parts of the world increasing their normal background levels. Further, food chain contamination through irrigation with untreated/partially treated water or contaminated groundwater, e.g., with As, also need attention.

Acknowledgment SM is grateful to SEED division of DST, New Delhi, for the project grant (SEED/TIASN/2018/74). SM and GS are also grateful to UCG, New Delhi, for the start-up grant (F.30-505/2020(BSR).

References

Ab Latif Wani AA, Usmani JA (2015) Lead toxicity: a review. Interdiscip Toxicol 8(2):55

Abbas M, Parveen Z, Iqbal M, Riazuddin M, Iqbal S, Ahmed M, Bhutto R (2010) Monitoring of toxic metals (cadmium, lead, arsenic and mercury) in vegetables of Sindh, Pakistan. Kathmandu Univ J Sci Eng Technol 6(2):60–65

Adepoju-Bello AA, Alabi OM (2005) Heavy metals: a review. Nig J Pharm 37:41–45

Adepoju-Belle AA, Ojomolade OO, Ayoola GA, Coker HAB (2009) Quantitative Analysis of some Toxic Metals in Domestic Water obtained from Lagos Metropolis. Nig J Pharm 42(1):57-60

Agarwal R, Kumar R, Behari JR (2007 Feb) Mercury and lead content in fish species from the river Gomti, Lucknow, India, as biomarkers of contamination. Bull Environ Contam Toxicol 78(2): 118–122

Agency for Toxic Substances and Disease Registry (ATSDR) (2008) Draft toxicological profile for cadmium. Atlanta, GA

Ahmed Q, Ali QM, Bat L, Ghory fs. (2017) Concentration of heavy metals in red, brown and green algae collected from the Karachi coast of Pakistan. Pak J Mar Sci 26(1):57–65

Ahmed T, Noman M, Manzoor N, Shahid M, Hussaini KM, Rizwan M, Ali S, Maqsood A, Li B (2021) Green magnesium oxide nanoparticles-based modulation of cellular oxidative repair mechanisms to reduce arsenic uptake and translocation in rice (Oryza sativa L.) plants. Environ Pollut 288:117785

Ali H, Khan E, Ilahi I (2019) Environmental chemistry and ecotoxicology of hazardous heavy metals: environmental persistence, toxicity, and bioaccumulation. J Chem 5:6730305

Alloway BJ, Jackson AP, Morgan H (1990) The accumulation of cadmium by vegetables grown on soils contaminated from a variety of sources. Sci Total Environ 91:223–236

Alloway BJ, Steinnes E (1999) Anthropogenic additions of cadmium to soils. In: Cadmium in soils and plants. Springer, Dordrecht, pp 97–123

Amasa SK (1975) Arsenic pollution at Obuasi Goldmine, town, and surrounding countryside. Environ Health Perspect 12:131–135

Amonoo-Neizer EH, Amekor EM (1993) Determination of total arsenic in environmental samples from Kumasi and Obuasi, Ghana. Environ Health Perspect 101(1):46–49

Amonoo-Neizer EH, Busari GL (1979) Arsenic status of ghana soils: contamination of soils near gold smelters. Ghana J Science 20(1-2):57–62

Appleton JD, Williams TM, Breward N, Apostol A, Miguel J, Miranda C (1999) Mercury contamination associated with artisanal gold mining on the Island of Mindanao, The Philippines. Sci Total Environ 228(2–3):95–109

Appleton JD, Williams TM, Orbea H, Carrasco M (2001) Fluvial contamination associated with artisanal gold mining in the Ponce Enriquez, Portovelo-Zaruma and Nambija areas, Ecuador. Water Air Soil Pollut 131(1):19–39

Arain MB, Kazi TG, Baig JA, Jamali MK, Afridi HI, Shah AQ, Jalbani N, Sarfraz RA (2009) Determination of arsenic levels in lake water, sediment, and foodstuff from selected area of Sindh, Pakistan: estimation of daily dietary intake. Food Chem Toxicol 47(1):242–248

Asante KA, Agusa T, Subramanian A, Ansa-Asare OD, Biney CA, Tanabe S (2007) Contamination status of arsenic and other trace elements in drinking water and residents from Tarkwa, a historic mining township in Ghana. Chemosphere 66(8):1513–1522

Asubiojo OI, Nkono NA, Ogunsua AO, Oluwole AF, Ward NI, Akanle OA, Spyrou NM (1997) Trace elements in drinking and groundwater samples in Southern Nigeria. Sci Total Environ 208(1–2):1–8

ATSDR (1992) Agency for toxic substances and disease registry. Case studies in environmental medicine - Lead Toxicity. Public Health Service, U.S. Department of Health and Human Services, Atlanta

ATSDR (1999) Agency for toxic substances and disease registry public health service. U.-S. Department of Health and Human Services. Toxicological Profile for Lead, Atlanta

Australia New Zealand Food Standards (2017). https://www.foodstandards.gov.au/Pages/default.aspx

Belkhiri L, Mouni L, Narany TS, Tiri A (2017) Evaluation of potential health risk of heavy metals in groundwater using the integration of indicator kriging and multivariate statistical methods. Groundw Sustain Dev 4:12–22

Berzas NJ, Garca BL, Rodrguez MD (2003) Distribution of mercury in the aquatic environment at Almadén, Spain. Environ Poll 122(2):261–271

Bhattacharya P, Samal AC, Majumdar J, Santra SC (2010) Arsenic contamination in rice, wheat, pulses, and vegetables: a study in an arsenic affected area of West Bengal, India. Water Air Soil Pollut 213(1):3–13

Birch G, Siaka M, Owens C (2001) The source of anthropogenic heavy metals in fluvial sediments of a rural catchment: Coxs River, Australia. Water Air Soil Pollut 126(1):13–35

BIS (2012) I. 10500 Indian standard drinking water–specification, second revision. Bureau of Indian Standards, New Delhi

Borah KK, Bhuyan B, Sarma HP (2010 Oct) Lead, arsenic, fluoride, and iron contamination of drinking water in the tea garden belt of Darrang district, Assam, India. Environ Monitor Assess 169(1):347–352

Brammer H, Ravenscroft P (2009) Arsenic in groundwater: a threat to sustainable agriculture in South and South-east Asia. Environ Int 35(3):647–654

Cadmium statistics and information US Geological survey (2022). https://www.usgs.gov/centers/national-minerals-information-center/cadmium-statistics-and-information

Cai SW (1989) Cadmium exposure and health effects among residents in irrigation area with ore dressing wastewater. Zhonghua Yu Fang Yi Xue Za Zhi 23(3):143–146

Carey M, Meharg C, Williams P, Marwa E, Jiujin X, Farias JG, De Silva PM, Signes-Pastor A, Lu Y, Nicoloso FT, Savage L (2020) Global sourcing of low-inorganic arsenic rice grain. Expos Health 12(4):711–719

Carr R, Zhang C, Moles N, Harder M (2008) Identification and mapping of heavy metal pollution in soils of a sports ground in Galway City, Ireland, using a portable XRF analyser and GIS. Environ Geochem Health 30(1):45–52

Cave MR, Wragg J, Harrison H (2013) Measurement modelling and mapping of arsenic bioaccessibility in Northampton, United Kingdom. J Environ Sci Health A 48(6):629–640

Chen MY, Chan BT, Lam CH, Chung SW, Ho YY, Xiao Y (2014) Dietary exposures to eight metallic contaminants of the Hong Kong adult population from a total diet study. Food Addit Contam A 31(9):1539–1549

Chiffoleau JF, Auger D, Chartier E, Michel P, Truquet I, Ficht A, Gonzalez JL, Romana LA (2001) Spatiotemporal changes in cadmium contamination in the Seine estuary (France). Estuaries 24(6):1029–1040

Chung MK, Cheung KC, & Wong MH. Pollutants in Hong Kong soils: As, Cd, Cr, Cu, Hg, Pb and Zn. bioRxiv. 2020 Jan 1

Clarkson TW, Magos L, Myers GJ (2003) The toxicology of mercury—current exposures and clinical manifestations. N Engl J Med 349(18):1731–1737

Cox MS, Bell PF, Kovar JL (1996) Arsenic supply characteristics of four cotton-producing soils. Plant Soil 180(1):1

Dabeka RW, McKenzie AD, Lacroix GM, Bowe CC, Graham RA, Conacher HB, Verdeer P (1993) Survey of arsenic in total diet food composites and estimation of the dietary intake of arsenic by Canadian adults and children. J AOAC Int 76(1):14–25

Dahal BM, Fuerhacker M, Mentler A, Karki KB, Shrestha RR, Blum WE (2008) Arsenic contamination of soils and agricultural plants through irrigation water in Nepal. Environ Pollut 155(1): 157–163

Davis HT, Aelion CM, McDermott S, Lawson AB (2009) Identifying natural and anthropogenic sources of metals in urban and rural soils using GIS-based data, PCA, and spatial interpolation. Environ Pollut 157(8–9):2378–2385

de Menezes MD, Bispo FH, Faria WM, Gonçalves MG, Curi N, Guilherme LR (2020 Apr) Modeling arsenic content in Brazilian soils: What is relevant? Sci Total Environ 10(712): 136511

Deepali KK, Gangwar K (2010) Metals concentration in textile and tannery effluents, associated soils and ground water. New York Sci J 3(4):82–89

Díaz OP, Leyton I, Muñoz O, Núñez N, Devesa V, Súñer MA, Vélez D, Montoro R (2004) Contribution of water, bread, and vegetables (raw and cooked) to dietary intake of inorganic arsenic in a rural village of Northern Chile. J Agric Food Chem 52(6):1773–1779

Djedjibegovic J, Marjanovic A, Tahirovic D, Caklovica K, Turalic A, Lugusic A, Omeragic E, Sober M, Caklovica F (2020) Heavy metals in commercial fish and seafood products and risk assessment in adult population in Bosnia and Herzegovina. Sci Rep 10(1):1–8

Dopp E, Hartmann LM, Florea AM, Rettenmeier AW, Hirner AV (2004) Environmental distribution, analysis, and toxicity of organometal (loid) compounds. Crit Rev Toxicol 34(3):301–333

Dwivedi S, Mishra S, Tripathi RD (2018) Ganga water pollution: a potential health threat to inhabitants of Ganga basin. Environ Int 1(117):327–338

Edition F (2011) Guidelines for drinking-water quality. WHO Chron 38(4):104–108

Edogbo B, Okolocha E, Maikai B, Aluwong T, Uchendu C (2020) Risk analysis of heavy metal contamination in soil, vegetables and fish around Challawa area in Kano State, Nigeria. Sci Afr 7:e00281

Elfidasari D, Nuril-Ismi L, Sugoro I (2020) Heavy metal concentration in water, sediment, and Pterygoplichthys pardalis in the Ciliwung River, Indonesia. AACL Bioflux 13(3):1764

Farooqi A, Masuda H, Firdous N (2007) Toxic fluoride and arsenic contaminated groundwater in the Lahore and Kasur districts, Punjab, Pakistan and possible contaminant sources. Environ Pollut 145(3):839–849

Fatima G, Raza AM, Hadi N, Nigam N, Mahdi AA (2019) Cadmium in human diseases: it's more than just a mere metal. Indian J Clin Biochem 34(4):371–378

Feeney S, Peterson JR, Zenz DR, Lue-Hing C (1984) National survey of the metals content of seven vegetable species. Department of Research and Development-Report (84-4)

Ferreccio C, González C, Milosavjlevic V, Marshall G, Sancha AM, Smith AH (2000) Lung cancer and arsenic concentrations in drinking water in Chile. Epidemiology 1:673–679

Friberg L, Nordberg GF, Vouk VB (1986) Handbook on toxicology of metals. Elsevier, Amsterdam; New York; Oxford, pp 130–184

Frisbie SH, Mitchell EJ, Mastera LJ, Maynard DM, Yusuf AZ, Siddiq MY, Ortega R, Dunn RK, Westerman DS, Bacquart T, Sarkar B (2009) Public health strategies for western Bangladesh that address arsenic, manganese, uranium, and other toxic elements in drinking water. Environ Health Perspect 117(3):410–416

Gabby PN (2006) Lead: in mineral commodity summaries. US Geological Survey, Reston, VA

Galal-Gorchev H, Ozolins G (1993) WHO guidelines for drinking-water quality. Water Supply 11(3):1–6

García-Sánchez A, Contreras F, Adams M, Santos F (2008) Mercury contamination of surface water and fish in a gold mining region (Cuyuni river basin, Venezuela). Int J Environ Pollut 33(2–3):260–274

Gray JE, Hines ME, Higueras PL, Adatto I, Lasorsa BK (2004) Mercury speciation and microbial transformations in mine wastes, stream sediments, and surface waters at the Almadén mining district, Spain. Environ Sci Technol 38(16):4285–4292

Guo Q, Planer-Friedrich B, Liu M, Li J, Zhou C, Wang Y (2017) Arsenic and thioarsenic species in the hot springs of the Rehai magmatic geothermal system, Tengchong volcanic region. Chem Geol 20(453):12–20

Guzzi G, La Porta CA (2008) Molecular mechanisms triggered by mercury. Toxicology 244(1):1–2

Hachiya N (2006) The history and the present of Minamata disease. Jpn Med Assoc J 49:112–118

Hackethal C, Kopp JF, Sarvan I, Schwerdtle T, Lindtner O (2021) Total arsenic and water-soluble arsenic species in foods of the first German total diet study (BfR MEAL Study). Food Chem 346:128913

Halder D, Biswas A, Šlejkovec Z, Chatterjee D, Nriagu J, Jacks G, Bhattacharya P (2014) Arsenic species in raw and cooked rice: Implications for human health in rural Bengal. Sci Total Environ 497:200–208

Hart BT, Lake PS (1987) Studies of heavy metal pollution in Australia with particular emphasis on aquatic systems. Wiley, Hoboken, NJ

Herath I, Vithanage M, Bundschuh J, Maity JP, Bhattacharya P (2016) Natural arsenic in global groundwaters: distribution and geochemical triggers for mobilization. Curr Poll Rep 2(1):68–89

Hernandez L, Probst A, Probst JL, Ulrich E (2003) Heavy metal distribution in some French forest soils: evidence for atmospheric contamination. Sci Total Environ 312(1–3):195–219

Holmgren GG, Meyer MW, Chaney RL, Daniels RB (1993) Cadmium, lead, zinc, copper, and nickel in agricultural soils of the United States of America. In: Crop science society of America, and soil science society of America. American Society of Agronomy, Madison, WI

Huang YC (1994) Arsenic distribution in soils. In: Arsenic in the environment, part I: cycling and characterization. Wiley, pp 17–49

Hurley JP, Cowell SE, Shafer MM, Hughes PE (1998) Partitioning and transport of total and methyl mercury in the lower Fox River, Wisconsin. Environ Scie Technol 32(10):1424–1432

IARC Working Group on the Evaluation of Carcinogenic Risks to Humans (2012) Pharmaceuticals. Volume 100 A. A review of human carcinogens. IARC monographs on the evaluation of carcinogenic risks to humans. 100(PT A):1

IARC Working Group on the Evaluation of Carcinogenic Risks to Humans (2006) Inorganic and organic lead compounds. In Inorganic and organic lead compounds. International Agency for Research on Cancer

Idrees N, Tabassum B, Abd Allah EF, Hashem A, Sarah R, Hashim M (2018) Groundwater contamination with cadmium concentrations in some West UP Regions, India. Saudi J Biol Sci 25(7):1365–1368

Igwilo IO, Afonne OJ, Maduabuchi UJ, Orisakwe OE (2006) Toxicological study of the Anam river in Otuocha, Anambra state, Nigeria. Arch Environ Occup Health 61(5):205–208

Ikem A, Egiebor NO (2005) Assessment of trace elements in canned fishes (mackerel, tuna, salmon, sardines and herrings) marketed in Georgia and Alabama (United States of America). J Food Compos Anal 18(8):771–787

International Agency for Research on Cancer (1993) Beryllium, cadmium, mercury, and exposures in the glass. Apresentado em: IARC Working Group on the Evaluation of Carcinogenic Risks to Humans: Beryllium, Lyon

Jacobs DE, Clickner RP, Zhou JY, Viet SM, Marker DA, Rogers JW, Zeldin DC, Broene P, Friedman W (2002) The prevalence of lead-based paint hazards in US housing. Environ Health Perspect 110(10):A599–A606

Javaid S, Shah SG, Chaudhary AJ, Khan MH (2008) Assessment of trace metal contamination of drinking water in the Pearl Valley, Azad Jammu and Kashmir. Clean—Soil Air Water 36(2): 216–221

Jiang GB, Shi JB, Feng XB (2006) Mercury pollution in China. Environ Sci Technol 40(12): 3672–3678

Jimoh WL, Mohammed MI (2016) Transfer of heavy metals from soil to cabbage (Brassica oleracea) grown in irrigated farmlands of Kaduna Metropolis. Niger J Chem Res 29(21):23–32

Joint FA (2001) WHO Expert Committee on Food Additives, World Health Organization. Evaluation of certain food additives and contaminants: fifty-fifth report of the Joint FAO/WHO Expert Committee on Food Additives. World Health Organization

Kabata-Pendias A, Pendias H (2001) Trace elements in soils and plants. CRC Press, Boca Raton, FL

Kambey JL, Farrell AP, Bendell-Young LI (2001) Influence of illegal gold mining on mercury levels in fish of North Sulawesi's Minahasa Peninsula (Indonesia). Environ Pollut 114(3): 299–302

Karunasagar D, Krishna MB, Anjaneyulu YA, Arunachalam J (2006) Studies of mercury pollution in a lake due to a thermometer factory situated in a tourist resort: Kodaikkanal. Environ Poll 143(1):153–158

Keshavarz M, Mohammadikia D, Gharibpour F, Dabbagh AR (2012) Accumulation of heavy metals (Pb, Cd, V) in sediment, roots and leaves of mangrove species in Sirik Creek along the sea coasts of Oman, Iran. J Appl Sci Environ Manag 16(4):323

Khan EA, Abbas Z (2021) A scoping review of sources of mercury and its health effects among Pakistan's most vulnerable population. Rev Environ Health 36(1):39–45

Khan MN, Wasim AA, Sarwar A, Rasheed MF (2011) Assessment of heavy metal toxicants in the roadside soil along the N-5, National Highway. Environ Monitor Assess 182(1):587–595

Khan S, Rehman S, Khan AZ, Khan MA, Shah MT (2010) Soil and vegetables enrichment with heavy metals from geological sources in Gilgit, northern Pakistan. Ecotoxicol Environ Saf 73(7):1820–1827

Khatiwada NR, Takizawa S, Tran TV, Inoue M (2002) Groundwater contamination assessment for sustainable water supply in Kathmandu Valley, Nepal. Water Sci Technol 46(9):147–154

Langford NJ, Ferner RE (1999) Toxicity of mercury. J Hum Hypertens 13(10):651–656

Leermakers M, Baeyens W, De Gieter M, Smedts B, Meert C, De Bisschop HC, Morabito R, Quevauviller P (2006) Toxic arsenic compounds in environmental samples: Speciation and validation. TrAC Trends Anal Chem 25(1):1

Li L, Chen F (2016) Oxidative stress, epigenetics, and cancer stem cells in arsenic carcinogenesis and prevention. Curr Pharmacol Rep 2(2):57–63

Li P, Feng XB, Qiu GL, Shang LH, Li ZG (2009) Mercury pollution in Asia: a review of the contaminated sites. J Hazard Mater 168(2–3):591–601

Lieber M, Welsch WF (1954) Contamination of ground water by cadmium. J Am Water Works Assoc 46(6):541–547

Lin K, Lu S, Wang J, Yang Y (2015) The arsenic contamination of rice in Guangdong Province, the most economically dynamic provinces of China: arsenic speciation and its potential health risk. Environ Geochem Health 37(2):353–361

Liu Y, Song S, Bi C, Zhao J, Xi D, Su Z (2019) Occurrence, distribution and risk assessment of mercury in multimedia of Soil-Dust-Plants in Shanghai, China. Int J Environ Res Public Health 16(17):3028

Ljujic B, Sundac L (1998) [[Council] Directive 98/83/EC [of 3 November 1998] on the quality of water intended for human consumption: review and integral translation [from English into Serbian]]. Voda i sanitarna tehnika (Serbia and Montenegro)

Lone MI, Saleem S, Mahmood T, Saifullah K, Hussain G (2003) Heavy metal contents of vegetables irrigated by sewage/tubewell water. Int J Agric Biol 5(4):533–535

Malmauret L, Parent-Massin D, Hardy JL, Verger P (2002) Contaminants in organic and conventional foodstuffs in France. Food Addit Contam 19(6):524–532

Mañay N, Goso C, Pistón M, Fernández-Turiel JL, García-Vallés M, Rejas M, Guerequiz R (2014) Groundwater arsenic content in Raigón aquifer system (San José, Uruguay)

Mandal BK, Suzuki KT (2002) Arsenic round the world: a review. Talanta 58(1):201–235

Mann SS, Rate AW, Gilkes RJ (2002) Cadmium accumulation in agricultural soils in Western Australia. Water Air Soil Pollut 141(1):281–297

Marchant BP, Saby NP, Arrouays D (2017 Aug) A survey of topsoil arsenic and mercury concentrations across France. Chemosphere 1(181):635–644

Marcovecchio JE, Botté SE, Freije RH (2007) Heavy metals, major metals, trace elements. Handb Water Anal 25(2):275–311

Mashiatullah A, Chaudhary MZ, Ahmad N, Javed T, Ghaffar A (2013) Metal pollution and ecological risk assessment in marine sediments of Karachi Coast, Pakistan. Environ Monitor Assess 185(2):1555–1565

Matschullat J (2000) Arsenic in the geosphere—a review. Sci Total Environ 249(1–3):297–312

Meharg AA, Williams PN, Adomako E, Lawgali YY, Deacon C, Villada A, Cambell RC, Sun G, Zhu YG, Feldmann J, Raab A (2009) Geographical variation in total and inorganic arsenic content of polished (white) rice. Environ Sci Technol 43(5):1612–1617

Menon M, Dong W, Chen X, Hufton J, Rhodes EJ (2021) Improved rice cooking approach to maximise arsenic removal while preserving nutrient elements. Sci Total Environ 755:143341

Mihucz VG, Virág I, Zang C, Jao Y, Záray G (2007) Arsenic removal from rice by washing and cooking with water. Food Chem 105(4):1718–1725

Mishra S, Dwivedi S, Kumar A, Chauhan R, Awasthi S, Mattusch J, Tripathi RD (2016) Current status of ground water arsenic contamination in India and recent advancements in removal techniques from drinking water. Int J Plant Environ 2(1 and 2):01–15

Mishra S, Mattusch J, Wennrich R (2017) Accumulation and transformation of inorganic and organic arsenic in rice and role of thiol-complexation to restrict their translocation to shoot. Sci Rep 7(1):1–3

Mishra S, Singh G, Dwivedi S, Singh K (2021) Role of edaphic factors in accumulation and chemical speciation of arsenic in agricultural food produce. Int J Adv Res Sci Commun Technol 12:213

Muhammad S, Shah MT, Khan S (2011) Heavy metal concentrations in soil and wild plants growing around Pb–Zn sulfide terrain in the Kohistan region, northern Pakistan. Microchem J 99(1):67–75

Mukherjee A, Sengupta MK, Hossain MA, Ahamed S, Das B, Nayak B, Lodh D, Rahman MM, Chakraborti D (2006) Arsenic contamination in groundwater: a global perspective with emphasis on the Asian scenario. J Health Popul Nutr 1:142–163

Mustafa HT, Hassan H, Abo-Melha A, Rihan TI (1988) Cadmium and zinc concentrations in the potable water of the Eastern Province of Saudi Arabia. Bull Environ Contam Toxicol 40(3):462

Mwale T, Rahman MM, Mondal D (2018) Risk and benefit of different cooking methods on essential elements and arsenic in rice. Int J Environ Res Public Health 15(6):1056

Naito S, Matsumoto E, Shindoh K, Nishimura T (2015) Effects of polishing, cooking, and storing on total arsenic and arsenic species concentrations in rice cultivated in Japan. Food Chem 168: 294–301

Ngoc NT, Chuyen NV, Thao NT, Duc NQ, Trang NT, Binh NT, Sa HC, Tran NB, Ba NV, Khai NV, Son HA (2020) Chromium, cadmium, lead, and arsenic concentrations in water, vegetables, and seafood consumed in a coastal area in northern Vietnam. Environ Health Insights 14: 1178630220921410

Nickson RT, McArthur JM, Ravenscroft P, Burgess WG, Ahmed KM (2000) Mechanism of arsenic release to groundwater, Bangladesh and West Bengal. Appl Geochem 15(4):403–413

Nriagu JO, Pfeiffer WC, Malm O, De Souza CM, Mierle G (1992) Mercury pollution in Brazil. Nature 356(6368):389
Olivero J, Solano B (1998) Mercury in environmental samples from a waterbody contaminated by gold mining in Colombia, South America. Sci Total Environ 217(1–2):83–89
Ong GH, Yap CK, Maziah M, Suhaimi H, Tan SG (2013) An investigation of arsenic contamination in Peninsular Malaysia based on Centella asiatica and soil samples. Environ Monit Assess 185(4):3243–3254
Otero XL, Tierra W, Atiaga O, Guanoluisa D, Nunes LM, Ferreira TO, Ruales J (2016) Arsenic in rice agrosystems (water, soil and rice plants) in Guayas and Los Ríos provinces, Ecuador. Sci Total Environ 573:778–787
Page AL, Chang AC, El-Amamy M (1987) Cadmium levels in soils and crops in the United States. In: Lead, mercury, cadmium and arsenic in the environment, vol 8. Wiley, New York, pp 119–146
Panaullah GM, Alam T, Hossain MB, Loeppert RH, Lauren JG, Meisner CA, Ahmed ZU, Duxbury JM (2009) Arsenic toxicity to rice (Oryza sativa L.) in Bangladesh. Plant Soil 317(1):31–39
Panwar RM, Ahmed S (2018) Assessment of contamination of soil and groundwater due to e-waste handling. Curr Sci 10:166–173
Park JD, Choi SJ, Choi BS, Lee CR, Kim H, Kim YD, Park KS, Lee YJ, Kang S, Lim KM, Chung JH (2016) Arsenic levels in the groundwater of Korea and the urinary excretion among contaminated area. J Expo Sci Environ Epidemiol 26(5):458–463
Pedron T, Segura FR, Paniz FP, de Moura SF, dos Santos MC, de Magalhães Júnior AM, Batista BL (2019) Mitigation of arsenic in rice grains by polishing and washing: evidencing the benefit and the cost. J Cereal Sci 87:52–58
Perera PC, Sundarabarathy TV, Sivananthawerl T, Kodithuwakku SP, Edirisinghe U (2016) Arsenic and cadmium contamination in water, sediments and fish is a consequence of paddy cultivation: evidence of river pollution in Sri Lanka. Achiev Life Sci 10(2):144–160
Pfeiffer WC, Malm O, Souza CM, de Lacerda LD, Silveira EG, Bastos WR (1991) Mercury in the Madeira river ecosystem, Rondonia, Brazil. For Ecol Manag 38(3–4):239–245
Podgorski J, Berg M (2020) Global threat of arsenic in groundwater. Science 368(6493):845–850
Pogoson E, Carey M, Meharg C, Meharg AA (2021) Reducing the cadmium, inorganic arsenic and dimethylarsinic acid content of rice through food-safe chemical cooking pre-treatment. Food Chem 338:127842
Qu R, Han G, Liu M, Li X (2019) The mercury behavior and contamination in soil profiles in Mun River Basin, Northeast Thailand. Int J Environ Res Public Health 16(21):4131
Rahman MA, Hasegawa H, Rahman MM, Rahman MA, Miah MA (2007) Accumulation of arsenic in tissues of rice plant (Oryza sativa L.) and its distribution in fractions of rice grain. Chemosphere 69(6):942–948
Rahman MA, Rahman MM, Reichman SM, Lim RP, Naidu R (2014) Heavy metals in Australian grown and imported rice and vegetables on sale in Australia: health hazard. Ecotoxicol Environ Saf 100:53–60
Rai PK, Lee SS, Zhang M, Tsang YF, Kim KH (2019) Heavy metals in food crops: health risks, fate, mechanisms, and management. Environ Int 125:365–385
Ramasamy EV, Jayasooryan KK, Chandran MS, Mohan M (2017) Total and methyl mercury in the water, sediment, and fishes of Vembanad, a tropical backwater system in India. Environ Monit Assess 189(3):1–9
Rehman K, Fatima F, Waheed I, Akash MS (2018) Prevalence of exposure of heavy metals and their impact on health consequences. J Cell Biochem 119(1):157–184
Riethmiller S (2005) From Atoxyl to Salvarsan: searching for the magic bullet. Chemotherapy 51(5):234–242
Robarge WP, Boos D, Proctor C (2004) Determination of trace metal content of fertilizer source materials produced in North America. In Hall WL Jr, Robarge WP (eds) Environmental impact of fertilizer on soil and water. ACS Symposium Series 872:75–89

Roberts TL (2014) Cadmium and phosphorous fertilizers: the issues and the science. Proc Eng 83: 52–59

Ros JPM, Slooff W (eds) (1987) Integrated criteria document. Cadmium. Bilthoven, National institute of public health and environmental protection (Report No. 758476004)

Roychowdhury T, Uchino T, Tokunaga H, Ando M (2002) Arsenic and other heavy metals in soils from an arsenic-affected area of West Bengal, India. Chemosphere 49(6):605–618

Rust DM, Soignet SL (2001) Risk/benefit profile of arsenic trioxide. Oncologist 6(S2):29–32

Rytuba JJ (2003) Mercury from mineral deposits and potential environmental impact. Environ Geol 43(3):326–338

Salman SA, Zeid SA, Seleem EM, Abdel-Hafiz MA (2019) Soil characterization and heavy metal pollution assessment in Orabi farms, El Obour, Egypt. Bull Natl Res Centre 43(1):1–3

Sankhla MS, Kumar R (2019) Contaminant of heavy metals in groundwater & its toxic effects on human health & environment. Available at SSRN 3490718

Sarwar N, Imran M, Shaheen MR, Ishaque W, Kamran MA, Matloob A, Rehim A, Hussain S (2017) Phytoremediation strategies for soils contaminated with heavy metals: modifications and future perspectives. Chemosphere 171:710–721

Shacklette HT, Boerngen JG (1984) Element concentrations in soils and other surficial materials of the conterminous United States. US Government Printing Office, Washington, DC

Sharafi K, Yunesian M, Mahvi AH, Pirsaheb M, Nazmara S, Nodehi RN (2019) Advantages and disadvantages of different pre-cooking and cooking methods in removal of essential and toxic metals from various rice types-human health risk assessment in Tehran households, Iran. Ecotoxicol Environ Saf 175:128–137

Sharma RK, Agrawal M, Marshall FM (2008) Heavy metal (Cu, Zn, Cd and Pb) contamination of vegetables in urban India: a case study in Varanasi. Environ Pollut 154(2):254–263

Sharma S, Prasad FM (2010) Accumulation of lead and cadmium in soil and vegetable crops along major highways in Agra (India). E-J Chem 7(4):1174–1183

Siddique A, Mumtaz M, Zaigham NA, Mallick KA, Saied S, Zahir E, Khwaja HA (2009) Heavy metal toxicity levels in the coastal sediments of the Arabian Sea along the urban Karachi (Pakistan) region. Mar Pollut Bull 58(9):1406–1414

Siddiqui SI, Chaudhry SA (2018) A review on graphene oxide and its composites preparation and their use for the removal of As3+ and As5+ from water under the effect of various parameters: application of isotherm, kinetic and thermodynamics. Process Saf Environ Prot 119:138–163

Siddiqui SI, Singh PN, Tara N, Pal S, Chaudhry SA, Sinha I (2020) Arsenic removal from water by starch functionalized maghemite nano-adsorbents: thermodynamics and kinetics investigations. Colloid Interface Sci Commun 36:100263

Signes-Pastor AJ, Carey M, Carbonell-Barrachina AA, Moreno-Jiménez E, Green AJ, Meharg AA (2016) Geographical variation in inorganic arsenic in paddy field samples and commercial rice from the Iberian Peninsula. Food Chem 202:356–363

Smedley PL (1996) Arsenic in rural groundwater in Ghana. J African Earth Sci 22:459–470

Smedley PL, Kinniburgh DG, Macdonald DM, Nicolli HB, Barros AJ, Tullio JO, Pearce JM, Alonso MS (2005) Arsenic associations in sediments from the loess aquifer of La Pampa, Argentina. Appl Geochem 20(5):989–1016

Smedley PL, Kinniburgh DG (2002) A review of the source, behaviour and distribution of arsenic in natural waters. Appl Geochem 17(5):517–568

Steinmaus C, Yuan Y, Bates MN, Smith AH (2003) Case-control study of bladder cancer and drinking water arsenic in the western United States. Am J Epidemiol 158(12):1193–1201

TatahMentan M, Nyachoti S, Scott L, Phan N, Okwori FO, Felemban N, Godebo TR (2020) Toxic and essential elements in rice and other grains from the United States and other countries. Int J Environ Res Public Health 17(21):8128

Tchounwou PB, Yedjou CG, Patlolla AK, Sutton DJ (2012) Heavy metal toxicity and the environment. Exp Suppl 101:133–164

USEPA (2009) National primary drinking water regulations. Arsenic and clarifications to 2009 May

USEPA (2022). https://www.epa.gov/system/files/documents/2022-03/fy-2022-2026-epa-strategic-plan.pdf

VanLoon GW, Duffy SJ (2005) Environmental chemistry: a global perspective. Oxford University Press, Oxford

Vodela JK, Renden JA, Lenz SD, McElhenney WH, Kemppainen BW (1997) Drinking water contaminants (arsenic, cadmium, lead, benzene, and trichloroethylene). 1. Interaction of contaminants with nutritional status on general performance and immune function in broiler chickens. Poult Sci 76(11):1474–1492

Wai KM, Umezaki M, Mar O, Umemura M, Watanabe C (2019) Arsenic exposure through drinking Water and oxidative stress Status: a cross-sectional study in the Ayeyarwady region, Myanmar. J Trace Elem Med Biol 54:103–109

Wang S, Mulligan CN (2006) Occurrence of arsenic contamination in Canada: sources, behavior and distribution. Sci Total Environ 366(2–3):701–721

Wen H, Zhang Y, Cloquet C, Zhu C, Fan H, Luo C (2015) Tracing sources of pollution in soils from the Jinding Pb–Zn mining district in China using cadmium and lead isotopes. Appl Geochem 52:147–154

Williams PN, Islam MR, Adomako EE, Raab A, Hossain SA, Zhu YG, Feldmann J, Meharg AA (2006) Increase in rice grain arsenic for regions of Bangladesh irrigating paddies with elevated arsenic in groundwaters. Environ Sci Technol 40(16):4903–4908

World Health Organization (1996) Trace elements in human nutrition and health. World Health Organization, Geneva

Wu C, Luo Y, Deng S, Teng Y, Song J (2014) Spatial characteristics of cadmium in topsoils in a typical e-waste recycling area in southeast China and its potential threat to shallow groundwater. Sci Total Environ 472:556–561

Wyatt CJ, Fimbres C, Romo L, Mendez RO, Grijalva M (1998) Incidence of heavy metal contamination in water supplies in Northern Mexico. Environ Res 76(2):114–119

Xia Y, Liu J (2004) An overview on chronic arsenism via drinking water in Pr China. Toxicology 198(1–3):25–29

Zamora-Ledezma C, Negrete-Bolagay D, Figueroa F, Zamora-Ledezma E, Ni M, Alexis F, Guerrero VH (2021) Heavy metal water pollution: a fresh look about hazards, novel and conventional remediation methods. Environ Technol Innov 22:101504

Zavala YJ, Duxbury JM (2008) Arsenic in rice: I. Estimating normal levels of total arsenic in rice grain. Environ Sci Technol 42(10):3856–3860

Zhang X, Wang X, Cao X, Xiao G, Miao H (2022) Heavy element contents of vegetables and health-risk assessment in China. Sci Total Environ 828:154552

Heavy Metal/Metalloid Contamination: Impact on Human Health and Mitigation Strategies

3

Seema Mishra, Apoorv Gupta, Ravi Kumar Tiwari, and Sanjay Dwivedi

Abstract

Heavy metals and metalloids (cumulatively referred to as metal(loid)s) are ubiquitous in the environment. They originate from Earth's crust, and their traces are present in all the environmental compartments. However, the concentration of many toxic heavy metal(loid)s has reached to a level causing serious health effects on humans. The sources and level of contamination of arsenic (As), cadmium (Cd), lead (Pb), and mercury (Hg) in water, soil, and food commodities have been discussed in details in the previous chapter. In the current chapter, the common and specific mode of toxicity through chronic dietary exposure of these priority heavy metal(loid)s to humans and target organs/systems has been discussed. The conventional and recent developments in mitigation techniques of these metal(loid)s from water and soil have also been included.

Keywords

Contamination in food · Heavy metal and metalloid; Impacts on human health · Provisional tolerable weekly intake · Mitigation strategies

S. Mishra (✉)
Department of Chemistry, University of Lucknow, Lucknow, India

A. Gupta
Department of Chemistry, Deen Dayal Upadhyay Gorakhpur University, Gorakhpur, India

R. K. Tiwari · S. Dwivedi
Plant Ecology and Environmental Science Division, CSIR-National Botanical Research Institute, Lucknow, India

R. P. Singh et al. (eds.), *Heavy Metal Toxicity: Environmental Concerns, Remediation and Opportunities*, https://doi.org/10.1007/978-981-99-0397-9_3

3.1 Introduction

Nowadays, heavy metal/metalloid contamination is a global issue which has severely perturbed the environment and plays a major role in poor human health. It has been observed that rapid urbanization and industrialization have led to high level of these elements in the environment and subsequently in food chain and food commodities. The problem is worst for the local residents in developing nations especially India, China, and Bangladesh where the compliances of environmental regulations are often compromised to keep lower costs of commodities. Further, many of the developing countries serve as global manufacturing hubs without stringent measures to protect their people and environment. At high concentrations, almost all heavy metals pose harmful effects; however, As, Cd, Pb, and Hg, being nonessential to plants, animals, and humans, are toxic even at low concentrations. Most of these heavy metal(loid)s have industrial significance; thus, they are extracted through mining and smelting and subsequently used in various manufacturing and domestic applications. In this way, mining and industrial effluents are major source of water and soil contamination leading to chronic metal poisoning episodes, such as itai-itai and Minamata disease in Japan due to Cd and methyl-Hg poisoning, respectively. Various domestic and industrial uses also contribute to significant amount of heavy metal(loid)s in the environment. Apart from these sources, the use of naturally As-enriched groundwater is a major cause of As contamination in water, soil, and subsequently food leading to chronic As poisoning in many parts of world. Water and food are the major ingestion routes of these heavy metal(loid)s in humans which leads to many serious health consequences including cancer (Adepoju-Bello and Alabi 2005). Occupational exposure, which involves working in industries like electroplating, mining, smelting, metallurgy, pesticides, or paint- and pigment-related occupation, also causes organ failures and serious illnesses. Since heavy metal(loid)s are nonbiodegradable, they tend to accumulate in various organs and adversely affect almost all vital processes and body organs, such as the skin, lungs, liver, kidneys, prostate, esophagus, stomach, central nervous system, brain, etc. Heavy metal(loid)s are also genotoxic, and their prolong exposure leads to various kinds of cancers. The general mechanism of metal(loid)s toxicity is the formation of metal-protein complexes, involving carboxylic acid, amine, and thiol functional groups. Upon binding to these metal(loid)s through the above mentioned functional groups, which have been linked to the catalytic or structural properties of enzymes and proteins, the enzymatic systems get inactivated, and protein structure gets affected. This results in redox imbalance and malfunctioning or death of the cells. Heavy metal(loid) toxicity also results in increased formation of free radicals which are generally dangerous causing the oxidation of biological molecules and results in genotoxicity. Arsenic, Cd, Pb, and Hg are metal(loid)s of major environmental and human health concern due to their high toxicity even at low concentration and/or high prevalence in the environment (Yedjou and Tchounwou 2007; Patlolla et al. 2009; Tchounwou et al. 2012). Depending on the direct or indirect evidence of genotoxicity or carcinogenicity on humans, they are classified as carcinogens (known or probable) by the International Agency for Research on Cancer (IARC) and USEPA. Arsenic and Cd are known carcinogens, and many of Pb and Hg compounds are also potential carcinogens (see Table 3.1 of the previous chapter). Though the general mechanism of toxicity of heavy metal(loid)s may resemble in terms of inactivation of vital enzymes and proteins and generation of reactive oxygen species (ROS), these may differ considerably in their specific mode and potentiality of toxicity depending on their mobility in environment and chemical form. For example, As, being a redox element, exists in several chemical forms which differ considerably in their toxicity (Leermakers et al. 2006). Mercury also exhibits different oxidation states and chemical forms with variable toxicity (García-Sánchez et al. 2008). Cadmium is toxic at very low concentration due to higher water and soil mobility, thus higher availability for plant accumulation and human exposure (Mishra et al. 2006; Fatima et al. 2019; Tchounwou et al. 2012). Cd exposure leads to kidney damage and hypertension, while Hg toxicity causes mental disturbance and motor dysfunction leading to impairment of speech, hearing, vision, and movement (Fields et al. 2017). Lead is a persistent toxicant and tends to accumulate in high concentrations in environment as well as in living systems causing chronic toxicity (Ab Latif Wani and Usmani 2015; Tchounwou et al. 2012). The central nervous system is most affected by lead toxicity, infants and children are particularly vulnerable to Pb even at low concentrations, causing intellectual disability and behavioral disorders. Thus, by consuming contaminated water, fishes, and food crops, humans get exposed to heavy metal(loid)s, often exceeding the maximum tolerable daily/weekly intake (MTDI/MDWI) values, leading to severe health hazards. They also hamper plant growth and productivity and hamper mineral nutrient acquisition, thus compromising quantity and nutritional quality of food. In this review, we discuss the effect of toxic heavy metal(loid)s on human health and their mitigation strategies employed globally for removal /reduction from water and soil and lowering the accumulation in plants and subsequently resulting in less exposure to humans.

3.2 Impact of Heavy Metal/Metalloid Intake on Human Health

In the view of the ubiquitousness and toxic nature of metal(loid)s, their consumption by human should be tightly regulated. Thus, depending on various studies, references doses (Rf) or provisional maximum tolerable daily or weekly intake values (MTDI/PTWI), which are considered nontoxic or have negligible toxicity on a lifetime exposure, has been set by different regulatory agencies (given in Table 3.1). Consumption of heavy metal(loid) beyond these limits can adversely affect different body organ and systems. Inactivation of enzymes, ROS generation, inhibition of antioxidant defense system, cellular redox imbalance, and multiple organ toxicity are the common mechanisms proposed for heavy metal(loid) toxicity. For example, As, Cd, Pb, and Hg all are reported to inactivate glutathione peroxidase and glutathione reductase enzymes. Some common and element-specific toxicity to human metabolic pathways/systems has been demonstrated in Fig. 3.1. The toxicity mechanism and diseases of each metal(loid)s have been elaborated below.

Table 3.1 Major food sources of heavy metal(loid)s to human, their permissible limits, and their toxicity to humans

S. no.	Element	PTWI (μg/kg BW)	Regulatory body	Food sources	Toxicity	References
1.	Arsenic	15	WHO/FAO 1989	Cereals like rice and wheat, vegetables, fish, shellfish, meat, poultry, dairy products	Encephalopathy, peripheral vascular disease, bone marrow depression, hepatomegaly, diarrhea, nephropathy, long QT syndrome, cancer (lungs, kidney, bladder, and skin), diabetes, cardiovascular diseases, hyperkeratosis, melanosis, speech and visual perception	Nickson et al. (2000), Smith et al. (2000), Riethmiller (2005), Leermakers et al. (2006), Rosado et al. (2007), Rai et al. (2019)
2.	Cadmium	2.5	EFSA 2011	Rice, shellfish, mussels, dry seaweed, shrimps, mushroom	Kidney failure, pneumonitis, inhibition of progesterone and estradiol, endocrine disruption proteinuria, osteoporosis, cancer (kidney, liver, prostate, hematopoietic system, stomach, lung, and breast), itai-itai	Waalkes et al. (1995), Baba et al. (2013), Rai et al. (2019)
3.	Lead	25	JECFA 2000	Fruits, vegetables, grains, seafood, red meat, wine, soft drinks	Encephalopathy, anemia, vomiting, nausea, nephropathy, abdominal pain, lack of attention, headache, memory loss, DNA breakage, and induced mutagenicity	Roy and Rossman (1992), Ab Latif Wani and Usmani (2015), Rai et al. (2019)

4.	Mercury	1.6	JECFA 2003	Fishes, crustaceans, pron fish, seafood, mushroom	Lung and kidney damage, hypersensitivity, amalgam disease, proteinuria, allergy, nephrotic syndrome, nephropathy, perturbed central nervous system (CNS), Minamata disease	Tchounwou et al. (2012), Rai et al. (2019)

PTWI, provisional tolerable weekly intake; EFSA, European Food Safety Authority; WHO/FAO, World Health Organization/Food and Agriculture Organization; JECFA, Joint FAO/WHO Expert Committee on Food Additives

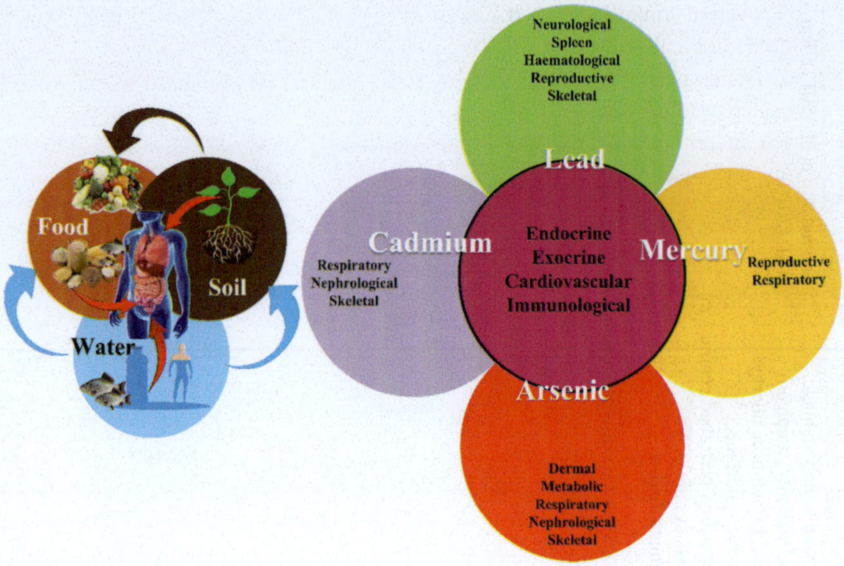

Fig. 3.1 Major routes of heavy metal(loid) exposure to humans and metabolic systems effected by their chronic exposure

3.2.1 Arsenic

Chronic As poisoning affects all body organs and systems. Inorganic As is classified as group 1 human carcinogen. Several carcinogenic and noncarcinogenic diseases, such as skin, lung, bladder, kidney, and liver cancer and neurological, respiratory, cardiovascular, and developmental disorders, have been reported as a result of chronic As exposure through drinking water (Smith et al. 2000; NRC 2013). It causes a characteristic pattern of dermal effects like melanosis, keratosis, and hyperkeratosis (Mandal et al. 1996). The appearance of keratosis and melanosis together is a typical indicator of arsenic toxicity. Recent studies have shown an inverse relationship with chronic As exposure and functions of the lung (Siddiqui et al. 2020). The accumulation of As in hair and nail and the level of As in urine are typical marker of As poisoning. The level of As in blood, urine, hairs, and nails has been found to be correlated with the level of As in drinking water (Ahamed et al. 2007). Chronic As toxicity has been reported to cause reproductive abnormalities like spontaneous abortion, stillbirth, preterm birth, low birth weight child, and neonatal death. Moreover, it has been reported that young adults who were exposed to As at fetus stage or during early life have increased risk of various types of cancers and other diseases and have higher risk of mortality (Smith and Steinmaus 2009; NRC 2013). A study conducted on Bangladeshi population concluded that As toxicity affects mental health (Brinkel et al. 2009). Arsenic toxicity may also cause reduction in locomotor activity that ultimately disturbs the central nervous system (Reiter et al. 2010). Children exposed to higher As concentration were found to show a low verbal intelligence (Calderon et al. 2001). Arsenic also causes neurotoxicological effects causing reduced cognitive performance and disturbances in speech and visual perception (Rosado et al. 2007). Arsenic is a known carcinogen, and an increased prevalence of skin, lung, bladder, kidney, and liver cancer has been reported in populations exposed to chronic As in India, Bangladesh, Vietnam, Taiwan, and elsewhere in the world (Mazumder et al. 2000; Smith et al. 2000; IARC 2017).

3.2.2 Cadmium

Bioaccumulation of cadmium in human and animals is comparatively higher than the other heavy metal(loid)s because it is assimilated rapidly and excreted slowly. Blood and urine Cd level is the primary marker of Cd exposure. Blood cadmium depicts recent exposure, while urine Cd indicates kidney burden, i.e., chronic exposure of Cd (Järup 2003). Cigarette smokers and metal workers contain higher blood and urine Cd level (Wittman and Hu 2002; Mannino et al. 2004). About 2.3% of the US population has been estimated to have elevated cadmium, i.e., >2 μg/g creatinine in urine (Tchounwou et al. 2012), while it was >15 μg/g creatinine in the Chinese population exposed to Cd for a long time. Oral ingestion or inhalation of Cd may lead to fatality because Cd is a severe gastrointestinal and pulmonary irritant. Chronic Cd exposure has been reported to cause kidney disease, osteoporosis, cardiovascular diseases, and cancer (Waalkes et al. 1995; Schutte et al. 2008; Gallagher et al. 2008). Cd also gets accumulated in the liver and kidney casing hepatic and renal injuries. Cd is highly hepatotoxic and causes irreversible renal injuries (Baba et al. 2013). The most severe chronic poisoning of Cd was identified in Japan after World War II. The inhabitants of the Jinzu River Basin in Toyama Prefecture of Japan were suffering with itai-itai disease due to the use of river water which was severely polluted by Cd from a Zn mine situated upstream of the river. In itai-itai disease, the renal dysfunction caused by Cd leads to insufficiency of active vitamin D which is characterized by fractures and severe pain (Aoshima 2012; Baba et al. 2013). Subsequently in the late 1980s, chronic Cd toxicity due to consumption of Cd-contaminated rice was observed in China (Cai 1989). Up to now, several studies have shown that chronic Cd exposure, even at low concentration, causes decrease in bone density and osteoporosis (Schutte et al. 2008; Gallagher et al. 2008). Some epidemiological studies have reported a potential association of Cd exposure and diabetes (Bernard 2004). Cadmium compounds are also classified as group 1 human carcinogens with pulmonary system being the most established target through rodent data and strong association observed between lung cancer and occupational Cd exposure (IARC 1993). Chronic Cd exposure causes pulmonary adenocarcinomas and has been also associated with kidney, liver, prostate, hematopoietic system, and stomach cancers (Waalkes et al. 1995). Cd-induced ROS generation leading to cell damage has been proposed as primary mechanism of Cd toxicity. It disrupts protein and nucleic acid synthesis and DNA damage (Tsuzuki et al. 1994; Stohs and Bagchi 1995; Fatima et al. 2019). Cadmium-induced cytotoxicity and DNA damage have been observed during in vitro studies, while in vivo studies have shown modulation of male reproductive organs in mice (Mukherjee and Das 2002). Earlier studies have reported that Cd affects signal transduction, blocking Ca channels, and causes increase in cytosolic-free Ca (Thevenod and Jones 1992).

3.2.3 Lead

Lead is highly persistent in the environment, and it is quickly absorbed in the bloodstream and affects vital organ systems like the cardiovascular system, the central nervous system, kidneys, and the immune system (Bergeson 2008; Ab Latif Wani and Usmani 2015). Drinking water and food are main sources of human Pb exposure in which drinking water accounts for 35–50% of Pb absorption in adults; it is even higher in children (ATSDR 1992, 1999). Pb is a known neurotoxin, reproductive toxin, and potential carcinogen. It gets accumulated in vital organs like the kidney, liver, heart, and brain tissues with highest proportion being in the kidney (Flora et al. 2006). The nervous system is the prime target of Pb poisoning leading to several neurological problems, such as lack of attention, headache, and memory loss. It causes several birth defects if ingested during pregnancy, like premature birth and low birth weight (Ab Latif Wani and Usmani 2015). Exposure during pregnancy causes transfer of Pb to the fetus and may cause abnormal brain development (Ong et al. 1985; Huel et al. 1992). The inhibition of NMDA receptor and the blockage of neurotransmitter release lead to Pb-induced neurotoxicity and cognitive impairment (Neal and Guilarte 2010). Children are more prone to Pb poisoning through deteriorated Pb paints on walls, soil surfaces, dust, and chips. Charney et al. (1980) reported that the level of Pb in blood was significantly higher in children who have damaged Pb paints in their homes. Pb easily gets transferred to the fetus if consumed by pregnant women (Ong et al. 1985). Pb exposure during pregnancy results in abnormal brain development in children (Huel et al. 1992). Pb reacts with proteins and hampers biological processes; it also inhibits Ca action (ATSDR 1999). Pb replaces Ca from biological molecules, thereby interfering with their normal actions (Flora et al. 2007). Apoptosis is also one of the toxicity mechanisms of Pb in which Bcl-2 gets inhibited and caspase-3 gets activated following the release of cytochrome C from mitochondria (Liu et al. 2012). Pb-induced oxidative stress has been reported as the major mechanism of toxicity leading to membrane damage. Some studies have reported gene expression alterations induced by Pb (Rossman 2000). Pb has been shown to interact with the Zn-binding site in protamine, an important human DNA-associated protein, and may cause altered gene expression (Quintanilla-Vega et al. 2000). Studies have indicated that Pb also affects activity of glucose-6-phosphate dehydrogenase (G6PD) (Cocco et al. 1995; Gürer et al. 1998). Pb interferes with the heme synthesis by inhibiting δ-aminolevulinic acid dehydratase (δ-ALAD) enzyme (Ahamed and Siddiqui 2007; Skoczynska 2008) and also interferes with the incorporation of iron in heme by inhibiting the enzyme ferrochelatase. Further Pb also disrupts the membrane integrity of RBCs (White et al. 2007). Thus, Pb-induced anemia is one of the prominent toxicity symptoms of Pb. The mechanism of neurotoxicity of Pb is proposed through the interference of Pb on Ca-regulated systems (Marchetti 2003; Toscano and Guilarte 2005). Another proposed mechanism for lead-induced neurotoxicity is calmodulin and cAMP phosphodiesterase-mediated interference in the release of neurotransmitter. Stimulation of cAMP phosphodiesterase and calmodulin has been reported in the presence of Pb (Verstraeten et al. 2008). Pb inhibits the level of glutathione reductase, thus reducing the level of glutathione, the main antioxidant in body. It also induces the generation of reactive oxygen and reactive nitrogen species (Patrick 2006). Thus, the disturbed redox state results in oxidative stress. Lead is classified as probable human carcinogen by IARC based on animal data for sufficient carcinogenicity; however, there is limited evidences on human (IARC 2006). Studies have demonstrated that lead compounds cause oxidative damage to genetic material. It inhibits the enzymes/proteins involved in DNA synthesis and repair and interferes with DNA-binding and tumor suppressor proteins. Toxic dose of lead acetate and lead nitrate has been shown to cause DNA breakage and induced mutagenicity (Roy and Rossman 1992). In vivo studies on rats showed that lead is potentially carcinogenic, inducing renal tumors in rats and mice (Waalkes et al. 1995). Pb-induced hematotoxicity and hepatotoxicity has been reported. Pb causes toxicity to respiratory and renal systems and may result in necrosis.

3.2.4 Mercury

The toxicity of Hg depends on its chemical form with elemental mercury, and methyl mercury being readily absorbed by the body are highly toxic (Clarkson et al. 2003). Humans are exposed to all forms of Hg through various ways like drinking Hg-contaminated water, consumption of food contaminated with Hg, occupational, dental care, preventive medical practices, industrial and agricultural operations, etc. Elemental Hg vapors get effectively absorbed through the mouth and lungs to the bloodstream and quickly pass to biological membranes. Inside cells, elemental mercury gets oxidized to Hg^{2+}, which is highly reactive and may cause oxidative damage at several levels. Methyl-Hg (CH_3-Hg) which is generally taken up by eating fishes is highly lipophilic and readily absorbed in the gastrointestinal tract. Both elemental Hg and CH_3-Hg cross the placental and blood-brain barriers. Excretion rate of Hg is extremely low; thus, once absorbed in the body, most of it accumulate in the kidneys, neurological tissue, and the liver and cause gastrointestinal toxicity, neurotoxicity, and nephrotoxicity (Tchounwou et al. 2012). The neurotoxicity incident due to chronic exposure to organic Hg by consuming contaminated fish in Minamata Bay, Japan, during the 1950s is well known. Later, the disease was known as Minamata disease. Several researches have suggested that oxidative stress is the main mechanism of Hg toxicity. Both ionic (Hg^{2+}) and CH_3-Hg have high affinity with sulfhydryl groups; thus, they form covalent bonds with cysteine residues of proteins and deplete cellular antioxidants. Increased ROS production has been observed during Hg toxicity (Shenker et al. 2000). Inorganic Hg is reported to cause increased production of ROS by interfering with oxidative phosphorylation and electron transport in mitochondria (Palmeira and Madeira 1997). Mercury compounds have been shown to cause lipid peroxidation and increase the level of MDA. Due to preferential accumulation of Hg in the kidneys, it causes renal dysfunction especially in the proximal tubules. The toxic effect of Pb on the intestine and liver has been also documented (Kim et al. 1995; Cheng et al. 2006). The carcinogenicity of Hg is controversial. Some studies report potential genotoxicity of Hg, while others show no association between mercury exposure and genotoxic damage (Valko et al. 2004).

The mechanism of carcinogenicity of heavy metal(loid)s is still elusive and intricate. It is assumed that heavy metals interfere the function of regulatory proteins involved in cell cycle regulation, DNA synthesis and repair, and apoptosis by binding to their active sites. Mutation in Ras proteins by heavy metals has been also proposed. Overexpression of Ras in human prostate epithelial cells has been shown upon As exposure. Activator protein 1 (AP-1), transcription factors like nuclear factor-kappa B (NF-κB), and tumor suppressor gene p53 are targets for Cd and As leading to uncontrolled cell growth and division (Valko et al. 2004). In vitro studies show that the level of ERK 1/2, a key protein for cell proliferation control and transcription factors jun and fos, was elevated by Cd. Arsenic inhibits poly (ADP-ribose) polymerase 1 (PARP-1), an enzyme responsible for DNA repair, thus causing DNA damage (Ding et al. 2009). Cd-induced malignant transformation of the prostate epithelial cell line through increased apoptotic resistance has been observed in vitro (Qu et al. 2007). Epigenetic alterations, including DNA methylation and histone modification, have been also observed by heavy metal(loid)s. DNA methylation inhibits the expression of some tumor suppressor causing malignant transformation of cells (Li and Chen 2016). Cd and As are reported to cause DNA methylation and specific histone modification (Ma et al. 2016). A recent study showed that chronic As exposure causes transgenerational effects through DNA methylation showing genotoxicity and reproductive effects up to F3 generation (Nava Rivera et al. 2022).

3.3 Mitigation Strategies of Heavy Metals/Metalloid

3.3.1 Heavy Metal Mitigation from Water

Access to clean water of the population is the main concern for the governments all over the world. Groundwater has been seen as a source of clean drinking water, particularly in developing countries. But contamination of groundwater by toxic heavy metal(loid)s has made the situation worst. Removal of heavy metal from water is expensive. Several methods have been developed including physical, chemical, and biological methods. However, until now, the environmentally friendly, cost-effective, and efficient water treatment process is a major challenge all over the world. Wastewater treatment is generally a three-step process. At the primary step, the organic matter and suspended solid are removed by using processes like filtration, sedimentation, gravity separation, and centrifugal separation. In the second step, soluble and insoluble inorganic and organic pollutants are removed via biological processes under aerobic and anaerobic conditions. Finally in the third step, various toxic heavy metal(loid)s are removed from the water. The third step is crucial and involves various chemical and physical methods like oxidation, adsorption, precipitation, ion exchange, etc. Some of the important heavy metal(loid) removal techniques have been discussed below in brief:

Oxidation/precipitation: This method is widely used in industries on a large scale for the treatment of industrial effluent contamination (Chen et al. 2013). The oxidation method was introduced in the 1980s utilizing a powerful oxidizing agent. The main principle of this method is based on production of hydroxyl group, a strong oxidant by using the oxidizing reagents like hydrogen peroxide, ozone, and oxygen in the presence of UV light, and a catalyst, normally TiO_2, is used. The bulk production of hydroxyl group oxidizes the water contaminants which are precipitated as metal oxides and are removed by filtration (Bora and Dutta 2019). The oxidation/precipitation method is also useful for inactivation of pathogens.

Ion exchange: This method is used to remove ionic impurities including toxic metals from water and is used in the reverse osmosis (RO) method to make potable water. The method was introduced by Thomas Way in 1850 for water treatment by using a solution of ammonium sulfate. In this method, synthetic resin like zeolites is used as medium for exchanging ionic impurities like heavy metals by similar charged ions (Kurniawan et al. 2006). During this procedure, dealkalization, deionization, and disinfection of water also happen. It is also used for removing hardness of water like calcium and magnesium ions and making the water soft.

Electrokinetic process: The electrokinetic method has been used to reduce the content of contaminants like heavy metals, pesticides, and polychlorinated biphenyls (Hamdan et al. 2014) This process was primarily used to remove heavy metal from contaminated fine grain soils through electrical adsorption. In electrokinetic process, low electricity current is used to mobilize the metals through electroosmosis (EO), electromigration (EM), and electrophoresis (EP). In addition to these transport mechanisms, electrochemical reactions such as electrolysis and electrodeposition may also take place. This process can be utilized both ex situ and in situ. During EO, fluid flows from anode to cathode. In EM process, reduction of cations at cathode and oxidation at anode occurs under electric field due to gain and loss of electrons at cathode and anode, respectively. This process is gaining more utility for soil remediation and can also be coupled with phytoremediation (Sanchez et al. 2019).

Membrane filtration/reverse osmosis: Currently, membrane filtration is the most used water purification technology and being more advertised for reducing heavy metal contamination. It also eliminates the dissolved solids and organic impurities from water. The membrane filtration technique involves two processes, i.e., adsorption and filtration. Depending on the size of impurities, ultrafiltration, nanofiltration, and RO method are introduced with specific pore size of the permeable membrane. In ultrafiltration, the pore size of membrane is 520 nm which can eliminate heavy metals, macromolecules, and suspended solids from contaminated water. The disadvantage in this method is that the filtration is limited to low molecular weight contaminants and nanoparticles. Thus, nowadays, polymer-supported ultrafiltration is gaining attention for eliminating the heavy metals from water (Bolisetty et al. 2019; Van der Bruggen and Vandecasteele 2002). However, for more efficient removal of heavy metals from water, the RO process and nanofiltration have also been tested and used. In the last decade, RO method has been more successfully used at community and domestic level to produce drinking water for direct consumption (Shahalam et al. 2002).

Adsorption method: Due to easy handling and low sludge generation, adsorption method is being increasingly used for purification of water. The vast availability of adsorbents, like nanomaterials, carbon materials, alumina, resins, etc., has made this method cost-effective too (Singh et al. 2018). Activated carbon is the most utilized adsorbent which eliminates both organic and inorganic pollutants including heavy metals (Abdulla et al. 2019; Siddiqui et al. 2019). The use of bio-based waste materials like coconut husk, peanuts, tree leaves, etc. to prepare activated carbon has further reduced its cost. In general, two types of adsorbent is being used, i.e., as powder and as a template.

In powder method, zeolite, iron core, resin, activated carbons, and graphene and carbon nanotubes are used to adsorb contaminants from water. Since ancient time, various powder materials, like iron and carbon-based materials, are used to remove heavy metals from water. However, still research is going on to improve properties of the adsorbents which can efficiently remove heavy metal(loid)s from water and can be recycled. Currently, graphene, carbon nanotubes, activated carbon, and carbon fiber are increasingly being used due to large surface area, thermal stability, and more active sites to adsorb contaminant (Wang et al. 2021). However, the separation of nanomaterial from water and the disposal of the used material are major limitations in nanomaterial-based powder method. In this backdrop, template materials prove better option.

In template method, a template of porous carbon materials is used to adsorb heavy metal from contaminated water. The porous carbon provides some unique properties like available adsorption sites on the adsorbent's surface, thermal stability, chemically inert, as well as easily separable from the purified water. In the recent years, considerable progress has been made in this direction, and several porous carbon materials have been made. For example, for the elimination of copper and lead from water, mesoporous carbon foam has been used (Burke et al. 2013; Lee et al. 2015), and for the chromium and arsenic decontamination, 3D magnetic graphene oxide foam (graphene oxide/Fe_3O_4) and 3D graphene oxide foam (graphene oxide/$CuFe_2O_4$), respectively, are used (Lei et al. 2014). Different types of carbon foams like carbon nanotube-embedded, resin-based, and functionalized carbon foams, graphite intercalation compound, etc., can also be used (Agrawal et al. 2019). These materials have improved efficiency and can also be recycled.

3.3.2 Heavy Metal Mitigation Strategies from Soil and Food Crops

Soil is the substrate from which toxic contaminants get transfer to crops, vegetables, and other agricultural produce. Soil gets loaded with heavy metal(loid)s, also from wastewater or from naturally contaminated irrigation water during years of applications. Therefore, the transfer of heavy metal(loid)s in the soil-crop system could be prevented by remediation of soil. Remediation technologies should be rapid, cost-effective, and environmentally friendly. The remediation techniques can be broadly categorized as physical, biological, ecological, and chemical approaches. Like in water purification, recent innovations in nanotechnology could also aid to soil remediation. Preventive measures to minimize further soil contamination is utmost important, for example, modifications in land-use policies to locate agricultural fields distant from municipal and industrial sources of heavy metals and reducing agrochemicals in fields, i.e., switching to organic cultivation. The methods for reduction of heavy metals from soil-crop system are summarized below:

Source reduction: These involve adequate treatment of industrial effluent, sewage, and sewage sludge before release in the environment. Strict air-quality management measures may significantly reduce contamination on food crop because most of the vegetables absorb metals from leaves. Roadside food crops are prone to metal contamination though deposition on leaves, e.g., Pb contamination in the leaves of *Amaranthus dubius* (Nabulo et al. 2006). Emission from urban traffic is also a source of heavy metal contamination in crop and vegetables grown in roadside areas. The area within 30 meters of the roadside is particularly prohibited for the cultivation of vulnerable crops (Liao et al. 2016). In a study, severe health risks to the population have been predicted due to the consumption of heavy metals contaminated with food crops like various vegetables, rice, and sugarcane near an acidic mining drainage area through atmospheric PM deposition (Liao et al. 2016). Heavy metal contamination in soil and plants has also been observed in green house through fertilizer and pesticide application. Root, vegetables, and fruit crops were the least affected by heavy metal(loid)s in comparison to leafy vegetables and thus can be grown in greenhouses (Hu et al. 2016; Zhang et al. 2017a). Hg and Pb are the main health risks from the products grown in greenhouses (Fan et al. 2017) resulting from insecticide application (Xu et al. 2015), while Cd and As can be brought into such controlled environment by the livestock manure (Li et al. 2009; Xu et al. 2015).

Eco-remediation: Biochar derived from waste has been widely used for heavy metal remediation from contaminated soil in different areas and has excellent immobilization effect (Wang et al. 2021). In addition to reduce heavy metal availability, biochar and composting can improve soil nutrient cycle, cation exchange capacity, and humidification, thus improving soil health (Beesley et al. 2014; Wu et al. 2017). Further, biochar in combination with metal-resistant bacteria has been found to result in remarkable minimization in Cd bioavalability in Cd-contaminated paddy soil (Li et al. 2018). In recent years, biochar amendment is gaining popularity for removal of As from soil and drinking water, as a low-cost, eco-friendly tool (Wang et al. 2017). However, several studies have shown that abiotic factors such as changes in soil condition (pH, redox, and dissolved organic matter), acid rain, flooded environment, and changes in biochar (Cl^- and alkali leaching) and biotic factors such as plant roots, soil microorganisms, and earthworms significantly limit and even reduce the effect of biochar with time (Wang et al. 2021). Further, the effect of biochar is not uniform for all heavy metal and metalloid. It has limited success for immobilization of As because the net negative charge present on most of the biochar surfaces restricts the sorption of negatively charged arsenate and arsenite onto biochar (pH 7–7.5) (Beesley and Marmiroli 2011). In contrast, the alkaline nature of biochar results in the increase of soil pH, which elevates As release, resulting in poor As removal efficiency from biochar (Qiao et al. 2018). Integrating phytoremediation with organic or chemical amendments may reduce the bioavailability of As in plant-soil systems. For instance, fern *Pteris vittata* in combination with biochar and activated carbon remarkably minimize As hazard in soil (Lessl et al. 2014; de Oliveira et al. 2017). Combination of *P. vittata* and activated carbon reduced plant available As in lettuce (*L. sativa*) by up to 22%. Biochar amendments are more effective for Pb and Cd than other metals, such as Co, Cr, Mn, Ni, and Zn, in the edible parts of crop (Rizwan et al. 2016; Peng et al. 2018). In a greenhouse study, application of biochar resulted in adverse effect on plants grown on slightly Cd-contaminated soil, i.e., it reduced the biomass and altered the nutritional content of Cd, whereas in heavily contaminated soil the Cd was effectively remediated through enhanced complexation of Cd (Zhang et al. 2017b). Thus, characteristics of soil and extent of heavy metal contamination are important factors impacting the efficacy of biochar. More research is needed to improve the efficiency of biochar in immobilization of heavy metals and its potentiality in soil remediation. Nanomaterials can remarkably increase the active sites and surface properties of biochar. Fortification of biochar with zerovalent iron nanomaterial has been promising (Ho et al. 2017); however, more research is needed to understand their sustainability and long-term environmental effects.

Phytoremediation: Phytoremediation is a green technology which involves the use of plants to remove contaminants from the substrate, i.e., soil and water. It comprises phytoextraction, i.e., removal of contaminant from soil by transferring them to above-ground part of the plants; rhizofiltration, i.e., removal of contaminant from water through plant roots; phytodegradation, i.e., degradation of contaminant from root secretions; phytostabilization, i.e., immobilization of contaminant in soil; and phytovolatilization, i.e., plant-mediated volatilization of contaminants (Tripathi et al. 2008). Phytoremediation technologies have been particularly applicable to wetland plants and have been successfully applied for removing heavy metals from rivers (Rai 2018a, 2018b). Phytoremediation can be an efficient technique for removal of heavy metal(loid) from soil and water by using appropriate plant for specific site. Hyperaccumulator plants also referred to as metallophytes are aquatic and terrestrial plants that can accumulate significantly high amounts of metal(loid)s from their substrate and, thus, can be used to phytoextract heavy metal(loid)s from water and soil. Hyperaccumulator plants have been successfully used for the extraction of different value metals, i.e., nickel phytomining by *Alyssum bertolonii* and *Pycnandra acuminata* (Robinson et al. 1997; Jaffré et al. 2018) and Hg phytostabilization by *Ipomoea reptans* (Sikanna et al. 2019). Some of the

Table 3.2 Some hyperaccumulator plants for potential use in phytoremediation of particular heavy metal/metalloid

S. no.	Element	Plant species	Metal accumulation (mgkg^{-1})	References
1.	Arsenic	*Isatis cappadocica*	>500	Kalve et al. (2011), Ma et al. (2001), Srivastava et al. (2006), Zhao et al. (2002), Karimi et al. (2009), Campos et al. (2015)
		Pteris vittata	>1000	
		Pteris cretica	>1800	
		Pteris longifolia	>2000	
		Pteris umbrosa	>2000	
		Pityrogramma calomelanos	>6000	
2.	Cadmium	*Thlaspi caerulescens*	>2000	Rai et al. (2009), Chehregani and Malayeri (2007), Wei and Zhou (2008), Lombi et al. (2001), Zhang et al. (2010), Stein et al. (2017), Zehra et al. (2020), Zheng et al. (2022)
		Azolla pinnata	740	
		Eleocharis acicularis	239	
		Rorippa globosa	100	
		Solanum photeinocarpum	158	
		Helianthus annuus	~ 40	
		Arabidopsis halleri	>100	
3.	Lead	*Euphorbia cheiradenia*	1138	Chehregani and Malayeri (2007), Mesjasz-Przybylowicz et al. (2001), Yoon et al. (2006)
		Minuartia bulgarica	>1500	
		Phyla nodiflora	>1000	
		Viola boashanensis	>1000	
		Gentiana pennelliana	>900	
		Noccaea rotundifolia	>1000	
4.	Mercury	*Medicago sativa*	>10	Shehu et al. (2014), Manikandan et al. (2015), Xun et al. (2017)
		Mentha arvensis	>1200	
		Cyrtomium macrophyllum	397	

hyperaccumulator/efficient accumulator plants for different heavy metal(loid)s which has been used/can be used for phytoremediation purposes have been listed in Table 3.2. In contrast to traditional chemical technologies, phytotechnologies are cost-effective and eco-sustainable. Although conventional physicochemical methods may be faster and transiently more effective for highly contaminated sites, but they are highly expensive and could induce secondary pollution by irreversibly changing soil properties (Mahar et al. 2016; Ashraf et al. 2019). Among different experimentally tested native plants in Florida, USA, the phytostabilization of Pb, Cu and Zn was effectively observed by *Phyla nodiflora*, and *Gentiana pennelliana*. Macrophytes and wetland plants are being used for the removal of heavy metals from the effluent from the sponge iron industries, and results are promising; thus, treated effluent can be used for irrigating agricultural fields (Gupta et al. 2008). Nonetheless, the effectiveness of phytotechnologies may be limited and time-consuming for reclamation of highly metal-contaminated sites. Further, despite immense potential of phytoremediation, there are certain limitations, for example, disposal of heavy metal-loaded biomass and transfer of contaminants from one compartment to another, e.g., phytovolatilization which transfers contaminants from soil to air (Muthusaravanan et al. 2018). However, for an integrated participatory approach of remediation soil scientists, biologists, farmers and site-specific management practices can be prepared, and sustainable utilization/safe disposal of metal-contaminated biomass, e.g., extraction of metals for industrial uses, can be done. In China, eco-remediation of heavy metal-contaminated soil through phytoextraction is assisted by scientific crop rotation systems for three types of oil crops. Yang et al. (2017) found the maximum efficiency for remediation of carcinogenic Cd by oilseed-rape-sunflower rotation. Heavy metal remediation can also be achieved by adding a beneficial microbial community to soil. For example, some plant growth-promoting bacteria have been found to enhance the phytoremediation of toxic metal(loid)s in crops like rice, wheat, maize etc. (Belimov et al. 2015; Hassan et al. 2016; Ashraf et al. 2017). Bioavailability/phytoavailability of heavy metals can be greatly reduced by plant-microbe interactions. Arbuscular mycorrhizal fungi have been found to reduce heavy metal accumulation and toxicity in plants (Hu et al. 2016).

Chemical or physicochemical remediation techniques: Although chemical remediation techniques are given lesser preference over biological processes due to cost intensiveness and alteration in soil characteristics, however, they have been used to restore many heavy metal-contaminated sites. These involve metal complexation in soil by using chemicals like EDTA to lower the health risks by making them less available to crops (Udom et al. 2004). In the recent years, researches are focused on developing green chemicals having no or less impact on soil health, like ferrate (Rai 2018a, 2018b). Synthetic zeolites, along with alkaline clay, have been found useful for remediation of heavy metal-polluted soil. Similarly, hematite, magnetite, zeolite, maghemite, hydrous manganese oxide, red mud, silicon calcium fertilizer and amended biochar have also demonstrated good heavy metal removal capacity in several studies (Yao et al. 2017; Rai 2018a). These amendments lead to immobilization of heavy metals, like Cd and As, in the soil in the form of hydroxides, phosphates and silicates which has been confirmed by state-of-the-art techniques like powder X-ray diffraction and SEM-EDS (Yao et al. 2017). Mandzhieva et al. (2017) practiced a 3-year field experiment on spring barley (*Hordeum sativum*) and found that amendments of chalk and manure (containing glauconite or natural zeolite, chalk and manure) reduced the mobility of heavy metals in soil as well as in *H. sativum*. The availability of Pb and Cd was reduced by ≥42.14% and ≥ 54.13%, respectively, in soil during wheat cropping by the use of single superphosphate, triple superphosphate, and calcium magnesium phosphate sepiolite in combination with $ZnSO_4$ (Guo et al. 2018). For leafy vegetables, phytoavailability of Pb and Cd was decreased by chemical amendment (phosphate containing diammonium phosphate and hydroxyapatite) (Waterlot et al. 2017). Among these, phosphate amendment was more effective in regulating the phytoavailability and bioavailability of Pb in comparison to Cd. In another study, As-induced toxicity in *Spinacia oleracea* (spinach) was greatly reduced upon addition of Ca and EDTA, thus minimizing the toxicity to humans also by reducing its bioavailability (Shahid et al. 2017).

Nanoparticle techniques: In the recent years, the area of agro-nanotechnology is being extensively researched. There are several studies showing the use of nanoparticles (NPs) for reducing heavy metal bioavailability (Shalaby et al. 2016; Rai 2018a). To ensure soil security, phytosynthesis of nanomaterials utilizing recent advances in genetic and protein engineering for the regulated biosynthesis on nonmaterial in soil remediation is being extensively studied (Kostal et al. 2005). These nano-tools might provide a cost-effective method for soil remediation. Current researches on nanotechnology should also focus on development of tools for mitigation of metal-contaminated wastewater and sludge to less hazardous for food crops as has been shown for improving the pesticide formulation through nanotechnology formulations (Hazra et al. 2017). Application of biochar nanosheets in industrially contaminated wheat field remarkably reduced the availability of carcinogenic metals in wheat due to adsorption of these metals (Yousaf et al. 2018). Though nanotechnology has been found useful for remediation of contaminants and also as fertilizer, but the adverse effects and fate of nanoparticles in environment and crops need a better understanding.

3.4 Conclusion

From the current review, it is evident that chronic poisoning of heavy metal(loid)s affects all body organs and systems. Although the mechanism of toxicity and primary target organ may be specific for each metal(loid), but in general, they directly or indirectly pose harmful effect on most of the biological pathways and organs. Arsenic is highly carcinogenic and cancer of the skin and internal organs has been widely documented around the world. Cadmium primarily causes hepatic and renal injury resulting in vitamin D deficiency and thus weak bones and factures. Children and developing fetus are most vulnerable to Pb toxicity leading to affect on nervous system and brain development. Mercury also passes the placenta and harms developing fetus. The pre-birth or early life exposure of these toxic metal(loid)s leads to several diseases in adults and causes early death. The level of As, Cd, Pb, and Hg is not only several folds higher than the maximum permissible limits in water, but also in various crops and sea foods their level has drastically surpassed the maximum tolerable daily intake values. Though water is the main route of human exposure, with the development of mitigation strategies (for domestic and community level) and people's awareness about clean drinking water, nowadays, food has become the major source of these heavy metal(loid)s in many areas. Further, the chronic toxicity of these heavy metal(loid)s is largely known; however, health impact of chronic exposure of multiple element is not much studied. Additionally, the total ingestion (from drinking water + food) can be several folds higher than can be assessed, and this is evident by the high global increase in incidents of different types of cancers and other diseases. In recent years, there has been a sufficient development in techniques for the removal of heavy metal(loid) from drinking water. The developed techniques have shown good efficiency, and many are cost-effective too; still, their benefits are not in the reach of everyone due to the lack of awareness and installation at community level. Further, food chain contamination through irrigation with untreated/partially treated water or contaminated groundwater, e.g., for As, also need attention. Several techniques/methods for mitigation of metal (loid)s from soil have been studied and employed; however, they have limited efficiency, and they do not provide long-term solutions, except soil reclamation by plants, i.e., phytoremediation. But later also has several limitations. In this backdrop, strict preventive measures and people's awareness and participation are of utmost importance along with mitigation strategies.

Acknowledgment SM is grateful to SEED division of DST, New Delhi, for the project grant (SEED/TIASN/2018/74) and to UCG, New Delhi, for the start-up grant (F.30–505/2020(BSR).

References

Ab Latif Wani AA, Usmani JA (2015) Lead toxicity: a review. Interdiscip Toxicol 8(2):55

Abdulla NK, Siddiqui SI, Tara N, Hashmi AA, Chaudhry SA (2019) Psidium guajava leave-based magnetic nanocomposite γ-Fe2O3@ GL: a green technology for methylene blue removal from water. J Environ Chem Eng 7(6):103423

Adepoju-Bello AA, Alabi OM (2005 Jan) Heavy metals: a review. Nig J Pharm 37:41–45

Agrawal PR, Singh N, Kumari S, Dhakate SR (2019) The removal of pentavalent arsenic by graphite intercalation compound functionalized carbon foam from contaminated water. J Hazard Mater 5(377):274–283

Ahamed M, Siddiqui MK (2007) Low level lead exposure and oxidative stress: current opinions. Clin Chim Acta 383(1–2):57–64

Ahamed S, Sengupta MK, Mukherjee A, Hossain MA, Das B, Nayak B, Pal A, Mukherjee SC, Pati S, Dutta RN, Chatterjee G (2007) Arsenic groundwater contamination and its health effects in the state of Uttar Pradesh (UP) in upper and middle Ganga plain, India: a severe danger. Sci Total Environ 370(2–3):310–322

Aoshima K (2012) Itai-itai disease: cadmium-induced renal tubular osteomalacia. Nihon eiseigaku zasshi. Jpn J Hyg 67(4):455–463

Ashraf MA, Hussain I, Rasheed R, Iqbal M, Riaz M, Arif MS (2017) Advances in microbe-assisted reclamation of heavy metal contaminated soils over the last decade: a review. J Environ Manag 198:132–143

Ashraf S, Ali Q, Zahir ZA, Ashraf S, Asghar HN (2019) Phytoremediation: environmentally sustainable way for reclamation of heavy metal polluted soils. Ecotoxicol Environ Saf 174: 714–727

ATSDR (1992) Agency for toxic substances and disease registry. Case studies in environmental medicine - Lead toxicity. Atlanta: Public Health Service, U.S. Department of Health and Human Services

ATSDR (1999) Agency for toxic substances and disease registry public health service. Atlanta: U.S. Department of Health and Human Services. Toxicological Profile for Lead

Baba H, Tsuneyama K, Yazaki M, Nagata K, Minamisaka T, Tsuda T, Nomoto K, Hayashi S, Miwa S, Nakajima T, Nakanishi Y (2013) The liver in itai-itai disease (chronic cadmium poisoning): pathological features and metallothionein expression. Mod Pathol 26(9):1228–1234

Beesley L, Inneh OS, Norton GJ, Moreno-Jimenez E, Pardo T, Clemente R, Dawson JJ (2014) Assessing the influence of compost and biochar amendments on the mobility and toxicity of metals and arsenic in a naturally contaminated mine soil. Environ Pollut 186:195–202

Beesley L, Marmiroli M (2011) The immobilisation and retention of soluble arsenic, cadmium and zinc by biochar. Environ Pollut 159(2):474–480

Belimov AA, Puhalsky IV, Safronova VI, Shaposhnikov AI, Vishnyakova MA, Semenova EV, Zinovkina NY, Makarova NM, Wenzel W, Tikhonovich IA (2015) Role of plant genotype and soil conditions in symbiotic plant-microbe interactions for adaptation of plants to cadmium-polluted soils. Water Air Soil Pollut 226(8):1–5

Bergeson LL (2008) The proposed lead NAAQS: is consideration of cost in the clean air act's future? Environ Qual Manag 18(1):79–84

Bernard A (2004) Renal dysfunction induced by cadmium: biomarkers of critical effects. Biometals 17(5):519–523

Bolisetty S, Peydayesh M, Mezzenga R (2019) Sustainable technologies for water purification from heavy metals: review and analysis. Chem Soc Rev 48(2):463–487

Bora AJ, Dutta RK (2019) Removal of metals (Pb, cd, cu, Cr, Ni, and co) from drinking water by oxidation-coagulation-absorption at optimized pH. J Water Process Eng 31:100839

Brinkel J, Khan MH, Kraemer A (2009) A systematic review of arsenic exposure and its social and mental health effects with special reference to Bangladesh. Int J Environ Res Public Health 6(5): 1609–1619

Burke DM, Morris MA, Holmes JD (2013) Chemical oxidation of mesoporous carbon foams for lead ion adsorption. Sep Purif Technol 104:150–159

Cai SW (1989) Cadmium exposure and health effects among residents in irrigation area with ore dressing wastewater. Zhonghua Yu Fang Yi Xue Za Zhi 23(3):143–146

Calderon J, Navarro ME, Jimenez-Capdeville ME, Santos-Diaz MA, Golden A, Rodriguez-Leyva I, Borja-Aburto V, Dıaz-Barriga F (2001) Exposure to arsenic and lead and neuropsychological development in Mexican children. Environ Res 85(2):69–76

Campos NV, Guerra MB, Mello JW, Schaefer CE, Krug FJ, Alves EE, Azevedo AA (2015) Accumulation and spatial distribution of arsenic and phosphorus in the fern Pityrogramma calomelanos evaluated by micro X-ray fluorescence spectrometry. J Anal At Spectrom 30(12): 2375–2383

Charney E, Sayre J, Coulter M (1980 Feb) Increased lead absorption in inner city children: where does the lead come from? Pediatrics 65(2):226–231

Chehregani A, Malayeri BE (2007) Removal of heavy metals by native accumulator plants. Int J Agric Biology (Pakistan) 9:462

Chen X, Huang G, Wang J (2013) Electrochemical reduction/oxidation in the treatment of heavy metal wastewater. J Metall Eng 2(4):161–164

Cheng JP, Wang WH, Jia JP, Zheng M, Shi W, Lin XY (2006) Expression of c-fos in rat brain as a prelude marker of central nervous system injury in response to methylmercury-stimulation. Biomed Environ Sci 19(1):67–72

Clarkson TW, Magos L, Myers GJ (2003) The toxicology of mercury—current exposures and clinical manifestations. N Engl J Med 349(18):1731–1737

Cocco P, Salis S, Anni M, Cocco ME, Flore C, Ibba A (1995) Effects of short-term occupational exposure to lead on erythrocyte glucose-6-phosphate dehydrogenase activity and serum cholesterol. J Appl Toxicol 15(5):375–378

Ding W, Liu W, Cooper KL, Qin XJ, de Souza Bergo PL, Hudson LG, Liu KJ (2009) Inhibition of poly (ADP-ribose) polymerase-1 by arsenite interferes with repair of oxidative DNA damage. J Biol Chem 284(11):6809–6817

European Food Safety Authority (2011) Use of the EFSA comprehensive European food consumption database in exposure assessment. EFSA J 9(3):2097

Fan Y, Li H, Xue Z, Zhang Q, Cheng F (2017) Accumulation characteristics and potential risk of heavy metals in soil-vegetable system under greenhouse cultivation condition in northern China. Ecol Eng 102:367–373

FAO W (1989) Evaluation of certain food additives and contaminants, FAO/WHO expert committee on food additives: 33rd report. Tech. Rep. Ser. 776. World Health Organization, Geneva

Fatima G, Raza AM, Hadi N, Nigam N, Mahdi AA (2019) Cadmium in human diseases: it's more than just a mere metal. Indian J Clin Biochem 34(4):371–378

Fields CA, Borak J, Louis ED (2017) Mercury-induced motor and sensory neurotoxicity: systematic review of workers currently exposed to mercury vapor. Crit Rev Toxicol 47(10):815–848

Flora SJ, Flora G, Saxena G (2006) Environmental occurrence, health effects and management of lead poisoning. In: Lead. Elsevier Science BV, Amsterdam, pp 158–228

Flora SJ, Saxena G, Gautam P, Kaur P, Gill KD (2007) Response of lead-induced oxidative stress and alterations in biogenic amines in different rat brain regions to combined administration of DMSA and MiADMSA. Chem Biol Interact 170(3):209–220

Gallagher CM, Kovach JS, Meliker JR (2008) Urinary cadmium and osteoporosis in US women≥ 50 years of age: NHANES 1988–1994 and 1999–2004. Environ Health Perspect 116(10): 1338–1343

García-Sánchez A, Contreras F, Adams M, Santos F (2008) Mercury contamination of surface water and fish in a gold mining region (Cuyuni river basin, Venezuela). Int J Environ Pollut 33(2–3):260–274

Guo G, Lei M, Chen T, Yang J (2018) Evaluation of different amendments and foliar fertilizer for immobilization of heavy metals in contaminated soils. J Soils Sediments 18(1):239–247

Gupta S, Nayek S, Saha RN, Satpati S (2008) Assessment of heavy metal accumulation in macrophyte, agricultural soil, and crop plants adjacent to discharge zone of sponge iron factory. Environ Geol 55(4):731–739

Gürer H, Özgünes H, Neal R, Spitz DR, Erçal N (1998) Antioxidant effects of N-acetylcysteine and succimer in red blood cells from lead-exposed rats. Toxicology 128(3):181–189

Hamdan SH, Molelekwa GF, Van der Bruggen B (2014) Electrokinetic remediation technique: an integrated approach to finding new strategies for restoration of saline soil and to control seawater intrusion. ChemElectroChem 1(7):1104–1117

Hassan W, Bashir S, Ali F, Ijaz M, Hussain M, David J (2016) Role of ACC-deaminase and/or nitrogen fixing rhizobacteria in growth promotion of wheat (Triticum aestivum L.) under cadmium pollution. Environ Earth Sci 75(3):1–4

Hazra DK, Karmakar R, Poi R, Bhattacharya S, Mondal S (2017) Recent advances in pesticide formulations for eco-friendly and sustainable vegetable pest management: a review. Arch Agric Environ Sci 2(3):232–237

Ho SH, Zhu S, Chang JS (2017) Recent advances in nanoscale-metal assisted biochar derived from waste biomass used for heavy metals removal. Bioresour Technol 246:123–134

Hu J, Tsang W, Wu F, Wu S, Wang J, Lin X, Wong MH (2016) Arbuscular mycorrhizal fungi optimize the acquisition and translocation of cd and P by cucumber (Cucumis sativus L.) plant cultivated on a cd-contaminated soil. J Soils Sediments 16(9):2195–2202

Huel G, Tubert P, Frery N, Moreau T, Dreyfus J (1992) Joint effect of gestational age and maternal lead exposure on psychomotor development of the child at six years. Neurotoxicology 13(1): 249–254

IARC (2006) Working group on the evaluation of carcinogenic risks to humans. Inorganic and organic lead compounds. In Inorganic and organic Lead compounds. International Agency for Research on Cancer

IARC (2017) Agents classified by the IARC Monographs, vol 1–120

International Agency for Research on Cancer. Beryllium, cadmium, mercury, and exposures in the glass. Apresentado em: IARC working group on the evaluation of carcinogenic risks to humans: Beryllium, Lyon. 1993

Jaffré T, Reeves RD, Baker AJ, Schat H, van der Ent A (2018) The discovery of nickel hyperaccumulation in the new Caledonian tree Pycnandra acuminata 40 years on: an introduction to a virtual issue. New Phytol 218(2):397–400

Järup L (2003) Hazards of heavy metal contamination. Br Med Bull 68(1):167–182

Joint FA, WHO Expert Committee on Food Additives (2003) Sixty-first meeting Rome, 10–19 June 2003. Summary and conclusions. http://www.who.int/pcs/jecfa/Summaries.htm

Joint FA, WHO Expert Committee on Food Additives, World Health Organization (2000) Evaluation of certain food additives and contaminants: fifty-third report of the Joint FAO/WHO expert committee on food additives. World Health Organization

Kalve S, Sarangi BK, Pandey RA, Chakrabarti T (2011) Arsenic and chromium hyperaccumulation by an ecotype of Pteris vittata–prospective for phytoextraction from contaminated water and soil. Curr Sci 25:888–894

Karimi N, Ghaderian SM, Raab A, Feldmann J, Meharg AA (2009) An arsenic-accumulating, hypertolerant brassica, Isatis capadocica. New Phytol 1:41–47

Kim CY, Watanabe C, Kasanuma Y, Satoh H (1995) Inhibition of γ-glutamyltranspeptidase decreases renal deposition of mercury after mercury vapor exposure. Arch Toxicol 69(10): 722–724

Kostal J, Prabhukumar G, Lao UL, Chen A, Matsumoto M, Mulchandani A (2005) Customizable biopolymers for heavy metal remediation. J Nanopart Res 7(4):517–523

Kurniawan TA, Chan GY, Lo WH, Babel S (2006) Physico–chemical treatment techniques for wastewater laden with heavy metals. Chem Eng J 118(1–2):83–98

Lee CG, Jeon JW, Hwang MJ, Ahn KH, Park C, Choi JW, Lee SH (2015) Lead and copper removal from aqueous solutions using carbon foam derived from phenol resin. Chemosphere 130:59–65

Leermakers M, Baeyens W, De Gieter M, Smedts B, Meert C, De Bisschop HC, Morabito R, Quevauviller P (2006) Toxic arsenic compounds in environmental samples: speciation and validation. TrAC Trends Anal Chem 25(1):1–10

Lei Y, Chen F, Luo Y, Zhang L (2014) Three-dimensional magnetic graphene oxide foam/Fe3O4 nanocomposite as an efficient absorbent for Cr (VI) removal. J Mater Sci 49(12):4236–4245

Lessl JT, Luo J, Ma LQ (2014) Pteris vittata continuously removed arsenic from non-labile fraction in three contaminated-soils during 3.5 years of phytoextraction. J Hazard Mater 279:485–492

Li L, Chen F (2016) Oxidative stress, epigenetics, and cancer stem cells in arsenic carcinogenesis and prevention. Curr Pharmacol Rep 2(2):57–63

Li L, Wang S, Li X, Li T, He X, Tao Y (2018) Effects of Pseudomonas chenduensis and biochar on cadmium availability and microbial community in the paddy soil. Sci Total Environ 640:1034–1043

Li W, Li YX, Lin CY, Dong YS, Han J, Han W (2009) Analysis of land suitability for manure application in Beijing. Acta Sci Circumst 29:882–889

Liao J, Wen Z, Ru X, Chen J, Wu H, Wei C (2016) Distribution and migration of heavy metals in soil and crops affected by acid mine drainage: public health implications in Guangdong Province. China Ecotoxicol Environ Saf 124:460–469

Liu CM, Ma JQ, Sun YZ (2012) Puerarin protects rat kidney from lead-induced apoptosis by modulating the PI3K/Akt/eNOS pathway. Toxicol Appl Pharmacol 258(3):330–342

Lombi E, Zhao FJ, Dunham SJ, McGrath SP (2001) Phytoremediation of heavy metal–contaminated soils: natural hyperaccumulation versus chemically enhanced phytoextraction. J Environ Qual 30(6):1919–1926

Ma L, Li J, Zhan Z, Chen L, Li D, Bai Q, Gao C, Li J, Zeng X, He Z, Wang S (2016) Specific histone modification responds to arsenic-induced oxidative stress. Toxicol Appl Pharmacol 1(302):52–61

Ma LQ, Komar KM, Tu C, Zhang W, Cai Y, Kennelley ED (2001) A fern that hyperaccumulates arsenic. Nature 409(6820):579

Mahar A, Wang P, Ali A, Awasthi MK, Lahori AH, Wang Q, Li R, Zhang Z (2016) Challenges and opportunities in the phytoremediation of heavy metals contaminated soils: a review. Ecotoxicol Environ Saf 126:111–121

Mandal BK, Chowdhury TR, Samanta G, Basu GK, Chowdhury PP, Chanda CR, Lodh D, Karan NK, Dhar RK, Tamili DK, Das D (1996) Arsenic in groundwater in seven districts of West Bengal, India–the biggest arsenic calamity in the world. Curr Sci 10:976–986

Mandzhieva SS, Minkina TM, Chaplygin VA, Bauer TV, Burachevskaya MV, Nevidomskaya DG, Sushkova SN, Sherstnev AK, Zamulina IV (2017) Protective mechanism of the soil–plant system with respect to heavy metals. J Soils Sediments 17(5):1291–1300

Manikandan R, Sahi SV, Venkatachalam P (2015) Impact assessment of mercury accumulation and biochemical and molecular response of Mentha arvensis: a potential hyperaccumulator plant. Sci World J 1:2015

Mannino DM, Holguin F, Greves HM, Savage-Brown A, Stock AL, Jones RL (2004) Urinary cadmium levels predict lower lung function in current and former smokers: data from the third National Health and nutrition examination survey. Thorax 59(3):194–198

Marchetti C (2003) Molecular targets of lead in brain neurotoxicity. Neurotox Res 5(3):221–235

Mazumder DN, Haque R, Ghosh N, De BK, Santra A, Chakraborti D, Smith AH (2000) Arsenic in drinking water and the prevalence of respiratory effects in West Bengal, India. Int J Epidemiol 29(6):1047–1052

Mesjasz-Przybylowicz J, Przybylowicz WJ, Pineda CA (2001) Nuclear microprobe studies of elemental distribution in apical leaves of the Ni hyperaccumulator Berkheya coddii. S Afr J Sci 97(11):591–593

Mishra S, Srivastava S, Tripathi RD, Kumar R, Seth CS, Gupta DK (2006) Lead detoxification by coontail (Ceratophyllum demersum L.) involves induction of phytochelatins and antioxidant system in response to its accumulation. Chemosphere 65(6):1027–1039

Mukherjee S, Das SK (2002) Acute cadmium toxicity and male reproduction. Adv Reproduct 6(3/4):76

Muthusaravanan S, Sivarajasekar N, Vivek JS, Paramasivan T, Naushad M, Prakashmaran J, Gayathri V, Al-Duaij OK (2018) Phytoremediation of heavy metals: mechanisms, methods and enhancements. Environ Chem Lett 16(4):1339–1359

Nabulo G, Oryem-Origa H, Diamond M (2006) Assessment of lead, cadmium, and zinc contamination of roadside soils, surface films, and vegetables in Kampala City, Uganda. Environ Res 101(1):42–52

National Research Council (2013) Critical aspects of EPA's IRIS assessment of inorganic arsenic: interim report. National Academies Press, Washington

Nava Rivera LE, Betancourt Martínez ND, Lozoya Martínez R, Carranza Rosales P, Guzmán Delgado NE, Carranza Torres IE, Delgado Aguirre H, Zambrano Ortíz JO, Morán MJ (2022) Author correction: transgenerational effects in DNA methylation, genotoxicity and reproductive phenotype by chronic arsenic exposure. Sci Rep 12(1):1

Neal AP, Guilarte TR (2010) Molecular neurobiology of lead (Pb2+): effects on synaptic function. Mol Neurobiol 42(3):151–160

Nickson RT, McArthur JM, Ravenscroft P, Burgess WG, Ahmed KM (2000) Mechanism of arsenic release to groundwater, Bangladesh and West Bengal. Appli Geochem 15(4):403–413

de Oliveira LM, Suchismita D, Gress J, Rathinasabapathi B, Chen Y, Ma LQ (2017) Arsenic uptake by lettuce from as-contaminated soil remediated with Pteris vittata and organic amendment. Chemosphere 176:249–254

Ong CN, Phoon WO, Law HY, Tye CY, Lim HH (1985) Concentrations of lead in maternal blood, cord blood, and breast milk. Arch Dis Child 60(8):756–759

Palmeira CM, Madeira VM (1997) Mercuric chloride toxicity in rat liver mitochondria and isolated hepatocytes. Environ Toxicol Pharmacol 3(3):229–235

Patlolla AK, Barnes C, Yedjou C, Velma VR, Tchounwou PB (2009) Oxidative stress, DNA damage, and antioxidant enzyme activity induced by hexavalent chromium in Sprague-Dawley rats. Environ Toxicol Int J 24(1):66–73

Patrick L (2006) Lead toxicity part II: the role of free radical damage and the use of antioxidants in the pathology and treatment of lead toxicity. Altern Med Rev 11(2):114

Peng X, Deng Y, Peng Y, Yue K (2018) Effects of biochar addition on toxic element concentrations in plants: a meta-analysis. Sci Total Environ 616:970–977

Qiao JT, Li XM, Li FB (2018) Roles of different active metal-reducing bacteria in arsenic release from arsenic-contaminated paddy soil amended with biochar. J Hazard Mater 344:958–967

Qu W, Ke H, Pi J, Broderick D, French JE, Webber MM, Waalkes MP (2007) Acquisition of apoptotic resistance in cadmium-transformed human prostate epithelial cells: Bcl-2 overexpression blocks the activation of JNK signal transduction pathway. Environ Health Perspect 115(7):1094–1100

Quintanilla-Vega B, Hoover D, Bal W, Silbergeld EK, Waalkes MP, Anderson LD (2000) Lead effects on protamine–DNA binding. Am J Ind Med 38(3):324–329

Rai PK (2018a) Heavy metal phyto-technologies from Ramsar wetland plants: green approach. Chem Ecol 34(8):786–796

Rai PK (2018b) Phytoremediation of emerging contaminants in wetlands. CRC Press, Boca Raton

Rai PK, Lee SS, Zhang M, Tsang YF, Kim KH (2019) Heavy metals in food crops: health risks, fate, mechanisms, and management. Environ Int 125:365–385

Rai R, Mishra D, Srivastav SK, Srivastav AK (2009) Ultimobranchial gland of a freshwater teleost, Heteropneustes fossilis, in response to cadmium treatment. Environ Toxicol 24(6):589–593

Reiter J, Manchester LC, Tan DX (2010) Neurotoxins: free radical mechanisms and melatonin protection. Curr Neuropharmacol 8(3):194–210

Riethmiller S (2005) From atoxyl to salvarsan: searching for the magic bullet. Chemotherapy 51(5): 234–242

Rizwan M, Ali S, Qayyum MF, Ibrahim M, Zia-ur-Rehman M, Abbas T, Ok YS (2016) Mechanisms of biochar-mediated alleviation of toxicity of trace elements in plants: a critical review. Environ Sci Pollut Res 23(3):2230–2248

Robinson BH, Chiarucci A, Brooks RR, Petit D, Kirkman JH, Gregg PE, De Dominicis V (1997) The nickel hyperaccumulator plant Alyssum bertolonii as a potential agent for phytoremediation and phytomining of nickel. J Geochem Explor 59(2):75–86

Rosado JL, Ronquillo D, Kordas K, Rojas O, Alatorre J, Lopez P, Garcia-Vargas G, del Carmen CM, Cebrián ME, Stoltzfus RJ (2007) Arsenic exposure and cognitive performance in Mexican schoolchildren. Environ Health Perspect 115(9):1371–1375

Rossman TG (2000) Cloning genes whose levels of expression are altered by metals: implications for human health research. Am J Ind Med 38(3):335–339

Roy NK, Rossman TG (1992) Mutagenesis and comutagenesis by lead compounds. Mutati Research/Genetic Toxicology 298(2):97–103

Sanchez V, López-Bellido FJ, Rodrigo MA, Rodríguez L (2019) Electrokinetic-assisted phytoremediation of atrazine: differences between electrode and interelectrode soil sections. Sep Purif Technol 211:19–27

Schutte R, Nawrot TS, Richart T, Thijs L, Vanderschueren D, Kuznetsova T, Van Hecke E, Roels HA, Staessen JA (2008) Bone resorption and environmental exposure to cadmium in women: a population study. Environ Health Perspect 116(6):777–783

Shahalam AM, Al-Harthy A, Al-Zawhry A (2002) Feed water pretreatment in RO systems: unit processes in the Middle East. Desalination 150(3):235–245

Shahid M, Rafiq M, Niazi NK, Dumat C, Shamshad S, Khalid S, Bibi I (2017) Arsenic accumulation and physiological attributes of spinach in the presence of amendments: an implication to reduce health risk. Environ Sci Pollut Res 24(19):16097–16106

Shalaby TA, Bayoumi Y, Abdalla N, Taha H, Alshaal T, Shehata S, Amer M, Domokos-Szabolcsy É, El-Ramady H (2016) Nanoparticles, soils, plants and sustainable agriculture. In: Nanoscience in food and agriculture 1. Springer, Cham, pp 283–312

Shehu J, Imeri A, Kupe L, Dodona E, Shehu A, Mullaj A (2014) Hyperaccumulators of mercury in the industrial area of a PVC factory in Vlora (Albania). Arch Biol Sci 66(4):1457–1464

Shenker BJ, Guo TL, Shapiro IM (2000) Mercury-induced apoptosis in human lymphoid cells: evidence that the apoptotic pathway is mercurial species dependent. Environ Res 84(2):89–99

Siddiqui SI, Singh PN, Tara N, Pal S, Chaudhry SA, Sinha I (2020) Arsenic removal from water by starch functionalized maghemite nano-adsorbents: thermodynamics and kinetics investigations. Colloid Interface Sci Commun 36:100263

Siddiqui SI, Zohra F, Chaudhry SA (2019) Nigella sativa seed based nanohybrid composite-Fe2O3–SnO2/BC: a novel material for enhanced adsorptive removal of methylene blue from water. Environ Res 178:108667

Sikanna R, Sutriono D, Prismawiryanti P (2019, June) The Phytostabilization of Mercury (Hg) in Ipomoea reptans Poir Plants from Polluted Soil. In: 1st International Conference on Science and Technology, ICOST 2019, 2–3 May, Makassar, Indonesia

Singh NB, Nagpal G, Agrawal S (2018) Water purification by using adsorbents: a review. Environ Technol Innov 11:187–240

Skoczynska A (2008) Genetyczne aspekty hipertensyjnego dzialania olowiu/genetic aspects of hypertensive effect of lead. Med Pr 59(4):325

Smith AH, Lingas EO, Rahman M (2000) Contamination of drinking-water by arsenic in Bangladesh: a public health emergency. Bull World Health Organ 78:1093–1103

Smith AH, Steinmaus CM (2009) Health effects of arsenic and chromium in drinking water: recent human findings. Annu Rev Public Health 30:107–122

Srivastava M, Ma LQ, Santos JA (2006) Three new arsenic hyperaccumulating ferns. Sci Total Environ 364(1–3):24–31

Stein RJ, Höreth S, de Melo JR, Syllwasschy L, Lee G, Garbin ML, Clemens S, Krämer U (2017) Relationships between soil and leaf mineral composition are element-specific, environment-dependent and geographically structured in the emerging model Arabidopsis halleri. New Phytol 213(3):1274–1286

Stohs SJ, Bagchi D (1995) Oxidative mechanisms in the toxicity of metal ions. Free Radic Biol Med 18(2):321–336

Tchounwou PB, Yedjou CG, Patlolla AK, Sutton DJ (2012) Heavy metal toxicity and the environment. In: Molecular, clinical and environmental toxicology. Springer, Basel, pp 133–164

Thevenod F, Jones SW (1992) Cadmium block of calcium current in frog sympathetic neurons. Biophys J 63(1):162–168

Toscano CD, Guilarte TR (2005) Lead neurotoxicity: from exposure to molecular effects. Brain Res Rev 49(3):529–554

Tripathi RD, Mishra S, Srivastava S (2008) Role of aquatic macrophytes in arsenic phytoremediation in wetlands. Proc Natl Inst Sci India B: Biol Sci 78:167–182

Tsuzuki K, Sugiyama M, Haramaki N (1994) DNA single-strand breaks and cytotoxicity induced by chromate (VI), cadmium (II), and mercury (II) in hydrogen peroxide-resistant cell lines. Environ Health Perspect 102(suppl 3):341–342

Udom BE, Mbagwu JS, Adesodun JK, Agbim NN (2004) Distributions of zinc, copper, cadmium and lead in a tropical ultisol after long-term disposal of sewage sludge. Environ Int 30(4): 467–470

Valko M, Izakovic M, Mazur M, Rhodes CJ, Telser J (2004) Role of oxygen radicals in DNA damage and cancer incidence. Mol Cell Biochem 266(1):37–56

Van der Bruggen B, Vandecasteele C (2002) Distillation vs. membrane filtration: overview of process evolutions in seawater desalination. Desalination 143(3):207–218

Verstraeten SV, Aimo L, Oteiza PI (2008) Aluminium and lead: molecular mechanisms of brain toxicity. Arch Toxicol 82(11):789–802

Waalkes MP, Diwan BA, Ward JM, Devor DE, Goyer RA (1995) Renal tubular tumors and atypical hyperplasias in B6C3F1 mice exposed to lead acetate during gestation and lactation occur with minimal chronic nephropathy. Cancer Res 55(22):5265–5271

Wang J, Shi L, Zhai L, Zhang H, Wang S, Zou J, Shen Z, Lian C, Chen Y (2021) Analysis of the long-term effectiveness of biochar immobilization remediation on heavy metal contaminated soil and the potential environmental factors weakening the remediation effect: a review. Ecotoxicol Environ Saf 207:111261

Wang N, Xue XM, Juhasz AL, Chang ZZ, Li HB (2017) Biochar increases arsenic release from an anaerobic paddy soil due to enhanced microbial reduction of iron and arsenic. Environ Pollut 220:514–522

Waterlot C, Pruvot C, Marot F, Douay F (2017) Impact of a phosphate amendment on the environmental availability and phytoavailability of cd and Pb in moderately and highly carbonated kitchen garden soils. Pedosphere 27(3):588–605

Wei S, Zhou Q (2008) Screen of Chinese weed species for cadmium tolerance and accumulation characteristics. Int J Phytoremediation 10(6):584–597

White LD, Cory-Slechta DA, Gilbert ME, Tiffany-Castiglioni E, Zawia NH, Virgolini M, Rossi-George A, Lasley SM, Qian YC, Basha MR (2007) New and evolving concepts in the neurotoxicology of lead. Toxicol Appl Pharmacol 225(1):1–27

Wittman R, Hu H (2002) Cadmium exposure and nephropathy in a 28-year-old female metals worker. Environ Health Perspect 110(12):1261–1266

Wu H, Lai C, Zeng G, Liang J, Chen J, Xu J, Dai J, Li X, Liu J, Chen M, Lu L (2017) The interactions of composting and biochar and their implications for soil amendment and pollution remediation: a review. Crit Rev Biotechnol 37(6):754–764

Xu L, Lu A, Wang J, Ma Z, Pan L, Feng X, Luan Y (2015) Accumulation status, sources and phytoavailability of metals in greenhouse vegetable production systems in Beijing, China. Ecotoxicol Environ Saf 122:214–220

Xun E, Zhang Y, Zhao J, Guo J (2017) Translocation of heavy metals from soils into floral organs and rewards of Cucurbita pepo: implications for plant reproductive fitness. Ecotoxicol Environ Saf 145:235–243

Yang Y, Zhou X, Tie B, Peng L, Li H, Wang K, Zeng Q (2017) Comparison of three types of oil crop rotation systems for effective use and remediation of heavy metal contaminated agricultural soil. Chemosphere 188:148–156

Yao A, Wang Y, Ling X, Chen Z, Tang Y, Qiu H, Ying R, Qiu R (2017) Effects of an iron-silicon material, a synthetic zeolite and an alkaline clay on vegetable uptake of as and cd from a polluted agricultural soil and proposed remediation mechanisms. Environ Geochem Health 39(2): 353–367

Yedjou CG, Tchounwou PB (2007) In-vitro cytotoxic and genotoxic effects of arsenic trioxide on human leukemia (HL-60) cells using the MTT and alkaline single cell gel electrophoresis (comet) assays. Mol Cell Biochem 301(1):123–130

Yoon J, Cao X, Zhou Q, Ma LQ (2006) Accumulation of Pb, cu, and Zn in native plants growing on a contaminated Florida site. Sci Total Environ 368(2–3):456–464

Yousaf B, Liu G, Abbas Q, Ullah H, Wang R, Zia-ur-Rehman M, Niu Z (2018) Comparative effects of biochar-nanosheets and conventional organic-amendments on health risks abatement of potentially toxic elements via consumption of wheat grown on industrially contaminated-soil. Chemosphere 192:161–170

Zehra A, Sahito ZA, Tong W, Tang L, Hamid Y, Wang Q, Cao X, Khan MB, Hussain B, Jatoi SA, He Z (2020) Identification of high cadmium-accumulating oilseed sunflower (Helianthus annuus) cultivars for phytoremediation of an Oxisol and an Inceptisol. Ecotoxicol Environ Saf 187:109857

Zhang H, Huang B, Dong L, Hu W, Akhtar MS, Qu M (2017a) Accumulation, sources and health risks of trace metals in elevated geochemical background soils used for greenhouse vegetable production in southwestern China. Ecotoxicol Environ Saf 137:233–239

Zhang MK, Liu ZY, Wang H (2010) Use of single extraction methods to predict bioavailability of heavy metals in polluted soils to rice. Commun Soil Sci Plant Anal 41(7):820–831
Zhang RH, Li ZG, Liu XD, Wang BC, Zhou GL, Huang XX, Lin CF, Wang AH, Brooks M (2017b) Immobilization and bioavailability of heavy metals in greenhouse soils amended with rice straw-derived biochar. Ecol Eng 98:183–188
Zhao FJ, Dunham SJ, McGrath SP (2002) Arsenic hyperaccumulation by different fern species. New Phytol 156(1):27–31
Zheng R, Teng W, Hu Y, Hou X, Shi D, Tian X, Scullion J, Wu J (2022) Cadmium uptake by a hyperaccumulator and three Pennisetum grasses with associated rhizosphere effects. Environ Sci Pollut Res 29(2):1845–1857

Heavy Metal Pollution in the Environment: Impact on Air Quality and Human Health Implications

4

Ritusmita Goswami and Nikita Neog

Abstract

Atmospheric pollution of heavy metals (HMs) has become a considerably huge cause of concern due to its greater toxicity, persistence nature, and bioaccumulative behavior. Although most of the HMs are natural environmental constituents, its biochemical equilibrium and geochemical processes have been altered by indiscriminate anthropogenic activities owing to the demand for better quality of life with modern conveniences. More than 80% of people who live in metropolitan areas are reportedly exposed to poor air quality, per the World Health Organization (WHO). As a consequence, the urban population has experienced a wide range of adverse health effects, including cancer and damage to essential organs as well as cardiovascular and pulmonary inflammation and acute respiratory disorders. This chapter presents an overview of the atmospheric heavy metal pollution, its impact on the air quality, and human health implications. A detailed discussion has been made on the sources, pathways, and fate of HMs in the environment followed by their toxicological effects on human health.

Keywords

Heavy metals · Air quality · Atmospheric pollution · Toxicology · Health impacts

R. Goswami (✉) · N. Neog
Centre for Ecology, Environment and Sustainable Development, Tata Institute of Social Sciences, Guwahati, India

R. P. Singh et al. (eds.), *Heavy Metal Toxicity: Environmental Concerns, Remediation and Opportunities*, https://doi.org/10.1007/978-981-99-0397-9_4

4.1 Introduction

Heavy metals (HMs) are a major environmental hazard across the world (Zhu et al. 2020). Because of their nonbiodegradability, bioaccumulation, environmental stability, persistence, and biotoxicity, HMs constitute a significant environmental danger to living beings and environments (Khan et al. 2019; Zhang et al. 2019). Natural (geological weathering, atmospheric precipitation, wave erosion, wind, and bioturbation) and anthropogenic (rapid industrialization, urbanization, agricultural runoff, and transportation) activities both contribute to the escalation of elevated air pollution levels (Nour et al. 2019; Kahal et al. 2020). However, rampant industrialization, rising energy demand, and the indiscreet utilization of the natural resources are the major causes of intensifying the global environmental pollution issues (Gautam et al. 2016; Briffa et al. 2020). Inorganic and organic constituents, gaseous pollutants, organometallic compounds, nanoparticles, and radioactive isotopes are major pollutants causing severe environmental contamination (Walker et al. 2012). The US Environmental Protection Agency (USEPA) listed HMs as a major category of pollutants under trace inorganic contaminants (Wuana and Okieimen 2011). Conventionally, HMs are defined as those having higher densities or high atomic weight. But now it also includes metallic chemical constituents, metalloids proven to be toxic (Lee et al. 2017). Few metalloids and lighter metals (selenium, arsenic, aluminum) are observed to be harmful in terms of the environment and human health and hence regarded as HMs (Tchounwou et al. 2012; Briffa et al. 2020).

Globally, a steady elevation of atmospheric HM concentrations has been observed over 30 years in few countries. Despite the fact that HMs are naturally found in the Earth's crust, the large increase in their use has led to an impending rise in the amount of metallic components in the environment (Gautam et al. 2016). Naturally, HMs are introduced through rock weathering, volcanic activities, soil erosion, metal corrosion, sediment resuspension, and metal evaporation from water and soil. The prime cause of HM pollution is contributed by anthropogenic activities such as mining activities, smelting, metal-based industries, and metal leaching from sources such as landfills, dumping sites, agricultural fields, livestock manure, etc. (Ali et al. 2021). The use of HM-based pesticides, fertilizers, and insecticides in the agriculture is considered the secondary source of HM pollution (Tchounwou et al. 2012; Gautam et al. 2016; Masindi and Muedi 2018). The widespread utilization in industrial, medical, and agricultural sectors along with others has led to its dispersal in the atmosphere, soil, and water (Wang 2009; Tchounwou et al. 2012; Lenntech 2018); however, HM emission by natural means cannot be overlooked as they contribute to huge quantities of dust comprising of HMs. The principal HMs estimated to have been released from natural resources are chromium (Cr), manganese (Mn), vanadium (V), copper (Cu), molybdenum (Mo), nickel (Ni), lead (Pb), selenium (Se), and zinc (Zn). Over 20% of cadmium (Cd), mercury (Hg), arsenic (As), chromium (Cr), copper (Cu), nickel (Ni), lead (Pb), and selenium (Se) in the atmosphere maybe attributed to volcanic activities (Allen et al. 2001). Of the total emissions into the atmosphere, sea salt aerosols contribute almost 10% of it (Allen et al. 2001). Fly ash (FA), released during coal combustion, is reported to contain

HMs. During 2001–2002, India was reported to emit >100 million tons of HM-enriched FA per year owing to the high ash content of coals (>32%) (Smith 2005). Although most of the HMs are natural environmental constituents, its biochemical equilibrium and geochemical processes have been altered by indiscriminate anthropogenic activities owing to the demand for better quality of life with modern conveniences (Briffa et al. 2020).

The contaminants are introduced into the atmosphere through a range of ways, including particles, droplets, and gases, or associated with other particles or droplets. Because they are too small to travel great distances, particles and droplets normally land on the ground after just a small distance (Ali et al. 2021). Particles smaller in size (<10 μm) and density persist for longer durations in the atmosphere and can be relocated over great distances. Air consistency along with vigorous vertical mixing circulation patterns promotes the distribution of pollutants in the troposphere layer. Once entering the air circulation, air pollutants can travel long distances and cause global environmental pollution issues. Soluble particulates react with rain and precipitate onto the water and land (Walker et al. 2012). Except for the metal Hg occurring in gaseous form, other atmospheric HMs of either natural or anthropogenic sources are chiefly associated with particulate matter. Metals of diameter ranging between 0.01 μm and 100 μm are adsorbed by particulate matter. The health impacts and haze episodes linked with atmospheric particle matter (PM) are of major concern to the public and government entities. PM has a high capacity for adsorbing hazardous metals, which can then enter the human body by breathing and cause negative physiological effects (Li et al. 2013). Wet and dry deposition of metal elements adsorbed to air PM can be seen in soils, aquatic bodies, and plant leaves. They may then accumulate in plants or animals as a result of metabolic processes, exposing people through the ingestion of contaminated plants or animals (Li et al. 2013). Sand storms, soil erosion, volcanic activities, and rock weathering are processes through which PMs are released. Meanwhile, burning of fossil fuels, smelting, vehicle exhaust, industrial activities, etc. are anthropogenic activities contributing to the PMs into the atmosphere (Briffa et al. 2020). As a result, a plethora of health implications including cancer and damage to essential organs as well as cardiovascular and pulmonary inflammation and acute respiratory disorders have been recorded mainly among urban populations (Fig. 4.1).

This chapter intends to provide a summary of HM pollution in the atmosphere and its consequences on air quality and the implications for human health. The sources and atmospheric pathway of some common HMs in the environment have been thoroughly discussed, as well as their toxicological impact on human health. The environmental bioaccumulation of hazardous components in human food chains and eventual human body contact are both facilitated by HM-contaminated media. In this chapter, the effects of common HMs on the human body are explored in terms of the amount and duration of exposure. To explore environmentally relevant HMs and their sinks and remobilization processes between air, soil, and water compartments, it is vital to analyze their transports together with fates and behaviors in air-water-soil biota. These concepts would be beneficial for future risk assessment, such as understanding ambient environmental variables; discovering, tracking, and

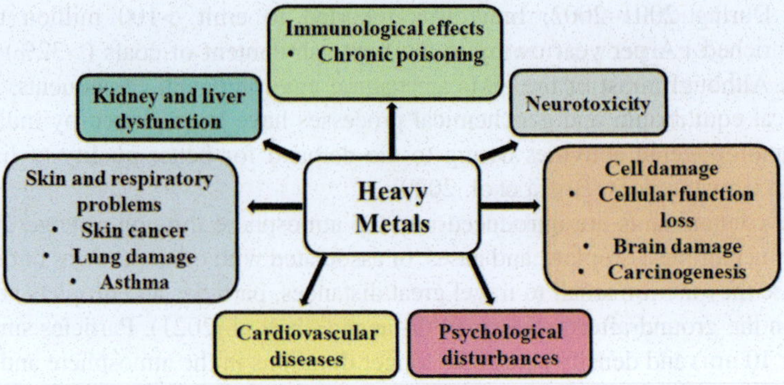

Fig. 4.1 The impact of heavy metals on the human body

analyzing problems that have occurred; and creating and ensuring command and control methods (Cheevaporn 2004).

4.2 Most Common Air Pollutants and Their Sources

Population expansion at a very uncontrollable rate is the chief cause of pollution with the peripheral causes of human activities, energy demands, and industrialization. However, besides this factor, there are several other contributing elements such as the use of biofuel, crop residue burning, biomass combustion, wood combustion, brick kilns, cremation, coal-based power plants, cyclones, forest fires, construction activities, long-range transport of dust, mining, and vehicular emissions (Ali et al. 2021). These activities have led to declining air quality, deteriorating human health, alterations in the meteorological conditions, precipitation rates and patterns, rate of snow melting, and causing acid rains. Moreover, atmospheric pollution along with degrading air quality can bring about ozone depletion, regression of biodiversity, economical loss, water, and land pollution, consequently acting in a comprehensive manner leading to a global change. The most commonly present atmospheric pollutants are Pb, Hg, and Cd; however, As, Mn, Cr, Co, Sb, Be, Se, and Ni are also observed to be present in high concentrations in the atmosphere.

Pb is a persistent contaminant which pollutes water, soils, and sediments by wet and dry deposition. Long-range wind transmission of soil particles, marine aerosol, volcanoes, biogenic materials, and forest fires are all natural sources of Pb. There is no volcanic component in India or its vicinity that can add to lead in the environment. In the mid-1990s, natural fluxes accounted for around 10% of total anthropogenic air emissions (Nriagu 1989). Coal, wood, and petroleum combustion, high-temperature metallurgy, the timber industry, and rubbish burning are all major anthropogenic contributions. Pb levels in the upper continental crust are at 170.5 ppm (Das 2022). As a result, crystal dust could be a major natural contributor of Pb in the atmosphere. In March 2012, Pb out of a strong Middle Eastern dust storm was detected over Delhi, India (Kumar et al. 2016). In India, coal burning is the most significant anthropogenic source of atmospheric Pb, which has expanded to fulfill the rising energy demand. Other source includes high-temperature enterprises like smelters and gasoline combustion, in addition to coal burning. India's high-temperature vehicle sector is rapidly growing, and the ever-growing fleet of automobiles in the cities is a significant Pb source in the atmosphere (Kumar et al. 2016). Organolead added to gasoline was a substantial source of exposure. Organic Pb was converted to inorganic lead oxide during combustion in car engines and released about completely in that form. This resulted in inorganic Pb exposure, particularly among those who lived in high-traffic regions. People who reside in urban regions have greater β-Pb levels than those who live in rural areas. In the Indian metropolis of Lucknow, Kaul et al. (2003) found 2000–3900 ng/m^3 in traffic zones. Residents in the vicinity may be exposed to industrial pollution and recycling processes. Furthermore, resuspension of polluted soils influences exposures due to environmental pollution from past industrial emissions. As a result, Pb exposure is enhanced in proximity to historical or present industrial sources, through inhalation and, probably more critically, consumption (due to children's hand-to-mouth activities). Weathering, chalking, and peeling paint releases environmental exposure to Pb-containing particles; Pb carbonate hydroxide has seen extensive use as a pigment in house paint in several nations. Dust in houses with a Pb-based paint and soil near Pb-emitting enterprises, for instance, can contain extremely high amounts of Pb (WHO/IPCS 1995; CDC 2002).

Hg exists naturally in the environment; however, in recent decades, human activity has profoundly altered its cycle. Volcanic eruptions and ocean discharges are both natural sources of Hg in the environment. Anthropogenic emissions include those from raw materials or fuels, applications in merchandise, and industrial processes. The major source of anthropogenic Hg emissions (37.7%) is artisanal and small-scale gold mining (ASGM), followed by stationary coal combustion (21%). Manufacturing of cement and nonferrous metals both account for 11% of total emissions. Since it was first released into the environment in small amounts, the Hg contained in it today can be transferred into the atmosphere through land, water, and other surfaces (US EPA).

Cd is released into the air by natural and artificial sources. Naturally, the chief sources of Cd in the atmosphere are windborne debris, biogenic emissions, and volcanoes (Nriagu 1989). The major contributor to airborne Cd is smelters. Additional sources of airborne Cd include burning fossil fuels, as well as incineration of municipal waste such as plastics and Ni-Cd batteries (ATSDR 1999). Iron and steel factories are another source of atmospheric Cd emissions. Smelting, soldering, and other high-temperature industrial activities produce Cd oxide as minute particles in the air (fume). When released into the environment, Cd compounds have the potential to bind with respirable airborne particles and travel long distances. Rain or falling from the sky deposits it on the ground below. Cd is rapidly absorbed into the food chain by plants such as green vegetables, root crops, cereals, and grains once it has fallen to the ground (ATSDR 1999).

Cr, a metallic element, is one of 129 priority pollutants identified by the USEPA and one of the 25 hazardous compounds that priority superfund sites have been found to contain the highest risk to human health. The largest natural emitter of Cr in the environment is continental dust flux. Anthropogenic sources provide more than 70% of Cr to the atmosphere, primarily from metal industry emissions and fuel combustion. Ferrochrome production, ore processing, chemical and refractory treatment, cement-making facilities, vehicle brake linings, and catalytic converters, leather tanneries, and chrome pigments are the most common sources of high atmospheric Cr exposure. The annual emissions of Cr from coal and oil burning are projected to be 1723 metric tons (ATSDR 1999).

As is a naturally occurring element that cycles through water, land, air, and living systems. High-temperature events such as coal-fired energy plants, burning vegetation, and volcanism release As into the atmosphere. As is liberated as highly soluble oxides as a result of combustion processes. As is also released by spontaneous biomethylation and reduction in low temperature to arsines. As, commonly discharged as As_2O_3, is mostly adsorbed onto a particulate matter. Arsines are released by microbial sources in soils or sediments which are oxidized in the air and reverted to nonvolatile forms, which are re-deposited on the soil. As a result of biogenic processes, gaseous alkyl As compounds are emitted from soil that has been treated with inorganic As compounds. Humans are predicted to breathe 40–90 ng of As per day, as recommended by the USEPA (Chung et al. 2014).

Mn is a necessary component of metabolic processes, but excessive quantities in the body can be hazardous, resulting in substantial neurologic and mental damage (Ranucci and Iorio 2019). In the environment, the occurrence of natural compounds of Mn is being steadily replaced by enrichment due to human activity. Mn emissions in the atmosphere are estimated to be 38.27 kt $yr.^{-1}$ by Livett (1992), with the biggest contributions coming from ferrous metal smelting and oil combustion. Mining, iron/steel manufacturing, ferro-/silico-Mn alloy and dry alkaline battery production, and welding are the main sources of Mn exposure. Low levels of Mn exposure in the environment are typically attributed to industrial sources, agricultural use, and the addition of Mn compounds to fuel. Since the addition of methylcyclopentadienyl Mn tricarbonyl (MMT) to unleaded gasoline, an increase in the atmospheric Mn content has been anticipated. Mn exists in atmospheric particles primarily as varying oxides that readily react with SO_2 and NO_2 and become water-soluble (Lucchini et al. 2014). On average, approximately 20 ng/m^3 Mn is reported in the air as per the ATSDR toxicological profile.

Sb ores are found in the crust of the Earth naturally. Sb oxides can be released into the environment by volcanoes. Anthropogenically, Sb oxides are a derivative of smelting lead and other metals, as well as coal-based power plants. Being a common component of coal and petroleum, emissions from car exhaust also contribute to Sb release. Refuse incinerators, minor industrial operations such as lead casting, and fossil fuel burning, for example, for home heating also cause emissions to air and land (NPI). Sb concentrations in ambient air range from >1 ng/m^3 to around 170 ng/m^3 according to the EPA's TTN ATW webpage; it may be observed at levels as high as 1000 ng/m^3 close to companies transforming Sb ores into metals or manufacture Sb oxide.

Pure Be, a hard gray metal, rarely exists naturally but is found in some types of rocks, coal and oil, soil, and volcanic dust as a chemical component. It can also be found in Be fluoride, Be chloride, Be sulfate, Be oxide, and Be phosphate, among other compounds. Be levels in the air next to the Be processing sector in Mumbai, India, were detected at 0.42–0.48 ng m^{-3} ($N = 397$) (Thorat et al. 2001).

Ni is the fifth most common metal. The Earth's surface and core are both rich in this silvery-white, lustrous metal. Wind-blown dust, generated from rock weathering and soils, forest fires, and volcanic activity are all natural producers of environmental Ni. Ni is also released due to the combustion of coal, diesel oil, and fuel oil, as well as the incineration of trash and sewage. The presence of Ni compounds in the air is primarily due to the combustion of fossil fuels (Genchi et al. 2020). Each cigarette has been observed to contain 1.1 to 3.1 g of Ni and may be available as Ni carbonyl in tobacco smoke, which is particularly dangerous to human health (Cempel and Nikel 2006).

Co exists as cobaltite, skutterudite, erythrite, spherocobaltite, and heterogeneity and can be found in both sedimentary and igneous rocks in excess. Co is found in 5–30% of cemented carbide. Co-leaching is caused by cement plants and carbide tool grinding operations (Lu et al. 2016). E-waste processing industries have also been discovered to leak Co at amounts exceeding the permitted limit (Lim and Schoenung 2010). Fine Co polishing discs are also utilized in diamond polishing. It also can generate Co dust (Leyssens et al. 2017). Co is also used as a siccative in the pigment and paint industries to speed up the drying process (Christensen and Poulsen 1994). Bottom ash from incinerators contains Co, which leaches into the soil and groundwater.

Se is a chalcogen element found in rocks and soil and is a widely dispersed element on the Earth's surface. Se is sometimes referred to as elemental Se or Se dust when existing in its purest form (Handa et al. 2016). Forest fires, volcanic activity, and soil erosion are all connected to Se origin. Se is also produced by human activities including the burning of trash, tires, paper, and other fossil fuels. The chief sources of atmospheric Se compounds are coal and oil combustion. According to their behavior in the atmosphere, three types of Se compounds can be distinguished: volatile organic molecules (DMSe, DMDSe, and methaneselenol), volatile inorganic compounds (selenium dioxide), and elemental Se attached to ashes or particles. In the atmosphere, dimethyl selenide is stable, whereas hydrogen selenide and Se dioxide are both unstable. Se and water are formed when hydrogen selenide is oxidized. In moist conditions, Se dioxide is converted to selenious acid (Mehdi et al. 2013).

Atmospheric particulate matter (PM) is of major concern for the public and government entities concerning the associated health impacts and haze events. Hazardous metals that can eventually enter the human body by breathing and create physiological issues have a high ability to adsorb in PM. PM can be released into the atmosphere directly or produced indirectly from gaseous precursors, primarily sulfur dioxide, nitrogen oxides, ammonia, and non-methane volatile organic compounds (Geiger and Cooper 2010).

4.3 Exposure Pathways of Atmospheric Heavy Metals

Inhalation and cutaneous contact are the two most common routes for metal intake in humans, both of which use air as the principal medium of interaction. The assessment of exposure is based on ambient and anthropogenic concentrations as well as numerous exposure routes (Geiger and Cooper 2010). People can risk their health by being exposed to dangerous air pollution in a variety of ways, including:

a. Inhaling contaminated air; consuming contaminated foods, such as fish sourced from polluted waters; meat, milk, or eggs from animals that consumed contaminated vegetation; and fruits and vegetables produced in contaminated soil that has been sprayed with air toxics.
b. Consuming polluted drinking water or soil contaminated by harmful air pollutants. Because they frequently swallow the dirt on their hands or the items they put in their mouths, young children are particularly vulnerable.
c. Contacting polluted dirt, dust, or water with one's skin (during recreational use of contaminated water bodies).

Air plays a role in many indirect pathways, including the deposition of metals to surface dust and intake through ingestion, inhalation, or dermal contact; deposition of metals to surface water and sediment and intake through ingestion and dermal contact; and absorption of deposited metals into aquatic and/or terrestrial biota, as well as entry into the food chain and ingestion. Although inhalation of surface dust is the primary contributor to human health risk in most cases where airborne elements have culminated in environmental contamination, this is not always the case. Inhaled metals have a substantially higher bioavailability than other forms of ingestion (Geiger and Cooper 2010). Even while inhaled doses are equivalent to intakes through other ways, this can result in extremely large internal dosages. Cigarette smoking, for example, contributes significantly to the bioaccumulation of Cd (Newman et al. 2004).

Due to differences in airway geometry and inflow velocities, which cause larger absorbed particulate deposition fractions in young children and infants than those in adults at the same exposure levels, infants and children are more vulnerable to airborne metal particles. Furthermore, studies demonstrate that particle pollution can limit children's lung function growth (C. EPA 2004).

4.4 Assessment of Heavy Metals in the Atmosphere

The current status of concentrations of HMs is crucial to present a good overview of ambient air quality in terms of heavy metal content in ambient air. Between 2000 and 2004, Fang et al. (2005) examined the atmospheric metallic elements throughout Asia. Fe was discovered to be the most prevalent element in Asian countries, varying in the value range from 165,000 to 265,500 ng m^{-3}. A significant concentration of Zn linked with total suspended particulate matter (TSP) was discovered in Taichung,

Taiwan, according to this study (1060 ng m^{-3}). Similar to TSP, a high quantity of Pb (820–1060 ng m^{-3}), Cu (370–1550 ng m^{-3}), and Mn (850 1470 ng m^{-3}) was found in Mumbai (India). At Sapporo and Tokyo, Ni (3.81–5.63 ng m^{-3}) and Cr (2.61–6.09 ng m^{-3}) associated with TSP were measured.

Wu et al. (2007) examined the trace metal contaminants related to atmospheric fine particulate matter of Asian countries from 1995 to 2005. According to this study, the Asian countries with the greatest mean concentrations of HMs connected with $PM_{2.5}$ were Mg (126.85 ng m^{-3}) > Fe (126.99 ng m^{-3}) > Zn (126.99 ng m^{-3}) (93.48 ng m^{-3}). Vehicle exhaust was the largest source of components such as Cr, Pb, Cu, Zn, and Cd in ambient particulate matter in Asian countries, according to the researchers. The primary source of elements like Zn and Pb is incineration, and the primary sources of Cr and Ni are coal combustion and industrial oil combustion, respectively. In the last 25 years, the toxicity of fine PM (1–2.5 μm) and ultrafine PM (0.1–1 μm) has been found to significantly increase human fatalities and morbidity (Geiger and Cooper 2010). However, emerging research during the last decade has indicated the toxicity of metallic constituents might be more hazardous as compared to other components of PM (Konkel 2009). The growing belief was that HMs adsorbing on PM is critical to its toxicity and detrimental health consequences. PM has been associated with a variety of illnesses, including cardiovascular and respiratory illnesses, as well as lung cancer (Li et al. 2013). Even at low levels, metals associated with PM present in ambient air have shown significant adverse health effects. Bushfires in Singapore resulted in substantial increases in Zn, Fe, and Cu concentrations. As a result, the number of patients seeking medical treatment for respiratory problems increased by 25% and the number of asthma cases increased by 29% (Karthikeyan et al. 2006).

The behavior of the PM in the mammalian respiratory system is determined by its aerodynamic properties and composition (Pendias and Mukherjee 2007). Studies have revealed that finer PM is more soluble and has high toxicity causing oxidative stress and inflammation (Sangani et al. 2010). In India's residential regions, the yearly mean concentration of PM_{10} (a primary source of HMs in the atmosphere) is three times greater than WHO standards. The standard yearly and 24-hour mean Pb values, according to India's National Ambient Air Quality Standards (NAAQS), were 500 ng/m^3 and 1000 ng/m^3, respectively. As and Ni had normal annual mean concentrations of 6 and 20 ng/m^3, respectively (The Gazette of India 2009). The ambient concentrations of As in residential neighborhoods of Hyderabad during the fireworks event were measured and reported by Kulshrestha et al. (2004) to be 25 ng/m^3, which is greater than the NAAQ guidelines. From March 2006 to May 2008, Kulshrestha et al. (2009) examined ambient Pb concentrations related to PM_{10} and $PM_{2.5}$ in urban and rural areas of Agra, finding two times higher Pb concentrations than the NAAQ requirements. In aerosols throughout the Delhi-Hyderabad-Delhi Road corridor, Singh et al. (2010) observed higher amounts of Zn, Mn, Cd, and Pb linked with PM_{10} in urban and semi-urban areas than in rural areas. The air quality in India is deteriorating as a result of urbanization and industrialization. HM concentrations in India are equivalent to those found in China and Pakistan, but greater than those reported in other developed Asian countries such as South Korea,

Japan, and Hong Kong. In China and India, fireworks during festivals raise the particulate matter and HM levels in the air. A major global risk factor is prolonged exposure to air pollution (Chen et al. 2022), arising from energy production, traffic, and industries which is a major consequence of the present industrial state and the increasing demand for a better quality of life with modern conveniences. Regulatory bodies at central, state, and local levels were set up to monitor and assess the air quality and categorize it as either safe or unsafe for breathing. The regulatory bodies are responsible for asserting standards or levels or goals for air quality (Geiger and Cooper 2010).

According to estimates from the World Health Organization, around 800,000 early deaths are brought on by urban air pollution which also reduces the life expectancy of 4.6 million people each year across the world. Developing nations are especially susceptible to rising amounts of airborne metals in the ambient atmosphere because they lack the regulatory infrastructure necessary to properly monitor and manage air toxins. PM pollution is responsible for 22,000 to 52,000 early deaths annually in the United States and 200,000 early deaths in Europe (Mokdad et al. 2004). Globally, 4–8% of premature deaths are owed to exposure to suspended PM, primarily in the indoor and ambient environment due to fine PM exposure. Country or state borders do not limit the emissions by the industries and other mobile sources. As per Health Effects Institute in Boston, Massachusetts, developing Asian nations account for two-thirds of the 800,000 early deaths brought on by urban air pollution (HEI 2004). Some of these nations, like some African nations, do not have sufficient industrial activity to pose concerns to human health, but others, like Mexico and China, have industrialization-related air toxics at levels considerably above what would be considered "safe" in more controlled systems (Geiger and Cooper 2010).

4.4.1 Metals in Particulate Matter

The concentration of HMs in particulate matter, particularly tiny particles such as $PM_{2.5}$, can endanger human health. The burning of garbage, high-temperature industrial processes, and the combustion of fossil fuels and timber all emit trace metals into the atmosphere. Volcanoes, wind erosion, forest fires, oceans, and other natural events are the primary sources of natural emissions (Nordberg et al. 2007). The major sources of Be, Co, Hg, molybdenum, Ni, Sb, Se, tin, and vanadium is specifically the burning of fossil fuels. Burning fossil fuels also adds to the emission of As, Cr, copper, Mn, and zinc by humans. A significant amount of As, Cd, Co, Ni, and zinc are also released during industrial metallurgical processes. Pb, Cu, Zn, Ni, and Cd were once present in varying degrees in gasoline exhaust fumes. Zn emission and tire rubber abrasion are connected (Councell et al. 2004) (Table 4.1).

The size of airborne particles determines the possibility for inflammatory inflammation, oxidative damage, and other biological impacts, according to several different groups of researchers (Costa and Dreher 1997; Ghio et al. 2002; Lippmann et al. 2006; Sangani et al. 2010). The deposited fraction of inhaled particles in distinct

Table 4.1 Industrial limits (μg m^{-3}) for hazardous metals in air and their adverse health effects

Metal	NIOSH REL (10 h TWA) (μg/m^3)	OSHA PEL (8 h TWA) (μg/m^3)	OSHA limit for work place air (μg/m^3)	Toxic effects
Sb	500	500		Affects skin and eyes, inflammation of lungs, chronic bronchitis, chronic emphysema, tuberculosis, cardiovascular effects, edema, and hemorrhage
Al				Dementia, central nervous system damage, kidney and liver dysfunction, colitis, lung damage, pulmonary fibrosis, hypoparathyroidism
Cd	N.E.	0.005	5	Kidney damage, prostate dysfunction, bone diseases and cancer, kidney dysfunction and proteinuria, lung cancer, osteoporosis
Cr	0.5	1	0.5–1000	Breathing problems, such as asthma, cough, shortness of breath, chronic poisoning, dermatitis, eczema, gingivitis, bronchitis, liver and kidney disease, sinusitis, pneumonia, lung cancer
Cr (III)	0.5	0.5		
Cr (VI)	0.001	0.005		
Co	0.05	0.1		Skin and respiratory problems, acute effects such as congestion, edema, lung hemorrhage, ventilator function reduction. Chronic inhalation may cause asthma, respiratory irritation, pneumonia, fibrosis, reduced lung functions, cardiac effects, nausea and vomiting, diarrhea, liver disorders
Cu	1	1		Liver cirrhosis, chronic anemia, brain and kidney dysfunction, stomach and intestine infection
As	2	10	10	Skin cancer and marked problems with circulatory system, developmental abnormalities, neurobehavioral sicknesses, cardiovascular diseases and hearing sickness, anemia, leukopenia, eosinophilia, and carcinoma. A sensation of "pins and needles" in hands and feet, darkening of the skin and the appearance of small "corns or warts" on the palms, feet, and torso, lung cancer, liver and kidney damage
Be	0.5	2		Inflammation of lungs, acute pneumonitis, berylliosis, shortness of breath, fatigue, immunological effects
Pb	50	50	50	Anemia, brain and kidney damage, affects the neurological system, impaired

(continued)

Table 4.1 (continued)

Metal	NIOSH REL (10 h TWA) (μg/m³)	OSHA PEL (8 h TWA) (μg/m³)	OSHA limit for work place air (μg/m³)	Toxic effects
				cognitive and motor function, miscarriage, infertility, carcinogen
Mn	1000	5000		Central nervous system dysfunction, manganism, weakness, lethargy, affects lungs
Hg	0.1	100	100 for organic Hg, 50 for metallic Hg	Tumor, memory loss, restlessness, depression, psychological disturbances, fatigue, insomnia, lung damage, kidney failures, death of fetus, gastrointestinal issues, increase in blood pressure
Ni	15	1000		Skin allergy, cancer of lungs, nose, sinuses, throat, immunotoxic, neurotoxic, genotoxic, loss of fertility
Se	200	200	200	Nausea, diarrhea, stomach pains, bronchitis, respiratory infection, selenosis, hepatoxicity
V	50	50		Throat, nose, and eye irritations, nose bleeding, dizziness, headaches, nausea, impairment to the nervous system, liver and kidney hemorrhage, paralyses and behavioral changes, cardiac diseases
Ag	10	10	10	Argyria, a blue-gray discoloration of the skin and other body tissues, breathing problems, lung and throat irritation, stomach aches, mild allergic reactions

NIOSH = National Institute for Occupational Safety and Health; OSHA = Occupational Safety and Health Administration; REL = recommended exposure limit; PEL = permissible exposure limit; TWA = time-weighted average

parts of the respiratory tract is determined by the particle size distribution of an aerosol (Oller and Oberdorster 2010). PM is a complex mixture of extremely small particles and liquid droplets that include acids (such as nitrates and sulfates), organic compounds, metals, and soil or dust particles, among other things. The EPA is particularly interested in particles having a size of 10 m or smaller since they enter the lungs through the nose and throat. If inhaled, these particles can damage the heart and lungs, travel throughout the body, accumulate in organs, pierce cell membranes, and result in serious health issues (Geiger and Cooper 2010). "Inhalable coarse particles," similar to that found near highways and dusty industries, are ≥2.5 m in diameter and < 10 m in diameter, according to the EPA. Fine particles, like those observed in smoke and haze, have a diameter of 2.5 meters or less. These particles may be released directly from events like forest fires or may develop in the atmosphere as a result of the reaction of gases produced by power plants, industries, and automobiles (Geiger and Cooper 2010).

As detailed by Nieboer et al. (2005), these particles can be further classified in terms of significant health impacts for specific regions:

a. The proportion of cumulative airborne particles entering the body through the mouth and/or nose during respiration is known as the "inhalable aerosol fraction." This fraction, which corresponds to particles with an aerodynamic equivalent diameter (dae) more than 100 m, is important for health impacts in the respiratory system, such as lung cancer, bronchitis, nasal irritation, and rhinitis. This proportion is also important in terms of systemic impacts.
b. The "thoracic aerosol fraction" [dae $<$ 30 m] is a subfraction of the inhalable fraction that is essential for the treatment of lung cancer, bronchitis, and asthma because it contains particles that can enter the tracheo-alveolar region of the lung.
c. The "respirable aerosol fraction" (also known as the "alveolar fraction") is a subfraction of inhaled particles [dae $<$ 10 m] that permeates into the alveolar surface of the lung and is crucial in the development of chronic diseases including pneumoconiosis and emphysema.

4.4.2 Characteristics of Fine Particulate Matter

The World Health Organization estimates that 2.4 million people each year die as a result of air pollution, primarily from small particles (WHO 2002). According to recent studies on human health, free radicals similar to those found in cigarettes are also common in airborne fine particles and may cause many of the same potentially fatal diseases (Dollemore 2008).

Utsunomiya et al. (2004) postulated homogeneously dispersed poisonous trace metals represent a concern to human health and the environment as pollutants in coarser, insoluble particles but the dangers are greatest if they are present as primary components in individual, trace metal, microscopic particles. The majority of studies generally indicate that the toxicity caused by oxidative stress and inflammatory processes increases with decreasing particle size and solubility (Valavandis et al. 2008). Metals were found to be a prominent source of cellular oxidant production and associated health consequences in research of $PM_{2.5}$ (Maciejczyk et al. 2010). Metal size distribution studies demonstrate that the majority of harmful metals collect in the tiniest particles ($PM_{2.5}$ or less) (Khaiwal et al. 2008). This small fraction may penetrate deeper into the respiratory tract, settling mostly in the alveolar region of the lungs, where trace element absorption effectiveness ranges from 60% to 80% (Pope and Dockery 2006). When a metallic ion comes into touch with lung tissue/cells, they are released into the biological system (Midander et al. 2007). The maximal residence time (100 days) for fine and ultrafine particulate matter in the atmosphere allows for a broad geographic dispersion (Utsunomiya et al. 2004). Ultrafine particles are considered to be more soluble than larger particles of the similar composition due to the high surface-to-volume ratio for small sizes (Navrotsky 2001). According to recent studies, the metal fraction of fine and ultrafine PM is extremely harmful and is the main cause of poor human health

(Magari et al. 2002). Additionally, these particles influence climate change and can be carried by strong winds over long distances (WHO 2007). To fully understand these findings' possible health effects, chemical characterization of the fine and ultrafine fractions of airborne particles must be prioritized (Khaiwal et al. 2008).

4.5 Public Health Concern

Industrialization, modern urbanization, and rapid economic development have all resulted from the increased use of HMs in industrial and agricultural activities around the world. These practices potentially pollute water, air, and soils with harmful HMs. HM-contaminated media lead to environmental bioaccumulation of toxic elements in food chains, ultimately impacting the human body. HM has a range of health concerns in humans, determined by the amount and duration of exposure (Ali et al. 2021).

4.5.1 Health Implications

The toxicity of an HM relies on the entry pathway of the metal, its entry rate, the distribution in tissue, the concentration retained, and the excretion rate. Inhibition of enzyme activities and protein synthesis, altering the functions of nucleic acid, and altering the cell membrane permeability are toxicity mechanisms induced by HMs (Bernardo 2021).

Cell organelles such as mitochondria, lysosomes, and nuclei along with cell membrane and enzymes are affected by HMs. The metal ions are associated with DNA and nuclear proteins leading to site-specific DNA damage (Valko et al. 2005; Tchounwou et al. 2012). The damages may be direct or indirect. Direct damages include structural alterations in the biomolecules due to association with the metal ions. Lipid peroxidation, DNA damage, altered sulfhydryl homeostasis, and free radical production are all effects of HM toxicity. The formation of reactive oxygen and nitrogen species containing hydroxyl and superoxide radicals, nitric oxide, hydrogen peroxide, and more endogenous oxidants is considered indirect damage (Valko et al. 2005). Copper, Cr, iron, Ni, and Cd have been observed to produce free radicals. Iron, copper, Cr, and Co follow Fenton reactions linked to peroxisomes, mitochondria, and microsomes (Valko et al. 2005). The metal-mediated free radicals result in DNA base mutation showing a vital linkage between carcinogenesis and oxidative stress. As, Cd, and Ni have been observed to inhibit DNA repair mechanisms (Valko et al. 2005; Briffa et al. 2020).

Recent studies have found that some metals can harm the nervous system and olfactory system. The neurological system and the outer world are directly connected by the olfactory system (Aschner et al. 2005). Foreign substances can reach the brain through the olfactory (nasal) neuron (Bondier et al. 2008). Metals that can travel through the olfactory nerve include aluminum, Cd, Co, Hg, Mn, Ni, and zinc.

4.5.2 Hazardous Air Pollutants (HAP)

The EPA's Health Effects Notebook for Hazardous Air Pollutants lists 11 distinct metals (EPA, Health Effects Notebook for HAPs 2010). HAPs, commonly referred to as harmful pollutants or air toxins, are contaminants that induce or may increase the risk of cancer as well as other serious health effects including birth defects or reproductive issues, as well as detrimental effects on the environment and the ecosystem. The information presented here is meant to serve as a basic overview rather than a full reference for metals' health impacts. The data for the Air Toxics Website (ATW) (EPA, Technology Transfer Network 2010), which incorporate information from the Agency for Toxic Substances and Disease Registry (ATSDR 1999), the Integrated Risk Information System (IRIS 2010), and the EPA, as well as recent air research, are presented in the following subsections. Additional sources of information on health effects include the California Office of Environmental Health Hazard Assessment, the Hazardous Substances Data Bank (HSDB 2010), and the American Conference of Governmental Industrial Hygienists (ACGIH 2010; OEHHA 2008).

4.5.2.1 Antimony (Sb)

Inhalation of Sb can cause skin and eye irritation, while ingestion might cause gastrointestinal problems. High acute exposure to Sb has caused respiratory consequences in animals, including a significant drop in ventilatory function, congestion, edema, and bleeding, in addition to liver and cardiovascular damage. Respiratory issues such as lung inflammation, chronic bronchitis, and chronic emphysema can be brought on by prolonged inhalation. Respiratory side effects include pleural adhesions, irritation, chronic bronchitis, chronic emphysema, latent tuberculosis, Sb pneumoconiosis (lung inflammation), and chronic bronchitis. There have also been reports of cardiovascular problems. Sb inhalation has been related to lung malignancies in animals, although no definitive link between cancer and Sb has been discovered in people. Sb has not been categorized as carcinogenic by the EPA.

4.5.2.2 Arsenic (As)

Inorganic As is most commonly absorbed through food, with smaller quantities absorbed through drinking water and breathing. Inhalation can happen in metal smelters and while burning wood that has been processed with an As preservative. Arsine poisoning is caused by inhaling it. As has no odor or flavor. Acute inorganic As inhalation and ingestion can cause gastrointestinal consequences (nausea, diarrhea, abdominal pain) as well as central and peripheral nervous system diseases. Acute inorganic As poisoning can end in death. Arsine is exceedingly poisonous, and exposure can cause abdominal discomfort, vomiting, and headaches within a few hours of exposure. Acute arsine poisoning might also result in mortality. Reduced production of red and white blood cells, an unsteady heartbeat, blood vessel damage, and a "pins and needles" feeling in the hands and feet are all symptoms of lower exposure levels. Skin and mucous membrane inflammation can occur after inhaling inorganic As. Long-term oral consumption can result in gastrointestinal problems, anemia, peripheral neuropathy, liver or renal damage, hyperpigmentation, and skin lesions. Reduced exposure over time can discolor the skin and create little corns or warts on the palms, soles, and torso. A kind of skin cancer and bladder, liver, and lung cancer have all been linked to oral exposure to inorganic As. Women who work in or live close to metal smelters may experience more spontaneous abortions than the general population, and their offspring may be born with lower weights. Inhalation investigations on humans have found that inorganic As exposure is highly linked to lung cancer. The EPA classifies inorganic As as a Group A human carcinogen, with the potential to cause cancer of the skin, lungs, liver, and bladder (ATSDR).

4.5.2.3 Beryllium (Be)

Be is most commonly found in or near facilities that mine, process, or convert it into alloys and chemicals. People can also become exposed through the inhalation of Be dust or fumes from the burning of coal, fuel oil, or tobacco. Be can also be consumed orally and is found in foods like fruits, vegetables, water, and soil.

During the 1980s, the average Be content in the air in the United States was 0.03 ng/m^3. Between 1977 and 1981, ambient values in 50 cities ranged from 0.1 to 0.4 ng/m^3. High levels of Be can cause acute pneumonitis or lung inflammation when inhaled for a brief period (reddening and swelling of the lungs). After exposure, symptoms may be reversible. Be compounds have been demonstrated to exhibit a wide spectrum of acute toxicity in animal tests, varying from mild to extreme acute toxicity when administered orally. Prolonged Be exposure can cause chronic Be illness (berylliosis), which causes noncancerous granulomatous lesions in the lungs. Chronic Be disease symptoms include mucous membrane irritation, decreased lung function, shortness of breath, malaise, fatigue, anorexia, dyspnea, and weight loss. In humans and animals, chronic inhalation exposure has had immunological consequences. Chronic pneumonitis, conjunctivitis, and skin allergies are some of other long-term consequences. The inhalation of Be is thought to increase the risk of developing lung cancer in humans, and animal studies have also linked the inhalation of Be to the disease. Be is a Group B1 probable human carcinogen according to the EPA.

4.5.2.4 Cadmium (Cd)

Based on the maximum concentrations of Cd detected in rural, urban, and industrialized sectors, the quantity of Cd inhaled does not surpass 0.04, 0.2, and 0.4 μg/day, respectively, assuming a daily inhalation of 20 m^3. Inhalation exposure is modest in the general population, but in locations with polluted soils, particularly where roadways have been coated with leftovers from nonferrous metal processing, home dust is a potentially substantial and persistent source of Cd exposure (e.g., zinc ashes or sintels as oven sludge). The cigarette is a potential source of Cd that is greater than that found in food. Cd levels in cigarettes range from 1 to 2 μg per cigarette. Approximately 10% of this is inhaled during the smoking process. Consequently, a person who smokes 20 cigarettes a day will ingest about 1 μg of Cd (Järup et al. 1998). According to Erzen and Kragelj (2006), light-to-moderate smokers

(>20 cigarettes per day) had a median blood Cd (B-Cd) concentration of 0.5 μg/L, heavy smokers (<20 cigarettes per day) had 1.5 μg/L, and nonsmokers had none.

Kidney and bone damage and cancer are important health endpoints. In Sweden, 820 women aged 53–64 years were studied to see if there was a link between Cd exposure and tubular and glomerular function. Increased B-Cd and U-Cd levels of 0.38 μg/l (median) and 0.52 μg/l (0.67 g/g creatinine), respectively, were associated with higher amounts of human complex-forming protein and N-acetyl-d-glucosaminidase in urine. The correlations persisted even at modest intake doses in women who never smoked. Inhalation exposure in work contexts has also been linked to an elevated risk of lung cancer. The inputs from transboundary air pollution and the usage of mineral and organic fertilizers in Europe are reasonably similar each year in Europe. They all continue to add to the relatively considerable Cd accumulations in the topsoil that already exist.

Cd exposure has been linked to prenatal deformities and other developmental problems in animals, but there is no solid evidence in humans. Cd toxicity from mining in Toyama Prefecture, Japan, causes itai-itai sickness. Cd toxicity resulted in bone softening (brittleness) and kidney failure. Cd is designated as a Group B1 probable human carcinogen by the Environmental Protection Agency.

4.5.2.5 Chromium (Cr)

With an average air level of 3 ng/m^3 observed in 13 cities in the United States, the average daily intake of Cr from air is between 200 and 400 ng (AIRS locations) (Chen and Lippmann 2009).

People who reside near Cr waste disposal facilities or Cr production and processing companies are more susceptible to high concentrations of Cr than the general population. Cr (VI) is far more hazardous than Cr (III), causing abdominal pain, vomiting, and hemorrhage in both inhalation and ingestion. Shortness of breath, coughing, wheezing, and other respiratory symptoms like asthma, nasal irritation, and discomfort are all side effects of Cr (VI) inhalation. It can result in pneumonia in the respiratory system, bronchitis, pulmonary dysfunction, and septal perforations and ulcerations. Cr exposure may also affect the liver, kidneys, and gastrointestinal and immunological systems, as well as the blood. Cr (VI) exposure can lead to difficulties throughout pregnancy and labor. Inhaled Cr (VI) is associated with an elevated risk of lung cancer, and Cr (VI) has been demonstrated to produce lung tumors in animal experiments. The Environmental Protection Agency has classified Cr (VI) as a Group A recognized human carcinogen when inhaled.

4.5.2.6 Cobalt (Co)

Respiratory consequences of Co inhalation include a considerable reduction in ventilatory function, lung congestion, edema, bleeding, wheezing, pneumonia, respiratory irritation, asthma, and fibrosis. Chronic exposure can have negative effects on the heart, the liver, the conjunctiva, and the immune system, including the development of Co sensitization. Inhalation exposure to Co has been linked to respiratory, cardiovascular, and neurological effects, as well as lower body weight, thymus necrosis, and blood, liver, and kidney consequences in animals.

4.5.2.7 Lead (Pb)

Pb is hazardous even in small amounts and elevated blood lead can result in fatalities in children. Pb poisoning can cause brain damage, renal damage, and gastrointestinal damage in people who are exposed to it even for a short time. Over time, Pb poisoning can also affect the kidneys, central nervous system, blood, blood pressure, blood, and vitamin D metabolism. Workers have observed neurological problems, as well as delayed nerve transmission in the peripheral nervous system in adults. Chronic Pb poisoning in children can result in IQ loss, slower cognitive development, stunted growth, loss of hearing, and other developmental problems. Possible negative effects of Pb exposure include decreased sperm count, accidental miscarriages, low birth weight, and sluggish postnatal neurobehavioral development. Pb is designated as a Group B2 probable human carcinogen by the EPA.

4.5.2.8 Mercury (Hg)

The lungs collect about 80% of Hg_0 vapor that is inhaled, and this substance is swiftly transported to the kidneys and brain, among other parts of the body. It easily passes through the blood-brain barrier and the placental barrier. The presence of elemental Hg vapors in pregnant women's blood can be transferred onto the growing fetus and deposited there (Geiger and Cooper 2010).

The majority of Hg_0 accumulates in the kidneys, and to a limited extent in the brain, where it is easily transformed to an inorganic form. When exposed to Hg_0, higher Hg is deposited within the brain than when exposed to inorganic Hg compounds. Hg_0's half-life in the body is expected to be around 60 days (WHO 2003). Acute poisoning from Hg vapor at increased concentrations (higher than 1000 g/m^3) for a limited time produces symptoms of lung injury such as extreme airway irritation, pneumonitis, and pulmonary edema. It can harm the brain, nerves, kidneys, and lungs, resulting in coma and/or death in severe cases. Chest discomfort, dyspnea, coughing, hemoptysis, and pulmonary function impairment were all symptoms of the injury among workers chronically exposed to Hg vapor for 4–8 hours (McFarland and Reigel 1978). After prolonged exposure to low amounts of Hg vapor (50–100 g/m^3), adverse effects on the kidneys, thyroid, and central nervous system might develop over time and manifest subtly. It's tough to tell the difference between Hg toxicity symptoms and those of other prevalent disorders. Tremors, muscle weakness, melancholy, behavioral modifications, and short-term memory loss are among the symptoms in adults, as are skin rashes, including redness and itching. In children, peeling of the hands and feet is common.

4.5.2.9 Manganese (Mn)

Health consequences of Mn are now being studied by scientists. Side effects of inhalation have been observed in several animal investigations. Based on short-term rat experiments, Mn is judged to have moderate acute toxicity. Mn poisoning affects the central nervous system in a number of ways, affecting functions like eye-hand coordination, hand steadiness, and visual reaction time. Inhalation exposure might have an impact on the respiratory system. When exposed to high quantities of manganese for an extended period of time, manganism, a disease that starts with feelings of weakness and lethargy, tremors, a mask-like face, and psychological issues, can progress to impotence and loss of libido (Geiger and Cooper 2010).

4.5.2.10 Nickel (Ni)

The lungs and kidneys were damaged in an instance of short-term inhalation exposure to a high quantity of Ni. Contamination of drinking water may cause neurological implications as well as gastrointestinal problems (such as nausea, vomiting, and diarrhea). Exposure to Ni carbonyl can lead to pulmonary fibrosis and renal edema. According to recent studies, Ni can, at ambient levels, cause abrupt changes in heart rate and have other negative effects on health, in part because it has the capacity to create reactive oxygen species (ROS) (Zelikoff et al. 2002).

Ni dermatitis, which causes itching in the fingers, hands, and forearms, can be caused by long-term skin contact. Ni refinery employees exposed to Ni refinery dust have been related to an elevated incidence of lung and nasal malignancies as a result of inhalation exposure (Geiger and Cooper 2010). The EPA classifies Ni refinery dust and nickel subsulfide as Group A human carcinogens, while nickel carbonyl is a Group B2 probable human carcinogen.

4.5.2.11 Selenium (Se)

Acute inhalation exposure to Se compounds such as Se dioxide and hydrogen selenide causes respiratory consequences in humans. Inhaling elemental Se dust for a short duration can induce irritation of the nose and throat mucous membranes, nosebleeds, bronchial spasms, dyspnea, bronchitis, and chemical pneumonia. Oral exposure to hydrogen selenide was found to be extremely hazardous during animal testing. High Se levels taken in over time can result in a variety of side effects, including hair loss, severe tooth decay, a strong garlic odor in the breath and urine, mental weariness, and listlessness (Geiger and Cooper 2010).

The EPA follows these processes to evaluate the remaining danger posed by HAPs:

- Evaluation of the degree of public exposure
- Evaluation of the nature and impact of negative consequences
- Dose-response assessment
- Characterizing overall risk

Thus, so far, there are currently no residual risk requirements involving sources that monitor metals or metal compounds; nevertheless, the next group of sources to have residual risk standards promulgated includes primary aluminum smelters and refineries that regulate metals (National Lime Association v US EPA 2000).

4.6 Long-Range Transboundary Air Pollution (LRTAP)

The issue of air pollution was seen as more of a national issue than a global one. The idea that one country's industrial pollution may contaminate another was controversial. Long-distance air pollution transfer has been identified as a significant element influencing ecosystems and human populations. The research report "Long-Range Transport of Air Pollutants: Measurements and Findings," which the Norwegian Institute for Air Research (NILU) produced for the Organization for Economic Co-operation and Development (OECD) in 1977, was a turning point. This study was critical in determining the extent to which air pollution may spread. The analysis showed that pollution could travel long distances and that national initiatives in a single country can only go so far in reducing acid rain. As proof of the importance of the long-distance transboundary air pollution movement grew, the affected states joined together to create international legislative procedures to lower emissions of the most dangerous pollutants. This led to the 1979 signing of the Convention on Long-Range Transboundary Air Pollution with the intended goal of limiting acid rain's impacts through sulfur and nitrogen emission management. The Convention's scope was later expanded to include ground-level ozone formation (Protocol to Abate Acidification, Eutrophication, and Ground-level Ozone, 1999), as well as persistent organic pollutants (Protocol on Persistent Organic Pollutants (POPs), 2001) and HMs (Protocol on Heavy Metals, 1998).

Long-range transboundary air pollution includes, but is not limited to, acidifying gases like sulfur oxides (SOx) and nitrogen oxides (NOx), ground-level ozone, particulate matter (dust), and environmental pollutants such HMs and persistent organic pollutants (POPs). The influence of long-distance air pollution transmission from far-off sources on global air quality is enormous. Changes in the meteorological conditions during long-distance travel allow for chemical transformations and the formation of secondary pollutants such as ozone, as well as the removal of pollutants by dry or wet processes. For instance, a study was conducted in the southernmost region of Norway to assess the deposition of the most common air pollutants in natural surface soils. These areas had less than a minimum contribution to air pollution; however, it was observed that the pollution levels concurred with the maximum precipitation zones. This might be due to orographic effects (Steinnes et al. 1987).

Based on new scientific data and revised estimates of emissions, air levels, deposition, and environmental destiny, the preliminary assessment of HM health concerns done by LRTAP in 2002 was revised. Data on the danger of these compounds, the propensity for some metals to travel great distances after being discharged into the environment, their persistence and accumulation in different environmental compartments, human environmental exposure pathways, and the results of research on the detrimental effects on human health have been summarized in several reports released by international organizations, like the World Health Organization (WHO) and the International Agency for Research on Cancer (IARC), or governmental organizations, like the Centers for Disease Control and Prevention (CDC) in the United States. Anthropogenic emissions generally outnumber natural emissions, and despite the present trend toward lower natural emissions, releases into the environment continue to cause modest increases in soil contamination across many regions of the world due to metal tenacity.

4.7 Fate and Behavior of HMs in the Environment

The environment is dynamic. The atmosphere, watershed, land, and living creatures are all impacted by physical changes or incidents that take place on Earth, whether they are natural or manmade. For instance, industrial sectors have been releasing a variety of pollutants, including HMs, into the air, watershed, and land. Multiple environmental compartments will experience infiltration, circulation, and finally agglomeration of contaminants (air, water, soil, and biota). These are referred to as contaminant pathways. They are also depicted as interconnected divisions of the atmosphere, hydrosphere, and lithosphere (Walker et al. 2006). In a short time, the HMs have significantly polluted and accumulated, as well as dispersed and transmitted across the compartments. Investigating their transportation, as well as their destinations and behaviors in the air, water, soil, and biota, is crucial (Pachana et al. 2010).

The localization of a large amount of HM causes an increase in its toxicity. The main channels for chemicals to enter and diffuse into the environment are air and water. The dispersion of lead off the coast of England revealed that it was produced on land, transported by air, and ingested by water and biota. These sources included industries, hazardous waste disposal sites, lead-suspended air conditions, and more (Cheevaporn 2004). Chemical pollution pathways include not only air and water but also soil and biota. Organisms harboring potential carcinogens can compartmentalize them into persistent deposits to stop them from interfering with cytoplasmic metabolic processes (Briffa et al. 2020) but remain persistent leading to bioaccumulation and biomagnification. Due to their nonbiodegradable and persistent nature, HMs cannot be broken down and exist in the sediments and soils for prolonged periods until transferred to some other compartments (Briffa et al. 2020). Transport can take place both within and between compartments. HMs are dispersed after infiltrating the environment and may be converted into different compounds. Photodegradation (e.g., UV), chemical degradation (e.g., hydrolysis), and biodegradation are examples of transformation processes in the environment (e.g., bacterial decomposition). Biotransformation is the process of HMs being converted within organisms (Boonsaner 2006).

As HMs enter compartments, their fates and behaviors have changed and moved between settings (Fig. 4.2). Based on their physicochemical features, the majority of dissolved HMs conveyed by natural water systems are quickly adsorbed by particulate matter (Forstner and Wittmann 1983). HMs can adhere to particulate matter or bond with organic groups to form organometallic complexes, resulting in lipophilic molecules and ions. As a result, they can be found in animals, plants, and sediments. Their destinies and behaviors are governed by sinks and remobilization processes, serving as entering compartments, as claimed by Forstner and Wittmann (1983) and

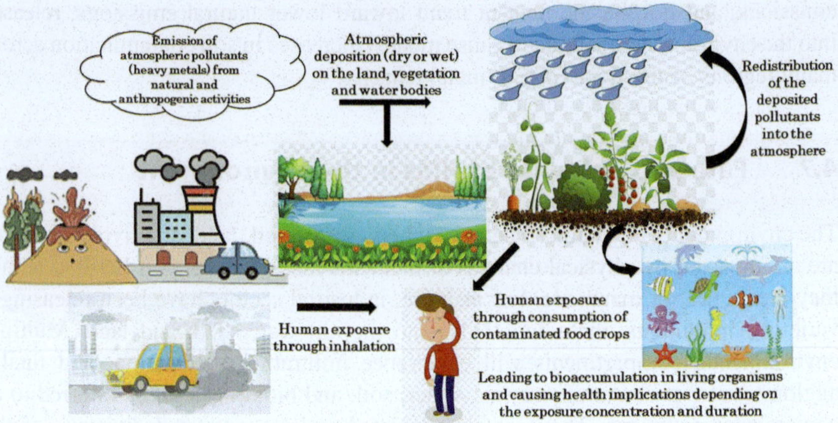

Fig. 4.2 Graphical illustration of the fate of atmospheric heavy metals

Cheevaporn (2004). Sink processes include adsorption and co-precipitation, precipitation, and absorption into biological activity, while remobilization mechanisms include increased salt concentration, redox condition, pH lowering, increased usage of organic complexing agents, and biochemical processes.

4.7.1 Sink Processes

4.7.1.1 Adsorption and Co-precipitation

HMs (Me^+) react in the form of positive ions. The sorptive properties of negatively charged ions, such as carboxyl and phenolic OH^- groups in organic materials, ferrous hydroxide ($FeOH^-$) and Mn hydroxide ($MnOH^-$) factions in hydrous Fe and Mn oxides, silicon hydroxides ($SiOH^-$), and aluminum hydroxide ($AlOH^-$) groups in clay minerals, are what cause their adsorption. The preferential adsorption of particular positive charges and the discharge of comparable charges associated with other species are made possible by the selective technique of balancing negative charges (Forstner and Wittmann 1983; Pachana et al. 2010).

Hydrous aluminum, iron, and Mn typically function as significant HM sinks in water systems, particularly the redox-sensitive Fe and Mn hydroxides and oxides under oxidative conditions. Such hydroxides and oxides, which both represent the Fe^{3+} ion and include ferric oxides ($Fe_2O_3^-$) and ferric hydroxide ($Fe(OH)_3$), efficiently bind or co-precipitate cations and anions. Additionally, hydrous Fe/Mn oxides could be a significant cause of dissolved metals since sorbed HMs are quickly mobilized in reducing conditions in natural waters (Pachana et al. 2010).

4.7.1.2 Precipitation

The equilibrium constant of solubility product (Ksp), which is used to describe a concentrated solution within which a dynamic equilibrium develops among a solid and its aqueous ions, is different for every metal (Purdue University 2008). When a chemical's solubility product is exceeded, the compound will precipitate until the product of the ionic levels surpasses the Ksp value. Precipitation will happen if the solubility product exceeds its Ksp value, which is determined by multiplying the results of the ion products (Ratchamongkol Technology University 2008). Precipitation will happen over the saturated solubility if the combined result of the ion products exceeds the Ksp value.

4.7.1.3 Integration of Biological Activity

Metals are nonbiodegradable, suggesting their inability to be broken down into less hazardous parts. Organisms detoxify by encapsulating active metal ions in proteins, such as metallothionein (which binds covalently to sulfur), or depositing them in insoluble form in intracellular granules for long-term storage or expulsion in the stool (Pachana et al. 2010).

4.7.2 Remobilization Processes

4.7.2.1 Increased Salt Concentrations

Due to increased salt content competing with HMs, desorption of HMs from solid materials occurs. Cations of alkali earth or alkali metals interact with immobilized metallic ions on solid materials or sediments before being released into the water. These adsorbed metals would be remobilized by desorption or dissolution, according to Forstner and Wittmann (1983).

4.7.2.2 Decrease in pH

The decline in pH causes an increase in solubility. Hydronium ions (H^+) are generated which interact with metallic ions to cause adsorption with sorptive particles, thus releasing metal ions (Pachana et al. 2010).

4.7.2.3 Redox Conditions

Eutrophication causes oxygen depletion in the aquatic systems; however, oxygen is an essential component for biodegradation by organic substances. As a result, the sediment serves as a sink for oxygen, which is supplied through the sediment surfaces at a rate controlled by the following factors: a biological oxygen demand resulting from respiration and metabolic activities in the sediment; a chemical oxygen demand resulting from inorganic elements, such as Fe2+ released to the sediment in the reduced state from decomposing biological matter accumulation; and diffusion, which helps control transportation (Pachana et al. 2010).

The oxidizing environment would typically have an Eh value, usually greater than 600 mv (Fitzpatrick et al. 2008). The Eh value drops dramatically in a decreasing environment at a sediment height of about 20 cm. At the same time, between 15 and 10 cm depth, there is a significant prevalence of iron and Mn in the soils (Forstner and Wittmann 1983). HMs are easily remobilized in this environment through the aqueous compartment.

4.7.2.4 Biochemical Process

Microbial processes facilitate the remobilization of HMs adsorbed onto particulate organic matters or sediments. Biodegradation, i.e., the destruction of organic matter into molecular weight components, is an example of a biochemical remobilization process (Pachana et al. 2010).

4.8 Conclusion

HM pollution of the environmental compartments is undoubtedly the most notable effect of our society's advancement. A chemical's hazard is determined by its toxicity, bioaccumulative potential, and persistence in the environment. HMs are deemed detrimental because of these three components: persistence, toxicity, and bioaccumulation. Soils contaminated by atmospheric HMs contain these contaminants on the surface layers of soil for long periods. It is critical to analyze and monitor the quantities of potentially harmful HMs and metalloids in various environmental segments as well as in the resident biota. Because of their high densities, the fates and behaviors would indicate which places should be examined. A rigorous environmental chemistry and ecotoxicology analysis of harmful HMs and metalloids shows that actions should be taken to limit the impact of these elements on human health and the environment and aid in determining the risk of exposure to vulnerable organisms. In terms of particulate matter and HMs, the implementation of Euro standards for automobile emissions and adequate management of industrial emissions will make Indian ambient air clean.

References

ACGIH, American Conference of Industrial Hygienists (2010) Threshold Limit Values (TLVs®) and Biological Exposure Indices (BEIs®). http://www.acgih.org/TLV/. Accessed May 2022

Ali MM, Hossain D, Al-Imran A, Khan MS, Begum M, Osman MH (2021) Environmental pollution with heavy metals: a public health concern. In: Nazal MK, Zhao H (eds) Heavy metals - their environmental impacts and mitigation. IntechOpen, London. https://doi.org/10.5772/intechopen.96805

Allen AG, Nemitz E, Shi JP, Harrison RM, Greenwood JC (2001) Size distribution of trace metals in atmosphere aerosols in the United Kingdom. Atmos Environ 35:4581–4591

Aschner M, Erickson K, Dorman D (2005) Manganese dosimetry: species differences and implications for neurotoxicity. Crit Rev Toxicol 35:1–32

ATSDR (1999) Toxicological profile for mercury – update. Atlanta, GA, Agency for Toxic Substances and Disease Registry. https://www.atsdr.cdc.gov/toxprofiledocs/index.html. Accessed 10 May 2022

Bernardo JF (2021) Aluminum toxicity: practice essentials, pathophysiology, etiology. https://emedicine.medscape.com/article/165315-overview. Accessed 13 May 2022

Bondier JR, Michel G, Propper A (2008) Harmful effects of cadmium on olfactory system in mice. Inhal Toxicol 20:1169–1177

Boonsaner M (2006) Environmental toxicology, 4th edn. Silpakorn University Press, Nakhon Prathom

Briffa J, Sinagra E, Blundell R (2020) Heavy metal pollution in the environment and their toxicological effects on humans. Heliyon 6(9):e04691. https://doi.org/10.1016/j.heliyon.2020.e04691

CDC (2002) *Managing elevated blood lead levels among young children: recommendations from the Advisory Committee on Childhood lead Poisoning Prevention.* Atlanta, GA, Centers for Disease Control and Prevention. CDC (2005c). *Preventing lead poisoning in young children. A statement by the Centers for Disease Control and Prevention,* August 2005. Atlanta, GA, Centers for Disease Control and Prevention

Cempel M, Nikel G (2006) Nickel: a review of its sources and environmental toxicology. Pol J Environ Stud 15:375–382

Cheevaporn V (2004) Analytical technique for environmental quality. Chonburi Karnpim Press, Chonburi

Chen L, Lippmann M (2009) Effects of metals within ambient air particulate matter (PM) on human health. Inhal Toxicol 21:1–31

Chen LC, Maciejczyk P, Thurston GD (2022) Chapter 6 - Metals and air pollution. In: Nordberg GF, Costa M (eds) Handbook on the toxicology of metals, 5th edn. Academic Press, London, pp 137–182, ISBN 9780128232927, https://doi.org/10.1016/B978-0-12-823292-7.00004-8. https://www.sciencedirect.com/science/article/pii/B9780128232927000048

Christensen JM, Poulsen OM (1994) A 1982–1992 surveillance programme on Danish pottery painters. Biological levels and health effects following exposure to soluble or insoluble cobalt compounds in cobalt blue dyes. Sci Total Environ 150:95–104

Chung JY, Yu SD, Hong YS (2014) Environmental source of arsenic exposure. J Prev Med Public Health 47(5):253–257. https://doi.org/10.3961/jpmph.14.036

Costa DL, Dreher KL (1997) Bioavailable transition metals in particulate matter mediate cardio-pulmonary injury in health and compromised animal models. Environ Health Perspect 105(105):1053–1060

Councell TB, Duckenfield KU, Landa ER, Callender E (2004) Tire-wear particles as a source of zinc to the environment. Environ Sci Technol 38:4206–4214

Das R (2022) Chapter 23 - Sources of lead (Pb) in atmosphere over Indian cities and health impacts. In: Singh RP (ed) Asian atmospheric pollution. Elsevier, Amsterdam, pp 435–452. ISBN 9780128166932, https://doi.org/10.1016/B978-0-12-816693-2.00014-7. https://www.sciencedirect.com/science/article/pii/B9780128166932000147

Dollemore D (2008) Newly detected air pollutant mimics damaging effects of cigarette smoke. *EurekAlert!, August 17, 2008.* Peer-Reviewed Publication, AMERICAN CHEMICAL SOCIETY

EPA, California (2004) Recent research findings: health effects of particulate matter and ozone pollution. Air Resources Board and American Lung Association of California

EPA. EPA Technology Transfer Network (TTN) (2010). National Air Toxics Assessments (NATA). May 10, 2022. http://www.epa.gov/ttn/atw/nata2002/. Accessed May 2022

EPA. Health Effects Notebook for HAPs (2010) Technology Transfer Network Air Toxics Web Site. May 10, 2022. http://epa.gov/ttnatw01/hlthef/hapindex.html. Accessed May 2022

Erzen I, Kragelj LZ (2006) Cadmium concentration in blood related to smoking habits in a group of males aged between 10 and 26 years. Trace Elem Electroly 23:60–65

Fang GC, Wu YS, Huang SH, Rau JY (2005) Review of atmospheric metallic elements in Asia during 2000-2004. Atmos Environ 39:3003–3013

Fitzpatrick R, Cass A and Besz W (2008). Measurement of redox potential (Eh) in periodically waterlogged viticultural soils. Retrieved October 23, 2009, from http://natres.psu.ac.th/Link/SoilCongress/bdd/symp35/2135-t.pdf

Forstner U, Wittmann GTW (1983) Metal pollution in the aquatic environment. Springer-Verlag, Berlin; Heidelberg

Gautam PK, Gautam RK, Chattopadhyaya MC, Banerjee S, Chattopadhyaya MC, Pandey JD (2016) Heavy metals in the environment: fate, transport, toxicity and remediation technologies.

thermodynamic profiling of pollutants view project materials for solid oxide fuel cells view project heavy metals in the environment: fate, Transport, Toxicity And Rem

Geiger A, Cooper J (2010) Overview of airborne metals regulations, exposure limits, health effects, and contemporary research. Environmental Protection Agency, Washington, pp 1–56. https://www3.epa.gov/ttnemc01/prelim/otm31appC.pdf

Genchi G, Carocci A, Lauria G, Sinicropi MS, Catalano A (2020) Nickel: human health and environmental toxicology. Int J Environ Res Public Health 17(3):679. https://doi.org/10.3390/ijerph17030679

Ghio AJ, Silbajoris R, Carson JL, Samet JM (2002) Biological effects of oil fly ash. Environ Health Perspect 110(1):89–94

Handa N, Bhardwaj R, Kaur H, Kapoor DP, Rattan A, Kaur S, Thukral AK, Kaur S, Arora S, Kapoor N (2016) Chapter 7 - Selenium: an antioxidative protectant in plants under stress. In: Ahmad P (ed) Plant metal interaction. Elsevier, Amsterdam, pp 179–207, ISBN 9780128031582, https://doi.org/10.1016/B978-0-12-803158-2.00007-2. https://www.sciencedirect.com/science/article/pii/B9780128031582000072

HEI, Health Effects Institute (2004) Health effects of outdoor air pollution in developing countries of Asia. Literature review. HEI International Scientific Oversight Committee, Boston, MA

HSDB, Hazardous Substances Data Bank (2010) TOXNET: Toxicology Data Network. May 10, 2022. http://toxnet.nlm.nih.gov/hsdb/. Accessed May 2022

IRIS, Integrated Risk Information System (2010) Environmental Protection Agency. May 10, 2022. http://www.epa.gov/IRIS/. Accessed May 2022

Järup L, Berglund M, Elinder CG, Nordberg G, Vahter M (1998) Health effects of cadmium exposure—a review of the literature and a risk estimate. Scand J Work Environ Health 24 (Suppl 1):1–51

Kahal A, El-Sorogy AS, Qaysi S, Almadani S, Kassem OM, Al-Dossari A (2020) Contamination and ecological risk assessment of the Red Sea coastal sediments, Southwest Saudi Arabia. Mar Pollut Bull 154:111125

Karthikeyan S, Balasubramanian R, Iouri K (2006) Particulate air pollution from bushfires: human exposure and possible health effects. J Toxicol Environ Health A 69:1895–1908

Kaul PP, Shyaml S, Srivastava R, Misra D, Salve PR, Srivastava SP (2003) Lead levels in ambient air and blood of pregnant mothers from the general population of Lucknow (U.P.), India. Bull Environ Contam Toxicol 71:1239–1243. https://doi.org/10.1007/s00128-003-8825-x

Khaiwal R, Stranger M, Grieken RV (2008) Chemical characterization and multivariate analysis of atmospheric PM2.5 particles. J Atmos Chem 59:199–218

Khan MB, Dai X, Ni Q, Zhang C, Cui X, Lu M, Deng M, Yang X, He Z (2019) Toxic metal pollution and ecological risk assessment in sediments of water reservoirs in Southeast China. Soil Sediment Contam Int J 28:695–715

Konkel L Heavy metal: some airborne particles pose more dangers than others. Environmental Health News, December 17, 2009

Kulshrestha A, Gursumeeran Satsangi P, Mashi J, Taneja A (2009) Metal concentration of PM2.5 and PM10 particles and seasonal variations in urban and rural environment of Agra, India. Sci Total Environ 407:6196–6204

Kulshrestha UC, Nageswara Rao T, Azhaguvel S, Kulshrestha MJ (2004) Emissions and accumulation of metals in the atmosphere due to crackers and sparkles during Diwali festival in India. Atmos Environ 38:4421–4425

Kumar S, Aggarwal SG, Malherbe J, Barre JPG, Berail S, Gupta PK, Donard OFX (2016) Tracing dust transport from Middle-East over Delhi in March 2012 using metal and lead isotope composition. Atmos Environ 132:179–187

Lee PK, Kang MJ, Yu S, Ko KS, Ha K, Shin SC, Park JH (2017) Enrichment and geochemical mobility of heavy metals in bottom sediment of the Hoedong reservoir, Korea and their source apportionment. Chemosphere 184:74–85

Lenntech BV (2018) Heavy metals. Available online https://www.lenntech.com/processes/heavy/heavy-metals/heavy-metals.htm. Accessed 10 May 2022

Leyssens L, Vinck B, Van Der Straeten C, Wuyts F, Maes L (2017) Cobalt toxicity in humans—a review of the potential sources and systemic health effects. Toxicology 387:43–56

Li H, Qian X, Wang Q (2013) Heavy metals in atmospheric particulate matter: a comprehensive understanding is needed for monitoring and risk mitigation. Environ Sci Technol 47(23): 13210–13211. https://doi.org/10.1021/es404751a

Lim SR, Schoenung JM (2010) Human health and ecological toxicity potentials due to heavy metal content in waste electronic devices with flat panel displays. J Hazard Mater 177:251–259

Lippmann M, Ito K, Hwang JS, Maciejczyk P, Chen LC (2006) Cardiovascular effects of nickel in ambient air. Environ Health Perspect 114:1662–1669

Livett EA (1992) Heavy metal pollution of the atmosphere. In: Nierenberg WA (ed) Encyclopedia of the earth system science. Academic Press, San Diego, pp 507–518

Lu H, Wei F, Tang J, Giesy JP (2016) Leaching of metals from cement under simulated environmental conditions. J Environ Manag 169:319–327

Lucchini RG, Guazzetti S, Zoni S, Benedetti C, Fedrighi C, Peli M, Donna F, Bontempi E, Borgese L, Micheletti S, Ferri R, Marchetti S, Smith DR (2014) Neurofunctional dopaminergic impairment in elderly after lifetime exposure to manganese. Neurotoxicology 45:309–317. https://doi.org/10.1016/j.neuro.2014.05.006

Maciejczyk P, Zhong M, Lippmann M, Chen LC (2010) Oxidant generation capacity of source-apportioned PM 2.5. Inhalation toxicology (early online), pp 1–8

Magari SR, Schwarrtz J, Williams PL, Hauser R, Smith TJ, Christiani DC (2002) The association of particulate air metals concentrations with heart rate variability. Environ Health Perspect 110(9): 875–880

Masindi V, Muedi KL (2018) Environmental contamination by heavy metals. In: Heavy metals. IntechOpen, London

McFarland RB, Reigel H (1978) Chronic mercury poisoning from a single brief exposure. J Occup Med 20:532–534

Mehdi Y, Hornick JL, Istasse L, Dufrasne I (2013) Selenium in the environment, metabolism and involvement in body functions. Molecules 18(3):3292–3311. https://doi.org/10.3390/molecules18033292

Midander K, Pan J, Wallinder IO, Leygraf C (2007) Metal release from stainless steel particles in vitro - influence of particle size. J Environ Monit 9:74–81

Mokdad AH, Marks JS, Stroup DF, Gerberding JL (2004) Actual causes of death in the United States, 2000. JAMA 291(10):1238–1245. https://doi.org/10.1001/jama.291.10.1238

National Lime Association v US EPA. 99-1325, Volume 233 F. 3d 625 (D.C. Circuit 2000, December 15, 2000

Navrotsky A (2001) Nanoparticles and the environment. In: Navrotsky JF, Banfield A (eds) Reviews in meralogy and geochemistry. Mineralogical Society of America, Washington, DC, pp 73–103

Newman MC, Diamond GL, Menzie C, Moya J, and Nriagu J (2004). Issue paper on metal exposure assessment. Submitted to Environmental Protection Agency by ERG

Nieboer E, Thomassen Y, Chashchin V, Odland JO (2005) Occupational exposure assessment of metals. J Environ Med 7:411–415

Nordberg GF, Fowler BA, Nordberg M, Friberg L (2007) Handbook on the toxicology of metals, 3rd edn. Academic Press, Burlington, MA

Nour HE, El-Sorogy AS, Abd El-Wahab M, Mohamaden M, Al-Kahtany K (2019) Contamination and ecological risk assessment of heavy metals pollution from the Shalateen coastal sediments, Red Sea, Egypt. Mar Pollut Bull 144:167–172

Nriagu JO (1989) A global assessment of natural sources of atmospheric trace metals. Nature 338: 47–49

OEHHA (2008) California Office of Environmental Health Hazard Assessment. Air toxicology and epidemiology. May 10, 2022. http://oehha.ca.gov/air/allrels.html. Accessed May 2022

Oller AR, Oberdorster G (2010) Incorporation of particle size differences between animal studies and human workplace aerosols for deriving exposure limits. Regul Toxicol Pharmacol 57:181–194

Pachana K, Wattanakornsiri A, Nanuam J (2010) Heavy metal transport and fate in the environmental compartments. NU Sci J 7(1):1–11. https://www.thaiscience.info/journals/Article/NUSJ/10888564.pdf

Pendias AK, Mukherjee AB (2007) Trace elements from soil to human. Springer, Berlin. https://doi.org/10.1007/978-3-540-32714-1

Pope CA, Dockery DW (2006) Health effects of fine particle air pollution: lines that connect. J Air Waste Manag 56:709–742

Purdue University (2008) Solubility product constant, Ksp. Purdue University. http://www.chem.purdue.edu/gchelp/howtosolveit/equilibrium/solubility_products.htm

Ranucci G, Iorio R (2019) Chapter 41 - Disorders that mimic Wilson disease. In: Kerkar N, Roberts EA (eds) Clinical and translational perspectives on Wilson disease. Academic Press, London, pp 419–426. ISBN 9780128105320. https://doi.org/10.1016/B978-0-12-810532-0.00041-0. https://www.sciencedirect.com/science/article/pii/B9780128105320000410

Ratchamongkol Technology University (2008) Chemical for engineer. Department of Physics, Faculty of Science, Ratchamongkol Technology University. http://www.rmutphysics.com/charud/scibook/chemistry/unit4.pdf

Sangani R, Soukup JM, Ghio AJ (2010) Metals in air pollution particles decrease whole-blood coagulation time. Inhal Toxicol 22(8):621–626

Singh MP, Singh VK, Patel DK, Tandon PK, Gaur JS, Behari JR, Yadav S (2010) Face mask application as a tool to diminish the particulate matter mediated heavy metal exposure among citizens of Lucknow, India. Sci Total Environ 408:5723–5728

Smith I (2005) Land uses of coal fly ash, benefits and barriers. IEA coal research, London, UK, CCC/96

Steinnes E, Hutchinson T, Meema KM (1987) Impact of long-range atmospheric transport of heavy metals to the terrestrial environment in Norway. Lead, mercury, cadmium and arsenic in the environment 1987. Scope 31:107–117

Tchounwou PB, Yedjou CG, Patlolla AK, Sutton DJ (2012) Heavy metal toxicity and the environment. Exp Suppl 101:133–164

The Gazette of India (2009) No.660, Ministry of Environment and Forest, Notification, New Delhi, 16th November, 2009, Environment (protection) seventh amendment rules

Thorat DD, Mahadevan TN, Ghosh DK, Narayan S (2001) Beryllium concentrations in ambient air and its source identification. Environ Monit Assess 69(1):49–61

Utsunomiya S, Jensen KA, Keeler GJ, Ewing RC (2004) Direct identification of trace metals in fine and ultrafine particles in the Detroit urban atmosphere. Environ Sci Technol 38(8):2289–2297

Valavandis A, Fiotakis K, Vlachogianni T (2008) Airborne particulate matter and human health: toxicological assessment and importance of size and composition of particles for oxidative damage and carcinogenic mechanisms. J Environ Sci Health C 26:339–362

Valko M, Morris H, Cronin M (2005) Metals, toxicity and oxidative stress. Curr Med Chem 12:1161–1208

Walker CH, Hopkin SP, Sibly RM, Peakall DB (2006) Principles of ecotoxicology, 3rd edn. Taylor & Francis, New York

Walker CH, Sibly RM, Hopkin SP (2012) D.B.P., in: Principles of ecotoxicology; Group, T. and F., Ed.; 4th Edition, CRC Press

Wang LK (2009) Heavy metals in the environment. CRC Press, Boca Raton

WHO (2002) World Health Organization. Estimated deaths and DAILYs attributable to selected environmental risk factors

WHO (2003) World Health Organization. Mercury. https://www.euro.who.int/__data/assets/pdf_file/0004/123079/AQG2ndEd_6_9Mercury.PDF

WHO (2007) World Health Organization Europe. In: Health risks of heavy metals from long-range transboundary air pollution. Regional Office for Europe of the World Health Organization, Copenhagen

WHO/IPCS (1995) Inorganic lead. Geneva, WHO World Health Organization Environmental Health Criteria, No. 165

Wu YS, Fang GC, Lee WJ, Lee JF, Chang CC, Lee CZ (2007) A review of atmospheric fine particulate matter and its associated trace metal pollutants in Asian countries during the period 1995-2005. J Hazard Mater 143:511–515

Wuana RA, Okieimen FE (2011) Heavy metals in contaminates soils: a review of sources, chemistry, risks and best available strategies for remediation. ISRN Ecol

Zelikoff JT, Schermerhorn KR, Fang K, Cohen MD, Schlesinger RB (2002) A role for associated transition metals in the immunotoxicity of inhaled ambient particulate matter. Environ Health Perspect 110(5):871–875

Zhang C, Shan B, Tang W, Wang C, Zhang L (2019) Identifying sediment- associated toxicity in rivers affected by multiple pollutants from the contaminant bioavailability. Ecotoxicol Environ Saf 171:84–91

Zhu G, Noman MA, Narale DD, Feng W, Pujari L, Sun J (2020) Evaluation of ecosystem health and potential health hazards in the Hangzhou Bay and Qianting Estuary region through multiple assessment approaches. Environ Pollut 264:114791

5

Heavy Metal Contamination in Surface Water Bodies Through Construction and Demolition Waste: A Case Study of City of Lakes – Bhopal, Madhya Pradesh, India

Amit Vishwakarma, Anupam Sharma, and S. S. Kushwah

Abstract

According to Chen et al., 2020, two billion tons of construction and demolition (C & D) wastes are produced each year globally. A report of CSE India, 2022, reveal the fact that out of 150 million tons of C & D waste annually generated in India, only 1% is treated in recycling capacity facilities. According to CPCB, 2017, Bhopal, the capital of Madhya Pradesh, India, generates 50 tons of C & D waste per day. According to Construction and Demolition Waste Management rules 2016, waste generators shall ensure that there is no littering or deposition so as to prevent obstruction to the traffic or the public or drains. The fact is that apart from rules and regulations, a lot of C & D waste is disposed in water streams. A report of national Hindi newspaper Dainik Bhaskar dated July 16, 2019, reveals that 16 major points which have a lot of C & D wastes directly meet the water stream. Gao et al., 2015, quoted "high risk of heavy metal contamination through C & D waste depending on the source type of C&D wastes". The present paper identifies heavy metal contamination zones through C & D waste in water streams of the City of Lakes Bhopal.

Keywords

Demolition waste · Polluted water · City of lakes · Heavy metal contamination · Bhopal lakes

A. Vishwakarma (✉) · A. Sharma · S. S. Kushwah
Department of Civil Engineering, University Institute of Technology, RGPV, Bhopal, Madhya Pradesh, India
e-mail: amitvishi@rgpv.ac.in

R. P. Singh et al. (eds.), *Heavy Metal Toxicity: Environmental Concerns, Remediation and Opportunities*, https://doi.org/10.1007/978-981-99-0397-9_5

5.1 Introduction

In urban and rural areas of India, a lot of infrastructure development activities are taking place. The construction industry in India generates 10–12 million tons of waste annually (CPHEEO, 2016). Construction and demolition wastes are generated through new or existing construction activities. It is estimated that two billion tons of construction and demolition wastes are produced each year globally (Chen et al. 2020). Out of 150 million tons of C & D wastes annually generated in India, only 1% are treated in recycling capacity facilities (CSE India, 2022). In 2018, construction and demolition (C & D) debris generation in the United States is more than twice the amount of municipal solid wastes (United States EPA, 2018). The Municipal Corporation of Gurugram (MCG) has collected close to 1.25 lakh tons of construction and demolition (C & D) wastes from across the city so far this year, which are to be processed at its treatment plant in Basai (Hindustan Times, 2022).

Bhopal is the capital of State Madhya Pradesh, India, with a population of 1,798,218 (Census, 2011). Bhopal City is also called the City of Lakes because of the many lakes in the city from ancient times. Bhopal generates 50 tons of C & D waste per day (CPCB, 2017). According to a Joint Inspection report of Madhya Pradesh Pollution Control Board (2020), the number of lakes in the city of Bhopal is seven: Upper Lake, Lower Lake, Shahpura Lake, Siddqui Hassan Talab, Motiya Talab, Bagmunsihussan Khan Lake, Sarang Pani Lake, '5' Number Talab, Saryoo Sarovar Talab, Lendia Talab and Jail Bagh. This report also reveals that 21 drains discharge untreated sewage water into the lower lake and cause pollution in the lake water (Joint Inspection Report, 2020).

The Upper Lake of Bhopal City is a freshwater lake of the city. It is said this lake was built up by Raja Bhojpal. An annual report of the Year (2020–2021) of "C & D Waste Management Rule 2016 of MPPCB", reveals a hierarchy of C & D waste management in a similar way as hierarchy in municipal solid waste management, i.e., reduction, reuse, recycling, recovery and disposal with reduction and reuse, and lastly landfill option. At present, Bhopal municipal corporation plant for C & D waste management is functional along with the functional plant of Indore, Ujjain, and Jabalpur cities. According to Swachh Bharat Mission (Urban) Portal, 338 ULBs have dedicated vehicles for collection and transportation of C & D wastes along with 333 ULBs C & D Waste Helpline in place. Moreover, according to Construction and Demolition Waste Management Rules 2016, waste generators shall ensure that there is no littering or deposition so as to prevent obstruction to the traffic or the public or drains. The fact is that apart from rules and regulations, a lot of C & D wastes are disposed of in water streams. A report of "Dainik Bhaskar dated July 16, 2019", a national Hindi newspaper, reveals that 16 major points which have a lot of C & D wastes directly meet the water stream (Dainik Bhaskar, 2019).

5.2 Study Area: Bhopal, Madhya Pradesh, India

Bhopal is the capital of the State of Madhya Pradesh, India, famous for its lakes, thus also known as the City of Lakes. Upper Lake of the city is situated at western side of the city and covers 361 Km2 catchment areas (Wikipedia, 2022). At present, along the periphery of Upper Lake of the city, a lot of construction activities are taking place. Without proper disposal and treatment of all C & D wastes, the quality of water of Upper Lake Bhopal is being affected. Present study focus on the field reality of pollution at identified locations of Upper Lake of the city.

5.2.1 Drainage Basin of Bhopal Region

Bhopal area has undulated hilly topography with highest elevation at airport of the city and lowest elevation at Phanda region of the city. Bhopal District covers two drainage basins: Betwa Basin and lower Chambal Basin. Figure 5.1 shows drainage basin in Bhopal District.

It is evident from Fig. 5.1 the maximum percentage area of the basin of Halali region of the city. The river in the city that has a maximum length is the Bah river. Bah region also has the second maximum percentage drainage area of the city. At present, Kerwa water reservoir has a source of water supply to some parts of the city. Sources of municipal water supply to the city are Narmada water supply, Kerwa water supply, Kolar reservoir water supply, and groundwater supply.

Figure 5.2 shows drainage area of lower Chambal Basin of Bhopal City. Parwati main river has a maximum 65% area of basin of lower Chambal Basin of Bhopal City.

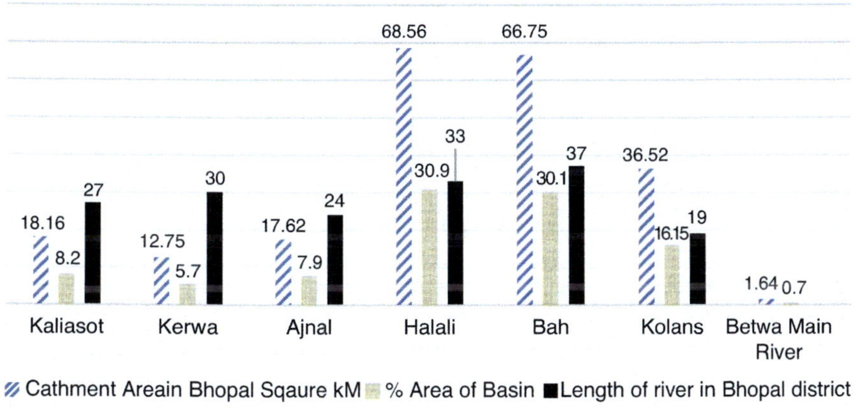

Fig. 5.1 Drainage area of Betwa Basin of Bhopal City (source: CGWB, 2013)

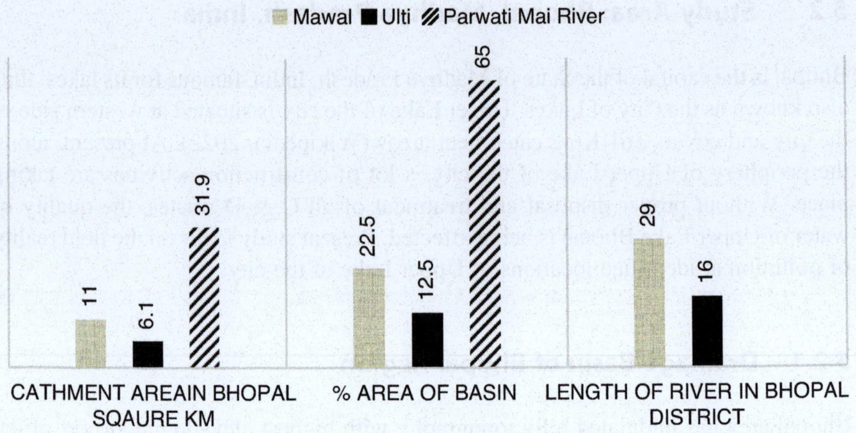

Fig. 5.2 Drainage area of lower Chambal Basin of Bhopal City (source: CGWB, 2013)

Table 5.1 Soil types of Bhopal District (source: CGWB, 2013)

Bhopal District regions	Area covered	Formation by
Black cotton soil	75% of total area	Weathering of basalt rocks
Yellowish-red mixed soil	25% of total area	Sandstone and shale
Alluvial soil	Along with river courses	Formation by watercourses and streams
Murrum	At hilly areas	Formed by weathering of basalt

5.2.2 Soil Classification of the City of Bhopal

Table 5.1 shows the soil types in and around the city of Bhopal. Almost 75% of the area has black cotton soil and in the remaining portion yellowish-red mixed soil, alluvial soil, and murrum. The city of Bhopal is famous for its "Bhojpur Temple" which has a very large Shiv Lingam made of single sandstone. In the city of Bhopal, sandstone and basalt rocks are present. Bhopal has hilly areas with a lot of greenery and has pleasant weather. Bhopal is also known as the City of Lakes.

5.3 Rainfall Data of Madhya Pradesh, India

It is evident from Fig. 5.3 that the maximum rainfall of Madhya Pradesh is 3.95 mm in the years 1998 to 2013.

Figure 5.4 shows the mean rainfall of Bhopal City of Madhya Pradesh, India.

It is evident from Fig. 5.4 that the average rainfall of Bhopal City is 2.67 mm in the years 1998 to 2013, which is 67.56% of the maximum rainfall in Madhya Pradesh.

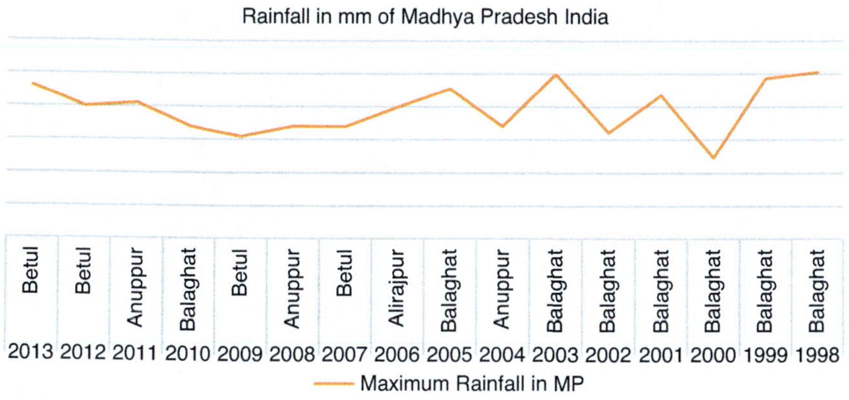

Fig. 5.3 Maximum rainfall of Madhya Pradesh, India (source: Rainfall Analytics, 2013)

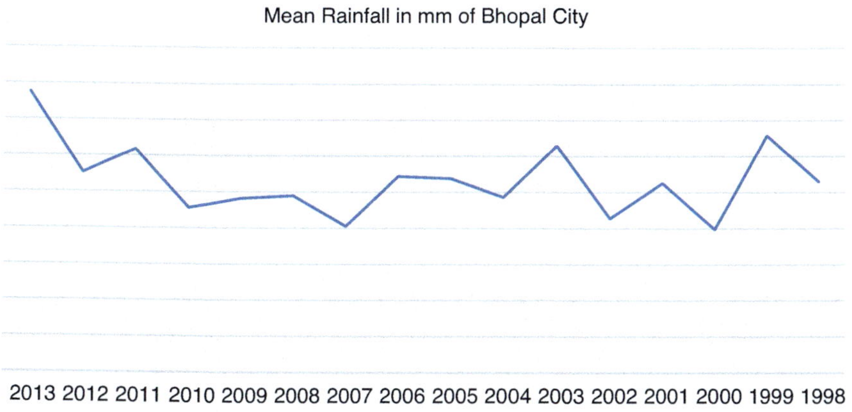

Fig. 5.4 Mean rainfall of Bhopal City (source: Rainfall Analytics, 2013)

5.4 Lakes of Bhopal City, Madhya Pradesh, India

Bhopal is the capital of Madhya Pradesh and known for its Bhopal gas tragedy and its lakes. The city is known as the City of Lakes. There are almost 17 lakes in the city of Bhopal. Figure 5.5 shows the spread area of lakes in the city of Bhopal.

It is evident from the figure that a major portion is covered by the Upper Lake of Bhopal. Upper Lake is man-made lake to store rainwater to fulfill the water requirement of the city.

All these lakes of the city have various ecological status and different purposes. Table 5.2 shows the ecological status and present use of the various lakes of the city of Bhopal. Out of 17 lakes of Bhopal, only Upper Lake, Kolar Reservoir, Kerwa Reservoir, and Damkheda water supplies are used for potable water supply. Other lakes are used for recreation and irrigation purposes.

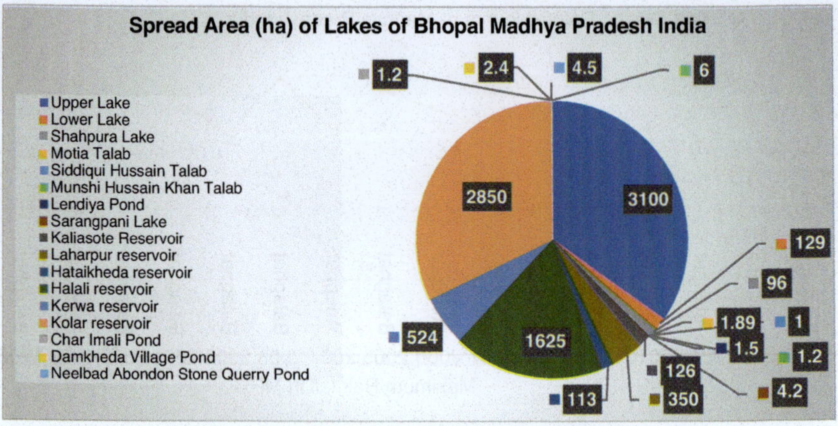

Fig. 5.5 Spread area (in hectare) of lakes of Bhopal Madhya Pradesh, India (source: City Development Plan: Bhopal; 2005)

Table 5.2 The ecological status and present use of lakes of the city of Bhopal (source: City Development Plan: Bhopal, 2005)

Name of water body	Ecological status	Present use
Upper Lake	Mesotrophic and part of the lake is eutrophic	Water supply and recreation
Lower Lake	Advance stage of eutrophic	Raw water supply recreation
Shahpura Lake	Advance stage of eutrophic	Recreation
Motia Talab	Advance stage of eutrophic	Recreation
Siddiqui Hassan Talab	Bog Lake	Recreation
Munshi Hussain Khan Talab	Eutrophic	Recreation
Lendiya pond	Advance stage of eutrophic	Recreation
Sarangpani Lake	Advance stage of eutrophic	Recreation
Kaliasote Reservoir	Mesotrophic and part of the lake is eutrophic	Irrigation
Laharpur reservoir	Advance stage of eutrophic	Irrigation
Hathaikheda reservoir	Mesotrophic	Irrigation
Halali reservoir	Mesotrophic	Irrigation
Kerwa reservoir	Mesotrophic	Irrigation
Kolar reservoir	Mesotrophic	Potable water supply and irrigation
Char Imali pond	Eutrophic	Recreation
Damkheda Village pond	Mesotrophic	Potable water and recreation
Neelbad Abondon stone Quarry pond	Mesotrophic	Recreation
Ayodhya Nagar abandon stone quarry ponds	Mesotrophic	Recreation

5.4.1 Upper Lake

It is evident from Table 5.2 that Upper Lake of the city serves water for drinking and other purposes too and also has ecological importance. Upper Lake of Bhopal City is a bigger lake of the city situated at western part. Upper lake is a man-made structure that serves as a water source for the people of Bhopal City. The effect of disposal of C&D Wastes in lakes and idol immersion in lakes, concentrations of phosphate, color, pH value, turbidity, and solids with heavy metals increase in water of lakes. Upper Lake of the city was constructed in the eleventh century on Kolans River by constructing earthen dam. Every year, this lake attracts and serves food to a lot of bird species with migratory birds. Hence, this lake plays an important role to have ecological balance of the city. Upper Lake covers a portion of agricultural land, forest areas including Van Vihar of the city, and urban population of the city.

Near and around the bank of the upper lake, a lot of construction activities are taking place. the demolition of old structures also produced lot of debris. The present work focused on affected water quality by C & D wastes near the bank of upper lakes of Bhopal City. Conservation of Upper Lake is required to save ecology and essential for human survival of the city of Bhopal.

Figure 5.6 shows the upper lake of the city and encircled study area polluted by C & D wastes

5.4.2 Lower Lake

Lower Lake of Bhopal City is also a man-made structure for recreation purposes situated at old Bhopal. It is mainly due to the seepage from upper lake and monsoon rainfall storage. It helps for ecological balance of the city and aesthetic and pleasant

Fig. 5.6 Upper lake of Bhopal, Madhya Pradesh, India (source: Googlemap, 2022)

atmosphere. Lower Lake of the city is a major place for idol immersion on festive occasions. At some places, its depth is greater than the Upper Lake of the city.

According to CGWB (2013), out of 241 MLD net water supplies to Bhopal City, 44.8% are released from Upper Lake and 55.2% from Kolar Dam. Twenty-two MLD is available from groundwater sources.

5.5 Construction and Demolition Waste

An annual report (2018–2019) of C & D Waste Management Rules 2016 (MPPCB, 2019) of Bhopal City regarding C & D waste reveals that C & D waste generated in Bhopal City is 655 TPD. only 336 TPD waste are processed out of the total generation of C&D wastes of the city Bhopal. According to this report Bhopal Municipal Corporation proposed a 100TPD capacity C & D processing Plant. This report also reveals the fact that 224 ULBs of Madhya Pradesh have started treating C & D waste in their ULBs through various means. C & D waste contains metal pieces, iron, steel, plastics, PVC pipes, construction building materials like broken bricks, aged concrete, paints and enamels, etc. Dilution & Dispersion of C & D wastes in lakes affect water quality.

The present work focus on testing the water of upper lakes at the possible identified contaminated locations where C & D wastes get mixed with water of Upper Lake. The possible contaminated locations of Upper Lakes were identified for collection of water samples and tested at laboratory. These identified locations are given in Table 5.3.

A brief introduction of sample sites is given below (Fig. 5.7):

Table 5.3 Identified locations of Upper Lake of the city at which C & D wastes mix with water

		Parameters			
S. no.	Location	Latitude	Longitude	Area (Sq. km)	Total population
1	Behata Bairagarh	23.265978°	77.342792°	–	–
2	Halalpura Lalghati	23.274051°	77.356022°	–	–
3	Khanugaon	23.258149°	77.372112°	0.81	4220
4	Intkedi	23.228135°	77.263653°	2.23	2447
5	Gora gaon	23.219768°	77.351012°	3.29	2710
6	Budhwara	23.253597°	77.408248°	0.28	2617
7	Jahagirabad	23.249992°	77.410035°	3.94	41,130
8	Near motia talab kohefiza	23.26422°	77.393642°	3.13	33,538
9	Idgah Hills	23.266103°	77.390816°	2.56	31,021
10	Manisha Market, Shapura	23.207058°	77.424158°	4.17	35,238
11	Manisha Market –2 near bansal	23.200517°	77.422222°	4.17	35,238

Marked Point on Map	Identified locations
1	Behata Bairagarh
2	Halalpura Lalghati
3	Khanugaon
7	Gora gaon
5	Budhwara
6	Jahagirabad
10	Near motia talab kohefiza
4	Idgah Hills
8	Manisha market shapura
9	Manisha market -2 near bansal

Fig. 5.7 Identified locations for water sample testing affected by C & D (source: Google Map, 2022)

5.5.1 Behata, Bairagarh

Behata, Bairagarh, has a latitude of 23.265978° and a longitude of 77.342792°. Behata is a part of Bairagarh of Bhopal City. It is near Lalghati, Gandhinagar, Idgah Hills, and Kohefiza. Upper Lake has its bank area near Idgah Hills and Kohefiza area of the city. Identified location of Behata Bairagarh is found and sample collected for testing its affected quality by C & D wastes.

5.5.2 Halalpur, Lalghati

Halalpur, Lalghati, has a latitude of 23.274051° and a longitude of 77.356022°. Halalpur, Lalghati, has a chartered bus terminal of the city. Heavy traffic existed around this area because it is connected to Bairagarh, Airport road, old city, and VIP road of the city. Some disposal points of C & D waste are located and samples collected from identified sites.

5.5.3 Khanugaon

Khanugaon of the city has a latitude of 23.258149° and a longitude of 77.372112°. The total population of this area is 4220. It is near Kohefiza, Idgah Hills, Shahjahanabad, and Peer Gate area. The debris of construction activity mixed with water quality of Upper Lake, hence sample collected from this site and tested at laboratory.

5.5.4 Intkhedi

Intkhedi has a population of 2447 with a latitude of 23.228135° and a longitude of 77.263653°. It is 12 km from Bhopal Station and located in Huzur Tehsil of Bhopal District. From identified locations, water samples were collected and tested at laboratory.

5.5.5 Goregaon

Goregaon has 2710 population with a latitude of 23.219768° and a longitude of 77.351012°. It is near Van Vihar, Prempura, Sair Sapata, Surajnagar, and Bhadbhada, Bhopal.

All these locations are very important as these locations are directly connected to animals, flora, and fauna of the city. At identified locations, samples were collected and tested at laboratory.

5.5.6 Budhwara

Budhwara is located near and around lower lake of the city. It has a latitude of 23.253597° and a longitude of 77.408248°. The population of Budhwara is 2617.

5.5.7 Jahangirabad

Jahangirabad is a center of the City of Bhopal. It is near the lower lake of the city. Jahangirabad has a latitude of 23.249992° and a longitude of 77.410035°. Famous Lal Pared ground situated in Jahangirabad. The population of Jahangirabad is 41,130.

5.5.8 Near Motia Talab, Kohefiza

Motia Talab, Kohefiza, has a latitude of 23.26422° and a longitude of 77.393642°. This area is situated at old Bhopal and has a population of 33,538. A famous Taj-ul-Masjid of the city is located in this area.

5.5.9 Idgah Hills

Idgah Hills of Bhopal City is located near the upper lake of the city. Its latitude is 23.266103° and longitude 77.390816°. The population of this area is 31,021. An intake structure and water treatment plant is also situated at Idgah Hills.

5.5.10 Manisha Market, Shahpura

Manisha Market is located at the bank of Shahpura Lake of the city Bhopal. Shahpura Lake is a man-made water reservoir for residential colonies, sewage treatment and irrigation purposes and fishing.

It is in posh localities of the city. It has a latitude of 23.207058° and a longitude of 77.424158°. The population of this area is 35,238.

Manisha Market −2 near Bansal has a latitude of 23.200517° and a longitude of 77.422222°.

5.6 Observation and Results

From sample site locations, water quality parameters are tested, and a brief discussion is given below:

Present work is focused on metal detection in water quality affected by C & D waste. TCLP analysis is done in the laboratory, and results are graphically presented in Fig. 5.8.

It is evident from Fig. 5.8 that manganese is found in most of the water samples in considerable amounts. At the sample of "Behata Bairagarh," an amount of manganese was found up to its highest level of 8.8 ppm. Small amount of manganese is essential for human health, but in excess, manganese can cause memory loss and attention loss in human health. According to IS 10500:2012, the permissible limit of manganese in drinking water is 0.3 ppm, while 0.1 ppm is acceptable limit. From Fig. 5.2, it is clearly represented that out of 17 samples, in 15 samples, manganese was found in excess amounts which is harmful to human health. Arsenic in two samples was found a little bit in excess although less than permissible limit. Boron in one water sample location was found in excess.

5.7 Conclusion

The present work is focused on detection of metal concentrations in possible contaminated water samples affected by construction and demolition (C & D) wastes. There is a high risk of heavy metal contamination through C & D waste depending on the type of source of C & D waste (Gao et al., 2015). The present work identifies heavy metal contamination zones through C & D waste in water streams of the City of Lakes, Bhopal. The guidelines of C & D waste disposal and the treatment process of C & D waste are already mentioned by municipalities with proper execution at their best level. Apart from a lot of work done by municipalities, government, and private agencies, proper monitoring on C & D waste disposal along the periphery of the lakes of the city is required. There is an urgent need of conservation of lakes of Bhopal as concentration of manganese and boron were found in excess in Upper Lake of Bhopal at identified location. Upper Lake is also

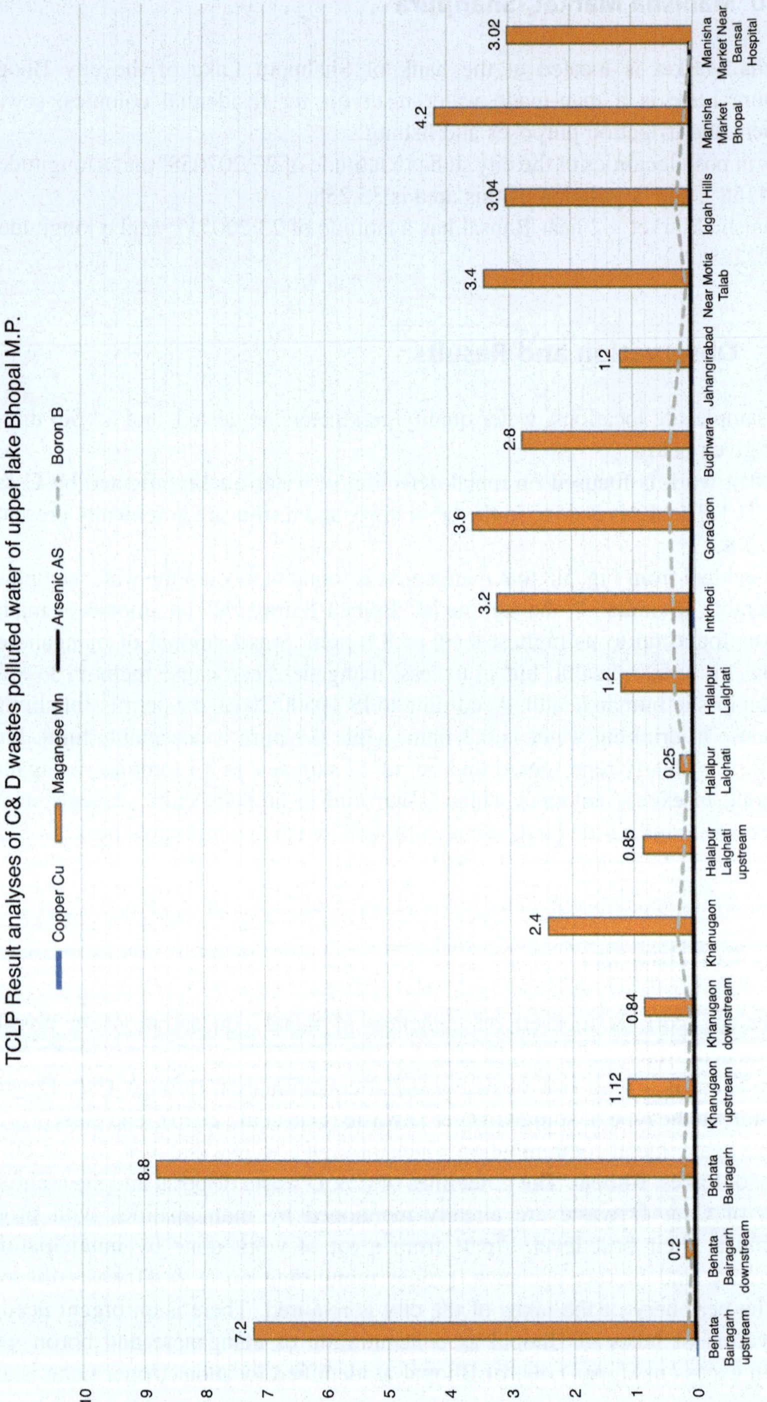

Fig. 5.8 TCLP result analyses of C & D wastes polluted water samples of Upper Lake Bhopal, Madhya Pradesh, India

fulfilling the water need of the people of the city with ecological balance and irrigation needs.

References

Census (2011) Population Census 2011. https://www.census2011.co.in

CGWB (2013) District ground water information booklet. Bhopal District Madhya Pradesh, Ministry of Water Resources, 2013

Chen P, Chen X, Wang Y, Wang P (2020) Preliminary study on the upcycle of non-structural construction and demolition waste for waste cleaning. Mater Constr 70(338):e220. https://doi.org/10.3989/mc.2020.13819

City Development Plan: Bhopal (2005) City Development Plan of Bhopal City 2005. Bhopal Municipal Corporation, JNNURM, pp 1–197

Construction and Demolition Waste Management rules 2016

CPCB (2017) Guidelines on environmental management of construction & demolition (C & D) wastes. Central Pollution Control Board, 2017

CPHEEO (2016) Municipal solid waste management manual part II: the manual. Central Public Health and Environmental Engineering Organisation (CPHEEO), Ministry of Urban Development

CSE India (2022) India manages to recover and recycle only about 1 per cent of its construction and demolition (C & D) waste, says new CSE analysis. https://www.cseindia.org/india-manages-to-recover-and-recycle-only-about-1-per-cent-of-its-construction-and-demolition-10326. Accessed 29 Jan 2022

Dainik Bhaskar (2019) A national Hindi News Paper of India. Accessed 16 July 2019

Gao X, Gu Y, Xie T (2015) Characterization and environmental risk assessment of heavy metals in construction and demolition wastes from five sources (chemical, metallurgical and light industries, and residential and recycled aggregates). Environ Sci Pollut Res 22(12):9332

Googlemap (2022). https://www.google.com/maps/place/Upper+Lake/@23.240931,77.2583943,12z/data=!3m1!4b1!4m5!3m4!1s0x397c5d7f3abf155f:0xec0ea7199923d63e!8m2!3d23.2532367!4d77.3381855. Accessed 03 May 2022

Hindustan Times (2022). https://www.hindustantimes.com/gurugram/1-25-lakh-tonnes-of-c-d-waste-collected-in-2020-fines-worth-rs-40-lakh-levied-for-violation/story-fk8EsIBqF1DpBFGqoxf7yJ.html

Joint Inspection Report (2020) Regional Office. M.P. Pollution Control Board, Bhopal

MPPCB (2019) Annual report 2018–19 C & D waste management rules 2016. Madhya Pradesh Pollution Control Board, Bhopal

Rainfall Analytics (2013) Madhya Pradesh rainfall data visualization. http://www.climatechange.mp.gov.in/en/rainfall. Accessed 20 July 2022

United States EPA (2018) Facts and figures about materials. Waste and recycling, construction and demolition debris: material-specific data; United States Environmental Protection Agency, 2018. https://www.epa.gov/facts-and-figures-about-materials-waste-and-recycling/construction-and-demolition-debris-material

Wikipedia (2022) Bhojtal. https://en.wikipedia.org/wiki/Bhojtal

Soil Deterioration and Risk Assessment of Heavy Metal Contamination

6

Akriti Ashesh and Ningombam Linthoingambi Devi

Abstract

Soils establish an understructure of ecological processes such as nutrient cycling, primary production, climate variables, biophysical habitats, species interaction, etc. Such an important part is vulnerable to various kinds of pollution, and one of them is heavy metal contamination. Soil gets contaminated with heavy metals through natural as well as anthropogenic sources. The former one includes weathering, volcanoes, wind, soil erosion, and wind storms, while the latter one includes mining, dumping of waste having heavy metals into landfills, fertilizer and pesticide application, coal combustion, sewage sludge, and petrochemicals. Almost 80% of heavy metal wastes in India are contributed by Gujarat, Andhra Pradesh, and Maharashtra. Heavy metals such as Fe, Zn, Cu, Mn, and Co are essential for improving soil health, but their excess amount could lead to contamination which affects soil hydrology, chemistry, and biota. Factors such as geoaccumulation index methods, ecological risk index (RI), and contamination factor (CF) are used to assess the anthropogenic influence of heavy metals on the soil. Physicochemical parameters such as pH, TOC, and texture govern the mobility and retention of heavy metals into agricultural soils. Phytoextraction is proven to be a realistic approach for mining heavy metals from soil. For instance, alkalinity decreases mobility and retention of heavy metals in soil, while loamy soils show better drainage and lower retention of heavy metals than clay soils. This chapter aims to yield decision-making information regarding agricultural soil quality in relation to risks associated with human health and the environment.

A. Ashesh · N. L. Devi (✉)
Department of Environmental Sciences, Central University of South Bihar, Gaya, Bihar, India
e-mail: nldevi@cub.ac.in

R. P. Singh et al. (eds.), *Heavy Metal Toxicity: Environmental Concerns, Remediation and Opportunities*, https://doi.org/10.1007/978-981-99-0397-9_6

Keywords

Agricultural soil · Pesticides · Ecological risk · Health risk

6.1 Introduction

The development of a man-made world comes at the cost of the degradation of environmental matrices. The air we breathe, the water we drink, and the Earth on which we live are immensely contaminated with heavy metals. Soil is characterized as nonrenewable resources that socialize ground, water, and air. Most of the soil is heavily contaminated with heavy metals that led to an increase in metal concentration in the environment (Cocârță et al. 2016). Agricultural soil contamination with heavy metals (Ennaji et al. 2020) has become a notable issue that needs not only our attention but a solution to work upon. The chemical way of defining heavy metal pollution is as follows: metal and metalloids that have an atomic mass more than 20 and specific gravity more than 5 are referred to as heavy metals, e.g., chromium (Cr), nickel (Ni), copper (Cu), zinc (Zn), arsenic (As), cadmium (Cd), mercury (Hg), and lead (Pb). On the other hand, in the perspective of biologists, heavy metals refer to those metals that are toxic for animals and plants when present even in less concentration (Li et al. 2019).

Heavy metals originate from anthropogenic as well as geogenic activities (Gayathri et al. 2021). Anthropogenic activities such as mining of gold, nickel, and copper, tailing, and dumping of wastes from mines have generated significant effects on ecology and human health (Hadzi et al. 2019). Detailed discussions on heavy metal contamination due to anthropogenic activities are covered in forthcoming sections. Natural sources of heavy metal pollution, despite being the primary source in soil, are less important when compared with anthropogenic sources (Li et al. 2019) as anthropogenic activities contribute more to metal pollution in soil (Zhuang et al. 2013). Natural sources arise from parent rock material which weathers and contributes to heavy metal pollution in soil. It is a slow geogenic process. Their efficiently persistent nature, irreversible quality, toxic characteristics, and tendency to bioaccumulate make them of utmost importance (Huang et al. 2017). The ability to persist in the environment for a long time is supported by their ability to resist biodegradation and thermodegradation (Kumar et al. 2019). Soil acts as the main sink for heavy metal accumulation (Yadav et al. 2018) which makes it the important matrix to be researched for remedial approaches and health risk assessment. Soil can hold heavy metals up to 1000 long years (Baharani et al. 2022). Soil matrix contaminated with heavy metals was studied for remediation using physical, chemical, and biological methods which is discussed in detail in later sections. Adults and children are exposed to soil dust from which heavy metals can accumulate into their tissues via ingestion, dermal contact, and inhalation (Ihedioha et al. 2017). Crops growing on such contaminated soil can be one of the major routes through which heavy metals can lead their way into the human intestine and result in fatal diseases (Fig. 6.1).

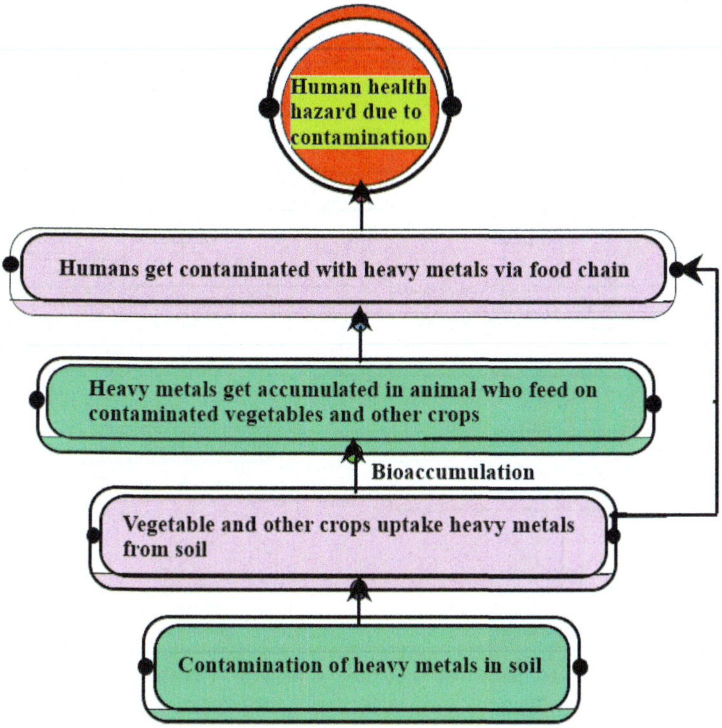

Fig. 6.1 Bottom-up approach showing the impact of heavy metals on food chain

6.2 Source of Heavy Metal in Agricultural Soils

Soils and plants need some of the essential nutrients for their steady growth. Nutrients like Zn, Cu, Fe, Mn, and Ni are referred to as essential micronutrients (Fageria et al. 2009). Hence, these nutrients when present in excess level impart their toxic characteristics. A few toxic characteristics such as growth suppression, cutback in crop yield, quality degradation, and health risk to humans and animals (Seth 2012) arise due to the accumulation of such metals in larger quantities. The entry of heavy metals in agricultural soils refers to both natural and anthropogenic origins (Fig. 6.2).

6.2.1 Natural Source

Earth's crust is made up of 95% igneous rocks and the rest 5% constitutes of sedimentary rocks. Igneous rocks are further divided into two types, that is, intrusive and extrusive igneous rocks. Intrusive igneous rocks encompass loads of heavy metals such as Ni, Cu, Zn, Cd, and Co (Sarwar et al. 2017; Li et al. 2019). Sedimentary rock mainly constitutes heavy metals such as Mn, Cu, Zn, Pd, and

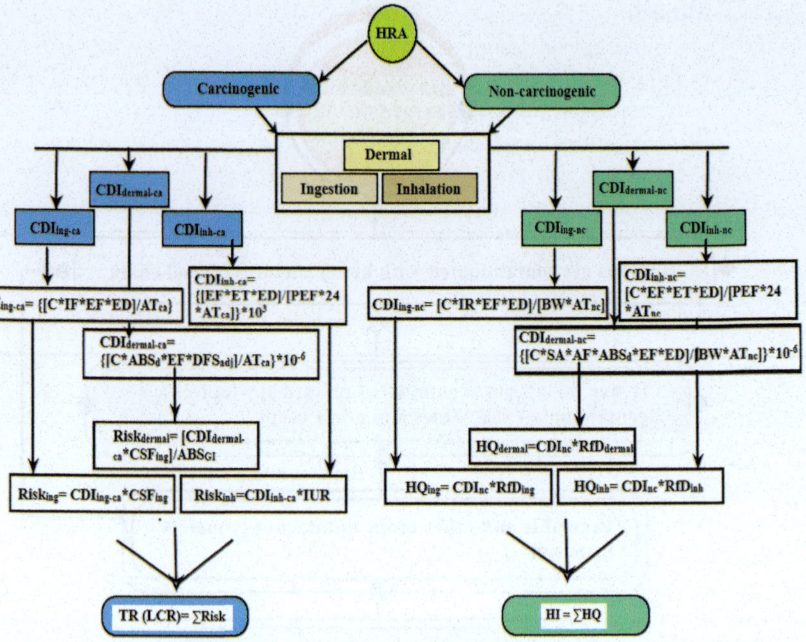

Fig. 6.2 Schematic representation of approaches followed in health risk assessment (HRA) C = metal concentration (mg/kg); IF = intake factor (mgyr/kg/d); EF = exposure frequency (d/yr); ED = exposure duration (yr); AT_{ca} = average time for carcinogens (d); IR = ingestion rate (kg/d); BW = body weight (kg); ET = exposure time (h/d); PEF = particulate emission factor ($m^3 kg^{-1}$); ABS_d = dermal absorption factor (no unit); DFS_{adj} = soil dermal contact, age-adjusted (mgyr/kg/d); SA = skin's surface area ($cm^2 event^{-1}$); AF = soil-skin adherence factor (mg/cm^2); CSF_{ing} = chronic oral slope factor; IUR = inhalation unit risk; ABS_{GI} = gastrointestinal absorption factor; HQ = hazard quotient; nc = noncarcinogenic; ca = carcinogenic; RfD = reference dose (mg/kgday); TR = total risk; LCR = lifetime cancer risk; HI = hazard index

Cd. Hence, soil when formed from igneous and sedimentary parent material constitutes a larger fraction of these metals. The movement of heavy metals from parent rock to soil felicitates through natural phenomena such as biogenic, meteoric, terrestrial, volcanic processes, wind, erosion, and leaching (Li et al. 2019). However, these processes occur very slowly in nature which got disturbed by anthropogenic activities as discussed in the next section.

6.2.2 Anthropogenic Source

Increased level of heavy metal contamination in soils is the result of upgradation in human lifestyle. The current lifestyle of rural as well as urban people, younger as well as older generation, demands advancement in the industrial and agricultural sector. These advancements lead to heavy metal contamination. Some anthropogenic sources are as follows:

6.2.2.1 Mining

Mining and connected smelting processes are major sources of heavy metals in soil. Extensive mine waste piles generated from milling accumulate on the ground surface. Particles of soil loaded with heavy metal have the tendency to move over 20 km from the starting point (Beattie et al. 2018). Sometimes, tailing may contaminate the surrounding soil through wind and water disposal as it contains metals (Hadzi et al. 2019). Other mining processes such as grinding and concentration of ores also release chunks of heavy metals into the atmosphere which pave their way to soil. Mining of gold through artisanal and small-scale gold mining releases more than 30% of mercury across the globe (Fernandez-F et al. 2022).

6.2.2.2 Sewage Irrigation

Urban sewage is a house to the majority of industrial and household discharges. These wastes are loaded with chemical and organic residues. The sewage coming out of industries and households is further utilized for irrigation of agricultural lands. The chemical and organic pollutants present in water used for irrigation infer with agricultural activities. This interference leads to yield reduction (Dorak 2020). Sewage coming out of stainless industries has a major content of Cr and Ni. Thus, paving their way to agricultural fields through sewage irrigation, these metals get concentrated in agricultural soil (Su et al. 2022). Zn, Al, Cr, and Fe residues were identified in the wastewater coming out of tannery industries (Appiah-Brempong et al. 2022). Therefore, agricultural land when irrigated with such wastewater tends to get contaminated with heavy metals.

6.2.2.3 Application of Pesticides

Organic fertilizer contains large amounts of copper (Su et al. 2022) and Zn (Ennaji et al. 2020). Application of fertilizer throughout the year has led to the accumulation of heavy metals such as Cd, As, and Pb, as these fertilizers contain heavy metals (Yokel and Delistraty 2003).

6.2.2.4 Traffic Emission

Heavy metals such as Pb enter the soil through vehicular emissions (Mielke et al. 2010). Fuel combustion in automobiles that run on the road along with the wearing of tires that produce cadmium powder moves toward soil causing heavy metal pollution in soil (Turer et al. 2001). The movement of heavy metal emission from road toward soil is supported via rainfall runoff.

6.2.2.5 Waste Dumping

Dumping of waste in India has no control. Wastes containing heavy metals are dumped so carelessly without any segregation. The expansion of cities has left no gap between these dumpsites, agricultural areas, and residential societies. The leachate from these dumpsites can travel via geological process carrying loads of heavy metals in it to agricultural areas contaminating the soil. Dumped stainless steel, rechargeable batteries, fabrics, leathers, electrical appliances, paint materials, gasoline, and discarded mechanical parts contribute to heavy metal contamination released from wasteyards (Ihedioha et al. 2017).

6.3 Current Status of Heavy Metal Contamination in Different Land Use Pattern

Soil is heavily polluted with heavy metals. All the above sources contribute to soil pollution at a global scale. The extent of the abovementioned activities is increasing continuously. The quality of the soil once degraded cannot regain its original strength. However, there are solutions available for reclamation of such degraded land through remediation approaches (Prathap et al. 2022). The soil from the coal mine dumpsite area was found to be heavily contaminated with heavy metals such as Fe, Mn, Zn, and Ni. The researchers are now trying to find out the removal strategies of over-dumped soil via plant management. Some of the plants that showed positive results are *C. dactylon*, *E. binata*, *L. indica*, and *C. oblongifolius* (Prathap et al. 2022). The agricultural soil irrigated with river water was found to be contaminated with heavy metals such as Fe, Mn, Zn, Cu, and Ni, but their concentrations were below the standard limit prescribed by regulatory bodies (Singh et al. 2021). However, metals such as As, Cr, and Pb were found to exceed their threshold in the hazardous waste disposal sites of Hyderabad (Parth et al. 2011). According to other reported studies, Pb makes one of the most potent toxic elements among other heavy metals such as Cd, Ni, Cr, Zn, and Cu. Pb has exceeded the permissible limits set by WHP/FAO 2007 and Indian standard in soil (Devi and Yadav 2018; Sonu et al. 2019). The kinds of research work that are published now are significantly different from the ones that used to be published 10 years ago. There is a huge sense of responsibility that can be seen in current research papers as they are not only monitoring the concentration levels but also adapting new approaches for remediation of contaminated sites. A study from agricultural soils in Kerala found elevated concentration of Pb in the soil and with their research work provided some of the plant species that could assist in the lead accumulation such as *D. chinensis* L., *C. indicum* L., *L. camara* L., and *R. simplex* C. (Arathi et al. 2021). Several other studies also worked on monitoring and remediation of heavy metals from soil (Adimalla et al. 2020; Bhat et al. 2021; Yadav et al. 2021; Vasudhevan et al. 2022). The concentrations of heavy metals in major parts of the country that are well researched are shown in Table 6.1.

Part of the lands located outside India has been well researched with the contamination of heavy metals (Table 6.2). The soils of the dumpsite were mentioned to be largely contaminated with Fe and Zn (Ihedioha et al. 2017). Agricultural soils were found to be contaminated with Cu and Zn because of the pesticide uses (Li et al. 2009; Ennaji et al. 2020). Urban soil showed extremely high concentration of Pb in Malaysia (Praveena et al. 2015). Agricultural soil of Romania region was found to have high concentration of Cd, whereas in Nigeria agricultural soils were found to have very less Cd concentration (Cocârță et al. 2016; Ogunlade and Agbeniyi 2011). This implies that the concentration of heavy metals varies from region to region

Table 6.1 Heavy metal contamination in soil from India (mg/kg)

Location	Soil type	Fe	Mn	Cu	Ni	Pb	Zn	Cd	As	Cr	Co	Hg	References
Bokaro, Jharkhand	Coal mine dumpsite	17817.8	278.513	–	8.37	4.6	38.07	–	5.76	–	-	–	Prathap et al. (2022)
Kali River, Uttar Pradesh	Agricultural	86.45	53.65	33.95	22.1	12.53	75.68	0.75	–	9.54	-	–	Singh et al. (2021)
Kazipalli, Hyderabad	Waste dumpsite	–	–	35.2	48	206.4	122.3	–	51.7	127.9	-	–	Parth et al. (2011)
Kopargaon area, Maharashtra	Agricultural	3.929	6.76	2.284	0.595	1.328	0.936	0.682	–	–	-	–	Patil et al. (2022)
Kozhikode, Kerala	Agricultural	–	–	–	–	0.19	–	–	–	–	-	BDL	Arathi et al. (2021)
Ludhiana, Punjab	Agricultural	19.6	8.61	1.16	0.61	5.72	4.36	0.2	–	0.39	0.23	–	Dhaliwal et al. (2021)
Patna, Bihar	Agricultural	–	417.73	49.604	–	24.454	136.952	–	–	–	–	0.445	Sonu et al. (2019)
Patna, Bihar	Roadside			143	165	984	713	9.50	–	55.5	–	–	Devi and Yadav (2018)
Ranga Reddy, Telangana	Urban	-	–	40.64	7.55	47.48	34.68	–	–	43.24	16.54	–	Adimalla et al. (2020)
Salem, Tamil Nadu	Agricultural	28.69	8.45	2.79	1.28	1.99	1.59	0.12	–	–	–	–	Vasudhevan et al. (2022)
Shahjahanpur, Uttar Pradesh	Agricultural	6.388	8.77	0.3	–	–	0.633	–	–	–	–	–	Yadav et al. (2021)
South Kashmir	Agricultural	–	–	29.46	41.46	22.6	82.87	0.128	8.99	110.06	18.71	–	Bhat et al. (2021)

Table 6.2 Heavy metal contamination in soil elsewhere in the world (mg/kg)

Location	Soil type	Fe	Mn	Cu	Ni	Pb	Zn	Cd	As	Cr	Co	Be	References
Akwa Ibom, Nigeria	Dumpsite	1804	91.2		11.82	11.8	146	12.2	–	4.05	–	–	Ihedioha et al. (2017)
Bahr El-Baqar, Egypt	–	–	–	122.96	67.83	43.096	133.25	12.702	–	111.57	–	–	Omran and Abd El Razek (2012)
Central Romania	Agriculture	–	–	–	50.68	704.22	–	23,083	–	27.29	–	0.15	Cocârță et al. (2016)
Cross River State, Nigeria	Agriculture	–	–		0.42	0.86	–	0.05	0.02	1.22	0.61	–	Ogunlade and Agbeniyi (2011)
Hunan Province, China		–	–	25.07	27.63	802.58	689.66	6.13	84.85	–	–	–	Huang et al. (2017)
Jilin Province, China	Metal dump, coal gangue	–	–	21.563	–	20.97	65.5	0.153	–	69.68	–	–	Jiang et al. (2014)
Nepal	Urban soil	1190	1870	277	121	137	1020	1.1	4.5	309	46	–	Yadav et al. (2018)
Neyshabur, Iran	Industrial	–	–		15.77	57.33	–	1.9	8.84	37.66	–	–	Mohammadi et al. (2020)
Seri Kembangan, Malaysia	Urban soil	–	–	–	–	2668.8	–	47.5	13.57	74	–	–	Praveena et al. (2015)
Shenyang, China	Agriculture	–	–	18.47	–	29.81	110.27	1.99	–	–	–	–	Li et al. (2009)
Tadla plain, Morocco	Agriculture	19249.3	–	138.1	–	31.72	162.11	0.92	–	32.0.72	–	–	Ennaji et al. (2020)

despite the same type of soil. A meta-analysis of heavy metal pollution in China showed the higher trend in southeast China whereas the lower trend in northwest China. The concentrations of Cd, Hg, Zn, Cu, and Pb were higher in vegetable and paddy sites (Huang et al. 2019).

6.4 Health Risk Assessment of Heavy Metals

Soil when contaminated with heavy metals has a high chance to pass it to higher organisms present at trophic level. Humans constitute the highest group of trophic level; hence, being the last repository, there is a maximum chance to get affected by harmful effects of heavy metals. Metals such as Cr, Cu, Pb, and Zn are known to cause detrimental health issues (Karim and Qureshi 2014). The severity of these issues can only be determined by performing the health risk assessment for available metals in the environment. Health risk assessment of humans can be defined as a process involving mathematical calculations that help in the estimation of likelihood of damages incurred to populations that had been exposed to pollution (Oves et al. 2012). In this process, the ways by which the population gets exposed to pollution and harmful effects of pollutants are assessed. The health risk assessment is an important parameter for decision-makers as it helps them give a new perspective for defining new policies or refining the existing ones. Now, different researchers followed different approaches for the assessment of health risks, but the parameters of calculating the risk is the same in every research article (Karim and Qureshi 2014; Qu et al. 2012; Adimalla and Wang 2018; Baltas et al. 2020; Kacholi and Sahu 2018). For assessing the human health risk of heavy metals in soil, four steps can be followed: the first one is to identify the hazard followed by dose-response assessment. The next step is the assessment of exposure, and the last is to characterize the risk (Koki et al. 2015). In most of the study, the population is exposed to heavy metals present in the soil through skin or dermal contact, inhalation, and ingestion of soil, dust, or agricultural crops and vegetables (Ma et al. 2018; Mohammadi et al. 2020; Praveena et al. 2015; Zheng et al. 2020). The human health risk assessment is divided into two pathways, carcinogenic and noncarcinogenic pathways. These two approaches are connected with the exposure pathways such as dermal, inhalation, and oral exposures. The schematic representation of calculating health risk is presented in Fig. 6.2.

6.4.1 Carcinogenic Risk

Carcinogenic risk is the risk associated with possible carcinogens that may lead to development of cancer in an individual exposed to it for an entire lifetime. The carcinogenic risk assessment can be calculated by multiplying chronic daily intake (CDI) with the individual metal's slope factor. Cumulative risks by dermal, inhalation, and ingestion exposure will result in lifetime cancer risk (LCR). The formula for calculation of CDI for each exposure phase is represented in Fig. 6.2.

6.4.2 Noncarcinogenic Risk

The noncarcinogenic risk is represented by hazard quotient (HQ). HQ is calculated by dividing chronic daily intake (CDI) (Fig. 6.2) by RfD of a specific chemical exposure. HQ can only be estimated for individual chemical exposure. But in the real environment, more than one exposure factor is responsible for the impact such as inhalation, ingestion, and dermal exposure. Therefore, concentration for single chemical exposure may be low within prescribed limits. But when two or more chemical's exposure combined together, the strength (impact in this case) increases. Hence, by combining individual HQ values, HI would be determined. If the overall value for HI is smaller than 1, this implies that the risk is not substantial, but if this value is greater than 1, it implies the occurrence of noncarcinogenic risk.

6.5 Ecological Risk Assessment of Heavy Metals

Ecological risk assessment is calculated to evaluate the affected biological communities in heavy metal-polluted areas (Kumar et al. 2019). The ecological risks imposed by heavy metals on soil and biological communities can be assessed by indices such as CF (contamination factor), EF (enrichment factor), and *I*geo (geoaccumulation) (Keshavarzi and Kumar 2020), P_N (Nemerow composite) and PERI (potential ecological risk indexes) (Cui et al. 2021), and PLI (pollution load index) and (bioconcentration factor) (Liu et al. 2014b). The indices mentioned above assess the extent by which the pollution level of heavy metals is spread over an area (Yahaya et al. 2021). These indices have different levels of classification according to their level of pollution and risk (Liu et al. 2014b; Mohseni-Bandpei et al. 2017; Olatunde et al. 2020; Cui et al. 2021) which is represented in Table 6.3. These indices are explained one by one in the following subsections.

6.5.1 Contamination Factor

The contamination factor is generally denoted by CF. CF is the measure of the pollution level of individual metal. It is the ratio of concentration of metal in the soil to the background concentration (concentration of metal in unpolluted soil) (Olatunde et al. 2020). CF value >1 indicates the contamination, and <1 indicates pollution of heavy metals (Yahaya et al. 2021):

$$\mathrm{CF} = \frac{\textit{Concentration of metal in soil}}{\textit{Background concentration}}$$

Table 6.3 Ecological risk assessment indices and their grades with the level of classification

Indices	Grades	Pollution level or risk level	References
Contamination factor	0	None polluted	Liu et al. (2014b)
	1	None to moderately polluted	
	2	Moderately pollution	
	3	Moderate to strongly polluted	
	4	Strongly polluted	
	5	Strong to very strongly polluted	
	6	Very strongly polluted	
Contamination factor	<1	Low risk	Mohseni-Bandpei et al. (2017)
	1–3	Moderate risk	
	3–6	Considerable risk	
	>6	High risk	
PLI (pollution load index)	<1	No pollution	Liu et al. (2014b)
	1–2	Moderate pollution	
	2–3	Heavy pollution	
	>3	Extremely heavy pollution	
Enrichment factor	<40	Low risk	Mohseni-Bandpei et al. (2017)
	40–80	Moderate risk	
	80–160	Considerable risk	
	160–320	High risk	
	>320	Very high risk	
PER (potential ecological risk index)	<65	Low risk	Mohseni-Bandpei et al. (2017)
	65–130	Moderate risk	
	130–260	Considerable risk	
	>260	High risk	
P_N index	<0.7	Safety state	Cui et al. (2021)
	0.7–1.0	Warning state	
	1.0–2.0	Slight pollution	
	2.0–3.0	Moderate pollution	
	>3.0	Heavy pollution	
I_{geo}	<0	Unpolluted	Olatunde et al. (2020)
	0–1	Unpolluted to moderately polluted	
	1–2	Moderately polluted	
	2–3	Moderately to strongly polluted	
	3–4	Strongly polluted	
	4–5	Strongly to very strongly polluted	
	≥5	Very strongly polluted	

6.5.2 Enrichment Factor

Not every researcher defines an enrichment factor in their research. Those who define it represent it as EF. EF is the ratio of proportion of metal in soil to the proportion of metal present in the Earth's crust (Yahaya et al. 2021):

$$\mathrm{EF} = \frac{Msample}{Resample} \Big/ \frac{Mreference}{Rereference}$$

where Msample and Resample are the contamination factors in polluted soil and Mreference and Rereference are the contamination factors in reference soil. EF assists in the estimation of potential pollution sources along with the impact of anthropogenic activities on contamination level in soil and associated human health (Kumar et al. 2019).

6.5.3 Pollution Load Index

The pollution load index, abbreviated as PLI, is used to find out the cumulative heavy metal pollution level of a particular site (Liu et al. 2014b). It is the geometric mean of CF of all the heavy metals studied for a site. Mathematically, it is expressed as:

$$\mathrm{PLI} = (CF1 * CF2 * CF3 * \ldots\ldots\ldots CFn)^{1/n}$$

where CF is the contamination factor and n is the number of metals identified for a particular site.

6.5.4 Nemerow Composite Index

Nemerow composite index also known as P_N index assesses the destruction in soil environmental quality due to heavy metal contamination (Cui et al. 2021). It is calculated using the following equation:

$$\mathrm{P_N} = \{[(Ci/Si)\max + (Ci/Si)ave]/2\}^{1/2}$$

where Ci = measured value of heavy metal i and Si = reference concentration of heavy metal i.

6.5.5 Potential Ecological Risk Index

PERI assists in environmental quality assessment. The effect of heavy metal pollution on biological communities can be assessed by potential ecological risk index and ecological risk factor (E_r). E_r is calculated for individual heavy metals, whereas PER is calculated for cumulative effect of heavy metals on soil quality (Mohseni-Bandpei et al. 2017):

$$E_r = \mathrm{T_r} \times \mathrm{CF}$$

$$\mathrm{PER} = \sum_{i}^{m} Er$$

where T_r = toxic factor for individual heavy metals and CF = contamination factor.

6.5.6 Geoaccumulation Index

This index is used to compare the present level of heavy metal contamination in soil with the past level concentrations. It is calculated for individual metal contamination (Tian et al. 2017; Yahaya et al. 2021). The geoaccumulation index represented as (I_{geo}) can be computed as:

$$I_{\mathrm{geo}} = \log_2[C_n/1.5B_n]$$

where C_n is the metal concentration in soil (mg/kg), B_n is the background concentration (mg/kg), and 1.5 is the constant that minimizes the lithogenic variation in background concentration. Anthropogenic contamination is depicted by the positive value of I_{geo}.

6.5.7 Bioconcentration Factor

Bioconcentration factor (BCF) determines the sharing amount of pollutant that the aerial part of the plant uptakes from soil. It is estimated by calculating the ratio of concentration of heavy metal present in plants to the concentration of heavy metals present in soil. Mathematically, it can be expressed as:

$$\mathrm{BCF} = C_{\mathrm{plant}}/C_{\mathrm{soil}}$$

where C_{plant} is the concentration of heavy metal in plant (mg/kg) and C_{soil} is the concentration of heavy metal in soil.

6.6 Heavy Metals and Soil Interaction

Soil is the end result of the weathering process. One of the natural sources of heavy metal is weathering of parent rock. During the soil development, a soil profile is formed which constitutes different soil layers starting with O followed by A, B, and

C. These layers are termed as horizons. The constituents of each horizon are different. For instance, O horizon consists of organic matter and other humic substances. A horizon (eluviation zone) is mainly composed of organic matter and minerals. The B horizon, often known as the alluviation zone, constitutes clay minerals. Fe oxyhydroxides are also present in the B horizon which has the ability to absorb heavy metals (Bradl 2005). C horizon is composed of weathered parent rock. Heavy metal's adsorption to the soil and its release from the soil depend upon conditions such as pH and redox (Bradl 2005). Metal sorption in the soil occurs at substantially lower pH as compared with the pH associated with metal hydroxide formation (form at pH 5.5 to 7.5). The presence of dissolved organic matter in soil helps in the sorption of heavy metals. From the soil, the metals can be moved to groundwater via leaching and to the atmosphere via erosion and colloid loss. A soluble metal-humate complex formed at high pH that lowers the metal precipitation. Carboxyl group present in the humic acid interacts with the metal cations (Spark et al. 1997). Heavy metals such as copper, chromium, cadmium, nickel, zinc, and lead are evidenced to reduce the potential of dehydrogenase enzymes present in the soil (Wyszkowska et al. 2006). The order of heavy metals according to their capability to reduce the dehydrogenase's activity in soil is highest in Cr (VI) followed by Cd; then Zn, Pb, Cu, and Ni have the weakest potential to reduce the enzyme's activity in soil. The activity of urease was reduced mostly by Cr (VI) and then Ni followed by other metals such as Cu, Cd, Zn, and Pb. All these metals when present in soil inhibit the activity of enzyme acid phosphatase. Cadmium produced the strongest harmful effect in inhibiting the activity of alkaline phosphatase, while copper produced the weakest effect, and Zn and Pb did not produce any harmful effect on this particular enzyme. The mobility of metals in the soil environment is dependent upon pH and varies from metal to metal. Low pH and CEC (cation exchange capacity) increase the mobility of heavy metal cations while decrease the anion (chromate and arsenate) mobility (Xu 2013). For instance, chromium and zinc are the most mobile heavy metals present in soil. Other metals except heavy metal reduce the adsorption of later in soil (Markiewicz-Patkowska et al. 2005). The heavy metals, when adsorbed by the soil, form complexes with organic material resulting in organomineral complexes. These complexes lead to the humic acid destruction and increase the aliphatic structures in soils (Minkina et al. 2006).

6.7 Heavy Metal and Human Interaction

Through the process of bioaccumulation, heavy metals make their way from soil to the human physiological system. All the heavy metals pose serious risks to human beings. For example, cadmium is a heavy metal that targets the liver, kidney, and vascular system of our body. Low concentration of Cd can cause Cd(II)-HSA (human serum albumin) complex and deform the original structure of HSA. Higher concentration of Cd(II) deforms the structure of protein and induces changes in size of HSA (Liu et al. 2014a). Heavy metals such as Au and Zn were proven to inhibit the activity of enzyme known as HNC (human neutrophil collagenase), which plays a substantial role in inflammatory disease by destructing tissues in human (Mallya and Van Wart 1989). Other heavy metals such as Cd(II), Cu(II), and Hg(II) also showed similar kinds of inhibition. Pb adsorption takes place in the gastrointestinal tract and then spreads into bone, soft tissue, and blood. Leads get attached with RBCs present in the blood. Being similar to calcium, lead has the tendency to mimic the pathway of calcium, while sometimes calcium also disturbs the movement of lead in the human body. Arsenic disturbs cellular respiration through inhibiting some of the important pathways such as glycolysis and gluconeogenesis. Therefore, acute exposure of arsenic is also related to the risk of diabetes (Alissa and Ferns 2011). Exposure of arsenic in deficiency of vitamin B and folic acid might impact the blood pressure by affecting the production of S-adenosylhomocysteine and homocysteine. Pb, Cd, and Ni are known carcinogens. Their chemical species such as Pb2+, Ni2+, and Cd2+ attach with dsDNA which leads to adulteration of the original structure of dsDNA. These changes make the helical structure unstable and further make the DNA vulnerable for reactions with oxidative agents (Oliveira et al. 2008).

6.8 Conclusion

Soil pollution through heavy pollution continued to be one of the major concerns to the rapidly growing human race. The increasing trend in contamination level has awakened the concern in the mind of industrialists, environmentalists, economists, and other classes of workers. The effect of heavy metal on human health, animal population, crop yield, and soil quality and land degradation has directed the researchers to think of strategies that can reduce the toxicity of metals from all our lives. The health risks and ecological risks are continuously increasing as the sources of heavy metal toxicity are vastly spread. Therefore, continuous monitoring and remediation approaches on the ground level would be one way to solve this drastically increasing problem. The uptake of metals from under the surface of soil to over the soil surface via plant uptake toward aerial parts is a growing concern that needs our attention. Hence, the monitoring of heavy metals needs to be supported by fresh threshold level and background values distinctly setup for agricultural fields.

References

Adimalla N, Wang H (2018) Distribution, contamination, and health risk assessment of heavy metals in surface soils from northern Telangana, India. Arab J Geosci 11(21):1–15

Adimalla N, Chen J, Qian H (2020) Spatial characteristics of heavy metal contamination and potential human health risk assessment of urban soils: a case study from an urban region of South India. Ecotoxicol Environ Saf 194:110406

Alissa EM, Ferns GA (2011) Heavy metal poisoning and cardiovascular disease. J Toxicol 2011: 870125

Appiah-Brempong M, Essandoh HMK, Asiedu NY et al (2022) Artisanal tannery wastewater: quantity and characteristics. Heliyon 8(1):e08680

Arathi T, Rahna KP, Sebastian DP et al (2021) Assessment of heavy metal and pesticide contamination in Banana fields and development of phytoremediation system in Kozhikode District, Kerala, India. Nat Environ Pollut Technol 20(3):1251–1256

Baharani A, Sentilraja K, Shahidha PA et al (2022) Rhizosphere microbes as biosorbents for heavy metal removal. Indian J Appl Res 12(3):45–48

Baltas H, Sirin M, Gökbayrak E, Ozcelik AE (2020) A case study on pollution and a human health risk assessment of heavy metals in agricultural soils around Sinop Province, Turkey. Chemosphere 241:125015

Beattie RE, Henke W, Campa MF et al (2018) Variation in microbial community structure correlates with heavy-metal contamination in soils decades after mining ceased. Soil Biol Biochem 126:57–63

Bhat NA, Bhat AA, Singh BP et al (2021) Heavy metal contamination in agricultural soils of NW Himalayas: with a perspective of spatial distribution, environmental contamination and health risk assessment. Arab J Geosci 14(23):1–14

Bradl H (ed) (2005) Heavy metals in the environment: origin, interaction and remediation. Elsevier, Amsterdam

Cocârță DM, Neamțu S, ReşetarDeac AM (2016) Carcinogenic risk evaluation for human health risk assessment from soils contaminated with heavy metals. Int J Environ Sci Technol 13(8): 2025–2036

Cui X, Geng Y, Sun R et al (2021) Distribution, speciation and ecological risk assessment of heavy metals in Jinan Iron & Steel Group soils from China. J Clean Prod 295:126504

Devi NL, Yadav IC (2018) Chemometric evaluation of heavy metal pollutions in Patna region of the Ganges alluvial plain, India: implication for source apportionment and health risk assessment. Environ Geochem Health 40:2343–2358

Dhaliwal SS, Setia R, Kumar V et al (2021) Assessment of seasonal variations and human health risks due to heavy metals in water, soils and food crops using multi-indices approach. Environ Earth Sci 80(11):1–11

Dorak S (2020) Determination of irrigation water quality parameters of Nilüfer stream and some of the discharged waste water treatment facilities in Nilüfer stream

Ennaji W, Barakat A, El Baghdadi M et al (2020) Heavy metal contamination in agricultural soil and ecological risk assessment in the northeast area of Tadla plain, Morocco. J Sediment Environ 5(3):307–320

Fageria NK, Filho MB, Moreira A et al (2009) Foliar fertilization of crop plants. J Plant Nutr 32(6): 1044–1064

Fernandez-F F, Lopez-C P, Febres-Molina C et al (2022) Identification and characterization of peruvian native bacterial strains as bioremediation of Hg-polluted water and soils due to artisanal and small-scale gold mining in the Secocha Annex, Arequipa. Sustainability 14(5): 2669

Gayathri S, Krishnan KA, Krishnakumar A et al (2021) Monitoring of heavy metal contamination in Netravati river basin: overview of pollution indices and risk assessment. Sustain Water Resour Manag 7(2):1–15

Hadzi GY, Ayoko GA, Essumang DK et al (2019) Contamination impact and human health risk assessment of heavy metals in surface soils from selected major mining areas in Ghana. Environ Geochem Health 41(6):2821–2843

Huang S, Yuan C, Li Q et al (2017) Distribution and risk assessment of heavy metals in soils from a typical Pb-Zn mining area. Pol J Environ Stud 26(3):1105–1112

Huang Y, Wang L, Wang W et al (2019) Current status of agricultural soil pollution by heavy metals in China: a meta-analysis. Sci Total Environ 651:3034–3042

Ihedioha JN, Ukoha PO, Ekere NR (2017) Ecological and human health risk assessment of heavy metal contamination in soil of a municipal solid waste dump in Uyo, Nigeria. Environ Geochem Health 39(3):497–515

Jiang X, Lu WX, Zhao HQ et al (2014) Potential ecological risk assessment and prediction of soil heavy-metal pollution around coal gangue dump. Nat Hazards Earth Syst Sci 14(6):1599–1610
Kacholi DS, Sahu M (2018) Levels and health risk assessment of heavy metals in soil, water, and vegetables of Dar es Salaam, Tanzania. J Chem 2018:1402674
Karim Z, Qureshi BA (2014) Health risk assessment of heavy metals in urban soil of Karachi, Pakistan. Hum Ecol Risk Assess Int J 20(3):658–667
Keshavarzi A, Kumar V (2020) Spatial distribution and potential ecological risk assessment of heavy metals in agricultural soils of northeastern Iran. Geol Ecol Landsc 4(2):87–103
Koki IB, Bayero AS, Umar A (2015) Health risk assessment of heavy metals in water, air, soil and fish. Afr J Pure Appl Chem 9(11):204–210
Kumar V, Sharma A, Kaur P et al (2019) Pollution assessment of heavy metals in soils of India and ecological risk assessment: a state-of-the-art. Chemosphere 216:449–462
Li C, Zhou K, Qin W et al (2019) A review on heavy metals contamination in soil: effects, sources, and remediation techniques. Soil Sediment Contam Int J 28(4):380–394
Li P, Wang X, Allinson G et al (2009) Risk assessment of heavy metals in soil previously irrigated with industrial wastewater in Shenyang, China. J Hazard Mater 161(1):516–521
Liu G, Yu Y, Hou J et al (2014b) An ecological risk assessment of heavy metal pollution of the agricultural ecosystem near a lead-acid battery factory. Ecol Indic 47:210–218
Liu Y, Chen M, Jiang L et al (2014a) New insight into molecular interaction of heavy metal pollutant—Cadmium (II) with human serum albumin. Environ Sci Pollut Res 21(11): 6994–7005
Ma W, Tai L, Qiao Z et al (2018) Contamination source apportionment and health risk assessment of heavy metals in soil around municipal solid waste incinerator: a case study in North China. Sci Total Environ 631-632:348–357
Mallya SK, Van Wart HE (1989) Mechanism of inhibition of human neutrophil collagenase by gold (I) chrysotherapeutic compounds: interaction at a heavy metal binding site. J Biol Chem 264(3): 1594–1601
Markiewicz-Patkowska J, Hursthouse A, Przybyla-Kij H (2005) The interaction of heavy metals with urban soils: sorption behaviour of Cd, Cu, Cr, Pb and Zn with a typical mixed brownfield deposit. Environ Int 31(4):513–521
Mielke HW, Laidlaw MA, Gonzales C (2010) Lead (Pb) legacy from vehicle traffic in eight California urbanized areas: continuing influence of lead dust on children's health. Sci Total Environ 408(19):3965–3975
Minkina TM, Motuzova GV, Nazarenko OG (2006) Interaction of heavy metals with the organic matter of an ordinary chernozem. Eurasian Soil Sci 39(7):720–726
Mohammadi AA, Zarei A, Esmaeilzadeh M et al (2020) Assessment of heavy metal pollution and human health risks assessment in soils around an industrial zone in Neyshabur, Iran. Biol Trace Elem Res 195(1):343–352
Mohseni-Bandpei A, Ashrafi SD, Kamani H et al (2017) Contamination and ecological risk assessment of heavy metals in surface soils of Esfarayen city, Iran. Health Scope 6(2). https://doi.org/10.5812/jhealthscope.39703
Ogunlade MO, Agbeniyi SO (2011) Impact of pesticides use on heavy metals pollution in cocoa soils of Cross-River State, Nigeria. Afr J Agric Res 6(16):3725–3728
Olatunde KA, Sosanya PA, Bada BS et al (2020) Distribution and ecological risk assessment of heavy metals in soils around a major cement factory, Ibese, Nigeria. Sci Afr 9:e00496
Oliveira SCB, Corduneanu O, Oliveira-Brett AM (2008) In situ evaluation of heavy metal–DNA interactions using an electrochemical DNA biosensor. Bioelectrochemistry 72(1):53–58
Omran ESE, Abd El Razek AA (2012) Mapping and screening risk assessment of heavy metals concentrations in soils of the Bahr El-Baker Region, Egypt. J Soil Sci Environ Manage 3(7): 182–195

Oves M, Khan MS, Zaidi A et al (2012) Soil contamination, nutritive value, and human health risk assessment of heavy metals: an overview. In: Toxicity of heavy metals to legumes and bioremediation. Springer, Vienna, pp 1–27

Parth V, Murthy NN, Saxena PR (2011) Assessment of heavy metal contamination in soil around hazardous waste disposal sites in Hyderabad city (India): natural and anthropogenic implications. J Environ Res Manage 2(2):027–034

Patil SN et al (2022) Magnetic susceptibility and heavy metals contamination in agricultural soil of Kopargaon Area, Ahmadnagar District, Maharashtra, India. In: Rai PK, Mishra VN, Singh P (eds) Geospatial technology for landscape and environmental management. Advances in geographical and environmental sciences. Springer, Singapore

Prathap A, Shaikh WA, Baudhh K et al (2022) Phyto-management potential of naturally thriving plants on the metal contaminated overburden dump of coal mines: a study from Jharkhand, India. Bioremediat J:1–11

Praveena SM, Ismail SNS, Aris AZ (2015) Health risk assessment of heavy metal exposure in urban soil from Seri Kembangan (Malaysia). Arab J Geosci 8(11):9753–9761

Qu C, Sun K, Wang S et al (2012) Monte Carlo simulation-based health risk assessment of heavy metal soil pollution: a case study in the Qixia mining area, China. Hum Ecol Risk Assess Int J 18(4):733–750

Sarwar N, Imran M, Shaheen MR et al (2017) Phytoremediation strategies for soils contaminated with heavy metals: modifications and future perspectives. Chemosphere 171:710–721

Seth CS (2012) A review on mechanisms of plant tolerance and role of transgenic plants in environmental clean-up. Bot Rev 78(1):32–62

Singh G, Patel N, Jindal T et al (2021) Heavy metal contamination in soils and crops irrigated by Kali River in Uttar Pradesh, India. Bull Environ Contam Toxicol 107(5):931–937

Sonu K, Yadav IC, Kumar A (2019) Dataset on assessment of heavy metals contamination in multi-environment samples from Patna, India. Data Brief 25:1–8

Spark KM, Wells JD, Johnson BB (1997) The interaction of a humic acid with heavy metals. Aust J Soil Res 35(1):89–102

Su H, Hu Y, Wang L et al (2022) Source apportionment and geographic distribution of heavy metals and as in soils and vegetables using kriging interpolation and positive matrix factorization analysis. Int J Environ Res Public Health 19(1):485

Tian K, Huang B, Xing Z et al (2017) Geochemical baseline establishment and ecological risk evaluation of heavy metals in greenhouse soils from Dongtai, China. Ecol Indic 72:510–520

Turer D, Maynard JB, Sansalone JJ (2001) Heavy metal contamination in soils of urban highways comparison between runoff and soil concentrations at Cincinnati, Ohio. Water Air Soil Pollut 132(3):293–314

Vasudhevan P, Manikandan E, Jonathan MP et al (2022) Pollution assessment and source apportionment of metals in paddy field of Salem, South India. Environ Earth Sci 81(6):1–15

Wyszkowska J, Kucharski J, Lajszner W (2006) The effects of copper on soil biochemical properties and its interaction with other heavy metals. Pol J Environ Stud 15(6):927

Xu RK (2013) Interaction between heavy metals and variable charge surfaces. In: Molecular environmental soil science. Springer, Dordrecht, pp 193–228

Yadav KK, David AA, Swaroop N et al (2021) Evaluation of Oudh sugar mill effluent impact on soil properties and heavy metal accumulation in the soil of Rosa, Shahjahanpur (UP). Pharma Innov J 10(9):1776

Yadav IC, Devi NL, Singh VP et al (2018) Spatial distribution, source analysis, and health risk assessment of heavy metals contamination in house dust and surface soil from four major cities of Nepal. Chemosphere 218:1100

Yahaya SM, Abubakar F, Abdu N (2021) Ecological risk assessment of heavy metal-contaminated soils of selected villages in Zamfara State, Nigeria. SN Appl Sci 3(2):1–13
Yokel J, Delistraty DA (2003) Arsenic, lead, and other trace elements in soils contaminated with pesticide residues at the Hanford site (USA). Environ Toxicol 18(2):104–114
Zheng S, Wang Q, Yuan Y et al (2020) Human health risk assessment of heavy metals in soil and food crops in the Pearl River Delta urban agglomeration of China. Food Chem 316:126213
Zhuang P, Zhi-An LI, BiZ et al (2013) Heavy metal contamination in soil and soybean near the Dabaoshan mine, South China , Pedosphere 23(3):298–304

Heavy Metal Contamination in Groundwater: Environmental Concerns and Mitigation Measures

7

J. Khan, R. Shrivastava, G. Gupta, and N. K. Singh

Abstract

Metal contamination in water especially in groundwater sources is increasing due to changes in geochemistry in the aquifers. Overexploitation of water with increasing human population and development leads to overextraction of more water from groundwater sources which change the level of groundwater table and its geochemistry. Heavy metal contamination in groundwater has become a global health concern due to its harmful impact on human health and other living beings. Contamination of groundwater with metal leads to human health hazards through contaminated drinking water. More reports are there on arsenic contamination in groundwater and its effect on plants and animals including humans. Monitoring of metal contamination and estimation of its concentration in groundwater will be helpful in investigating contaminated sites and understanding its route and fate in the environment. Understanding metal contamination and its environmental concerns is important for assessing health hazards in developing mitigation strategies for health and environmental safety.

Keywords

Metal · Bioaccumulation · Biomagnification · Water treatment · Water quality

J. Khan · G. Gupta · N. K. Singh (✉)
Environmental Science Discipline, Department of Chemistry, Manipal University Jaipur, Jaipur, Rajasthan, India

R. Shrivastava
Environmental Science Discipline, Department of Chemistry, Manipal University Jaipur, Jaipur, Rajasthan, India

Poornima College of Engineering, Jaipur, Rajasthan, India

R. P. Singh et al. (eds.), *Heavy Metal Toxicity: Environmental Concerns, Remediation and Opportunities*, https://doi.org/10.1007/978-981-99-0397-9_7

7.1 Introduction

Water is the most crucial natural resource for life. Vorosmarty et al. (2010) reported that about 80% of the world's people use water as a critical commodity and are now facing water shortage and health issues. Even more, the surface of the Earth is covered with water; however, less than 3% of the water is available as freshwater, and 97% of the remaining water is available in bigger oceans as marine water with excess salt which is unsuitable for human use. In fact, less than 1% of freshwater is only available on Earth for human use, while 1.96% of fresh Earth's water is locked in polar ice caps and glaciers. Furthermore, freshwater resources slowly become unavailable for drinking water supply owing to contamination due to negligence by the public and industrial authorities (Kumar Reddy and Lee 2012). Groundwater resources play a significant role for maintaining the Indian economy, environment sustainability, and standard of living. Basically, we mostly depend on groundwater sources for domestic, agricultural, and industrial use in India. Overextraction and overexploitation of groundwater lead to water scarcity and contamination particularly by toxic metals due to geological changes and processes. Metals like Fe, Cu, Zn, Co, Cr, and Mn are very essential for growth and development of living being; however, other metals considered as nonessential metals have toxic effects on the plants and animals like lead (Pb), arsenic (As), mercury (Hg), and cadmium (Cd) even at very low concentration in water. Bradl (2002) reported that natural phenomena such as volcanic eruptions and geogenic processes significantly contribute to heavy metal contamination in soil and water. Therefore, Pb, Ni, As, Cd, Cr, and Hg are the most common metal contamination in water due to geogenic and anthropogenic activities. Metal contamination in soil and water causes threats to human health including other animals and plants. Continuous exposure and accumulation of metals through drinking water and food contamination cause harmful human health impacts leading to death. Toxic level of mercury and lead can lead to immune system suppression, arthritis, and circulatory and respiratory problems in humans (Rajendran et al. 2003; Johnson and Hallberg 2005; Oelofse et al. 2007). Even at low concentration, some metals or metalloids cause toxic effects in the plants and animals (Rascio and Navariizzo 2011). Water contamination with metal has become a significant environmental and health concern. Rapid urbanization and extensive developmental activities with increasing human population intensified the metal contamination and depletion of water quality (Sarwar et al. 2016).

7.2 Metal Contamination in Groundwater

7.2.1 Sources of Contamination

Metal contamination of water occurs by natural and anthropogenic sources. Extensive studies have shown that natural and geogenic processes are usually less concerned as compared with anthropogenic activities for metal contamination (Dixit et al. 2015). Metals present in the soils are mostly derived from the parental

Table 7.1 Sources of metal contamination

Heavy metal sources		References
Nonessential metals		
Cd	Fertilizers, pesticides, and plastics	Mandour and Azab (2011); Thambavani and Mageswari (2013)
Hg	Mining, paint and paper industry, volcanic eruption, coal burning	Virha et al. (2011)
As	Pesticides, atmospheric deposition mining	Krishna et al. (2009)
Pb	Paint industry, batteries, coal combustion	Mandour and Azab (2011)
Cr	Tanneries and electroplating industry	Krishna et al. (2009)
Essential metals		
Zn	Brass manufacturing, plumbing, oil refining	Krishna et al. (2009); Gowd and Govil (2008)
Cu	Printing, copper polishing, plating	Mandour and Azab (2011)
Fe	Intake of iron supplements	Usman et al. (2021)

rock material. Ingenious rocks cover 95% of the Earth's crust, whereas sedimentary rocks make up 5% (Sarwar et al. 2016). Metals like Cu, Cd, Ni, and Co are naturally found in basaltic igneous rocks, while Pb, Cu, Zn, Mn, and Cd are present in shale rock. Muradoglu et al. (2015) reported heavy metals mobilized in the environment by biogenic processes, volcanic eruptions, soil erosion, leaching, and seepage. Anthropogenic sources of metal contamination are particularly mining, fossil fuel combustion (Muradoglu et al. 2015), discharge of urban waste (Khan et al. 2016), agrochemical application (Ogunlade and Agbeniyi 2011), and irrigation with wastewater and sewage (Sun et al. 2013). Most cities of India have been affected with groundwater metal contamination due to municipal solid waste disposal in unsanitary landfills, farming, and industrial and other activities (Ravindra et al. 2019; Mor et al. 2006a, 2006b). Table 7.1 summarizes the main sources of heavy metals.

7.2.2 Route and Fate of Metal Contamination

Groundwater contamination with heavy metals has become a major concern in recent years due to increased industrialization and urbanization. Metals bioaccumulate in plants and animals throughout the food chain, eventually finding their way to higher trophic level animals including humans (Singh et al. 2006). Certain metals are toxic to living organisms as they accumulate in tissues via food chain and cause harmful effect. Many researchers have gained global interest in assessing groundwater heavy metal contamination to estimate groundwater quality from various parts of the globe (Qian and Li 2011; Malassa et al. 2013; Khan et al. 2013a, 2013b). Qian et al. (2012, 2013) reported that metal contamination in groundwater has the potential to cause noncarcinogenic health impacts in humans. More than five million people die by waterborne disease due to contaminated water with increasing heavy metal contamination at alarming rate. Geological composition of aquifers typically allows the leaching of toxic elements during extensive groundwater extraction and supply. Arsenic, copper, chromium, and selenium are main metal contaminants in groundwater. Arsenic as one of the metal contaminants of groundwater causes serious human health hazards and risks to people's lives in various countries. In Bangladesh, more than 95% of the population uses groundwater from ten million tube wells. About 60% of deep-water wells throughout the Ganga-Brahmaputra River system contain arsenic over the permissible level (Ahmed et al. 2004). High arsenic concentrations are caused by the weathering of arsenic-containing rocks in mountain ranges, the deposition of organics in river floodplains, and the long residence durations of water in groundwater aquifers, which create anoxic conditions for arsenic release. Groundwater contamination is related to the natural formation of aquifers when toxins join the subsurface water table and release to other drinking water. Contamination of water and soil with heavy metal occurs from urban solid waste disposal sites, radioactive waste sites, hazardous waste sites, accidental leaks, farmland drainage, and industrial waste discharge. Many abandoned sites where waste materials are illegally stored and disposed contribute to groundwater contamination with heavy metals. Groundwater is the main source of drinking water; therefore, monitoring and assessment of metal concentration are essential to determine health risk of humans. Nonessential heavy metals impact biological functions directly or indirectly in the living being. Raj et al. (2006) reported metal pollution in untreated wastewater caused by industrial, agricultural, and geologic processes.

7.2.3 Factors Affecting Metal Contamination

Physical, chemical, and biological processes result to the release of heavy metals in the aquatic environments (Bazrafshan et al. 2015). Different environmental effects are caused by heavy metal contamination (Torres-Cruz et al. 2018). Environmental effects of heavy metal contamination depend on the background condition of the environment, and changes occur in the diversity, composition of species and community with respect to the concentration of metals in water and soil (Pendergast and Hoek 2011; Xu et al. 2018), and redox conditions and densities of living life-forms (Gurung et al. 2018). In water, the fate of metals is mostly dependent on different variables, for example, in methylation and reduction to the metallic form, produce environmental effects on metals. Complex of metals with organic ligands and chlorides cause a decrease in sorption cycle and an increase in housing time in the water. Effects of heavy metal on aquatic plants are highly complex and toxic at higher concentration (Mishra and Dubey 2006). Main responses of plants under heavy metal stress are decrease in diversity and species richness (Yadav 2010; Aihemaiti et al. 2018). Heavy metal release in the aquatic environment changes the physical condition by changing pH, ground organic content, and particle size in water (Jortner 2008; Street 2012; Martinez-Cortijo and Ruiz-Canales 2018). Heavy metals like Cr, Cd, Pb, Hg, Sb, Ag, and As are toxic to plants even at low concentrations (Mustafa and Komatsu 2016; Amari et al. 2017; Muszynska et al. 2018).

7.3 Environmental Concern of Metal in Groundwater

Water is one of the primary life-sustaining resources on Earth (Vanloom and Duffy 2005). Heavy metals can be defined as any metallic compound that has a high density especially greater than 4 g/cm^2 (Duruibe et al. 2007; Nagajyoti et al. 2010; Sujitha et al. 2014). Heavy metals when present in a minute amount play a vital role in human body metabolism, while high amounts cause acute and chronic disease (Bodaghpour et al. 2012). Metals typically present in the water either in colloidal particulates or in dissolved states. Although some metals play an important role for sustaining life, several metals, like sodium, potassium, magnesium, and calcium, must be present in optimum amounts for proper human growth and metabolic functions. Metals like zinc, iron, manganese, copper, and cobalt, if present in less amount, are used as the catalyst for enzyme activities, however, at high concentration, such metals cause toxicity and serious human health impacts. Toxic heavy metals directly react with the proteins to form more toxic elements in which carboxylic acid (–COOH), amine ($–NH_2$), and thiol (–SH) groups are included. The ability of biological molecules to operate effectively is lost as a result of their change, which leads to cell dysfunction and death. When metal react with the above groups, they directly affect the protein structure and inactivate the whole enzyme system, as it's directly linked to the catalytic properties of enzymes. The formation of these toxic radicals is very dangerous as it causes the oxidation of biological molecules. Heavy metals in concentrations above critical levels can cause major health problems. Toxicity of heavy metal can harm or reduce brain and central nervous system activities and affect the lungs, kidney, liver, blood components, and other vital organs. Multiple sclerosis, muscular dystrophy, Alzheimer's disease, and several types of cancer may be induced by long-term exposure to harmful heavy metals.

7.3.1 Effect of Metal Contamination on Water Quality

The geology and hydrogeology of the aquifer play a significant role in the natural sources of heavy metals in water (Wang and Mulligan 2006). Overpopulation and municipal growth have degraded groundwater quality, where heavy metal contamination is one of the most evolving harmful substances for aquatic pollution. These compounds influence living species' metabolism by entering the food chain through water (Popovic et al. 2001). An aquatic ecosystem has been used as a dumping place for the disposal of heavy metal effluents. Both household and industrial pollution affect equally the quality of water (Amalia et al. 2018). Characteristics of water are directly depicted by various contaminants, i.e., inorganic or organic substances like pesticides contain a high amount of heavy metals, although some heavy metals if present in less amount are essential for metabolism activity in living organisms but high amount reacts with As toxic element and gives adverse impacts on water quality (Permanawati et al. 2013). Due to the contamination of the whole aquatic ecosystem, i.e., river, lakes, basin, etc., by heavy metals, all living and nonliving microorganisms are dealing with the current scenario. Due to the presence of the nonessential heavy metals in high amounts, at a certain point, they act as essential metals and disturb the major key cellular functions which are the common reason for acute toxic bioaccumulation (Huat et al. 2019). It has also been found that the excess amount of dissolved metals in water makes it more toxic that can easily enter into the cell body by crossing the membrane. The accumulation of these heavy metals drastically affects the aquatic environment. Rapid population increase and anthropogenic activities have led to the increased discharge of more hazardous materials into water river worldwide (Islam et al. 2015), where if it is discharged into a water body without any treatment it affects the whole aquatic environment (Venugopal et al. 2009). Due to the recalcitrant nature of metals, it is not easy to treat, so it's directly disposed of into the water and gives high toxicity. Accumulation of these heavy metals into the aquatic environment changes the physicochemical properties of water that directly affect the quality of water and make it unfit for further human consumption. The concentration of Zn, Cr, and Cd in the atmosphere is also increased through atmospheric deposition, which is comparatively high in rural areas (Pandey et al. 2014), and effect of temperature of the water. Heavy metals precipitate and form undissolved oxide and carbonate in an alkaline medium. Due to the depletion of dissolved oxygen and the presence of a high concentration of heavy metals in the water body, the amount of dissolved oxygen decreases; therefore, it also increases the energy demand for aquatic microorganisms. Copper is considered one of the most important elements for life which pollute water and is found in water due to the agricultural runoff and by using copper sulfate to control the growth rate of algal blooms (Hong et al. 2020). The high toxicity of copper into the water directly affects its hydrochemical properties. Zinc is also an important trace element for life. A high concentration of Zn directly affects aquatic organisms (Anzecc 2000). Mainly, zinc intake from dietary sources is 10–15 mg on a daily basis; when this concentration increases into the water body, it directly produces toxic effects on water quality. The quality of freshwater can be evaluated by the toxicity percentage. At very low concentration in the water body, lead can directly affect the water's physical property which can directly damage the nerve especially in infants and children (Lin et al. 2017). High amounts of lead also decrease the amount of dissolved oxygen in the water which affects the physicochemical property of water. Chromium is generally present in polluted water bodies as hexavalent chromium (Cr^{+6}), which directly comes from electroplating, tanning, or pigment manufacturing (Testa et al. 2004; Rengaraj et al. 2007) which causes serious pollution effects in the wastewater. The concentration of heavy metal discharge is higher than the regulatory limit; hence, effluents must be treated before being released into the water. Figure 7.1 summarizes the heavy metal effects on living organisms.

Industrial effluent and urban sewage are being disposed of directly to land as an alternative means of treatment and disposal, by giving the logic of a useful source of plant nutrients; however, these effluents contain a high amount of many organic and inorganic materials and heavy metals, which depends on the type of industrial effluents (Khalil et al. 2008). In the last 20 years, the concentration of metals in soil has increased considerably as they are nondegradable (Singh and Kalamdhad

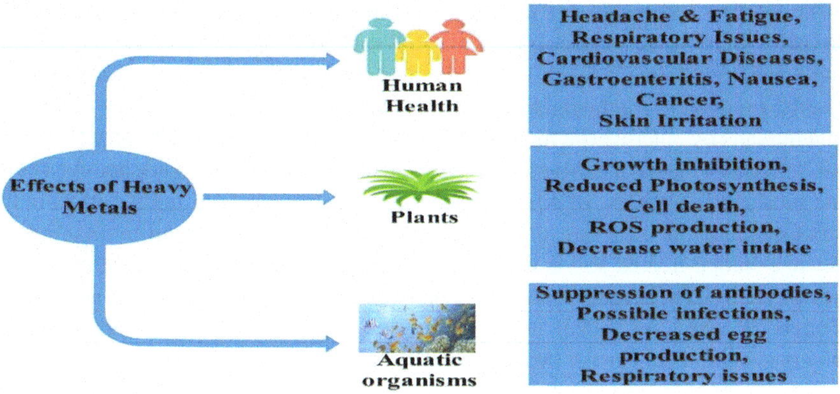

Fig. 7.1 Effect of heavy metals on soil and agro-economic system

2011; Sardar et al. 2013). Sewage containing metals leads to plant uptake, accumulation, and biomagnification (Maric et al. 2013), and its continued usage in farming could be a serious threat to the health and life of living organisms because these heavy metals are transferred via plants to human beings. Mainly, crops and leafy vegetables are used as rich sources of nutrients (116 g/day) because they are rich in minerals and vitamins. Due to excessive metal deposition in soil, soil microorganisms die, and soil quality declines (Oves et al. 2012). Agricultural use of sewage water in cities must be done carefully because it directly impacts human health (Wuana and Okieimen 2011). All contaminants are transported to crops and harm consumer health. Therefore, the existence of contaminants in treated sewage sludge should be considered, and its use in agriculture should be debated scientifically (Shuaib et al. 2021). There are many factors that affect the mobility of metals in soil to land, including pH, clay and organic matter content, and plant characteristics (Stefanowicz et al. 2020). Heavy metals are a cause of soil pollution, which is increasing worldwide as a result of various human activities. One of the most prominent anthropogenic sources of metal is municipal solid waste (MSW) as metals are not degraded; they go into deeper layers of the soil at groundwater or can be incorporated into the food chain and show a negative impact on environmental quality. Sewage and sludge application to soils cause heavy metal contamination which causes usable land to become limited (Singh and Kalamdhad 2011). Heavy metal uptake by plants from soil varies depending on the components of the crop plant. For example, in wheat, the maximum quantity of heavy metals is taken by roots, while the lowest amount is absorbed by stems and grains. Wang et al. (2021) reported the concentration of Pb (16%), Cd (26%), and Cr (0.56%) in store vegetables. Crops are generally grown in urban or suburban areas which get contaminated by various heavy metals that mainly depend on the exposure time of heavy metals, air, water, species, their concentration in soil, and their morphological matter (Cheng et al. 2021). Due to the nondegradable nature of metals, their effective cleanup requires removing toxicity. In recent years, research has started to develop cost-effective techniques for cleaning contaminated areas utilizing microbes, biomass, or living plants (Karimian et al. 2021). Heavy metal concentrations in soils have increased to toxic levels due to the discharge of untreated industrial effluents. Plants' productivity is being reduced since these contaminants are risky to animals and humans by entering the food chain via soil-plant-animals/human route (Lin et al. 2021). Heavy metals in sewage sludge can accumulate in plant bodies and then enter the food chain. Amount of a heavy metal taken up by plants from the soil is, in the strictest sense, the criterion of its availability. Uptake of cadmium by plants from sewage-sludge-treated soil is usually substantially greater than from untreated soil, but phytotoxicity is rare. Zinc, which is usually present in the largest amounts in sludge, copper, and nickel are responsible either singly or in combination for phytotoxic effects in crops (Mandal et al. 2003; Mudhoo et al. 2011).

7.3.2 Effects of Metals on Plants

Metals are one the main pollutants which affect plant growth and cause phytotoxicity. Some metals, particularly essential metals, are required for plant growth; however, at higher concentration, they cause toxicity. Other Metals like As, Cd, Pd, and Hg, which do not give any essential worthwhile effects on the environment and organisms, are therefore considered a "primary threat" as it is more dangerous to plants as well as for animals. In addition, contaminants in the environment cause harm to humans. Metals are taken up by plants at the level of primary producers and subsequently accumulate in the organisms at higher trophic level through food chain contamination. Human water consumption and plant roots are the main exposure sites for metal ions, whereas, in water, plants directly uptake metal ions and/or directly absorbed by leaves (Peralta-Videa et al. 2009). Metals like copper, zinc, and iron are acute for all living and nonliving organisms, and their availability varies with the place, and metals such as Cu, Mn, Zn, Mo, Fe, Ni, and Co are essentially required as micronutrients (Reeves and Baker 2000); however, in excess, it can lead to toxic effects (Monni et al. 2000). Some metals are required in less concentration for plant growth and maintenance, while excessive concentration can be more damaging to plants, allowing them to accumulate nonessential heavy metals (Djingova and Kuleff 2000). Since the metal cannot be decomposed, when the concentration in the plant exceeds optimal levels, which directly and indirectly adversely affect plants, as well as some direct toxic effects caused by plants, high metal concentrations include inhibition of cytoplasmic enzymes and damage to cellular structures due to oxidation pressure. Indirect toxic effects are the substitution of essential nutrients at plant cation exchange sites (Schaller and Diez 1991). Negative effects of heavy metals on soil microbial growth and activity also indirectly affect the growing plant. A decrease in the population of beneficial soil microbes due to excessive metal concentrations may result in a decrease in the decomposition of organic matter, resulting in a decrease in soil fertility. Enzyme activity is very useful for plant metabolism and is hampered by metals interfering with the activity of soil microorganisms. These toxic effects (directly and indirectly) result in reduced plant growth and ultimately plant death (Kibra 2008). Metal toxicity affects plant growth and development based on the amount of metal accumulation. Metals do not have any beneficial effect on plant growth, such as Pb, Cd, Hg, and As. Low metal concentrations in the growth medium are harmful. Cd reduces wheat shoot and root growth at 5 mg/L in soil, according to Ahmad et al. (2012). Heavy metal ions are essential micronutrients; however, at large amounts and with additional ions like Cd, Hg is hazardous to metabolic activity. The effects of harmful heavy metals on plants have been studied worldwide (Reeves and Baker 2000). Metal contamination in agricultural soil has a serious environmental impact. Toxic metals are soil and water contaminants due to their widespread prevalence and acute and chronic impacts on plants. Increasing mental concentration exposure leads to more toxicity in the cells and tissues (Rizvi and Khan 2018). Metals interfere with ionic homeostasis and activities of enzymes, nutrient uptake, seed germination, growth, photosynthesis, and plant water balance. Metal toxicity causes chlorosis, senescence, leaf rolling and putrefaction, low biomass production, shrivelling and stunted growth, few seeds, and death (Shahid et al. 2017; Shi et al. 2018).

7.3.3 Effects of Metals on Aquatic Animals

Heavy metals have become one of the critical environmental problems all over the world due to biological toxicity, nondegradability, and capability to enter into the food web. Heavy metals tend to react with environmental organic substances and produce even more harmful contaminants in a certain amount, but exceeding the level of these pollutants adversely affects all aquatic animals. Some aquatic organisms directly absorb the heavy metals by their gills; after this, heavy metals directly accumulate on the skin surface, various organs, and tissues of the body. Other aquatic organisms directly uptake the heavy metals through food intake. In some cases, heavy metals are directly absorbed through the animal's subcutaneous layers which are then entered into the body (Long et al. 2016). Metals once enter into aquatic organisms, and then they do not get easily decomposed, metabolized, and excluded from the aquatic organisms and accumulate in the liver and kidney of the aquatic animals (Gaete Olivares et al. 2016). In early development of toxicity of organisms, a high amount of metals also affects overall development of aquatic animals. Metals affect the embryonic stage of fishes through interference with the larvae, growth, biological activities, and chemical structure of the organisms. Metal toxicity affects the development of the central nervous system, genetic development, and endocrine which can lead to the disease and death of the affected organisms (Inesta-Vaquera et al. 2021). Cu, Zn, and Pb are the metals that also affect the hatching rate of early embryos of aquatic animals. Metals also interfere with the function of aquatic organisms and induce immune stimulation effects at high metal accumulation that leads to increased phagocytic activities of the blood cells in organisms (Andre et al. 2022). Continuous exposure and accumulation of metals into the aquatic animals lead to genetic mutations and damage the tissue and organs (Gaete Olivares et al. 2016). High concentration of metal can easily damage the endocrine system of aquatic organisms and also lead to metabolic diseases like hypertension and diabetes where some metalloids like polychlorinated biphenyl directly interfere with hormone synthesis and its secretion (Razmara 2021). Cd and Hg enters into organisms directly competing with other substances through the sulfhydryl of metallothionein (MT) discharged, the harmful substances from the body. It has also been found that aquatic animals secrete high mucus after being affected by metals (Mao et al. 2012).

7.4 Mitigation Measures

Leachate from improperly managed wastes is one of the primary sources of groundwater contamination and a serious concern for mitigating groundwater contamination (Gupta and Bharagava 2021). In developing countries, wastes from homes, agriculture, and industries are disposed in an open system without an appropriate liner. A number of mitigating measures have been proposed in order to prevent toxic metal transmission to groundwater.

7.4.1 Treatment Technology

Several technologies exist for the remediation of heavy metal-contaminated groundwater:

(i) Physicochemical treatment technologies
(ii) Chemical treatment technologies
(iii) Biological treatment technologies

The overall classification has been pictorially presented in Fig. 7.2.

Contamination of groundwater is frequently spread in plumes across wide areas deep under the surface, making traditional remediation procedures challenging to implement. In some cases, chemical treatment methods may be the best alternative. Metal contaminants are reduced into inactive forms using chemicals, which reduce their toxicity and mobility. For this, oxidation, reduction, and neutralization reactions can be utilized (Capodaglio 2020). Table 7.2 summarizes chemical treatment techniques covered in this section. According to Hashim et al. (2011), dithionites decrease redox-sensitive metals like Cr, U, and Th. Dithionites, which convert Fe(III) to Fe(II) in clay minerals, can be injected directly downstream of the polluting plume to give a reduced treatment zone. Thornton and Jackson (1994) studied H_2S gas for in situ immobilization of chromate-contaminated soils, although supplying the gas was proven to be difficult. Nitrogen could transport and regulate H_2S gas during treatment and remove unreacted chemicals from the soil. Iron-based technology to treat contamination of groundwater and soil is highly effective. Iron can reduce redox-sensitive elements like Fe(0) and Fe(II) in lab and field testing (Ludwig et al. 2007). CL:AIRE (2007) observed that reductive precipitation reduced

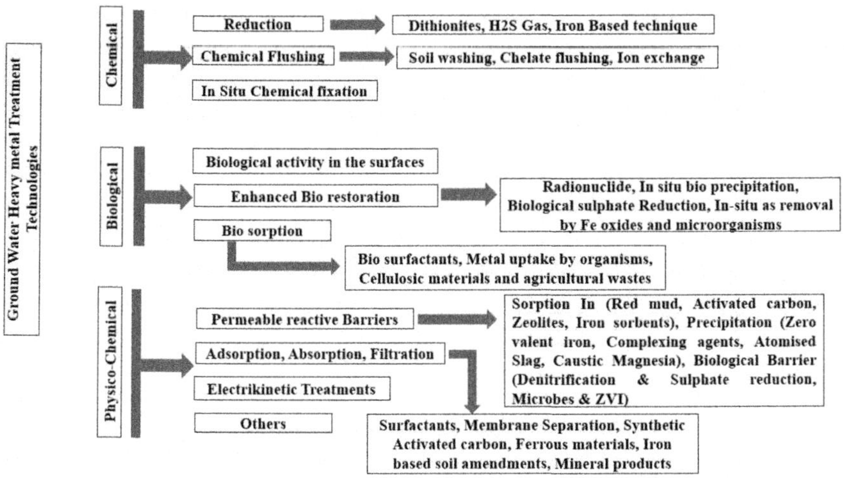

Fig. 7.2 Chemical treatment technologies

Table 7.2 Chemical treatment techniques for water treatment

Treatment technology	Scope	Mechanism and process	References
1. In situ reduction process			
1.1 Reduction by dithionites	Redox-sensitive elements (Cr, U, Th)	Reductive precipitation at alkaline pH	Jemison et al. (2020)
1.2 Reduction by H_2S (g)	Redox-sensitive metals (Cr)	Sulfide oxidized to sulfate and metal is precipitated as hydroxide	Hashim et al. (2011)
2. Soil washing			
2.1 In situ soil Flushing	A wide range of heavy metals, e.g., Cr, Fe, Cu, Co, Al, Mn, Mo, Ni	Desorption of metals at lower pH and recovering of leachate by pump-and-treat system from aquifer	Palansooriya et al. (2020)
2.2 In situ chelate Flushing	Pb, Cd, Cr, Hg, Cu, Zn, Fe, As	Formation of stable chelate complexes between chelate and contaminants	Blue et al. (2008), Hong et al. (2008), Lim et al. (2004), and Warshawsky et al. (2002)
2.3 In situ remediation by selective ion exchange	Heavy metals and transition metals	Liquid-liquid extraction and ion exchange process involving a separate solid phase	Warshawsky et al. (2002)
3. In situ chemical fixation	Pb, As, and other metals in agricultural soil	Stabilization of metals by oxidizing and trapping in the structure	Yang et al. (2007)

Cr(VI) in groundwater from 85,000 mg L^{-1} to 50 mg L^{-1} at a former paper mill site on the Delaware River in the United States. Soil washing is used in the first method as a fractionating process for extracting smaller particles such as clay, silt, or humic compounds, which entrap contaminants in the soil. Dermont et al. (2008) investigated the basic concept, application, benefits, and limitations, as well as methodologies for forecasting and improving the performance of physical and chemical soil washing systems.

When washing, users should think about the type of leaching agent, concentration, pH, time, and solid-to-liquid ratio. To reduce groundwater contamination, future soil washing studies should focus on producing ecologically friendly washing agents, optimizing washing conditions while decreasing economic expenses, and preventing damage of the soil's function, complicated pollution, and the overall soil-ground ecology. Navarro and Martínez (2010) utilized water to flush metals out of an ancient mining region of 0.9 ha that had been contaminated by uncontrolled disposal of base-metal smelting slags. Groundwater geochemical modeling revealed the existence of ferrihydrite, which could have resulted in the mobilization of As, Sb, and Se. Hong et al. (2008) used 100-mM EDTA and 150-psi (10 atm) pressure to extract Pb from 3300 mg kg^{-1} soil. Pociecha and Lestan (2010) used an EDTA solution to extract 67.5% of the Pb from the contaminated soil, generating a washing solution containing 1535-mg Pb per liter and 33.4-mM EDTA. After washing, chelating agents like [S,S]-ethylenediaminedisuccinic acid (EDDS), methylglycinediacetic acid (MGDA), and citric acid were used to remove Cu, Pb, and Zn from contaminated soil samples. Both EDDS and MGDA got rid of Cu, Pb, and Zn in 10 minutes to an hour, but EDDS worked best after 10 days. Di Palma et al. (2005) recovered Cu 93.3% from a contaminated soil using Na_2-EDTA and evaluated percolation speed and chelating agent concentration at pH 7.3. Ion exchange resins have a big advantage over other adsorbents in that they can be effectively regenerated up to 100% of the time (Warshawsky et al. 2002). Lombi et al. (2002) used red mud, lime, and beringite to evaluate heavy metals and metalloids in agricultural soil (a modified aluminosilicate). Yang et al. (2007) studied in situ chemical As fixation with $FeSO_4$, $CaCO_3$, and $KMnO_4$. Due to iron compounds' strong interaction with arsenic and ferric arsenate's poor solubility, $FeSO_4$ was used in fixation solutions. To oxidize any As(III) in the soil samples, KMnO4 was used in two of the treatment solutions.

7.4.1.1 Biological, Biochemical, and Biosorptive Treatment Technologies

Biological treatment methods take advantage of natural biological processes that allow specific plants and microorganisms to assist in metal removal from soil and groundwater. Adsorption, oxidation, reduction, and methylation are used to remove heavy metals from groundwater or subsurface soil (Dixit et al. 2015). According to Boopathy (2000), land farming, composting, bioreactors, oxygen bioventing, biofilters, microbial culture bioaugmentation, and nutrient biostimulation are heavy metal bioremediation techniques (2000). Bioaccumulation, bioleaching, and phytoremediation are some of other processes. Wang and Zhao (2009) studied the viability of using biological approaches to clean up As-contaminated soils and groundwater. Salati et al. (2010) reported a strategy for enhancing maize shoot heavy metal uptake from contaminated soil. According to Yong and Mulligan (2004), the natural attenuation of many heavy metals decreased with time. Kim and Owens (2010) evaluated phytoremediation in polluted areas or landfills using biosolids. Moreno-Jiménez et al. (2011) investigated the phytostabilization of heavy metals such as As, Zn, Cu, Cd, and Al in the Guadiamar River Valley in southern Spain and discovered that *Retama sphaerocarpa*, a native Mediterranean shrub, has promising phytostabilization ability in heavy metal-contaminated soils. In recent years, more bioprocesses for heavy metal bioremediation have been developed. Michalsen et al. (2009) neutralized pH and removed nitrate and radionuclides from groundwater over 21 months. The addition of ethanol stimulated the establishment of a denitrifying community, elevated pH from 4.7 to 6.9, facilitated the elimination of 116-mM nitrate, and immobilized 94% of total U. (VI). Groudev et al. (2010) used the indigenous soil microflora to treat experimental plots extensively contaminated with radionuclides (mostly U and Ra) and nonferrous heavy metals (mostly Zn, Cd, Pb, and Cu). Heavy metals in groundwater are immobilized as precipitates (mainly sulfides) by in situ bioprecipitation (ISBP). The ISBP technique was found to be an effective strategy for improving groundwater quality for heavy metal stabilization, including Cu, Zn, Cd, Ni, Co, Fe, Cr, and As. Sulfate-reducing bacteria (SRB) catalyze the reduction of sulfate to sulfide by using sulfate as an electron acceptor (Gibson 1990). Sen Gupta et al. (2009) successfully used this idea in the field by putting a set amount of oxygenated water (DO >4 mg L^{-1}) back into the aquifer to make an oxidized zone. This started to turn the process of bacteria reducing arsenic reverse without using chemicals. This field of remediation technology is still in its early stages of development, with limited field applicability. Experiments with different biosorbents yielded positive results. Biosorption has a variety of advantages over traditional treatment procedures, including low cost, minimal biochemical sludge, biosorbent regeneration, high efficiency, and recovery of metal. Surfactin, rhamnolipids, and sophorolipids are biosurfactants that can remove Zn, Cu, Cd, and Ni from polluted soil (Mulligan et al. 1999a, 1999b; Wang and Mulligan 2004). Rangsayatorn et al. (2002) studied the ability of *Spirulina platensis* TISTR 8217 to remove Cd from water (>100 mg L-1). According to Pandey et al. (2008), *Calotropis procera*, a wild perennial plant, has a high Cd(II) absorption capability between pH 5.0 and 8.0. Regardless of the Cd ion concentration, the adsorption equilibrium of 90% elimination was reached in less than 5 minutes. The biological, biochemical, and biosorption treatment processes are summarized in Table 7.3.

7.4.1.2 Physicochemical Treatment Technologies

This section describes civil barrier construction, physical absorption, mass transfer, and harnessed biochemical processes. To manage the contamination problem, two or more processes are usually connected together. Table 7.4 shows the physicochemical treatment technologies.

According to the USEPA (1989), a permeable reactive barrier (PRB) is "a subsurface emplacement of reactive media designed to intercept a contaminated

Table 7.3 The biological, biochemical, and biosorption treatment processes

Treatment technology	Scope	Mechanism and process	References
1. Biological activity in the subsurface	Cr, Co, Cd, Ni, Zn, Pb, Cu	Oxidation, precipitation, bioaccumulation	Baker (1995), Salati et al. (2010), and Yong and Mulligan (2004)
2. Enhanced biorestoration			
2.1. Immobilization of radionuclides by microorganisms	U, Ra, Tc	Reduction, agglomeration, absorption of U(IV) into sediments	Anderson et al. (2003)
2.2. BSR	Divalent metal cations	Reduction of sulfate to metal sulfide ppts, catalyzed by the activity of SRB	Gibert et al. (2002) and Waybrant et al. (1998)
2.3. In situ As removal by ferrous oxides and microorganisms	As, Fe, Mn	Oxidation of Fe(II) and As (III) by elevating Eh and boosting microbial growth and then co-precipitating As, Fe, and Mn	Camacho et al. (2011), Katsoyiannis and Zouboulis (2004), Leupin and Hug (2005), and Sen Gupta et al. (2009)
3. Biosorption of heavy metals			
3.1. Biosurfactants	Cd, Zn, Ni	Bioadsortion through metal complex forming with surfactants due to lowering of interfacial tension	AsçI et al. (2010), Mulligan and Wang (2006), and Ron and Rosenberg (2001)
3.2. Uptake by organisms	Cd, Cr, Zn, As, Fe, Ni	Bacteria, fungus, plants, and DNA aptamers uptook metals in cell cytoplasm or stabilized them	Pandey et al. (2008), Prakasham et al. (1999), and Srivastava et al. (2011)

plume, provide a flow path through the reactive media, and transform the contaminant(s) into environmentally acceptable forms to achieve remediation concentration goals downgradient of the barrier." PRBs are made to be more permeable than the aquifer materials that surround them, allowing water to easily flow through while contaminants are removed and maintaining groundwater hydrogeology (Yin and Allen 1999). According to the separation process used, permeable barrier technologies are classified into three groups by Scherer et al. (2000) as sorption (oxides, zeolites, etc.), chemical reaction (minerals, zero-valent metals, etc.), and biological treatment (organic materials). Many different types of adsorbents and filtering processes mentioned in this study are utilized to treat contaminated water or wastewater, but not in situ groundwater. Most of them can be utilized to generate a reactive zone in the PRB technique. Surfactants initiate a number of good processes that help get rid of contaminants, such as making things more soluble, lowering surface tension, dissolving particles, making things wet, and making foam (Mulligan et al. 2001). Doong et al. (1998) found that complexing agents and surfactants perform together to remove Pb, Zn, and Cd from soil. Membrane technology can

Table 7.4 Physicochemical treatment technologies

Technology	Scope	Mechanism and process	References
1. Permeable reactive barriers			
1.1. Red mud	Pb, As, Cd, Zn	Sorption of metal cations in the channels of negatively charged cancrinite framework	Apak et al. (1998), Gupta and Sharma (2002)
1.2. Activated carbon	Cr, Cd, and other heavy metals	Adsorption by high surface area (about 1000 $m^2 g^{-1}$) and presence of surface functional groups	Fine et al. (2005), Huttenloch et al. (2001), and Thiruvenkatachari et al. (2008)
1.3. Zeolites (clinoptilolite, chabazite-phillipsite, clinoptilolite, fly ash zeolites)	Cd, Cu, Ni, Cr, As	Adsorption, ion exchange, catalytic and molecular sieving through 3D aluminosilicate structure	Roehl et al. (2005), Ruggieri et al. (2008), and Xenidis et al. (2010)
2. Adsorption, filtration, and absorption mechanisms			
2.1. Absorption by using inorganic surfactants	Cd, Pb, Zn, As, Cd, Cu, Ni	Metal sorption depending on charge of surfactant	Mulligan et al. (2001) and Scherer et al. (2000)
2.2. Membrane and filtration technology	Cu, Cd, Pb, Cr, Hg, Pb, Zn, U, Tc, As	All the membranes and filters have separate mechanisms, e.g., electrostatic capture, complexation, dialysis, micellar capture in 3-D structure	Hsieh et al. (2008), Sang et al. (2008), and Sikdar et al. (1998)
2.3. Use of ferrous materials as adsorbents	As(V), Cr, Hg, Cu, Cd, Pb	Sorption by Fe oxides, oxyhydroxides, and sulfides and microbe-mediated reactions involving Fe as an electron acceptor	Chowdhury and Yanful (2010), Rao and Karthikeyan (2007), Ruiping et al. (2009), Smedley and Kinniburgh (2002), and Sylvester et al. (2007)
3. Electrokinetic remediation	As, Cd, Cr, Co, Hg, Ni, Mn, Mo, Zn, Sb, Pb	Process involves electro-osmosis, electromigration, and electrophoresis	Colacicco et al. 2010; Giannis et al. 2007; Scullion 2006; Virkutyte et al. 2002

be combined with soil washing or flushing procedures to concentrate contaminants in the wash liquid, allowing the raffinate to be treated appropriately. Membranes can be electrodialytic membranes, liquid membranes, polymer membranes, ultrafiltration membranes, nanofiber membranes, and many other varieties. According to Sikdar et al. (1998), an effective membrane technology should decrease contaminated water volume while producing clean water that meets effluent standards. The Palmerton zinc superfund site in Palmerton, Pennsylvania, used this method to remove suspended Zn particles with greater than 99% efficiency. Ultrafiltration may be used to remove dissolved metal ions from the filtrate (James and Stacy 1993). Sang et al. (2008) used a nanofiber membrane named M^{-1} to remove Cu^{2+}, Pb^{2+}, and Cd^{2+} from simulated groundwater. Activated carbon absorbed 2860 mg g-1 As (Rajakovic 1992). Gu et al. (2005) developed iron-containing granular activated carbon as an adsorbent (GAC). Singh et al. (2008) chemically treated tamarind wood to make activated carbon. This material has a BET surface area of 612 $m^2\ g^{-1}$ and a total pore volume of 0.508 $cm^3\ g^{-1}$. It was tested for Pb(II) adsorption from a dilute aqueous solution with an initial concentration of 40 mg L^{-1}, a concentration of 3 g L^{-1}, and a pH of 6.5. The maximum removal rate was 97.95% (experimentally) and 134.22 mg g^{-1} (Langmuir isotherm model). Heavy metals can be absorbed or immobilized by Fe oxides, oxyhydroxides, and sulfides, according to Contin et al. (2007), Heal et al. (2003), and Kumpiene et al. (2006). Most iron-based treatments remove As (V) more effectively than As (III). As (III) can be oxidized to As (V) (Rao and Karthikeyan 2007).

7.4.2 Safe Water Storage and Supply

More than two-thirds of the Earth's surface is covered in water. Only 2.7% of the world's available freshwater is accessible, and only1% of that (in lakes, rivers, and groundwater) is drinkable (Glavan 2018). Water is intimately connected to every aspect of human life, whether directly or indirectly. Everyone needs safe water for drinking, cooking, personal hygiene, and sanitation without sacrificing health or dignity. Therefore, humans have a basic right to reliable and safe (fresh and clean) water (Samra and Fawzi 2011). Unsafe drinking water is a problem for 1.1 billion people worldwide, according to a UNDP report (UNDP 2015). In some countries, especially Africa, nearly half the population lacks adequate drinking water, leading to poor health (Davis 2013). For safe drinking water, there are basic standards, norms, criteria, and indications. Bos et al. (2016) have clearly defined these concepts. Water laws are necessary for the provision of sufficient quantities of safe, accessible, acceptable, economical, and dependable drinking water. Water source; water treatment, distribution, and consumption; wastewater; and gray water are all regulated by drinking water regulations. A sufficient, clean, and safe drinking water supply must be provided for a different type of users as a criterion (Bos et al. 2016). The principle of sustainability, system robustness, and resilience includes the concept of reliability, which refers to the continuity of service provision for current and future generations. Acceptability relates to the aesthetic value of water, which includes its sight, taste, and odor. Water resources are intertwined with other industries; therefore, the world's long-term development is heavily reliant on them. It requires the implementation in all three aspects of long-term growth (environmental, economic, and social) (UNESCO 2015). According to Berg and Danilenko, WSS has encountered various global difficulties in the twenty-first century (Berg and Danilenko 2011). Population growth, climate change, socioenvironmental issues, limited water, economic problems, and aging are all big challenges. Aging causes low pressure and loss of water quality degradation (Alegre et al. 2006). Worldwide primary challenge is providing safe water storage and sanitation (Moe and Rheingans 2006). In many communities, especially in developing countries, water storage is still necessary for home. Water storage, which is commonly done in tanks/vessels, and related user activities are regarded to be sources of pollution. Several studies have examined this phenomenon, albeit in isolation (Manga et al. 2021). Water storage has long been recognized as a source of domestic water contamination, as it is a critical component of the indirect cold water delivery system and many other un-piped water supply systems. Many challenges face household water storage, all of which inevitably degrade water quality (Nnaji et al. 2019). Several studies, mostly from developing countries, have examined the impact of water storage on water quality. Schafer (2010), Ziadat (2005), Mohanan et al. (2017), and Douhri et al. (2015) studied the effects of storage material on water quality, whereas Holt (2005) and Agensi et al. (2019) studied the effects of user practices on stored home water quality. It is vital to mention that in comparison to other tank materials, plastic tanks were widely used in various regions. Tank material was discovered to be a primary source of water contamination. Microbiological contamination occurred at different rates based on the type of water tank (Al-Bahry et al. 2013). Akuffo et al. (2013) reported that water storage tank materials increase the pollutants' contact with construction/production or inside coatings. Temperature affects water quality as different materials have varied thermal conductivities; for example, steel has a higher thermal conductivity and cools faster than plastic under the same weather conditions (Schafer 2010; Schafer and Mihelcic 2012); Akuffo et al. 2013, Al-Bahry et al. 2013). Schafer and Mihelcic (2012) reported that polyethylene tank temperatures are higher than fiberglass and fiber cement materials. Polyethylene tanks have higher microbial activity than other tank materials. Mohanan et al. (2017) found that brass, copper, and clay water storage containers have more antimicrobial than steel, plastic, and aluminum jars. Water quality is affected by tank size and capacity by extending the retention or residency time. Ogbozige et al. (2018) analyzed water quality in different containers during storage and reported that all container materials can store water for up to 21 days, except clay pots, which can only hold water for 6 days. Uncoated steel tanks weren't permitted. Black plastic containers protect water quality better than other colored containers during storage. Large tanks allow for longer water storage, which can increase pollution and chlorine volatilization (Graham and VanDerslice 2007). The color of the tanks also affects the temperature of the water in storage tanks as different colors absorb heat at different rates and change the temperature. Heat is absorbed more effectively by darker colors than by lighter color water storage tanks. According to Schafer (2010), black polyethylene tanks are prevalent in Bolivia and revealed that tank design has an indirect effect on water quality through affecting user practices.

7.5 Conclusion

Metal contamination of water caused by both natural and anthropogenic sources is a health issue. Because billions of people utilize contaminated groundwater for drinking, groundwater remediation is a top priority. Groundwater treatment processes have been evaluated and categorized in this work into three broad categories: physicochemical, biological, and chemical. Water treatment methods have progressed significantly with more advancement to ensure removal of metal and other contaminants and supply of safe drinking water for environmental and health safety. Numerous technologies have been reviewed from ex situ physical separation to in situ microbiological and adsorption. As water is a precious resource for life and environmental sustainability on the Earth planet, it is essential to upkeep water resources with natural quality of water by preventing metal contamination and conserving its whole entities by sustainable utilization. Accessing clean and safe drinking water for the growing human population could be achieved by implementing more sustainable water treatment technologies. Improving access to safe water in underdeveloped nations necessitates the supply of high-quality education and the formation of effective governance for more awareness and water resource management through best practices of recycling, reuse, and safe storage.

Acknowledgment Thanks are due to the Department of Chemistry, Environmental Science Discipline, Manipal University Jaipur, India, for providing facilities.

References

Agensi A, Tibyangye J, Tamale A, Agwu E, Amongi C (2019) Contamination potentials of household water handling and storage practices in Kirundo subcounty, Kisoro District, Uganda. J Environ Public Health 2019:1–8

Ahmad I, Akhtar MJ, Zahir ZA, Jamil A (2012) Effect of cadmium on seed germination and seedling growth of four wheat (*Triticum aestivum* L.) cultivars. Pak J Bot 44:1569–1574

Ahmed KM, Bhattacharya P, Hasan MA, Akhter SH, Alam SM, Bhuyian MH, Sracek O (2004) Arsenic enrichment in groundwater of the alluvial aquifers in Bangladesh: an overview. Appl Geochem 19:181–200

Aihemaiti A, Jiang J, Li D, Liu N, Yang M, Meng Y, Zou Q (2018) The interactions of metal concentrations and soil properties on toxic metal accumulation of native plants in vanadium mining area. J Environ Manag 222:216–226

Akuffo I, Cobbina S, Alhassan E, Nkoom M (2013) Assessment of the quality of water before and after storage in the Nyankpala community of the Tolon-Kumbungu district, Ghana. Int J Sci Technol Res 2:221–227

Al-Bahry S, Al-Hinai J, Mahmoud I, Al-Musharafi S (2013) Opportunistic and microbial pathogens in municipal water distribution systems. APCBEE Proc 5:339–343

Alegre H, Baptista JM, Cabrera EJ, Cubillo F, Duarte P, Hirner W, Merkel W, Parena R (2006) Performance indicators for water supply services. IWA, London

Amalia WR, Hariyadi S, Anwar S (2018) Heavy metal contamination as water quality test of cengkareng drain estuary. In proceeding of international symposium for sustainable humanosphere, pp 93–103

Amari T, Ghnaya T, Abdelly C (2017) Nickel, cadmium, and lead phytotoxicity and potential of halophytic plants in heavy metal extraction. South Afr J Bot 111:99–110

Anderson RT, Vrionis HA, Ortiz-Bernad I, Resch CT, Long PE, Dayvault R, Karp K, Marutzky S, Metzler DR, Peacock A, White DC, Lowe M, Lovley DR (2003) Stimulating the in-situ activity of geobacter species to remove uranium from the groundwater of a uranium-contaminated aquifer. Appl Environ Microbiol 69:5884–5891

Andre C, Pilote M, Gagnon C, Gagne F (2022) Ecotoxicological impacts of oil sand mining activity to endemic caged mussels Pyganodon grandis. Comp Biochem Physiol C Toxicol Pharmacol 251:109–193

Anzecc A (2000) Australian and New Zealand guidelines for fresh and marine water quality. Australian and New Zealand Environment and Conservation Council and Agriculture and Resource Management Council of Australia and New Zealand, Canberra, pp 1–103

Apak R, Tutem E, Hugul M, Hizal J (1998) Heavy metal cation retention by unconventional sorbents (red muds and fly ashes). Water Res 32:430–440

AsçI Y, Nurbas M, Sag AçIkel Y (2010) Investigation of sorption/desorption equilibria of heavy metal ions on/from quartz using rhamnolipid biosurfactant. J Environ Manag 91:724–731

Baker AJM (1995) Metal hyperaccumulation by plants: our present knowledge of Ecophysiological phenomenon. Abstract book: fourteenth annual symposium in current topics in plant biochemistry. Physiology and Molecular Biology, p. 7

Bazrafshan E, Mohammadi L, Ansari-Moghaddam A (2015) Heavy metals removal from aqueous environments by electrocoagulation process—a systematic review. J Environ Health Sci Eng 13:74

Berg C, Danilenko A (2011) The IBNET water supply and sanitation performance blue book, the international benchmarking network for water and sanitation utilities data book, water and sanitation program. The World Bank, Washington DC, p 58849

Blue LY, Van Aelstyn MA, Matlock M, Atwood DA (2008) Low-level mercury removal from groundwater using a synthetic chelating ligand. Water Res 42:2025–2028

Bodaghpour S, Biglarijoo N, Ahmadi S (2012) A review on the existence of chrome in cement and environmental remedies to control its effects. Int J Geol 2:62–67

Boopathy R (2000) Factors limiting bioremediation technologies. Biores Technol 74:63–67

Bos R, Alves D, Latorre C, Macleod N, Payen G, Roaf V, Rouse M (2016) Manual on the human rights to safe drinking water and sanitation for practitioners. IWA Publishing, London

Bradl H (2002) Heavy metals in the environment: origin, interaction and remediation, vol 6. Academic Press, London

Camacho LM, Gutiérrez M, Alarcón-Herrera MT, Villalba MDL, Deng S (2011) Occurrence and treatment of arsenic in groundwater and soil in northern Mexico and southwestern USA. Chemosphere 83:211–225

Capodaglio AG (2020) Critical perspective on advanced treatment processes for water and wastewater AOPs, ARPs, and AORPs. Appl Sci 10:4549

Cheng P, Zhang S, Wang Q, Feng X, Zhang S, Sun Y, Wang F (2021) Contribution of nano-zero-valent iron and arbuscular mycorrhizal fungi to phytoremediation of heavy metal-contaminated soil. Nanomaterials 11:1264

Chowdhury SR, Yanful EK (2010) Arsenic and chromium removal by mixed magnetite-maghemite nanoparticles and the effect of phosphate on removal. J Environ Manag 91:2238–2247

CL:AIRE (2007) Treatment of Chromium contamination and chromium ore processing residue. Technical Bulletin (TB 14). Contaminated Land: Applications in Real Environment. http://www.claire.co.uk. London, UK

Colacicco A, De Gioannis G, Muntoni A, Pettinao E, Polettini A, Pomi R (2010) Enhanced electrokinetic treatment of marine sediments contaminated by heavy metals and PAHs. Chemosphere 81:46–56

Contin M, Mondini C, Leita L, Nobili MD (2007) Enhanced soil toxic metal fixation in iron (hydr)oxides by redox cycles. Geoderma 140:164–175

Davis ML (2013) Water and wastewater engineering: design principles and practice. McGraw-Hill Education, New York

Dermont G, Bergeron M, Mercier G, Richer-Laflèche M (2008) Soil washing for metal removal: a review of physical/chemical technologies and field applications. J Hazard Mat 152:1–31
Di Palma L, Ferrantelli P, Medici F (2005) Heavy metals extraction from contaminated soil: recovery of the flushing solution. J Environ Manag 7:205–211
Dixit R, Wasiullah, Malaviya D, Pandiyan K, Singh UB, Sahu A, Shukla R, Singh BP, Rai JP, Sharma PK, Lade H, Paul D (2015) Bioremediation of heavy metals from soil and aquatic environment: an overview of principles and criteria of fundamental processes. Sustainability 7 (2):2189–2212. https://doi.org/10.3390/su7022189
Djingova R, Kuleff I (2000) Instrumental techniques for trace analysis. In: Vernet JP (ed) Trace elements: their distribution and effects in the environment. Elsevier, Amsterdam
Doong RA, Wu YW, Lei WG (1998) Surfactant enhanced remediation of cadmium contaminated soils. Water Sci Technol 37:65–71
Douhri H, Raissouni I, Tazi S, Badia D (2015) Effect of house storage on water's quality in rural areas of Tangier-Tetuan Region (Morocco). Larhyss J 24:301–314
Duruibe JO, Ogwuegbu MOC, Egwurugwu JN (2007) Heavy metal pollution and human biotoxic effects. Int J Phys Sci 2:112–118
Fine P, Scagnossi A, Chen Y, Mingelgrin U (2005) Practical and mechanistic aspects of the removal of cadmium from aqueous systems using peat. Environ Pollut 138:358–367
Gaete Olivares H, Moyano Lagos N, Jara Gutierrez C, Carrasco Kittelsen R, Lobos Valenzuela G, Hidalgo Lillo ME (2016) Assessment oxidative stress biomarkers and metal bioaccumulation in macroalgae from coastal areas with mining activities in Chile. Environ Monit Assess 188:1–11
Giannis A, Gidarakos E, Skouta A (2007) Application of sodium dodecyl sulfate and humic acid as surfactants on electrokinetic remediation of cadmium contaminated soil. Desalination 211:249–260
Gibert O, de Pablo J, Cortina JL, Ayora C (2002) Treatment of acid mine drainage by sulphate-reducing bacteria using permeable reactive barriers: a review from laboratory to full-scale experiments. Rev Environ Sci Biotechnol 1:327–333
Gibson GR (1990) Physiology and ecology of the sulphate-reducing bacteria. J Appl Microbiol 69: 769–797
Glavan M (ed) (2018) Water challenges of an urbanizing world. InTech
Gowd SS, Govil PK (2008) Distribution of heavy metals in surface water of Ranipet industrial area in Tamil Nadu, India. Environ Monit Assess 136:197–207
Graham J, VanDerslice J (2007) The effectiveness of large household water storage tanks for protecting the quality of drinking water. J Water Health 5:307–313
Groudev S, Spasova I, Nicolova M, Georgiev P (2010) In situ bioremediation of contaminated soils in uranium deposits. Hydrometallurgy 104:518–523
Gu Z, Fang J, Deng B (2005) Preparation and evaluation of GAC-based iron-containing adsorbents for arsenic removal. Environ Sci Technol 39:3833–3843
Gupta PK, Bharagava RN (2021) Mitigation of groundwater pollution: heavy metal retention characteristics of fly ash based liner materials. In: Fate and transport of subsurface pollutants, Microorganisms for sustainability, vol 24. Springer, Singapore, pp 79–104
Gupta VK, Sharma S (2002) Removal of cadmium and zinc from aqueous solutions using red mud. Environ Sci Technol 36:3612–3617
Gurung SB, Geronimo FK, Hong J, Kim LH (2018) Application of indices to evaluate LID facilities for sediment and heavy metal removal. Chemosphere 206:693–700
Hashim MA, Mukhopadhyay S, Sahu JN, Sengupta B (2011) Remediation technologies for heavy metal contaminated groundwater. J Environ Manag 92:2355–2388
Heal K, Younger PL, Smith K, Glendinning S, Quinn P, Dobbie K (2003) Novel use of ochre from mine water treatment plants to reduce point and diffuse phosphorus pollution. Land Contam Reclam 11:145–152
Holt S (2005) A survey of water storage practices and beliefs in households in Bonao, Dominican Republic in 2005. Springer, Berlin

Hong PKA, Cai X, Cha Z (2008) Pressure-assisted chelation extraction of lead from contaminated soil. Environ Pollut 153:14–21

Hong YJ, Liao W, Yan ZF, Bai YC, Feng CL, Xu ZX, Xu DY (2020) Progress in the research of the toxicity effect mechanisms of heavy metals on freshwater organisms and their water quality criteria in China. J Chem 2020:9010348

Hsieh LHC, Weng YH, Huang CP, Li KC (2008) Removal of arsenic from groundwater by electro-ultrafiltration. Desalination 234:402–408

Huat TJ, Camats-Perna J, Newcombe EA, Valmas N, Kitazawa M, Medeiros R (2019) Metal toxicity links to Alzheimer's disease and neuroinflammation. J Mol Biol 431:1843–1868

Huttenloch P, Roehl KE, Czurda K (2001) Sorption of nonpolar aromatic contaminants by chlorosilane surface modified natural minerals. Environ Sci Technol 35:4260–4264

Inesta-Vaquera F, Navasumrit P, Henderson CJ, Frangova TG, Honda T, Dinkova-Kostova AT, Wolf CR (2021) Application of the in vivo oxidative stress reporter Hmox1 as mechanistic biomarker of arsenic toxicity. Environ Pollut 270:116053

Islam MS, Ahmed MK, Raknuzzaman M, Habibullah-Al-Mamun M, Masunaga S (2015) Metal speciation in sediment and their bioaccumulation in fish species of three urban rivers in Bangladesh. Arch Environ Contam Toxicol 68:92–106

James SV, Stacy GL (1993) Overview of separation technologies for EPA's superfund innovative technology evaluation program. Gulf Publishing, Houston, TX

Jemison N, Reimus P, Harris R, Boukhalfa H, Clay J, Chamberlain K (2020) Reduction and potential remediation of U (VI) by dithionite at an in-situ recovery mine: insights gained by δ238U. Appl Geochem 115:104560

Johnson DB, Hallberg KB (2005) Acid mine drainage remediation options: a review. Sci Total Environ 338(1-2):3–14. https://doi.org/10.1016/j.scitotenv.2004.09.002

Jortner BS (2008) Effect of stress at dosing on organophosphate and heavy metal toxicity. Toxicol Appl Pharmacol 233:162–167

Karimian S, Shekoohiyan S, Moussavi G (2021) Health and ecological risk assessment and simulation of heavy metal-contaminated soil of Tehran landfill. RSC Adv 11:8080–8095

Katsoyiannis IA, Zouboulis AI (2004) Application of biological processes for the removal of arsenic from groundwaters. Water Res 38:17–26

Khalil N, Sinha R, Raghav AK, Mittal AK (2008) UASB technology for sewage treatment in India: experience, economic evaluation and its potential in other developing countries. In Twelfth international water technology conference, vol 12, pp 1411–1427

Khan I, Ghani A, Rehman AU, Awan SA, Noreen A, Khalid I (2016) Comparative analysis of heavy metal profile of *Brassica campestris* (L.) and *Raphanus sativus* (L.) irrigated with municipal wastewater of Sargodha city. J Clin Toxicol 6:1–4

Khan K, Lu Y, Khan H, Zakir S, Khan S, Khan AA, Wei L, Wang T (2013a) Health risks associated with heavy metals in the drinking water of Swat, northern Pakistan. J Environ Sci 25:2003

Khan S, Shahnaz M, Jehan N, Rehman S, Shah MT, Din I (2013b) Drinking water quality and human health risk in Charsadda district. Pak J Clean Prod 60:93–101

Kibra MG (2008) Effects of mercury on some growth parameters of rice (*Oryza sativa* L.). Soil Environ 27:23–28

Kim KR, Owens G (2010) Potential for enhanced phytoremediation of landfills using biosolids - a review. J Environ Manag 91:791–797

Krishna AK, Satyanarayanan M, Govil PK (2009) Assessment of heavy metal pollution in water using multivariate statistical techniques in an industrial area: a case study from Patancheru, Medak District, Andhra Pradesh, India. J Hazard Mater 167:366–373

Kumar Reddy DH, Lee SM (2012) Water pollution and treatment technologies. J Environ Anal Toxicol 2:103

Kumpiene J, Ore S, Renella G, Mench M, Lagerkvist A, Maurice C (2006) Assessment of zerovalent iron for stabilization of chromium, copper, and arsenic in soil. Environ Pollut 144: 62–69

Leupin OX, Hug SJ (2005) Oxidation and removal of arsenic (III) from aerated groundwater by filtration through sand and zero-valent iron. Water Res 39:1729–1740

Lim TT, Tay JH, Wang JY (2004) Chelating-agent-enhanced heavy metal extraction from a contaminated acidic soil. J Environ Eng 130:59–66

Lin H, Liu C, Li B, Dong Y (2021) *Trifolium repens* L. regulated phytoremediation of heavy metal contaminated soil by promoting soil enzyme activities and beneficial rhizosphere associated microorganisms. J Hazard Mat 402:123829

Lin WC, Li Z, Burns MA (2017) A drinking water sensor for lead and other heavy metals. Anal Chem 89:8748–8756

Lombi E, Zhao FJ, Zhang G, Sun B, Fitz W, Zhang H, McGrath SP (2002) In situ fixation of metals in soils using bauxite residue: chemical assessment. Environ Pollut 118:435–443

Long Y, Luo Y, Xiao J, Guo Z, Chen L, Xiao Y (2016) Research advances in effects of heavy metals stress on fish. J Southern Agric 47:1608–1614

Ludwig RD, Su C, Lee TR, Wilkin RT, Acree SD, Ross RR, Keeley A (2007) In situ chemical reduction of Cr(VI) in groundwater using a combination of ferrous sulfate and sodium dithionite: a field investigation. Environ Sci Technol 41:5299–5305

Malassa H, Al-Qutob M, Al-Khatib M, Al-Rimawi F (2013) Determination of different trace heavy metals in groundwater of southwest bank/Palestine by ICP/MS. J Environ Pollut 4:818–827

Mandal BK, Ogra Y, Suzuki KT (2003) Speciation of arsenic in human nail and hair from arsenic-affected area by HPLC-inductively coupled argon plasma mass spectrometry. Toxicol Appl Pharmacol 189:73–83

Mandour RA, Azab YA (2011) The prospective toxic effects of some heavy metals overload in surface drinking water of Dakahlia Governorate, Egypt. Int J Occup Environ Med 4:245–253

Manga M, Ngobi TG, Okeny L (2021) The effect of household storage tanks/vessels and user practices on the quality of water: a systematic review of literature. Environ Syst Res 10:18

Mao H, Tan FQ, Wang DH, Zhu JQ, Zhou H, Yang WX (2012) Expression and function analysis of metallothionein in the testis of stone crab Charybdis japonica exposed to cadmium. Aqu Toxicol 124:11–21

Maric M, Antonijevic M, Alagic S (2013) The investigation of the possibility for using some wild and cultivated plants as hyperaccumulators of heavy metals from contaminated soil. Environ Sci Pollut Res 20:1181–1188

Martinez Cortijo J, Ruiz-Canales A (2018) Effect of heavy metals on rice irrigated fields with wastewater in high pH Mediterranean soils: the particular case of the Valencia area in Spain. Agric Water Manag 210:108–123

Michalsen MM, Peacock AD, Smithgal AN, White DC, Spain AM, SanchezRosario Y, Krumholz LR, Kelly SD, Kemner KM, McKinley J, Heald SM, Bogle MA, Watson DB, Istok JD (2009) Treatment of nitric acid, U(VI)-, and Tc(VII)-contaminated groundwater in intermediate-scale physical models of an in situ biobarrier. Environ Sci Technol 43:1952–1961

Mishra S, Dubey RS (2006) Heavy metal uptake and detoxification mechanisms in plants. Int J Agri Res 1:122–141

Moe CL, Rheingans RD (2006) Global challenges in water, sanitation and health. J Water Health 4: 41–57

Mohanan N, Manju E, Sonia J (2017) The effect of different types of storage vessels on water quality. Int J Innov Res Sci Eng Technol 6:20362–20368

Monni S, Salemaa M, Millar N (2000) The tolerance of Empetrum nigrum to copper and nickel. Environ Pollution 109:221–229

Mor S, Ravindra K, Dahiya RP, Chandra A (2006a) Leachate characterization and assessment of groundwater pollution near municipal solid waste landfill site. Environ Monit Assess 118:435–456

Mor S, Ravindra K, Visscher AD, Dahiya RP, Chandra A (2006b) Municipal solid waste characterization and its assessment for potential methane generation: a case study. Sci Total Environ 371:1–10

Moreno-Jiménez E, Vázquez S, Carpena-Ruiz RO, Esteban E, Peñalosa JM (2011) Using Mediterranean shrubs for the phytoremediation of a soil impacted by pyritic wastes in southern Spain: a field experiment. J Environ Manag 92:1584–1590

Mudhoo A, Sharma SK, Garg VK, Tseng CH (2011) Arsenic: an overview of applications, health, and environmental concerns and removal processes. Criti Rev Environ Sci Technol 41:435–519

Mulligan CN, Wang S (2006) Remediation of a heavy metal-contaminated soil by a rhamnolipid foam. Eng Geol 85:75–81

Mulligan CN, Yong RN, Gibbs BF (1999a) On the use of bio surfactants for the removal of heavy metals from oil-contaminated soil. Environ Prog 18:50–54

Mulligan CN, Yong RN, Gibbs BF (1999b) Removal of heavy metals from contaminated soil and sediments using the bio surfactant. J Soil Contam 8:231–254

Mulligan CN, Yong RN, Gibbs BF (2001) Remediation technologies for metal contaminated soils and groundwater: an evaluation. Eng Geol 60:193–207

Muradoglu F, Gundogdu M, Ercisli S, Encu T, Balta F, Jaafar HZE, Zia-Ul-Haq M (2015) Cadmium toxicity affects chlorophyll a and b content, antioxidant enzyme activities and mineral nutrient accumulation in strawberry. Biol Res 48:1–7

Mustafa G, Komatsu S (2016) Toxicity of heavy metals and metal-containing nanoparticles on plants. Biochim Biophys Acta Protein Proteomics 8:932–944

Muszynska E, Labudda M, Rozanska E, Hanus-Fajerska E, Znojek E (2018) Heavy metal tolerance in contrasting ecotypes of *Alyssum montanum*. Ecotoxicol Environ Saf 161:305–317

Nagajyoti PC, Lee KD, Sreekanth TVM (2010) Heavy metals, occurrence and toxicity for plants: a review. Environ Chem Lett 8:199–216

Navarro A, Martínez F (2010) The use of soil-flushing to remediate metal contamination in a smelting slag dumping area: column and pilot-scale experiments. Eng Geol 115:16–27

Nnaji C, Nnaji I, Ekwule R (2019) Storage-induced deterioration of domestic water quality. J Water Sanit Hyg Dev 9:329–337

Oelofse SHH, Hobbs PJ, Rascher J, Cobbing JE (2007) The pollution and destruction threat of gold mining waste on the Witwatersrand – A West Rand case study. Symposium on environmental issues and waste management in energy and mineral production (SWEMP 2007), 11–13 December, Bangkok

Ogbozige F, Ibrahim F, Adie D (2018) Effect of storage time and container material on potable water quality. Life J Sci Technol 1:59–71

Ogunlade MO, Agbeniyi SO (2011) Impact of pesticides use on heavy metals pollution in cocoa soils of Cross-River State, Nigeria. Afr J Agri Res 6:3725–3728

Oves M, Khan MS, Zaidi A, Ahmad E (2012) Soil contamination, nutritive value, and human health risk assessment of heavy metals: an overview. In: Toxicity of heavy metals to legumes and bioremediation. Springer, Vienna, pp 1–27

Palansooriya KN, Shaheen SM, Chen SS, Tsang DCW, Hashimoto Y, Hou D, Bolan NS, Rinklebe J, Ok YS (2020) Soil amendments for immobilization of potentially toxic elements in contaminated soils: a critical review. Environ Int 134:105046

Pandey J, Pandey U, Singh AV (2014) Impact of changing atmospheric deposition chemistry on carbon and nutrient loading to Ganga River: integrating land–atmosphere–water components to uncover cross-domain carbon linkages. Biogeochemistry 119:179–198

Pandey PK, Verma Y, Choubey S, Pandey M, Chandrasekhar K (2008) Biosorptive removal of cadmium from contaminated groundwater and industrial effluents. Bioresour Technol 99:4420–4427

Pendergast MM, Hoek EMV (2011) A review of water treatment membrane nanotechnologies. Energy Environ Sci 4:1946–1971

Peralta-Videa JR, Lopez ML, Narayan M, Saupe G, Gardea-Torresdey J (2009) The biochemistry of environmental heavy metal uptake by plants: implications for the food chain. Int J Biochem Cell Biol 41:1665–1677

Permanawati Y, Zuraida R, Ibrahim A (2013) Heavy metal content (Cu, Pb, Zn, Cd, and Cr) in seawater and sediment in Jakarta Bay. J Geologi Kelautan 11:9–16

Pociecha M, Lestan D (2010) Electrochemical EDTA recycling with sacrificial Al anode for remediation of Pb contaminated soil. Environ Pollut 158:2710–2715

Popovic A, Djordjevic D, Polic P (2001) Trace and major element pollution originating from coal ash suspension and transport processes. Environ Int 26:251–255

Prakasham RS, Merrie JS, Sheela R, Saswathi N, Ramakrishna SV (1999) Biosorption of chromium VI by free and immobilized *Rhizopus arrhizus*. Environ Pollut 104:421–424

Qian H, Li P (2011) Hydrochemical characteristics of groundwater in Yinchuan plain and their control factors. Asian J Chem 23:2927–2938

Qian H, Li P, Howard KWF, Yang C, Zhang X (2012) Assessment of groundwater vulnerability in the Yinchuan plain, Northwest China using OREADIC. Environ Monit Assess 184:3613–3628

Qian H, Li P, Wu J, Zhou Y (2013) Isotopic characteristics of precipitation, surface and ground waters in the Yinchuan plain, Northwest China. Environ Earth Sci 70:57–70

Raj BG, Patnaik MC, Babu SP, Kalakumar B, Singh MV, Shylaja J (2006) Heavy metal contaminants in water, soil, plant and animal continuum due to pollution of Musi River around Hyderabad in India. Indian J Anim Sci 76:131–133

Rajakovic LV (1992) The sorption of arsenic onto activated carbon impregnated with metallic silver and copper. Separat Sci Technol 27:1423–1433

Rajendran P, Muthukrishnan J, Gunasekaran P (2003) Microbes in heavy metal remediation. Ind J Exp Biol 41:935–944

Rangsayatorn N, Upatham ES, Kruatrachue M, Pokethitiyook P, Lanza GR (2002) Phytoremediation potential of *Spirulina (Arthrospira) platensis*: biosorption and toxicity studies of cadmium. Environ Pollut 119:45

Rao TS, Karthikeyan J (2007) Removal of As(V) from water by adsorption on to low-cost and waste materials. Progress in Environmental Science and Technology. Science Press, Beijing, China, pp 684–691

Rascio N, Navariizzo F (2011) Heavy metal hyperaccumulating plants: how and why do they do it, and what makes them so interesting. Plant Sci 18:169–181

Ravindra K, Mor S, Pinnaka VL (2019) Water uses, treatment and sanitation practices in rural areas of Chandigarh and its relation with water-borne diseases. Environ Sci Pollut Res 26:19512–19522

Razmara P (2021) Effect of copper nanoparticles on rainbow trout olfaction and recovery (Doctoral dissertation). University of Lethbridge

Reeves RD, Baker AJM (2000) Metal accumulating plants. In: Raskin I, Finsley BD (eds) Phytoremediation of toxic metals: using plants to clean up the environment. Wiley, New York, pp 193–229

Rengaraj S, Venkataraj S, Yeon JW, Kim Y, Li XZ, Pang GKH (2007) Preparation, characterization, and application of Nd–TiO_2 photocatalyst for the reduction of Cr (VI) under UV light illumination. Appl Catal B Environ 77:157–165

Rizvi A, Khan MS (2018) Heavy metal induced oxidative damage and root morphology alterations of maize (*Zeamays* L.) plants and stress mitigation by metal tolerant nitrogen fixing *Azotobacter chroococcum*. Ecotoxicol Environ Saf 157:9–20

Roehl KE, Meggyes T, Simon FG, Stewart DI (2005) Long-term performance of permeable reactive barriers. Elsevier Publishers, Amsterdam

Ron EZ, Rosenberg E (2001) Natural roles of biosurfactants. Environ Microbiol 3:229–236

Ruggieri F, Marín V, Gimeno D, Fernandez-Turiel JL, García-Valles M, Gutierrez L (2008) Application of zeolitic volcanic rocks for arsenic removal from water. Eng Geol 101:245–250

Ruiping L, Lihua S, Jiuhui Q, Guibai L (2009) Arsenic removal through adsorption, sand filtration and ultrafiltration: in situ precipitated ferric and manganese binary oxides as adsorbents. Desalination 249:1233–1237

Salati S, Quadri G, Tambone F, Adani F (2010) Fresh organic matter of municipal solid waste enhances phytoextraction of heavy metals from contaminated soil. Environ Pollut 158:1899–1906

Samra SCJ, Fawzi SCM (2011) The right to water in rural Punjab: assessing equitable access to water through the Punjab rural water supply and sanitation project. Health Hum Rights 13:36–49

Sang Y, Li F, Gu Q, Liang C, Chen J (2008) Heavy metal-contaminated groundwater treatment by a novel nanofiber membrane. Desalination 223:349–360

Sardar K, Ali S, Hameed S, Afzal S, Fatima S, Shakoor MB, Tauqeer HM (2013) Heavy metals contamination and what are the impacts on living organisms. Green J Environ Manag Pub Saf 2: 172–179

Sarwar N, Imran M, Shaheen MR, Ishaque W, Kamran MA, Matloob A, Rehim A, Hussain S (2016) Phytoremediation strategies for soils contaminated with heavy metals: modifications and future perspectives. Chemosphere 171:710–721

Schafer C, Mihelcic J (2012) Effect of storage tank material and maintenance on household water quality. J Am Water Works Assoc 104:E521–E529

Schafer CA (2010) Impact of tank material on water quality in household water storage systems in Cochabamba. University of South Florida, Bolivia

Schaller A, Diez T (1991) Plant specific aspects of heavy metal uptake and comparison with quality standards for food and forage crops. Der Einfluß von festen Abfällen auf Böden, Pflanzen: Sauerbeck D, Lübben S, Eds, KFA, Jülich, Germany 92-125

Scherer MM, Richter S, Valentine RL, Alvarez PJJ (2000) Chemistry and microbiology of permeable reactive barriers for in situ groundwater cleanup. Crit Rev Environ Sci Technol 30: 363–411

Scullion J (2006) Remediating polluted soils. Naturwissenschaften 93:51–65

Sen Gupta B, Chatterjee S, Rott U, Kauffman H, Bandopadhyay A, DeGroot W, Nag NK, Carbonell-Barrachina AA, Mukherjee S (2009) A simple chemical free arsenic removal method for community water supply - a case study from West Bengal, India. Environ Pollut 157:3351–3353

Shahid M, Dumat C, Khalid S, Schreck E, Xiong T, Niazi NK (2017) Foliar heavy metal uptake, toxicity and detoxification in plants: a comparison of foliar and root metal uptake. J Hazard Mater 325:36–58

Shi T, Ma J, Wu X, Ju T, Lin X, Zhang Y, Li X, Gong Y, Hou H, Zhao L, Wu F (2018) Inventories of heavy metal inputs and outputs to and from agricultural soils: a review. Ecotoxicol Environ Saf 164:118–124

Shuaib M, Azam N, Bahadur S, Romman M, Yu Q, Xuexiu C (2021) Variation and succession of microbial communities under the conditions of persistent heavy metal and their survival mechanism. Micro Pathogen 150:104713

Sikdar SK, Grosse D, Rogut I (1998) Membrane technologies for remediating contaminated soils: a critical review. J Membr Sci 151:75–85

Singh CK, Sahu JN, Mahalik KK, Mohanty CR, Mohan BR, Meikap BC (2008) Studies on the removal of Pb(II) from wastewater by activated carbon developed from tamarind wood activated with sulphuric acid. J Hazard Mater 153:221–228

Singh J, Kalamdhad AS (2011) Effects of heavy metals on soil, plants, human health and aquatic life. Int J Res Chem Environ 1:15–21

Singh KP, Malik A, Mohan D, Singh VK, Sinha S (2006) Evaluation of groundwater quality in Northern Indo-Gangetic alluvium region. Environ Monit Assess 112:211–230

Smedley PL, Kinniburgh DG (2002) A review of the source, behaviour and distribution of arsenic in natural waters. Appl Geochem 17:517–568

Srivastava PK, Vaish A, Dwivedi S, Chakrabarty D, Singh N, Tripathi RD (2011) Biological removal of arsenic pollution by soil fungi. Sci Total Environ 409:2430–2442

Stefanowicz AM, Kapusta P, Zubek S, Stanek M, Woch MW (2020) Soil organic matter prevails over heavy metal pollution and vegetation as a factor shaping soil microbial communities at historical Zn–Pb mining sites. Chemosphere 240:124922

Street RA (2012) Heavy metals in medicinal plant products - an African perspective. S Afr J Bot 82: 67–74

Sujitha K, Shankaret R, Gajendran T, Vasanthi B (2014) Invitro reduction, kinetic modelling and optimisation of parameters for biosorption of Cr (VI) using an ecological sorbent. IOSR J Environ Sci Toxicol Food Technol 8:1–7

Sun WH, Jiang YX, Li X (2013) Research of the evaluation on heavy-metal pollution in rice by sewage irrigation. Appl Mech Mater 295:1594–1599

Sylvester P, Westerhoff P, Möller T, Badruzzaman M, Boyd O (2007) A hybrid sorbent utilizing nanoparticles of hydrous iron oxide for arsenic removal from drinking water. Environ Eng Sci 24:104–112

Testa JJ, Grela MA, Litter MI (2004) Heterogeneous photocatalytic reduction of chromium (VI) over TiO_2 particles in the presence of oxalate: involvement of Cr (V) species. Environ Sci Technol 38:1589–1594

Thambavani DS, Mageswari TSRU (2013) A comprehensive geochemical evaluation of heavy metals in drinking water. J Chem Biol Phys Sci 3:2942–2956

Thiruvenkatachari R, Vigneswaran S, Naidu R (2008) Permeable reactive barrier for groundwater remediation. J Ind Eng Chem 14:145–156

Thornton EC, Jackson RL (1994) Laboratory and field evaluation of the gas treatment approach for in situ remediation of chromate-contaminated soils. Department of Energy, USA

Torres-Cruz TJ, Hesse C, Kuske CR, Porras-Alfaro A (2018) Presence and distribution of heavy metal tolerant fungi in surface soils of a temperate pine forest. Appl Soil Ecol 13:66–74

UNDP (2015) Sustainable development goal. 17 goals to transform our world. Goal 6: ensure access to water and sanitation for all. UNDP Report Available from: http://www.un.org/sustainabledevelopment/water-and-sanitation/

United Nations Scientific and Cultural Organization (UNESCO) (2015) Water for a sustainable world. The United Nations world water development report. UNESCO, Paris

USEPA (1989) Evaluation of groundwater extraction remedies. EPA Office of Emergency and Remedial Responses, Washington, DC

Usman UA, Yusoff I, Raoov M (2021) Natural sources of iron and manganese in groundwater of the lower Kelantan River Basin, North-eastern coast of peninsula Malaysia: water quality assessment and an adsorption-based method for remediation. Environ Earth Sci 80:425

Vanloom GW, Duffy SJ (2005) The hydrosphere in environmental chemistry: a global perspective, 2nd edn. Oxford University Press, New York

Venugopal T, Giridharan L, Jayaprakash M, Velmurugan PM (2009) A comprehensive geochemical evaluation of the water quality of River Adyar, India. Bull Environ Contam Toxicol 82:211–217

Virha R, Biswas VK, Qureshi TA, Borana K, Malik N (2011) Seasonal variation in physicochemical parameters and heavy metals in water of upper lake of Bhopal. Bull Environ Contam Toxicol 86:168–174

Virkutyte J, Sillanpää M, Latostenmaa P (2002) Electrokinetic soil remediation—critical overview. Sci Total Environ 289:97–121

Vorosmarty CJ, McIntyre PB, Gessner MO, Dudgeon D, Prusevich A, Green P, Davies PM (2010) Global threats to human water security and river biodiversity. Nature 467:555–561

Wang Z, Chen X, Yu D, Zhang L, Wang J, Lv J (2021) Source apportionment and spatial distribution of potentially toxic elements in soils: a new exploration on receptor and geostatistical models. Sci Total Environ 759:143428. https://doi.org/10.1016/j.scitotenv.2020.143428

Wang S, Mulligan CN (2004) An evaluation of surfactant foam technology in remediation of contaminated soil. Chemosphere 57(9):1079–1089. https://doi.org/10.1016/j.chemosphere.2004.08.019

Wang S, Mulligan CN (2006) Occurrence of arsenic contamination in Canada: sources, behavior, and distribution. Sci Total Environ 366:701–721

Wang S, Zhao X (2009) On the potential of biological treatment for arsenic contaminated soils and groundwater. J Environ Manag 90:2367–2376

Warshawsky A, Strikovsky AG, Vilensky MY, Jerabek K (2002) Interphase mobility and migration of hydrophobic organic metal extractant molecules in solvent-impregnated resins. Separ Sci Technol 37:2607–2622

Waybrant KR, Blowes DW, Ptacek CJ (1998) Selection of reactive mixtures for use in permeable reactive walls for treatment of mine drainage. Environ Sci Technol 32:1972–1979

Wuana RA, Okieimen FE (2011) Heavy metals in contaminated soils: a review of sources, chemistry, risks and best available strategies for remediation. International Scholarly Research Notices 2011

Xenidis A, Stouraiti C, Papassiopi N (2010) Stabilization of Pb and As in soils by applying combined treatment with phosphates and ferrous iron. J Hazard Mater 177:929–937

Xu J, Cao Z, Zhang Y, Yuan Z, Lou Z, Xu X, Wang X (2018) A review of functionalized carbon nanotubes and graphene for heavy metal adsorption from water: preparation, application, and mechanism. Chemosphere 195:351–364

Yadav SK (2010) Heavy metals toxicity in plants: an overview on the role of glutathione and phytochelatins in heavy metal stress tolerance of plants. S Afr J Bot 76:167–179

Yang L, Donahoe RJ, Redwine JC (2007) In situ chemical fixation of arsenic contaminated soils: an experimental study. Sci Total Environ 387:28–41

Yin Y, Allen HE (1999) In-situ chemical treatment. Technology Evaluation Report, TE-99e01. Ground-Water Remediation Technologies Analysis Center, Pittsburg, PA

Yong RN, Mulligan CN (2004) Natural attenuation of contaminants in soils. CRC Press, Boca Raton, FL

Ziadat A (2005) Impact of storage tanks on drinking water quality in Al-Karak Province-Jordan. J Appl Sci 5:634–638

Effect of Heavy Metals on Roadside Vegetation

8

Naresh Kumar and Suman

Abstract

Pollution, along with the rapidly growing number of vehicles, is a major problem worldwide. Auto pollution is known primarily for emitting various forms of living and nonliving pollutants into the atmosphere, but recent research shows that vehicular exhaust is a key source of heavy metal toxicity in municipal areas through processes such as diesel and oil combustion, corrosion of metal parts, engine aging, tire and brake wear, and road damage. Heavy metal pollution caused by vehicle emissions has a detrimental effect on the environment and vegetation along roadsides. The physical characteristics of plants are affected by vehicular pollution. The absorption of heavy metals gradually alters soil pH, affecting anatomical, physiological, and plant fertility characteristics. The seed germination rate is reduced because of the high toxicity of lead. Excessive concentration of thrash metals reduces metabolic activity and plant growth. Even though some heavy metals are important micronutrients in low-grade plants, their high concentration causes metabolic disorders and reduced growth in a variety of plants, especially those growing along the road. Due to this, there is an urgent necessity to formulate an effective environmental management system in urban areas that includes promoting new technologies, using biofuels, developing green belts, and public participation. This chapter summarizes the adverse consequences of heavy metal pollution triggered by the growing number of vehicles on the vegetation along the roadside.

N. Kumar · Suman (✉)
Department of Environmental Science, School of Basic Science and Research, Sharda University Greater Noida, Noida, Uttar Pradesh, India
e-mail: suman.kumar@sharda.ac.in

R. P. Singh et al. (eds.), *Heavy Metal Toxicity: Environmental Concerns, Remediation and Opportunities*, https://doi.org/10.1007/978-981-99-0397-9_8

Keywords

Vehicular pollution · Heavy metal pollution · Urban environment · Roadside · Physiological aspects

8.1 Introduction

Environmental pollution is caused by toxic compounds and energy released by the different components of the environment, which affects the Earth's ecosystem (Altaf et al., 2021; Santra, 2001). Air impairment is the utmost harmful of all types of pollution (Asthana, 2006). There is a significant increase in pollution in current years due to increase in man-made activities, such as the flaming and manufacturing of nonrenewable sources of fuels like petroleum and kerosene and car exhaust (Amusan, Bada, and Salami, 2003). Environmental degradation due to hazardous discharges from motor carriage is presently a universal issue. Every year, the use of automobiles for different purposes increases around the world, increasing environmental pollution, especially air pollution (Maksimtsev, Dudarets and Yukhnovskyi, 2021). The global population has grown steadily in recent decades, from 7.4 billion to 7.9 billion in the last three decades. By the year 2050, the population of the world is likely to touch several 990 crores people. To fulfill the needs of this rapidly growing population (Muthu et al., 2021), there is also a hike in transport to meet the needs of people with increased development and industrialization. The use of private motors has also quietly amplified due to the scarcity of public transport. Wheeler being a convenient and comparatively cheap mode of transport as compared to four-wheeled vehicles has grown in popularity, mostly in low economic countries like India (Yunus, Singh and Iqbal, 1996).

Universally, approximately 53 million autos were estimated five decades ago; in 2000, this number increased to 500 million, and approximately 19 million vehicles are added per year. With the sales of 349 crores units of private and commercial automobiles, India was at the fifth position in the automobile market in the financial year 2020 (IBEF, 2017). The global production data of automobiles in the last three decades are shown in Fig. 8.1.

Urbanization and the rapid increase in registered motor vehicles have resulted in many different types of pollutants. Automobiles are one of the many causes of pollution in large cities, and they play a significant role in degrading the environment (Ghosh, Maiti, and Singh, 2009).

8.2 Vehicular Pollution

Vehicle emissions are caused by the burning of nonrenewable sources of energy like petrol and diesel, abrasion of vehicle parts, and the use of lubricants (De Silva et al., 2021). Automobiles emit volatile organic compounds (VOCs), particulate matter, oxides of sulfur and nitrogen, carbon monoxide, and potentially hazardous

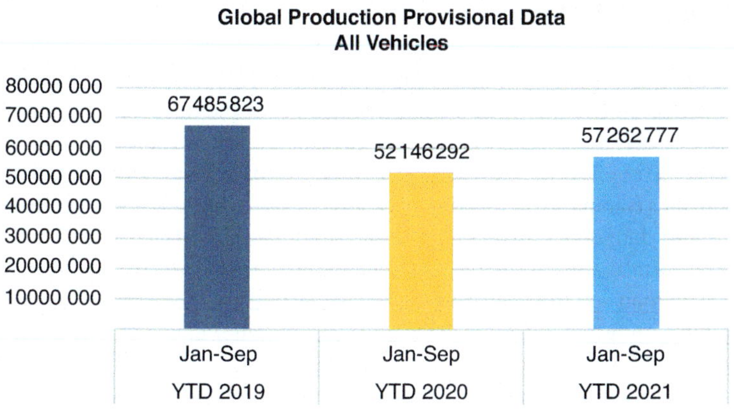

Fig. 8.1 Global production provisional data (source: International Organization of Motor Vehicle Manufacturers)

Table 8.1 Sources of vehicle emissions and contamination with potentially harmful materials

Sources	Metals															
	Ag	As	Cd	Co	Cr	Cu	Mn	Mo	Ni	Pb	Pt	Rh	Sb	Se	Ti	Zn
Break wear		■	■		■	■			■	■			■		■	■
Tyre Wear			■	■	■	■			■	■						■
Engine Oil	■		■		■	■	■	■	■	■						■
Road abrasions					■				■							
VEC										■	■					
Fuel		■	■		■		■		■	■				■		■

Colored boxes denote the accessibility of potentially toxic metals in the various vehicular sources

substances such as zinc, cadmium, mercury, and others which damage roadside vegetation (Wagh et al., 2006). Due to their everlasting nature and long continuance, heavy metals are most closely related to many of the contaminants associated with road dust in metropolitan settings (Masih et al., 2019; Gautam et al. 2019). Particles emitting from vehicular exhaust, lubricating oil deposits, tire and brake abrasion, abrasion of motor and engine parts, refined paint particles, and weathering of an asphalt roadway and road exteriors are all anthropogenic sources of heavy metals (Roy et al., 2019). The following table (Table 8.1) shows the presence of different heavy metals in the various sources of automobile exhaust.

8.3 Heavy Metals

Heavy metals are those metals that have a density of 4–5 grams per centimeter or more (Nagajyoti, Lee and Sreekanth, 2010). Some of the heavy metals, like Cd, Cr, As, and Pb, are more noxious and are present in the Earth's crust (Wuana and Okieimen, 2011). They can cause dangerous impairment to the ecosystem and environment (Babula et al., 2009). Because heavy metals are not biodegradable, they are also dangerous in small amounts (Brodin et al., 2017; Ferrey et al., 2018). They are naturally present in the atmosphere, but current industry and development, as well as human activities and fertilizer, have resulted in increased amounts of these metals in the environment, as well as exposure to higher levels of living things (Ali, Khan and Ilahi, 2019).

Metal and metalloid ions are divided into three categories. Mercury, cadmium, and lead are among the first group of metals that are dangerous in small amounts. The second group of metals is less harmful than the third group which is involved in several metabolic reactions. The methods occur in the body and are noxious only in high concentrations (Odobašić, Šestan, and Begić, 2019).

Sources

Mining, manufacturing units, unprocessed sewage slush, combustion by-products from metal pipelines, traffic, and power plants that use coal as an energy source are also examples of dispersal causes of potentially hazardous metal (www.unep.org, n. d.). According to the UNEP/GPA (2006), the supervision of e-waste, especially the dumping of discarded computers and mobile phones, contains more than 1000 different elements, many of which are harmful to people (UNEP/GPA, 2006). Toxic chemical emissions into the environment have spread from developed countries. Heavy metals can be introduced into the air by many industrial units along with heavy traffic.

Lead, cadmium, and zinc are among the potentially hazardous substances found in traffic (Viard et al. 2004; Onder and Dursun, 2006). Levels of heavy metals added to different components of the environment from anthropogenic sources increase as a result of lifestyle changes (Olowoyo, van Heerden and Fischer, 2010; Odukoya et al. 2000). Figure 8.2 shows the various major sources of heavy metals in the ambient atmosphere.

Generally, gaseous contaminants and atmospheric aerosols comprising heavy metals are out into the air by motor movement and industrial activity. For those

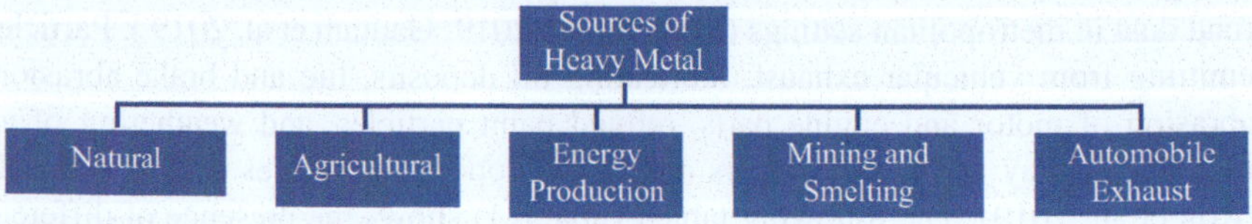

Fig. 8.2 Sources of heavy metals

who are more likely to live close to vehicular fumes, pollutants generated by traffic movement are extra harmful than pollutants emitted from other origins (Khalid et al., 2018). Greenery along the road acts as a barrier to the pollutants caused by vehicles. The surface area of the leaves allows particulates and heavy metals to settle and remove gaseous pollutants (Singh et al., 2018). Plants exposed to airborne pollution, on the other hand, exhibit rapid changes in physiological response just before visual cues appear on leaves (Rai and Panda, 2015).

8.4 Impact of Heavy Metals on Roadside Vegetation

Heavy metal pollution is caused by vehicular emissions, which have adverse consequences on the ambient environment and roadside vegetation. Plant growth in areas around roadways is affected by automobile exhaust emissions (Shakeel et al., 2022). Plants growing along roadsides and near bus stops are particularly vulnerable to air pollution (Rai, 2016; Krishnaveni, et al., 2015). Physiological, structural, and biochemical properties of plant types are all harmed by increased air pollution (Karmakar and Padhy, 2019; Kaur and Nagpal 2017). The absorption of heavy metals changes soil pH over time, affecting plant morphological, physiological, and reproductive characteristics (Ramos-Montaño, 2020; Banerjee, Palit, and Banerjee, 2021). Automobile pollutants hurt plant breeding parts (Farahzadi et al., 2020; De Silva et al., 2020). Roadside vegetation seedling growth and germination are affected by heavy metal contamination (Weitekamp et al., 2020; Azab and Hegazy, 2020). Increased Pb toxicity results in reduced seed germination (Bai et al., 2020; Wu et al., 2020). Increasing traffic pollution is negatively correlated with species diversity and richness (Karmakar, Ballav and Hazra, 2021). The effect of traffic pollution on roadside plants is shown in Fig. 8.3.

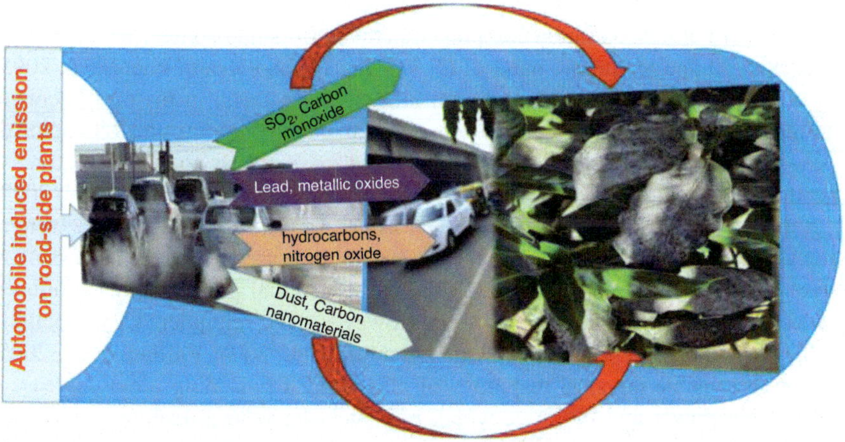

Fig. 8.3 Impact of vehicular emission on roadside vegetation

In addition, due to the ubiquitous occurrence of heavy metals on roadsides, urban parks, and green areas, roadside vegetation, which is accomplished by particulate material and allied heavy metals, has been used for potential toxic element poisoning in urban settings (Martin et al., 2018; Xu et al., 2019).

8.4.1 Impact of Copper (Cu) on Plants

Copper is a vital component for the evolution and expansion of vegetation, but at elevated concentrations, it successfully prohibits plant growth and has antiseptic, antimycotic, and insecticidal properties (Jampílek and Kráľová, 2022). Copper is a vital micronutrient for all plant species at optimal levels. In plants, copper is essential for cell breathing, photosynthesis, cell wall metabolism, and lignin production, as well as the oxidative stress response and hormone communication (Printz et al., 2016). Copper poisoning prevents the decomposition of reserve food supplies like cellulose by obstructing the events of alpha-amylase and invertase isoenzymes. Cu toxicity at the seed germination stage is also associated with changes in normal metabolism and water transport (Sethy and Ghosh 2013). Copper poisoning inhibits PIN1-mediated auxin transport in cells by increasing nitric oxide levels (Yuan et al., 2011). Surplus Cu damages the structure of the plastid, leading to alter in the structure of the thylakoid membrane (Rehman et al., 2019). Cu poisoning restricts root growth, which interferes with root assimilation and prevents critical minerals from being absorbed through diffusion (Marschner 2011). Cu poisoning results in reduced levels of absorption of essential micronutrients for plants (Li et al., 2019). There are various adverse effects of Cu deficiency and overabundance on vegetation which was shown in Fig. 8.4.

8.4.2 Impact of Zinc (Zn) on Plants

Zn is the 23rd utmost common component in the Earth's lithosphere, with a nucleon number of 30 and an atomic mass of 65.37. (Zn: Human Health Fact Sheet 2005). Unadulterated zinc is white-tinged with blue, lustrous metal in texture (Escobedo Monge et al., 2019). Zinc (Zn) is crucial in plant physiology and biochemistry because of its acknowledged essentialness and noxiousness to alive creatures at specific Zn concentrations, i.e., fatal beyond acceptable limits (Natasha et al. 2022).

Stagnant growth, reduced leaf size, leaf hysteresis, and spikelet sterility were all seen as clear indicators of Zn deficiency in plants. Increased Zn levels in the soil are phytotoxic, causing a variety of structural and functional problems that ultimately reduce plant routine (Gondal et al., 2021). However, these rejoinders vary depending on the plant class and progressive stage (Khudsar et al., 2004; Vaillant et al., 2005). The utmost visible physical responses of plants to Zn poisoning are a decrease in germination power and biomass, which ultimately leads to a decline in yield and product quality, as shown in Fig. 8.5 (Garg and Singh, 2018; Zhang et al., 2019). Furthermore, under Zn stress, higher pectin content has been found, with increased

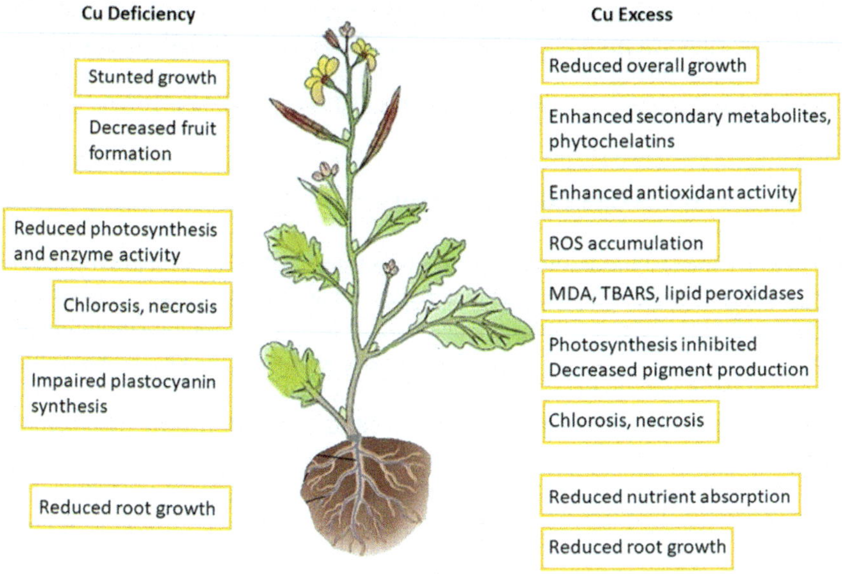

Fig. 8.4 Impact of copper (Cu) on plants

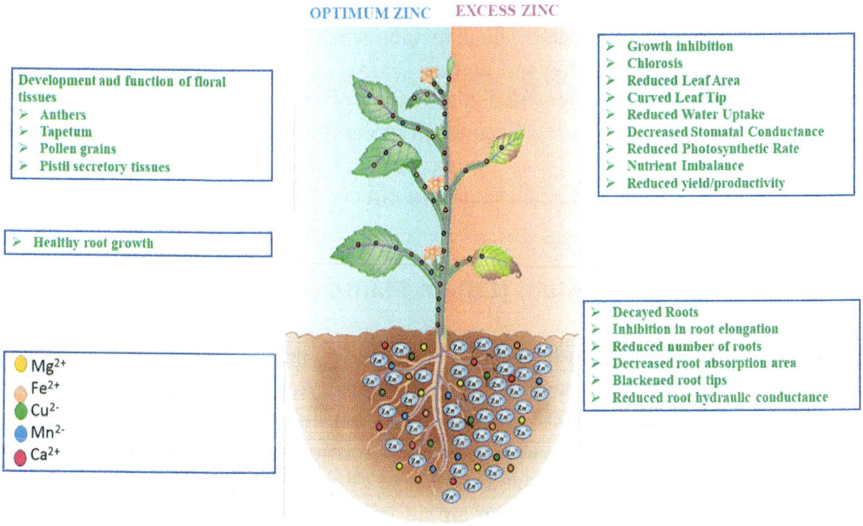

Fig. 8.5 Impact of zinc (Zn) on plants

callose and pectin content that seems to accompany everyone (Kaur and Garg, 2021). The enlarged pectin binds to extra zinc in the cell wall, and glucose deposits stabilize it, preventing Zn from entering the cytoplasm (Feigl et al., 2019). The adverse consequences of optimal and surplus zinc are shown in Fig. 8.5.

8.4.3 Impact of Lead (Pb) on Plants

It is one of the most hazardous heavy metals present in the environment. Because of its significant physicochemical characteristics, it has been used since ancient times. On a worldwide scale, it is a widely distributed, significant, yet dangerous environmental chemical (Mahaffey, 1990). It is an example of a possible heavy metal that is easily absorbed and stored in many areas of a plant (Nas and Ali, 2018). It is neither a necessary element nor does it play a role in cell metabolism. Leaded gasoline is the most common source of lead pollution. Lead inhibits photosynthesis, as is the nutrient cycle, moisture balance, and enzyme activity (Sharma and Dubey, 2005). Lead affects seed incubation and slows sapling growth, lowering germination percentage, germination index, root/shoot length, tolerance index, and root and shoot dry mass (Mishra et al., 2006).

Lead has been shown to affect floral output, with plants producing fewer flowers at high lead concentrations (Opeolu, Adenuga, and Ndakidemi, 2010). The activity and number of key enzymes in metabolic processes such as the photosynthetic Calvin cycle (Stevens, Creissen, and Mullineaux, 1997), nitrogen metabolism (Kumar and Dubey, 1999), and sugar metabolism (Verma and Dubey, 2003) can all be affected by lead (Verma and Dubey, 2003).

Whole plants, detached leaves, isolated protoplasts, and mitochondria all show stimulation of respiration at lower concentrations (Parys et al., 1998). Lead affects the growth of roots and aerial plant components at low concentrations (Sengar et al., 2008). Figure 8.6 depicts the many harmful effects of lead.

8.4.4 Impact of Cadmium (Cd) on Plants

It is a chemical element with nucleon number 48 and the symbol Cd. Cadmium (Cd^{2+}) is a bivalent heavy metal cation that causes cytotoxic effects in plants and is accepted and accreted deep into the food chain, causing disease and damage (Wang et al., 2020). There are various man-made sources of cadmium which account for about 90% of Cd released into the atmosphere (Yuan et al., 2019).

Chlorosis, leaf roll, and dwarfism are all indicators of Cd toxicity, and they are most clearly evident on plant parts (Kadioglu et al., 2012). Cd toxicity can reduce nitrogen fascination and primary ammonia integration in soybean nodules (Hussain et al., 2020). Under Cd stress, physiological and metabolic changes are common. Changes in gas conversation property, proline, malondialdehyde (MDA), sugar,

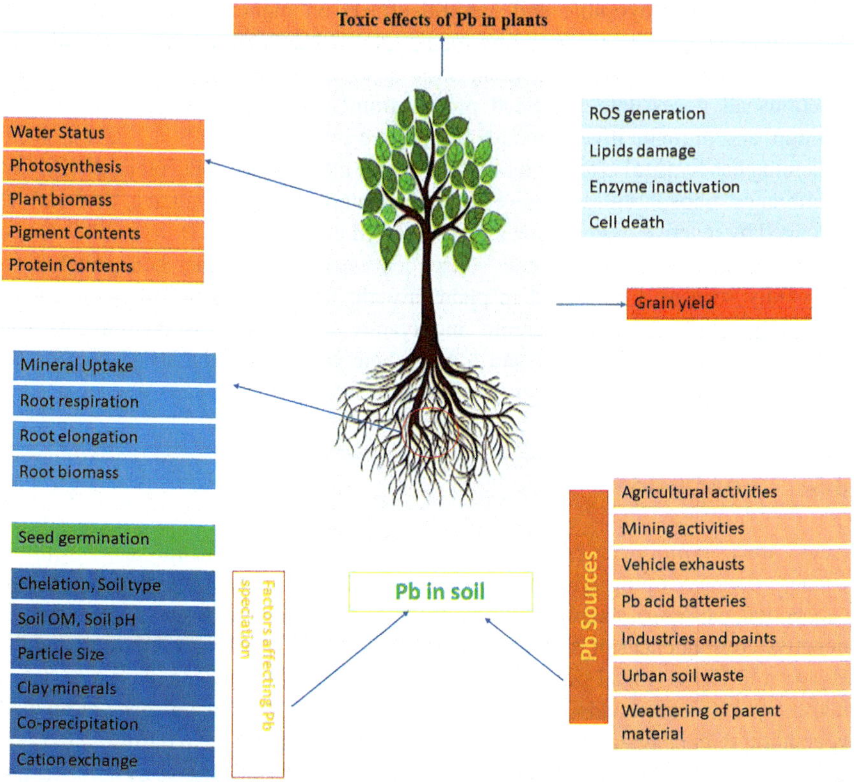

Fig. 8.6 Impact of lead (Pb) on plants

protein, and enzyme activity are among the most common modifications (Anjum et al., 2016; Marzban et al., 2017). The change is caused by an excess of free radicals, enzyme inhibition, and/or a lack of food (Hasanuzzaman, Nahar, and Fujita, 2018; Rizwan et al., 2017). Cadmium toxicity has effects on plants by preventing carbon fixation and reducing chlorophyll concentration and photosynthetic activity (Gallego et al., 2012). Toxicity of cadmium results in an excess of reactive oxygen species (ROS), which damage plant crusts and destroy cell macromolecules and organelles (Abbas et al., 2018). Cd normally interferes with Ca, P, Mg, K, and Mn (Nazar et al., 2012).

8.4.5 Impact of Nickel (Ni) on Plants

Plants use nickel (Ni) as one of the major micronutrient heavy metals. It is the 24th utmost frequent element in the Earth's crust, accounting for roughly 3% of its total mass.

This has both favorable and harmful effects on plant evolution liable to the concentration of Ni in the growing media. Normal development, enzyme activity (such as urea), nitrogen digestion, iron absorption, and specialized metabolic reactions all necessitate Ni. Seed propagation, root and shoot growth, biomass buildup, and ultimate invention are all inhibited by Ni (Hassan et al., 2019). Nickel poisoning has harmful effects on other nonliving factors (Gill and Tuteja, 2010). Cell membranes, phospholipids, stains, enzymes, chloroplasts, and nucleic acids are all damaged by reactive oxygen species in plants (Maheshwari and Dubey, 2009; Gopal and Nautiyal, 2012). The adverse consequences of Ni and other heavy metals are frequently seen as a reduction in plant growth, which is a standard indicator for assessing pollution levels (Seregin and Ivanov, 2001). Excess Ni prevents seed germination and development and lowers their yields significantly (Boominathan and Doran, 2002). The toxicity of nickel is depicted in Fig. 8.7.

8.5 Conclusion

Heavy metal pollution is rapidly becoming a global environmental issue of public concern. In developing countries like India, vehicular exhaust is one of the most substantial causes of heavy metal pollution. This chapter aims to delineate the negative consequences of heavy metal pollution from car exhaust on the surrounding vegetation. Heavy metal pollution from vehicle exhaust has been shown to have a detrimental effect on roadside vegetation around the world. The interface of roadside vegetation with heavy metals has two aspects. Heavy metals have harmful effects on plants, on the one hand, and, on the other hand, have their defense systems against toxic effects and remove heavy metal pollution. As shown in this chapter, the presence of heavy metals affects all elements of plant physiology and morphology.

A country such as India has strict regulations on various fire festivals to manage the number of heavy metals in the ambient atmosphere. In addition, diesel cars must be controlled to limit particulate matter, which transfers most of the heavy metals to the ambient air.

This can be accomplished by requiring automobiles to meet Euro standards. Most low-income countries have not yet adopted the modern Bharat stage emission standards or Euro standards. Lower heavy metal emissions can also be helped by proper road maintenance and vehicle maintenance. If the pollution control rules are strictly implemented, then they will be able to regulate the level of particulate and heavy metals in the atmosphere at the same time.

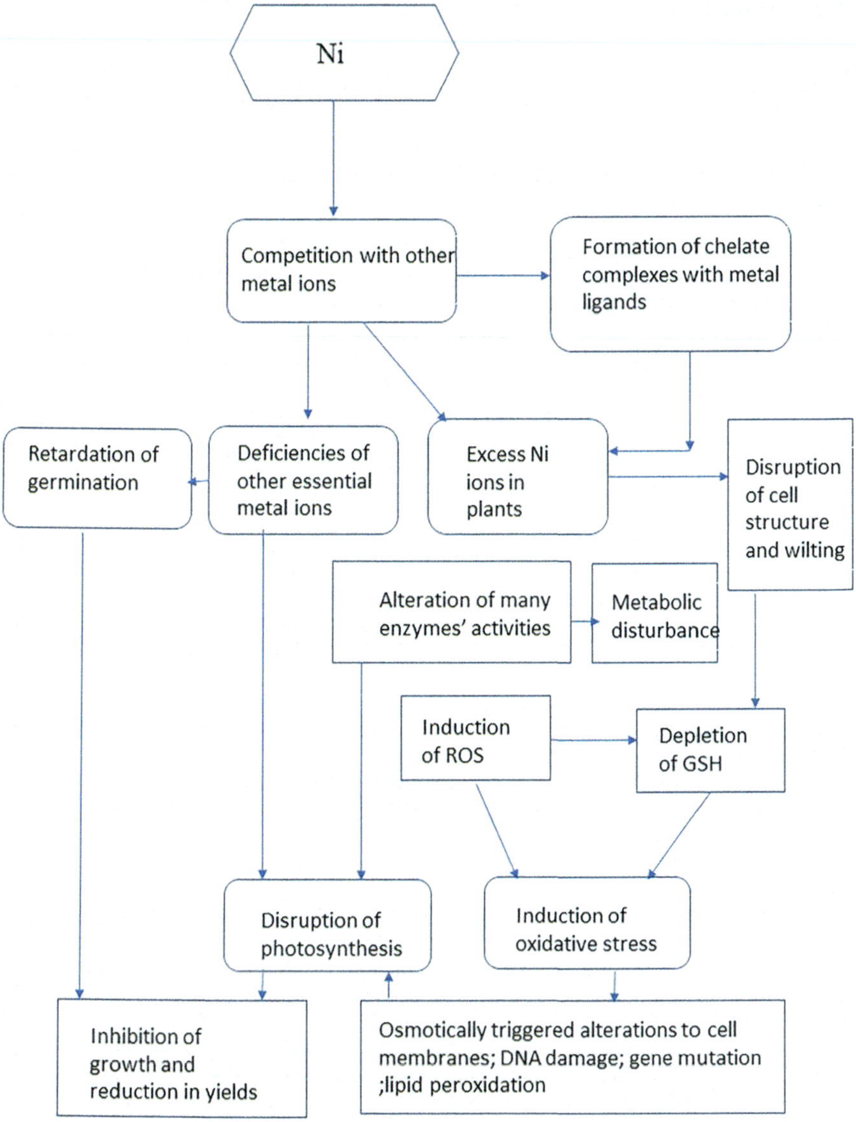

Fig. 8.7 Impact of nickel (Ni) on plants

References

Abbas T, Rizwan M, Ali S, Adrees M, Zia-ur-Rehman M, Qayyum MF, Ok YS, Murtaza G (2018) Effect of biochar on alleviation of cadmium toxicity in wheat (*Triticum aestivum* L.) grown on Cd-contaminated saline soil. Environ Sci Pollut Res 25(26):25668–25680

Ali H, Khan E, Ilahi I (2019) Environmental chemistry and ecotoxicology of hazardous heavy metals: environmental persistence, toxicity, and bioaccumulation. J Chem 2019:6730305. https://doi.org/10.1155/2019/6730305

Altaf R, Altaf S, Hussain M, Shah RU, Ullah R, Ullah MI, Rauf A, Ansari MJ, Alharbi SA, Alfarraj S, Datta R (2021) Heavy metal accumulation by roadside vegetation and implications for pollution control. PLoS One 16(5):0249147. https://doi.org/10.1371/journal.pone.0249147

Amusan AA, Bada SB, Salami AT (2003) Effect of traffic density on the heavy metal content of soil and vegetation along roadsides in Osun state, Nigeria. West Afr J Appl Ecol 4(1):107

Anjum SA, Ashraf U, Khan I, Tanveer M, Ali M, Hussain I, Wang LC (2016) Chromium and aluminum phytotoxicity in maize: morpho-physiological responses and metal uptake. Clean - Soil, Air, Water 44(8):1075–1084. https://doi.org/10.1002/clen.201500532

Asthana D (2006) Textbook of environmental studies. S. Chand Publishing, New Delhi

Azab E, Hegazy AK (2020) Monitoring the efficiency of Rhazya stricta L. plants in phytoremediation of heavy metal-contaminated soil. Plan Theory 9:1057. https://doi.org/10.3390/plants9091057

Babula P, Adam V, Opatrilova R, Zehnalek J, Havel L, Kizek R (2009) Uncommon heavy metals, metalloids, and their plant toxicity: a review. In: Organic farming, pest control and remediation of soil pollutants, vol 1. Springer, Dordrecht, pp 275–317

Bai T, Liu Y-Y, Muhammad I, Yang X, Yin X-J, Bai L, Wang YJ (2020) Mixed nitrogen form addition facilitates the growth adaptation of legume plant to heavy metal contamination in degraded mining areas. Glob Ecol Conserv 24:e01387

Banerjee S, Palit D, Banerjee A (2021) Variation of tree biochemical and physiological characters under different air pollution stresses. Environ Sci Pollut Res 28:17960–17980. https://doi.org/10.1007/s11356-020-11674-3

Boominathan R, Doran PM (2002) Ni-induced oxidative stress in roots of the Ni hyperaccumulator, Alyssum bertolonii. New Phytol 156(2):205–215

Brodin M, Vallejos M, Opedal MT, Area MC, Chinga-Carrasco G (2017) Lignocellulosic as sustainable resources for the production of bioplastics—a review. J Clean Prod 162:646–664

De Silva S, Ball AS, Indrapala DV, Reichman SM (2020) Review of the interactions between vehicular emitted potentially toxic elements, roadside soils, and associated biota. Chemosphere 263:128135. https://doi.org/10.1016/j.chemosphere.2020.128135

De Silva S, Ball AS, Indrapala DV, Reichman SM (2021) Review of the interactions between vehicular emitted potentially toxic elements, roadside soils, and associated biota. Chemosphere 263:128135

Escobedo Monge MF, Barrado E, Alonso Vicente C, Redondo del Río MP, Marugán M, de Miguelsanz J (2019) Zinc nutritional status in patients with cystic fibrosis. Nutrients 11(1):150

Farahzadi HN, Arbabian S, Majd A, Tajadod G (2020) Long-term effect different concentrations of Zn (NO3) 2 on the development of male and female gametophytes of *Capsicum annuum* L. var California wonder. Caryologia 73(1):145–154. https://doi.org/10.13128/caryologia-174

Feigl G, Molnár A, Szőllősi R, Ördög A, Törőcsik K, Oláh D, Bodor A, Perei K, Kolbert Z (2019) Zinc-induced root architectural changes of rhizotron-grown B. napus correlate with a differential nitro-oxidative response. Nitric Oxide 90:55–65

Ferrey ML, Hamilton MC, Backe WJ, Anderson KE (2018) Pharmaceuticals and other anthropogenic chemicals in atmospheric particulates and precipitation. Sci Total Environ 612:1488–1497

Gallego SM, Pena LB, Barcia RA, Azpilicueta CE, Iannone MF, Rosales EP, Zawoznik MS, Groppa MD, Benavides MP (2012) Unraveling cadmium toxicity and tolerance in plants: insight into regulatory mechanisms. Environ Exp Bot 83:33–46

Garg N, Singh S (2018) Arbuscular mycorrhiza Rhizophagus irregularis and silicon modulate growth, proline biosynthesis and yield in *Cajanus cajan* L. Millsp. (pigeonpea) genotypes under cadmium and zinc stress. J Plant Growth Regul 37:46–63

Gautam S, Patra AK, Kumar P (2019) Status and chemical characteristics of ambient PM2. 5 pollutions in China: a review. Environ Dev Sustain 21(4):1649–1674

Ghosh SP, Maiti SK, Singh G (2009) Heavy metal contamination in roadside soil and vegetation: a review. Indian J Environ Protect 29(4):334–345

Gill SS, Tuteja N (2010) Reactive oxygen species and antioxidant machinery in abiotic stress tolerance in crop plants. Plant Physiol Biochem 48(12):909–930

Gondal AH, Zafar A, Zainab D, Toor MD, Sohail S, Ameen S, Ijaz AB, Ch BI, Hussain I, Haider S, Ahmad IA (2021) A detailed review study of zinc involvement in animal, plant and human nutrition. Ind J Pure App Biosci 9(2):262–271

Gopal R, Nautiyal N (2012) Growth, antioxidant enzymes activities, and proline accumulation in mustard due to nickel. Int J Vet Sci 18:223–234

Hasanuzzaman M, Nahar K, Fujita M (2018) In: Hasanuzzaman M, Nahar K, Fujita M (eds) Plants under metal and metalloid stress—responses, tolerance, and remediation. Springer Nature, Singapore. https://doi.org/10.1007/978-981-13-2242-6

Hassan MU, Chattha MU, Khan I, Chattha MB, Aamer M, Nawaz M, Ali A, Khan MAU, Khan TA (2019) Nickel toxicity in plants: reasons, toxic effects, tolerance mechanisms, and remediation possibilities—a review. Environ Sci Pollut Res 26(13):12673–12688

Hussain S, Khaliq A, Noor MA, Tanveer M, Hussain HA, Hussain S, Shah T, Mehmood T (2020) Metal toxicity and nitrogen metabolism in plants: an overview. In: Datta R, Meena R, Pathan S, Ceccherini M (eds) Carbon and nitrogen cycling in soil. Springer, Singapore. https://doi.org/10.1007/978-981-13-7264-3_7

IBEF (2017) Automobile industry in India, Indian Automobile Industry, sector, trends, statistics. [online] Ibef.org. Available at https://www.ibef.org/industry/india-automobiles.aspx

Jampílek J, Kráľová K (2022) Impact of copper-based nanoparticles on economically important plants. In: Copper nanostructures: next-generation of agrochemicals for sustainable agroecosystems. Elsevier, pp 293–339

Kadioglu A, Terzi R, Saruhan N, Saglam A (2012) Current advances in the investigation of leaf rolling are caused by biotic and abiotic stress factors. Plant Sci 182:42–48

Karmakar D, Padhy PK (2019) Air pollution tolerance, anticipated performance, and metal accumulation indices of plant species for greenbelt development in urban industrial area. Chemosphere 237:124522. https://doi.org/10.1016/j.chemosphere.2019.124522

Karmakar NC, Ballav J, Hazra A (2021) Roadside vegetation in and around the megacity of Kolkata along an urbanisation gradient. Acta Bot Hung 63(1–2):77–86

Kaur H, Garg N (2021) Zinc toxicity in plants: a review. Planta 253(6):1–28

Kaur M, Nagpal A (2017) Evaluation of air pollution tolerance index and anticipated performance index of plants and their application in development of green space along the urban areas. Environ Sci Pollut Res 24(23):18881–18895. https://doi.org/10.1007/s11356-017-9500-9

Khalid N, Hussain M, Ashraf M, Masood A, Akhtar Y (2018) Spatiotemporal variation in cadmium released by automobiles along two roads in Pakistan. Pak J Bot 50(2):529–536

Khudsar T, Mahmooduzzafar Iqbal M, Sairam RK (2004) Zinc-induced changes in morpho-physiological and biochemical parameters in Artemisia annua. Biol Plant 48(2):255–260

Krishnaveni M, Silambarasan V, Kumar RS, Sabari M, Eswari V, Silpavathi G (2015) Air pollution tolerance index of plants studied near Omalur bus stand, soil nutrient analysis. Int J Pharm Sci Res 31:154–157

Kumar RG, Dubey RS (1999) Glutamine synthetase isoforms from rice seedlings: effects of stress on enzyme activity and the protective roles of osmolytes. J Plant Physiol 155(1):118–121

Li Q, Chen HH, Qi YP, Ye X, Yang LT, Huang ZR, Chen LS (2019) Excess copper effects on growth, uptake of water and nutrients, carbohydrates, and PSII photochemistry revealed by OJIP transients in citrus seedlings. Environ Sci Pollut Res 26:30188–30205

Mahaffey KR (1990) Environmental lead toxicity: nutrition as a component of the intervention. Environ Health Perspect 89:75–78
Maheshwari R, Dubey R (2009) Nickel-induced oxidative stress and the role of antioxidant defense in rice seedlings. Plant Growth Regul 59:37–49
Maksimtsev S, Dudarets S, Yukhnovskyi V (2021) Accumulation of heavy metals in soil and litter of roadside plantations in western Polissia of Ukraine. Folia For Pol A 63(3):232–242
Marschner H (2011) Marschner's mineral nutrition of higher plants. Academic Press, Cambridge
Martin JAR, Gutierrez C, Torrijos M, Nanos N (2018) Wood and bark of Pinus halepensis as archives of heavy metal pollution in the Mediterranean Region. Environ Pollut 239:438–447
Marzban L, Akhzari D, Ariapour A, Mohammadparast B, Pessarakli M (2017) Effects of cadmium stress on seedlings of various rangeland plant species (Avena fatua L., Lathyrus sativus L., and Lolium temulentum L.): growth, physiological traits, and cadmium accumulation. J Plant Nutr 40(15):2127–2137. https://doi.org/10.1080/01904167.2016.1269347
Masih J, Nair A, Gautam S, Singhal RK, Basu H, Dyavarchetty S, Uzgare A, Tiwari R, Taneja A (2019) Chemical characterization of submicron particles in indoor and outdoor air at two different microenvironments in the western part of India. SN Appl Sci 1(2):1–12
Mishra S, Srivastava S, Tripathi RD, Kumar R, Seth CS, Gupta DK (2006) Lead detoxification by coontail (*Ceratophyllum demersum* L.) involves induction of phytochelatins and antioxidant system in response to its accumulation. Chemosphere 65(6):1027–1039. https://doi.org/10.1016/j.chemosphere.2006.03.033
Muthu M, Gopal J, Kim DH, Sivanesan I (2021) Reviewing the impact of vehicular pollution on road-side plants—future perspectives. Sustainability 13(9):5114
Nagajyoti PC, Lee KD, Sreekanth TVM (2010) Heavy metals, occurrence, and toxicity for plants: a review. Environ Chem Lett 8(3):199–216
Nas FS, Ali M (2018) The effect of lead on plants in terms of growing and biochemical parameters: a review. MOJ Ecol Environ Sci 3(4):265–268
Natasha N, Shahid M, Bibi I, Iqbal J, Khalid S, Murtaza B, Bakhat HF, Farooq ABU, Amjad M, Hammad HM, Niazi NK (2022) Zinc in the soil-plant-human system: a data-analysis review. Sci Total Environ 808:152024
Nazar R, Iqbal N, Masood A, Khan MIR, Syeed S, Khan NA (2012) Cadmium toxicity in plants and role of mineral nutrients in its alleviation. Am J Plant Sci 03(10):1476–1489. https://doi.org/10.4236/ajps.2012.310178
Odobašić A, Šestan I, Begić S (2019) Biosensors for determination of heavy metals in waters. In: Biosensors for environmental monitoring. IntechOpen, London
Odukoya OO, Arowolo TA, Bamgbose O (2000) Pb, Zn, and Cu levels in tree bark are indicators of atmospheric pollution. Environ Int 26(1–2):11–16
Olowoyoz JO, Van Heerden E, Fischer JL (2010) Investigating Jacaranda mimosifolia tree as a biomonitor of atmospheric trace metals. Environ Monit Assess 164(1):435–443
Onder S, Dursun S (2006) Airborne heavy metal pollution of Cedrus libani (A. Rich.) in the city center of Konya (Turkey). Atmos Environ 40(6):1122–1133
Opeolu BO, Adenuga PA, Ndakidemi Olujimi OO (2010) Assessment of phytotoxicity potential of lead on tomato (*Lycopersicon esculentum* L) planted on contaminated soils. Int J Phys Sci 5(2): 68–73
Parys E, Romanowska E, Siedlecka M, Poskuta JW (1998) The effect of lead on photosynthesis and respiration in detached leaves and mesophyll protoplasts of Pisum sativum. Acta Physiol Plant 20:313–322
Printz B, Lutts S, Hausman JF, Sergeant K (2016) Copper trafficking in plants and its implication on cell wall dynamics. Front Plant Sci 7:601
Rai PK (2016) Biodiversity of roadside plants and their response to air pollution in an Indo-Burma hotspot region: implications for urban ecosystem restoration. Asia Pac Bus Rev 9(1):47–55
Rai PK, Panda LS (2015) Roadside plants as bioindicators of air pollution in an industrial region, Rourkela, India. Int J Adv Res Technol 4(1):14–36

Ramos-Montaño C (2020) Vehicular emissions effect on the physiology and health status of five tree species in a Bogota, Colombia urban forest. Rev Biol Trop 68
Rehman M, Liu L, Wang Q, Saleem MH, Bashir S, Ullah S, Peng D (2019) Copper environmental toxicology, recent advances, and future outlook: a review. Environ Sci Pollut Res 26:18003–18016
Rizwan M, Ali S, Adrees M, Ibrahim M, Tsang DCW, Zia-Ur-Rehman M, Zahir ZA, Rinklebe J, Tack FMG, Ok YS (2017) A critical review on effects, tolerance mechanisms, and management of cadmium in vegetables. Chemosphere 182:90–105. https://doi.org/10.1016/j.chemosphere.2017.05.013
Roy S, Gupta SK, Prakash J, Habib G, Baudh K, Nasr M (2019) Ecological and human health risk assessment of heavy metal contamination in road dust in the National Capital Territory (NCT) of Delhi, India. Environ Sci Pollut Res 26(29):30413–30425
Santra SC (2001) Environmental science. New Central Book Agency, Kolkata, pp 301–314
Sengar RS, Gautam M, Sengar RS, Garg SK, Sengar K, Chaudhary R (2008) Lead stress effects on physiochemical activities of higher plants. Rev Environ Contam Toxicol 196:73–93
Seregin IV, Ivanov VB (2001) Physiological aspects of cadmium and lead toxic effects on higher plants. Russ J Plant Physiol 48(4):523–544
Sethy SK, Ghosh S (2013) Effect of heavy metals on germination of seeds. J Nat Sci Biol Med 4:272
Shakeel T, Hussain M, Shah GM, Gul I (2022) Impact of vehicular emissions on anatomical and morphological characteristics of vascular plants: a comparative study. Chemosphere 287:131937
Sharma P, Dubey RS (2005) Lead toxicity in plants. Brazilian J Plant Physiol 17(1):35–52
Singh H, Sharma R, Savita SM, Kumar M, Verma A, Ansari MW, Negi M, Sharma SK (2018) Adaptive physiological response of Parthenium hysterophorus to elevated atmospheric CO2 concentration. Indian For 144(1):6–19
Stevens RG, Creissen GP, Mullineaux PM (1997) Cloning and characterization of a cytosolic glutathione reductase cDNA from pea (*Pisum sativum* L.) and its expression in response to stress. Plant Mol Biol 35(5):641–654
UNEP/GPA (2006) The state of the marine environment: trends and processes. UNEP/GPA Coordination Office, The Hague
Vaillant N, Monnet F, Hitmi A, Sallanon H, Coudret A (2005) Comparative study of responses in four Datura species to zinc stress. Chemosphere 59:1005–1013
Verma S, Dubey RS (2003) Lead toxicity induces lipid peroxidation and alters the activities of antioxidant enzymes in growing rice plants. Plant Sci 164(4):645–655
Viard B, Pihan F, Promeyrat S, Pihan JC (2004) Integrated assessment of heavy metal (Pb, Zn, Cd) highway pollution: bioaccumulation in soil, Graminaceae, and land snails. Chemosphere 55(10):1349–1359
Wagh ND, Shukla PV, Tambe SB, Ingle ST (2006) Biological monitoring of roadside plants exposed to vehicular pollution in Jalgaon city. J Environ Biol 37(2):419–421
Wang YM, Tang DD, Yuan XY, Uchimiya M, Li JZ, Li ZY, Luo ZC, Xu ZW, Sun SG (2020) Effect of amendments on soil Cd sorption and trophic transfer of Cd and mineral nutrition along the food chain. Ecotoxicol Environ Saf 189:110045
Weitekamp CA, Stevens T, Stewart MJ, Bhave P, Gilmour MI (2020) Health effects from freshly emitted versus oxidatively or photochemically aged air pollutants. Sci Total Environ 704:135772. https://doi.org/10.1016/j.scitotenv.2019.135772
Wu Z-z, Zhang Y-x, Yang J-y, Zhou Y, Wang C-q (2020) Effect of vanadium on Testa, seed germination, and subsequent seedling growth of alfalfa (*Medicago sativa* L.). J Plant Growth Regul 40:1566–1513
Wuana RA, Okieimen FE (2011) Heavy metals in contaminated soils: a review of sources, chemistry, risks and best available strategies for remediation. Int Sch Res Notices 2011:402647

Xu H, Wang W, Wang H, Sun Y, Zhong Z, Wang S (2019) Differences in quantity and composition of leaf particulate matter and morphological structures in three evergreen trees and their association in Harbin, China. Environ Pollut 252:1772–1790

Yuan M, Li X, Xiao J, Wang S (2011) Molecular and functional analyses of COPT/Ctr-type copper transporter-like gene family in rice. BMC Plant Biol 11:69

Yuan Z, Luo T, Liu X, Hua H, Zhuang Y, Zhang X, Zhang L, Zhang Y, Xu W, Ren J (2019) Tracing anthropogenic cadmium emissions: from sources to pollution. Sci Total Environ 676: 87–96

Yunus M, Singh N, Iqbal M (1996) Global status of air pollution: an overview. Wiley, Chichester

Zhang H, Yang J, Li W, Chen Y, Lu H, Zhao S, Li D, Wei M, Li C (2019) PuHSFA4a enhances tolerance to excess zinc by regulating reactive oxygen species production and root development in Populus. Plant Physiol 180:2254–2271

Heavy Metal Pollution in Atmosphere from Vehicular Emission

9

Shreya Singh and Ningombam Linthoingambi Devi

Abstract

The transport sector contributes a significant source of air pollution in metro cities. It emits various types of organic and inorganic pollutants in the form of gaseous or particulate matter into the atmosphere. Heavy metals generated by automobiles can be dangerous to people and the environment because they disrupt ecosystems and contaminate air, water, and soil, and some heavy metals can cause toxicity even at a low amount of exposure. Chromium, nickel, and cadmium can cause cancer in living beings, and their distribution and depositions are heavily influenced by meteorological parameters and topographical factors. Zinc, copper, and lead were released from vehicles, approximately 90% of the overall emitted quantity. So, there is a need for the creation of an efficient sustainable policy and control plan for metropolitan zones, which includes the upgradation of new advanced technologies to reduce the pollutant emission, the implementation of sustainable biofuels, the expansion of the greenbelt development, and the public engagement to increase the quality of air. This chapter will be emphasizing the emission level of heavy metals from (a) various transportation sectors, (b) the type of fuels used in heavy and light vehicles, and (c) the type of tires used in a vehicle and also will extend the health risk assessment of people located near road sites/cities.

Keywords

Heavy metal · Air pollution · Vehicular exhaust · Human health

S. Singh · N. L. Devi (✉)
Department of Environmental Science, Central University of South Bihar, Gaya, India
e-mail: nldevi@cub.ac.in

R. P. Singh et al. (eds.), *Heavy Metal Toxicity: Environmental Concerns, Remediation and Opportunities*, https://doi.org/10.1007/978-981-99-0397-9_9

9.1 Introduction

The ambient air quality deteriorated due to automobile exhaust emissions with increasing transportation facilities and rapid urbanization. Vehicle emission levels have risen in response to the growing vehicle population, serving as the primary driver of increased heavy metals in the atmosphere. Although the importance of motor vehicles cannot be denied for general socioeconomic development, it led to various harmful emissions and is identified with many environmental pollution problems (Nagpure et al. 2016). Road transport contributes significant emissions of heavy metals, resulting in serious health and environmental impacts. Transportation influences the regional and global environment in several ways, and for some contaminants like heavy metals, the public vehicle sector adds the largest to environmental externalities (Button and Nijkamp 1997). Globally, the number of vehicles is increasing about 5% per year faster than the world's population, compared to 2% of the population (Sharma et al. 2008). The number of vehicles registered has increased significantly in India. The overall figure of registered vehicles raised from 300,000 to 295.8 million from 1951 to 2019 (Ministry of Road Transport and Highways Yearbook 2021). Similarly, the overall figure of growth in the human population is also observed from 361 million to over 1000 million during this phase (Ministry of Road Transport and Highways Yearbook 2021). The increasing trend of motor vehicles registered from the year 1951 to 2019 is clearly shown in Fig. 9.1. India consumes 25% of energy in the road sector, and it is the main contribution of metal pollution through vehicular exhaust (Ministry of Road Transport and Highways Yearbook 2021). Motor vehicle exhaust contributes more to air pollution, and it's the main culprit to harm human health and the surrounding environment (Meister et al. 2012). Vehicular air contamination is well known for releasing numerous types of volatile organic compounds into the air, but

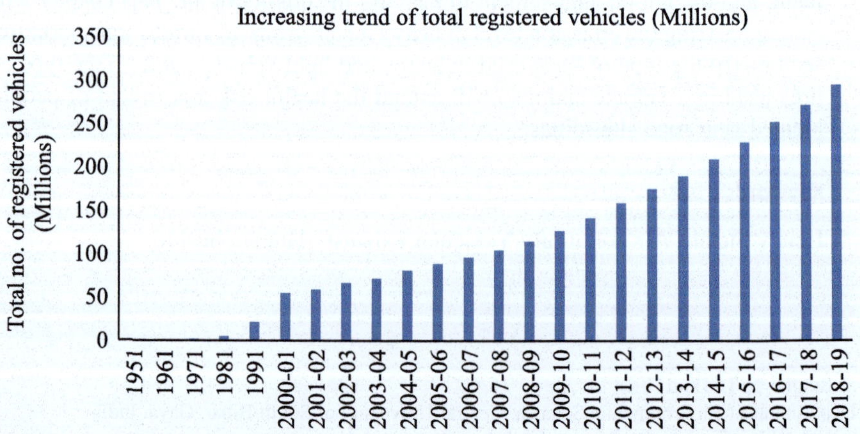

Fig. 9.1 Increasing trend of motor vehicles registered for the year 1951 to 2019 in India (source: Ministry of Road Transport and Highways Yearbook 2021)

new research identifies that the transport sector plays a significant role in heavy metal contamination in metropolitan areas via processes such as gasoline and diesel exhaust, engine wear, tire and brake pad wear, decay of metallic parts, and low quality of road caused by vehicular transit (Acosta et al. 2014). Motor vehicles are one of the most remarkable sources of particulate heavy metal emissions into the atmosphere (Suryawanshi et al. 2016). Various studies reveal that other than fuel combustion, brake and tire wear particles contribute heavy metals to an urban environment (Tanushree et al. 2011). Metropolitan areas have a large population and high levels of motor traffic, so publics are exposed to high levels of heavy metal pollution, leading to serious health-related complications. Environmental distress has become one of the most important problems in transportation policy discussions, as in several other portions in the world air pollution from motor vehicles has become one of the most intense and rapidly growing issues. In developing countries like India, where urbanization trends are increasing and air emissions from the automobile sector will increase further in the future, strict regulations and standards for heavy metals are needed for better air quality. There is a need for the creation of an environment sustainable management policy and control plan to combat air pollution in metropolitan areas, which includes the advancement of new technologies to reduce pollutant emissions. A global step is taken for the implementation of sustainable biofuels in the transport sector and expansion toward greenbelt development and public engagement for the betterment of air quality. Bica et al. (2014) reported a case study of greenbelt implementation in Timisoara, Romania, and its surroundings and found that its successful implementation leads to improvement in air quality. Greenbelts bring several profits to the local environment, it conveys a positive inspiration to the surroundings by lowering the daytime temperature, and they can also purify the germs from the air (Bica et al. 2014).

9.2 Heavy Metal Pollution from Automobiles

The sources of heavy metals through vehicular emissions include lead-acid batteries, fuel, and road dust associated with a tire, brakes, and clutch wearing. Various heavy metals were released from different parts of vehicles through abrasion. Heavy metals such as Cd, Cu, Ni, Fe, Zn, and Pb are released through different parts of vehicles. Lead is employed in fuel as an anti-knocking agent to extend their octane levels in previous days; however, it is banned in India due to its negative consequences (Suryawanshi et al. 2016). Although leaded fuel was phased out a decade ago, it was found in resuspended road dust, indicating a prior use of lead in fuel and its use in other vehicle parts (Hwang et al. 2016). Sources of different heavy metal pollution from automobiles are given in Table 9.1. Automobile emission is the major source of Zn within the metropolitan atmosphere because zinc is employed in tire tread as zinc oxide and its use in other vehicle parts makes its concentration high in non-exhaust emissions (Gunawardena et al. 2012). Furthermore, various heavy metals, such as cadmium and zinc, mostly arise from automobile exhaust in urban areas (Lv et al. 2018). Lead is added to gasoline and other automotive parts of heavy and light

Table 9.1 Sources of heavy metal pollution from automobiles

Heavy metal	Sources	References
Fe	Tire tread and tire dust pollution	Fukuzaki et al. (1986)
Ba	Brake pedal dust	Sternbeck et al. (2002)
Sb	Brake linings	Hjortenkrans et al. (2008)
Mo	Tire tread and tire dust pollution	Fukuzaki et al. (1986)
Cu	Automobile emissions	Xia et al. (2011)
Pb	Lead gasoline	Yang et al. (2003)
Cd	Tire wear	Hjortenkrans et al. (2008)
Zn	Vehicle emissions, rubber tire wear	Hjortenkrans et al. (2008)
Ni	Petroleum burning	Tian et al. (2012)
Cr	Oil burning	Tian et al. (2010)
V	Burning of fossil fuels such as oil	Lin and Lin (2005)
Mn	Gasoline antiknock additive	Loranger and Zayed (1995)
Ti	Brake wear	Adamiec et al. (2016)

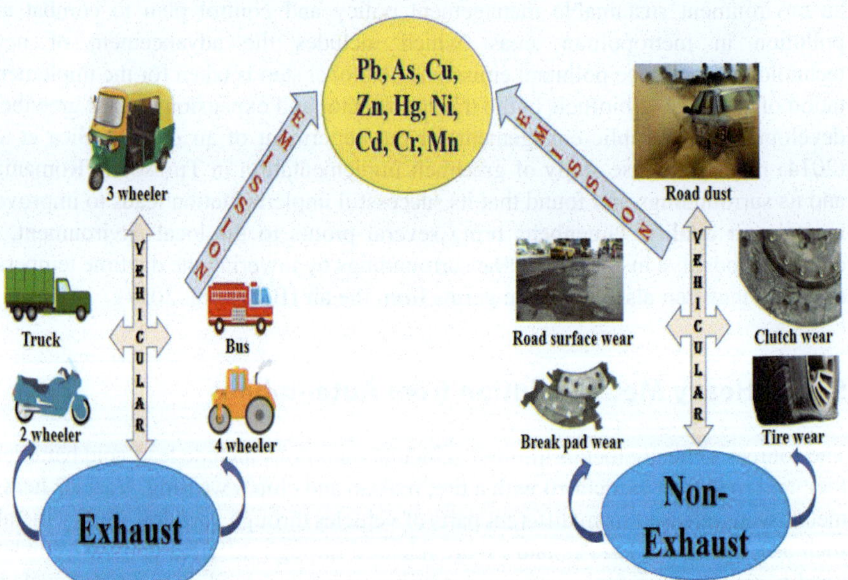

Fig. 9.2 Heavy metal pollution from exhaust- and non-exhaust-mediated automobile emission

vehicles, including wheel balancing weights, wheel rims, and batteries, responsible for lead pollution in metropolitan cities (Hwang et al. 2016). To make smooth braking and avoid brakes from trembling, copper has been used in brake pads, responsible for the heavy metal emission from automobiles (Hwang et al. 2016). Figure 9.2 shows heavy metal pollution from exhaust- and non-exhaust-mediated automobile emissions.

9.2.1 Exhaust-Generated Heavy Metal Pollution

During the combustion process, diesel engine exhaust produces various heavy metals, trace elements, and other toxins absorbed into the surface of the particulate matter. Engine speed and load, type of engine used, fuel composition, lubrication oil type, and emission control technologies all influence the particle extent distribution and chemical makeup of diesel exhaust giving out (Environmental Protection Agency 1998). A study by the EPA found that approximately 90% of particles less than 1 μ and 94% of particles less than 2.5 μ have adverse health consequences when released from diesel fuel (Health Effects Institute 1995). Additionally, the EPA noted that it is unclear if the threat posed by diesel emissions has diminished over time as engine technology has advanced (USEPA 2000). The extent of diesel exhaust's toxicity depends on the particle size and chemical composition of the emission, both of which change as technology advances. Other fuels such as petrol and gasoline are equally responsible for heavy metal pollution from vehicles by the combustion process, resulting in serious health effects. It is necessary to investigate both emissions of mass concentration of particulate matter and chemical composition from diesel exhaust, as well as other fuels utilized in a vehicle to track the pollution of heavy metal from the automobile sector. Heavy metal exposure from the automobile sector has negative health consequences, including poisoning (Chow et al. 1994). Calcium, phosphorus, and zinc are commonly found as additives in engine lubricating oil. Some of the metals like zinc, iron, phosphorus, lanthanum, and barium are also detected more in exhaust at an emissions rate of 1.65 mg/km (Lowenthal et al. 1994). Maher et al. (2008) studied that Pb-enriched resuspended soil dust is found along city roadside, indicating the use of leaded fuel in prior decades. Cd, Mn, and Pb are mostly linked to motor fuel leakage, as well as oil lubricants and vehicle abrasion (Bourliva et al. 2017).

9.2.2 Non-Exhaust-Generated Heavy Metal Pollution

The complexity of non-exhaust vehicle emissions is exacerbated by the large range of materials used. Dust is produced by a variety of causes during vehicle operation, including the wear of the braking system, wear of clutch drive plates, degradation of the dynamic parts of the catalytic converter, and road dust resuspension (Adachi and Tainosho 2004). The amounts of major heavy metals in roadside dust have been studied and found Mn, As, Zn, Pb, Cr, Cd, Hg, and Ni levels ranging from 227.9 to 2765, 0.1 to 0.5, 19 to 1540, 20.1 to 123, 0.3 to 0.7, 37.4 to 398.6, 0.1 to 0.9, and 0.1 to 0.7 mg/kg, respectively (Al-Taani et al. 2019). Tire wear and road surface abrasion are well-known contributors to road dust pollution (Apeagyei et al. 2011). The primary causes of non-exhaust vehicular emissions that cause road dust are tire and brake wear, clutch wear, road surface wear, other vehicle parts, and road component deterioration (Adamiec et al. 2016). Metals, particularly Cu, Ti, and Cr, are well-known non-exhaust brake wear key tracers, revealing that road dust pollution was largely produced by brake lining and tire wear. A natural source of fine particles includes the physical and chemical destruction of shoulders and pavement, as well as background dust and windblown dirt. A wide range of materials, including glass and polymer fibers and brass chips that conduct heat, are used to make brake linings for cars, causing heavy metal pollution (Chan and Stachowiak 2004). Some of the meals were significantly released from tire wear dust. During vulcanization, ZnS and ZnO were added to the tire, making Zn the most prevalent heavy metal in tire wear (Fauser et al. 1999). According to Councell et al. (2004), wear particles are another important source of Zn in metropolitan areas. Pb is produced by tire and brake pad wear (Sansalone et al. 1996). Fukuzaki et al. (1986) also investigated the metals emitted from tire tread and tire particulate matter added Mn, Zn, Ni, Cd, Pb, and Fe into the air in the form of airborne particulate matter. Wear from brake systems is another cause of road dust. Brake in vehicles is subjected to a lot of heat produced from friction during quick braking; it is transferred to the brake discs and causes particles to be added to air (Sternbeck et al. 2002). The source of Cu, Sb, and Ba in aerosol has been identified as brake dust, which accounts for 47% of the total Cu loading in a metropolitan area runoff (Davis et al. 2001). Metal emissions through brake dust and tire tread are suspended in road dust, responsible for non-exhaust heavy metal contamination. This demonstrates that road abrasion and vehicle activity have a significant influence on the levels of heavy metals in road dust, which can have a variety of negative health impacts.

9.3 Factors Affecting Heavy Metal Pollution Through Vehicular Emissions

Numerous factors are directly or indirectly influencing the heavy metal concentrations in the atmosphere generated by vehicular emissions. The degree of pollution is mainly dependent on the various meteorological parameters, climatic conditions of the area, geographical locations (latitude and altitude), topography (relief, plain,) and environmental conditions (stability, instability). Several factors influence road abrasion, including vehicle speed, climate, road moistness, asphalt type, and the percentage of heavy-duty vehicles on the road (Gupta 2020). Heavy metal concentration is also depending upon the type of fuel used in a vehicle (gasoline, petrol, diesel, etc.), the type of vehicle used for road transportation (two-wheeler, three-wheeler, four-wheeler, light-duty, and heavy-duty automobiles, etc.), the type of material used for the road construction (asphalt, concrete), the type of engine used in a vehicle (two strokes, four strokes), vehicle speed, traffic congestion, average vehicle kilometers travelled for different vehicle types, etc.

9.3.1 Climatic and Environmental Factors

Meteorological variables such as relative humidity (RH), wind speed (WS), wind direction (WD), and temperature are responsible for the local weather pattern, and they can influence the levels of heavy metals in the atmosphere. Long-term variation in the meteorological conditions governs climatology, which directly influences the fate of heavy metal pollution in the atmosphere. Wind speed, wind direction, rainfall frequency, intensity profiles, and fog duration have been shown to play a role in the transportation of pollutants (Piron-Frenet et al. 1994; Melaku et al. 2008). The spatial dispersion of heavy metal is heavily influenced by wind speed and direction. The higher the wind speed, the higher will be the dispersion of pollutants in ambient air. Rainfall frequency also determines the pollution level; rain sweeps away contaminants in the atmosphere, lowering heavy metal concentrations in the process. Vehicle emissions are concentrated around the sources during foggy conditions and stable environmental conditions so the heavy metal distribution is negligible while there is more dispersion of heavy metal pollution in unstable environmental conditions (Gupta 2020). The "canyon" effect of buildings in the city's center and the dilution of heavy metals released by autos and the paths for their dispersion are more affected by the inner reaches (Namdeo et al. 1999).

9.3.2 Geographical and Topographical Factors

Geographical conditions of a particular area are responsible for the fate and long-range transport of contamination of heavy metal in ambient air. The distribution and deposition of metals are strongly influenced by meteorological and topographical factors (Suryawanshi et al. 2016). The topographical feature has directly affected the transportation of pollutants in the air. Due to the mountain chain, dispersal is less common in hilly areas, while it is more common in plain and desert-like conditions where relief is not higher. The heavy metal pollution load on highways is higher in mountainous terrain, which has more twists and tunnels in comparison to the routes in flat areas.

9.3.3 Types of Fuel Used in a Vehicle

Different types of fuel can be used in the transportation sector such as diesel, petrol, biofuel, and gaseous fuel. The two leading fossil fuels used to power automobiles on the road are gasoline and diesel and contain several potentially hazardous heavy metals. According to published data by the Press Information Bureau, the Government of India, the automobile sector is the major user of petrol in India accounting for 99.6% of the total petrol sales (Ministry of Petroleum and Natural Gas 2014). Additionally, a major increase in fuel consumption by the road transportation sector is caused by the shifting of passengers from train to roads, which leads to increased emissions of various pollutants (Ministry of Road Transport and Highways Yearbook 2021). Sector-wise consumptions of petrol for two-wheeler, three-wheeler, four-wheeler, and others are given in Fig. 9.3. The percentage share of sector-wise consumption of diesel for different vehicles is shown in Fig. 9.4.

Automobiles powered by gasoline are reported to be the main source of lead in non-rural areas (Kannan 1991; Jeba Rajasekhar et al. 2001). Pb was the second most

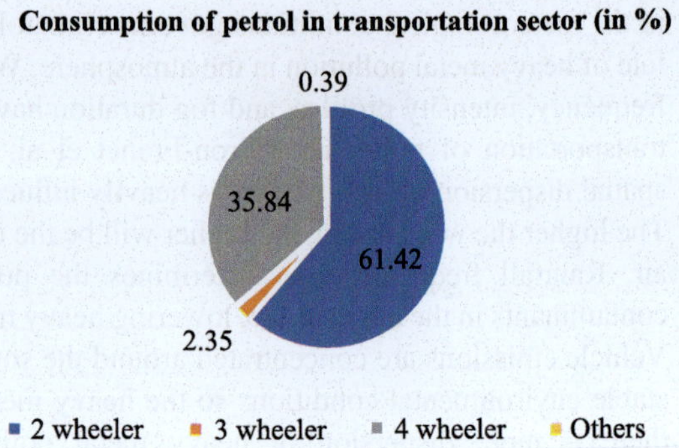

Fig. 9.3 The percentage share of petrol consumption for different types of vehicles (source: Ministry of Petroleum and Natural Gas 2014)

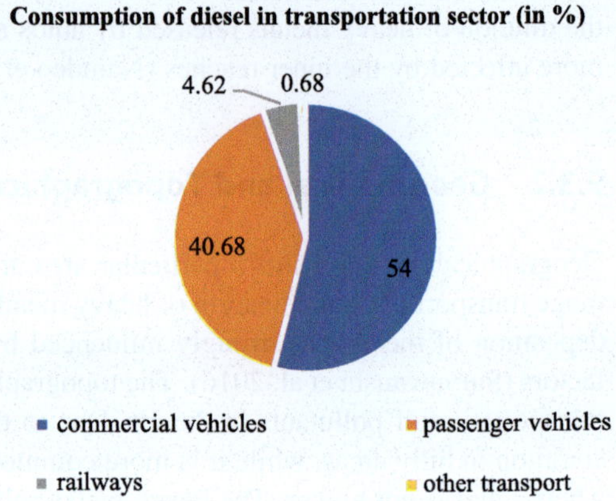

Fig. 9.4 The percentage share of diesel consumption for different types of vehicles (source: Ministry of Petroleum and Natural Gas 2014)

often found heavy element, despite the fact that the use of leaded gasoline was dropped a period ago. Previously produced lead is linked with pavement soil and is resuspended into the air, as well as vehicular-related scrape products (Gunawardena et al. 2012). Apart from engine improvements, catalytic converters and unleaded gasoline have contributed to minimizing car emissions, yet diesel exhaust is the most common source of heavy metals from vehicle transport, accounting for between 62% and 93% of total emissions (Liu et al. 2018a, 2018b). Vehicle fuel burning processes may produce Cr (Song and Gao 2011), and vehicle emissions may also contain Ni and Pb (Song and Gao 2011; Zheng et al. 2013). Copper, cadmium, chromium, arsenic, selenium, zinc, lead, mercury, and nickel are the principal heavy metals contributed by the combustion of fossil fuel (Pulles et al. 2012). Pulles et al. (2012) estimated heavy metal concentration in diesel and petrol, which is used as a fuel for automobiles in Europe as given in Fig. 9.5. Zinc is the major heavy metal emitted from both fuels; however, other heavy metals like Cr, Cu, and Hg are also emitted in greater concentration in both fuels (Pulles et al. 2012). Due to increasing morning rush hours with a majority of diesel vehicles, diesel-fueled vehicles are the major contributor to toxic heavy metal generation (Silva 2020).

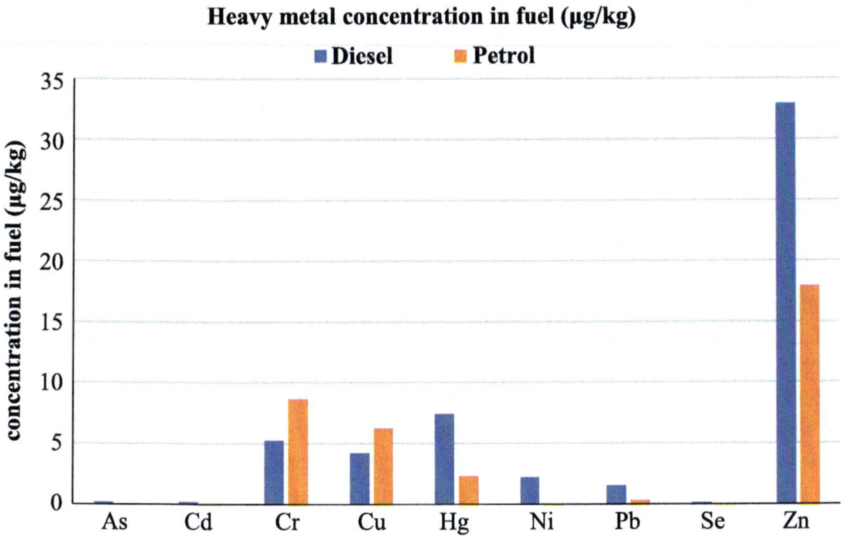

Fig. 9.5 Heavy metal concentration in petrol and diesel oil in Europe (source: Pulles et al. 2012)

9.3.4 Traffic Density Factor

High traffic density is a prominent factor responsible for vehicular-mediated emission of metal contamination in the atmosphere. In urban areas, traffic congestion is a common issue, contributing to a major fraction of heavy metal pollution through mobile and stationary traffic. Three of the most common heavy metals emitted by vehicular traffic are zinc, copper, and lead, accounting for at least 90% of the overall emitted quantity (Popescu et al. 2011). Some other heavy metals are also correlated with traffic density, such as nickel and cadmium (Gupta 2020). When traffic volume is higher, the increased vehicular load results in higher vehicle emissions of heavy metals. High traffic places, such as bus stops, petrol stations, and road crossroads and junctions, frequently saw peaks in heavy metal pollution. High traffic regions have been identified as a significant source of pollutants: chromium, antimony, nickel, vanadium, cobalt, antimony, vanadium, and chromium, (Silva 2020). Lead content in metropolitan areas is higher during working days, although cadmium content is stable (Popescu et al. 2011). Regrettably, there is only a rudimentary acceptance of the traffic characteristics and its effect from transportation (Gunawardena et al. 2012). Vehicle traffic and wear particles from tires are the main sources of Zn in the urban air (Kemp 2002; Councell et al. 2004). Heavy-duty traffic sources are more likely to generate Cu and Cr (Gunawardena et al. 2012). The surge in vehicle traffic may be to blame for Nagpur's ambient air's variable concentrations of heavy metals (Chaudhari et al. 2012).

9.3.5 Road-Associated Factors

Heavy metal emissions from automobiles also depend upon the road surface type and age of the road. Silva et al. (2015) reported a significant relationship between metal accumulation in Melbourne roadside soil and different road age (New, medium, and old). Generally, the road surface is made up of asphalt or concrete. Driving on a concrete highway causes greater tire abrasion, resulting in more heavy metal emissions compared to driving on a smooth highway because the concrete surface is much more uneven than the smooth surface, so automobiles on a concrete roadway experience more tire abrasion than vehicles on an asphalt highway (Duong and Lee 2011). In addition, cars travelling on a bumpy surface consume more energy, so there is more fuel consumption, and eventually the heavy metal emissions are also high. The width of the road, surface type, lateral clearance, and smoothness of the road affects the pollution level, which has a direct impact on vehicular-mediated heavy metal pollution (Gupta 2020). Nowadays, waste material is used for the construction of roads, which can contribute to the heavy metal emissions in the environment through road abrasion and leaching. Schwab et al. (2014) studied the impact of heavy metal leaching from road construction made up of industrial waste material and found that potentially toxic heavy metals are released from the site. Chowdhury et al. (2010) used various leaching data for heavy metal (Pb, Cd, Cr, Hg, Cu, As, Co) in mg/kg for industrial waste products such as the bottom, recycled concrete pavement, and fly ash, which is used as a construction material in road making and reported lifecycle-based environmental impact assessment of this construction material used in road making.

9.3.6 Vehicle-Associated Factors

Heavy metal pollution is further influenced by factors such as vehicle age, the type of vehicle being used for transportation, its condition and frequency of service, engine type (two-stroke vs. four-stroke), and vehicle speed. If older automobiles are not properly maintained, they will produce more pollutants than modern vehicles. Servicing frequency of the vehicles contributes to the emission level, and maintained vehicles generate less pollution. Vehicles with four-stroke engines emit fewer pollutants than those with two-stroke engines. Vehicles equipped with a catalytic converter will also generate less pollution. Silicon, zinc, calcium, copper, and phosphorus are among the metallic elements emitted by four-stroke heavy-duty engines, whereas lead, manganese, calcium, zinc, and chromium are among the metallic elements emitted by two-stroke engines (Hare 1977). Kebede et al. (2022) reported on several types of public transportation vehicles and investigated the impact of vehicle mileage and age on contribution of particulate matter and other pollutants, reporting that emissions from old minibus are four times more in comparison to old large buses, promoting large bus with high passenger-carrying capacity because it can significantly reduce the pollution levels as per passenger kilometers travelled in Addis Ababa region. There is a remarkable difference observed in emission levels concerning age groups of vehicles (year of manufacture) and size of the vehicles (Kebede et al. 2022).

9.3.6.1 Type of Tires Used in a Vehicle

The road-tire contact could be a major source of particle emissions like heavy metal from automobiles. Dahl et al. (2006) calculated emission factors for particles in the size 15–700 nm generated by the road-tire interface for studded and non-studded tires as 3.8×10^{11} and 6.1×10^{11}, respectively, on the same road surface at 50 km/h speed. The overall number of emission factors was more affected by vehicle speed than by tire or pavement type, with greater speeds creating more particles (Dahl et al. 2006). Hussein et al. (2008) estimated road particle emissions in a different type of tires (studded, friction, summer) in the Stockholm region and found that studded tires, regardless of the asphalt type, produce more pollution than friction and summer tires due to more road surface wear. Studded tires produce more particles than friction tire, while Zn is more enriched in friction tires for all particle sizes (Gustafsson et al. 2008).

9.3.6.2 Vehicle Speed

When calculating emission factors, one of the crucial variables is the average speed of each kind of vehicle (Sun et al. 2016; Zhang et al. 2014). The high quantities of heavy metals in road dust along highways are most likely due to the increased emissions by tire and road abrasion and by high-speed vehicles (Duong and Lee 2011). Silva et al. (2015) studied roadside soil of heavy metals in western Melbourne and found the speed of the vehicles is the prominent factor governing heavy metal pollution. Speed limits have an impact on pollution levels since quicker driving increases fuel consumption. As a result, lowering speed restrictions on highways is predicted to reduce both fuel consumption and pollution emissions. The amount of heavy metals of vehicle speed affects the quantity of heavy metals in road dust, resulting in increased tire wear and increased fuel combustion due to higher speeds. Sb release from vehicles, accumulated in roadside soil, is due to the acceleration and speed of the vehicles (Iijima et al. 2008). Speed bumps are installed in a city to control vehicle speed, which is a reason for increasing emission levels of particulate matter, during braking (Baltrėnas et al. 2017).

9.4 Emission Level of Heavy Metals from the Various Transportation Sectors

According to vehicle categorization (Kenya National Bureau of Statistics 2017), the Kenya National Bureau of Statistics (NBSC) in China utilizes the National Statistical Yearbook to characterize the yearly vehicle population in the following categories: light, medium, and heavy passenger automobile vehicles. Different categories of vehicles led to various heavy metal emission factors, which had serious environmental impacts. Two-wheelers lead the road traffic and become a main source of atmospheric heavy metal contamination in India and most developing countries due

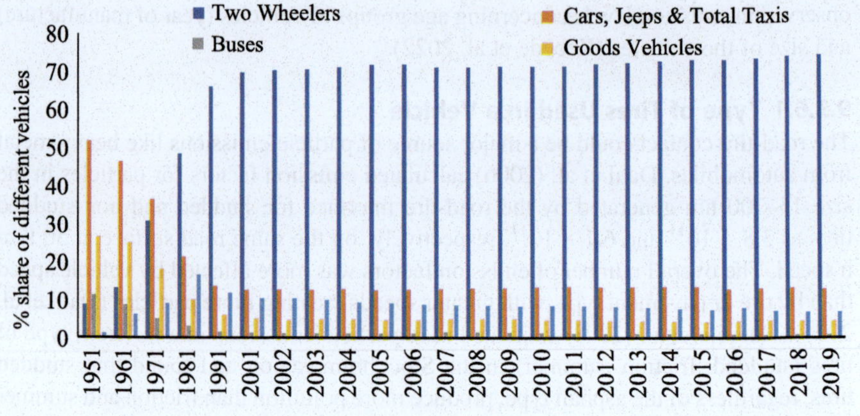

Fig. 9.6 Trends in the category-wise share of vehicles (in %) in total registered vehicles further on years 1951 to 2019 in India (source: Ministry of Road Transport and Highways Yearbook 2021)

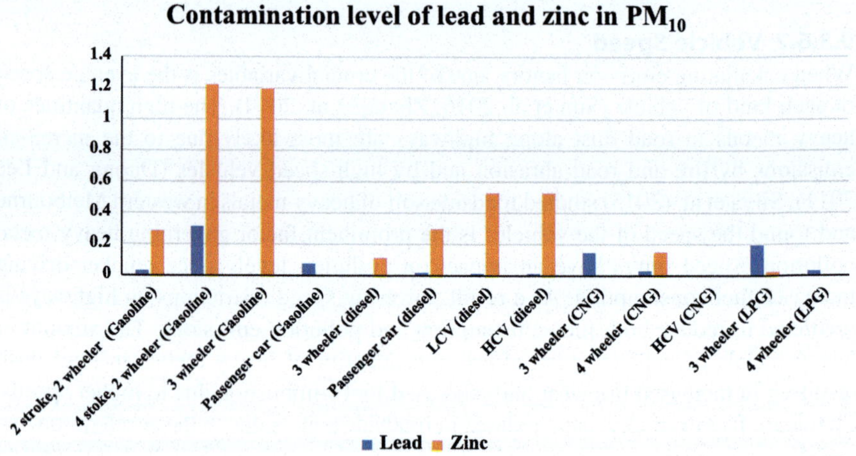

Fig. 9.7 Percentage distribution of heavy metals in PM_{10} for different categories of vehicles (source: CPCB national summary report, 2011)

to their simple availability and low cost. Figure 9.6 shows trends in the category-wise share of vehicles (%) in total registered vehicles from the year 1951 to 2019 in India, which indicates two-wheeler trends over these years (source: Ministry of Road Transport and Highways Yearbook 2021).

In India, the number of two-wheelers accounted for 70% of the total vehicle production, responsible for vehicular emissions (Ministry of Road Transport and Highways Yearbook 2021). Different vehicle categories, types of fuel used, and engine types are the prominent factors responsible for vehicular heavy metal pollution. Figure 9.7 shows heavy metal percentage distribution in PM_{10} for different vehicles, for different engines, and for different fuels used in a vehicle. Two- and three-wheeler gasoline-used vehicles dominate the heavy metal pollution followed

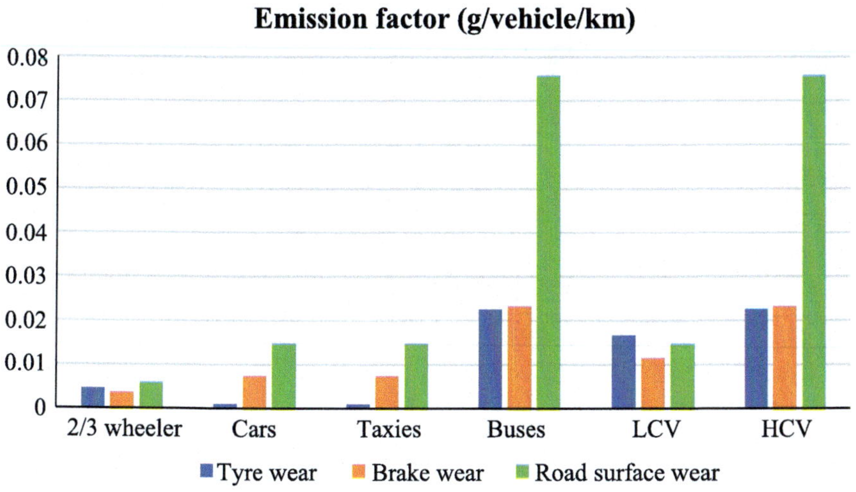

Fig. 9.8 Non-exhaust emission factor (g/vehicle/km) from different vehicle categories of megacity Delhi (Source: Kumari et al. 2013)

by light and heavy commercial diesel-used vehicles. In gasoline exhaust, elements such as wear metal (Fe, Pb, and Cu) are found to be higher than in diesel emissions, and the percentage distribution of heavy metal (Zn and Pb) is higher in gasoline fuel as compared to the other fuel in different vehicle category (CPCB national summary report, 2011).

Figure 9.8 shows non-exhaust emission factors for two- and three-wheelers, cars, taxis, busses, light commercial vehicles (LCV), and heavy commercial vehicles (HCV) in the megacity of Delhi. Road surface wear is the major non-exhaust vehicle emission factor responsible for heavy metal pollution in the megacity of Delhi, followed by brake wear and tire wear for different vehicle categories (Kumari et al. 2013).

9.5 Global Contamination Level of Heavy Metals in Air Through Vehicular Emission

Table 9.2 summarizes the heavy metal pollution in road dust and brake pad dust in various transportation-related regions. The amount of Cr and Hg in road dust was much higher, indicating that they were primarily anthropogenic in origin (Al-Taani et al. 2019). A total of seven different heavy metal species, including Cu, Pb, Mn, Cd, Ni, Zn, and Cr, were examined in the Gold Coast region of Australia due to automotive emissions, and zinc was found to be present at higher concentrations in each of them (Gunawardena et al. 2012). Although leaded fuel usage is banned across the world, its concentration in roadside soil is more due to the previous usage of leaded fuel in resuspended road dust (Maher et al. 2008). One reason for decreasing Pb levels in the city's air may be the introduction of unleaded gasoline

Table 9.2 Concentration (mg/kg) of heavy metal emission from the transportation sector

Location	Type	Pb	Zn	Ni	Cr	Cd	Mn	As	References
Bahrain	Road dust	742	67	12	9.6	1.5	–	–	Madany et al. (1994)
Iran, Shiraz	Road dust	115.7	403.5	77.5	67.2	0.5	438.5	6.58	Keshavarzi et al. (2015)
Jordan, Aqaba	Road dust	93–212	103–160	51–115	–	1.9–2.9	–	–	Al-Khashman (2007)
Oman, Muscat	Road dust	65	124	930	64	–	–	–	Abdul-wahab and Yaghi (2004)
Pakistan, Islamabad	Road dust	104	116	23	–	5	–	–	Faiz et al. (2009)
Poland, Katowice	Brake pad dust	2897	8310	369	789	2.34	–	–	Adamiec et al. (2016)
UAE, Abu Dhabi	Road dust	50.05	173.01	0.3	306.33	0.48	1158.5	0.23	Al-Taani et al. (2019)

in Nagpur in 1999 and the following phase-out of leaded fuel (Chaudhari et al. 2012). Kumari et al. (2013) estimated heavy metals from different mobile sources for both exhaust and non-exhaust emissions from megacity Delhi and found that Pb emits a major fraction from tire wear, despite the introduction of unleaded gasoline, while Cu was most significantly found in brake wear. There is more concern about heavy metal pollution such as Cr and Cd in China, while V was found to be much lower than the WHO limits (Duan and Tan 2013). In comparison to other cities across the world, in comparison to technologically sophisticated nations like the United States and the European Union, China's cities have significantly higher levels of atmospheric heavy metals, whereas cities in India and Pakistan have somewhat lower amounts (Duan and Tan 2013). Wang et al. (2019) estimated that $PM_{2.5}$ is associated with heavy metal contamination and suggested that vehicle exhaust and traffic-related dust contribute 11.26% of the emissions. The occurrence of heavy metals in the air of Nagpur, Maharashtra, India, between 2001 and 2006 is estimated by Chaudhari et al. (2012), and because of the significant rise in the number of cars, a total of six heavy metals in the following order, Fe > Zn > Cd > Pb > Ni > Cr in the year 2001 and Zn > Fe > Pb > Ni > Cd > Cr in the year 2006, results in significant vehicular pollution (Chaudhari et al. 2012)

9.6 Health Risk Assessment of People Located near Road Site

Health risk assessment is a method for calculating the quantifiable danger to human health presented by a variety of pollutants like heavy metals via multiple exposure paths (Kampa and Castanas 2008; Luo et al. 2012). Risk analysis is crucial in toxicological and public health investigations. To evaluate the health risk, the hazard quotient (HQ) and health index (HI) were utilized. The "total of more than one hazard quotient (HQ) for numerous toxic compounds or various exposure paths" is denoted as the "hazard index (HI)." As a result, by putting the HI of each acquaintance channel (dermal contact, ingestion, and inhalation) together, it is possible to establish a grouping of noncarcinogenic hazards for individuals from several exposure paths (US EPA 1989). If the HQ or HI number is lower than 1, it indicates that there is no considerable danger of noncarcinogenic consequences. If the value of HQ or HI is more than 1, it indicates that high likelihood of noncarcinogenic effects (USEPA 2001; Zheng et al. 2010). The most common route to be exposed to heavy metals is ingestion (Jena and Singh 2016). The primary routes of trace element exposure in humans were discovered to be ingestion and inhalation (Huang et al. 2018). According to the health risk assessment model, ingesting dust particles was responsible for more than 83% of the total noncarcinogenic risk (Najmeddin et al. 2018). Roy et al. (2019) calculated health risk based on HQ and suggested that assimilation was the main route of heavy metal exposure to humans from road dust. The danger level of metal exposure is more among children, especially in urban and suburban areas with a large population and high traffic density (Lv et al. 2018). In the human health risk evaluation, the noncarcinogenic values were reported to be below the threshold levels, and ingestion is the main way that both children and adults are exposed to heavy metals in Kermanshah province, Iran, for atmospheric dust generated by vehicular exhaust (Doabi et al. 2018). However, the hazard index value of Pb (0.459), a notably poisonous metal, came close to the safe limit in the city of Thessaloniki, Greece, and all metal concentrations in road dust were lower than the tolerable threshold for children (Bourliva et al. 2017). In Tianjin, China, the US EPA standard reported that the nine heavy metals that pose health risks when inhaled through the respiratory system, such as Cr, Pb, Mn, Zn, Cd, As, Ni, Cu, and Hg, are in decreasing order of risk (Chen et al. 2014). All heavy metals have deteriorating health impacts in the following order: High traffic density areas have lower carcinogenic risks for Pb, Cr, Ni, and Zn metals in PM_{10} via inhalation and derma for children and adults in Southwest, Iran (Goudarzi et al. 2018). Noncarcinogenic health hazard study indicated a high likelihood of deleterious effects on children, whereas adults are far below the acceptable limit in Kanpur, India (Izhar et al. 2016). Cancer risk decreased in the same order for both adults and children in the following order, Cr > Cd > Ni > As>Pb, which denotes a value higher than the permissible limits (Izhar et al. 2016). Huang et al. (2018) showed that Shanghai's air needed to be cleaned up and that the health dangers associated with ingesting atmospheric Pb must be taken into account. The carcinogenic risk assessment revealed that the residents of Dhanbad, India, have a very high risk of cancer, particularly among youngsters (10–18 times higher than adults) (Jena and Singh 2016). The amplification of HI values (10.8) for the transport sector, as well as the noticeably high HQ values provided by $PM_{2.5}$-bound Co and Ni, indicates a greater noncarcinogenic health risk due to inhalation exposure in Dhanbad, India (Jena et al. 2019). The carcinogenic health risk in adults in the transportation sector (1.57×10^{-4}) was considerably higher than the threshold value, demonstrating that adults are vulnerable to $PM_{2.5}$-bound trace elements (Jena et al. 2019). Kumari et al. (2021) performed a health risk associated with heavy metal concentrations in $PM_{2.5}$, PM_{10}, and road dust in Dhanbad, India, suggesting that automobiles are the major sources of heavy metal pollution and skin contact and ingestion were the key ways. Children were more impacted than adults in this study in terms of the noncarcinogenic threats presented by heavy metals, while the carcinogenic risk posed by heavy metals was negligible. However, the effects on children and adults that were verified to be noncarcinogenic are in the acceptable limits (Kumari et al. 2021). Higher levels and orally solubilization of heavy metals from traffic areas of dust indicated that there were more health concerns to residents in the Chengdu area, China, and found that for both adults and children due to ingestion and dermal contact (Li et al. 2017). According to an assessment based on the hazard index in Nanjing, China, the noncarcinogenic health hazard related to exposure to potentially toxic metals in road dust was within the acceptable range, with the exception of the polluted hotspots where exposure with Pb, Cr, and Cu may be dangerous for children (Liu et al. 2014). Liu et al. (2018a, 2018b) calculated the health risk assessment of heavy metals by traffic emissions in Beijing, China, suggesting that arsenic has the highest carcinogenic risk among all heavy metals and exceeded the accepted level. Peng et al. (2017) evaluated danger of heavy metals in automobiles being carcinogenic and not carcinogenic in a southern Chinese city, suggesting that health risk for heavy metals is within the acceptable limits; however, Cr had the highest risk among all heavy metals. Based on calculations of the hazard index (HI) for both carcinogenic and noncarcinogenic risks of toxic substances in street dust samples in Dhaka City, it was discovered that the noncarcinogenic effect of dermal contact between children and adults through exposure to street dust causes a greater risk for heavy metals (Rahman et al. (2019). Due to the high HI for Cr (1.04), which was slightly over the acceptable threshold 1, and Cd (0.69), which was near to the safe limit 1, children were more likely to have health issues. The cancer risk for As (9.59×10^{-7}) was closer to the upper limit of threshold values than the cancer risk for Cr (4.27×10^{-6}) which was within the range of threshold values (10^{-4} to 10^{-6}) (Rahman et al. 2019). Roy et al. (2019) calculated health risk based on hazard quotient and found that the toxic substances in the area of road dust in Delhi, India, had no possible danger of being noncarcinogenic; however, in road dust, metals have been proven to be dangerous to children. Ingestion of the material, followed by dermal contact, was the route of exposure that raised the risk the most for people exposed to road dust in China; however, adults faced larger health hazards than children, except in a few sites (Shi et al. 2011). The only carcinogenic metal with a mean risk greater than 10^{-6} is Cr and Cd which account for a 95% of the sum of the total cancer risk (Shi et al. 2011).

9.7 Impacts of Heavy Metal Pollution Through Vehicular Emissions

Impacts of heavy metals from vehicular emissions are seen in all environmental segments, by degrading their quality and contamination. It has diverse impacts on the environment, and it can cause harmful effects not only on humans but also on plants, animals, and microorganisms.

9.7.1 Impacts on the Environment

Heavy metals discharged into the atmosphere are eventually deposited on other environmental components such as the hydrosphere and lithosphere affecting the species of flora and fauna living in it (Gawade et al. 2016). Heavy metal released from vehicular emission gets deposited nearby roadside soil from the atmosphere by dry and wet deposits. Before depositing in the soil, it may travel far distances by long-range transport through air masses. From soil, it may get deposited in water bodies through surface runoff, and eventually it may also contaminate groundwater. Therefore, vehicular-mediated heavy metal pollution can contaminate the different environmental compartments. Pb shows a high binding capacity with soil and sediments (Union Europian 2002). Apart from the contamination of environmental media, it equally affects plants, animals, and microorganisms. For plants, the extent of heavy metal toxicity is different for different varieties of plants (Bajgai et al. 2013).

9.7.2 Impacts on Human Health

The detrimental impacts of traffic-related pollution on health have been widely recognized by researchers (Lim et al. 2005). Even at low concentrations in the air, heavy metals are harmful to human beings. After inhalation, they form complexes or interactions with necessary protein molecules, denaturing them and leading to cell malfunction or death (USEPA 1996). Heavy metals present in PM_{10}, and $PM_{2.5}$, are harmful to human health (Kumari et al. 2021). Increased morbidity and death in the community due to cancer, neurotoxicity, immunotoxicity, and cardiotoxicity are due to heavy metal contamination in the atmosphere (Dockery et al. 1993; Silbergeld 1995, 1996). Heavy metal pollution also has an impact on the human body's other organs, including the kidney, liver, lungs, and central nervous system. The impact of tire wear particles on human lung cells and macrophages was estimated by Karlsson et al. (2008). Gualtieri et al. (2005) reported the impact of tire breaking on human alveolar lung cells and found a substantial rise in DNA damage and cell death. Karlsson et al. (2011) estimated the impact of particles created by studded tires colliding with pavement on bronchial epithelia, macrophages, and nasal epithelia in human cell cultures. Cadmium pollution causes acute pulmonary effects, damages the kidney, and also affects the liver and gastrointestinal tract (Kolonel 1976). Mercury poisoning can harm the lungs, kidneys, and neurological system in people who are exposed to it for a short period of time (Abosede et al. 2017). Being the most dangerous of all the heavy metals, lead can have teratogenic consequences when it is ingested through food and drink as well as breathed (Ferner 2001). Acute and chronic central nervous system damage, peripheral nervous system damage, cardiovascular dysfunction, and the suppression of hemoglobin production are all additional effects of lead poisoning (Ogwugbu and Muhanga 2005). Lead and zinc poisoning has been shown to induce similar effects, which makes it simple to misdiagnose (McCluggage 1991). Arsenic coagulates the proteins to form complexes with coenzymes and prevents adenosine triphosphate (ATP) generation during respiration (INECAR 2000). According to the International Agency for Research on Cancer, Ni and Cr have a causative influence on cancer (Cao et al. 2014).

9.8 Regulation and Standards

It is crucial to examine vehicle emission inventories as well as automobile emission inclinations and characteristics over time to create more efficient automotive emission control rules and emission reduction plans (Lv et al. 2018). For private autos and conventional energy vehicles, vehicle population control, such as purchase restrictions, will help to decrease vehicle emissions and mitigate heavy metal pollution (Lv et al. 2018). Vehicle emission reduction for heavy metals is improved by the use of clean fuel, the best technological invention for reduction of atmospheric pollutants emitted from the exhaust, and the application of electric vehicles for public transportation, which can improve the air quality. In India, customers have switched to CNG or electric cars; therefore, the country's transportation industry uses very little gasoline and diesel in the states including Delhi, Haryana, Gujarat, and Odisha. By reducing the usage of diesel fuel, the detrimental effects on the respiratory system can be successfully reduced (Silva 2020). In India, standards for Pb, Ni, and As have been set up by the CPCB in the national ambient air quality standards, but not for other compounds, indicating a major gap for policymakers to improve the pollution level. For outdoor air quality improvement in India, various Bharat stage norms were implemented from time to time for attaining cleaner fuels. Two-wheelers dominate the road traffic in India, and Bharat stage III regulation norms were introduced to control emissions from two-wheeler vehicles (Suvarapu et al. 2018). India is shifting toward Bharat stage VI regulation norms by 2020 to achieve cleaner fuels and to regulate the emissions derived from motor vehicles. The upcoming Bharat stage VI regulation norms reduced the PM levels by around 82%, which can also cut down the heavy metals associated with PM (Pothumsetty et al. 2020). Table 9.3 presents the World Health Organization, the European Union, and national ambient air quality regulations in India and China's limits and requirements for several heavy metals in ambient air. Policymakers are advised to continue efforts to reduce heavy metals to enhance the quality of the air. The local government administration and related authorities must establish rigorous legislation to define local ambient limits for the presence of heavy metal pollution in the air in order to improve air quality.

9.9 Conclusion

Vehicular discharges are the major contributing source of heavy metals in the ambient atmosphere, through exhaust and non-exhaust emissions. In the air, it can travel a large distance by long-range transport. Meteorological parameters influence the dispersion and fate of pollution of heavy metals in the atmosphere. Geographical, topographical, vehicle speed, road pavement type, tire type, road congestion, type of vehicle, and type of fuel are some of the major factors, governing the pollution scenario of heavy metals in the atmosphere by vehicular emissions. Heavy metals generated by vehicles can contaminate different environmental media and deteriorate their quality. It can adversely affect public health and surroundings. There is an urgent need to set up a limit for individual heavy metals in the atmosphere for regulation and control. For better air quality, it is necessary to create an effective environmental management strategy for urban areas that incorporates the development of new technologies and the use of biofuels.

Table 9.3 Limits/standards for different heavy metals in ambient air by different agencies

Limits/standards	Mercury	Lead	Vanadium	Arsenic	Manganese	Nickel	Chromium	Cadmium
WHO, 2000	1000	500	1000	6.6	150	25	0.25	5
EU, (directive), 2004	–	–	–	6	–	20	–	5
India (NAAQS), 2009	–	500	–	6	–	20	–	–
China (NAAQS- GB3095), 2012	50	500	–	6	–	–	0.025	5

References

Abdul-Wahab SA, Yaghi B (2004) Total suspended dust and heavy metal levels emitted from a workplace compared with nearby residential houses. Atmos Environ 38(5):745–750

Abosede IA, Peter OA, Adunola AA et al (2017) Biomass valorization: agricultural waste in environmental protection, phytomedicine and biofuel production. In: Biomass volume estimation and valorization for energy. InTech, Rijeka

Acosta JA, Faz A, Kalbitz K et al (2014) Partitioning of heavy metals over different chemical fraction in street dust of Murcia (Spain) as a basis for risk assessment. J Geochem Explor 1(144): 298–305

Adachi K, Tainosho Y (2004) Characterization of heavy metal particles embedded in tire dust. Environ Int 30(8):1009–1017

Adamiec E, Jarosz-Krzemińska E, Wieszała R et al (2016) Heavy metals from non-exhaust vehicle emissions in urban and motorway road dusts. Environ Monit Assess 188(6):1–11

Al-Khashman OA (2007) Determination of metal accumulation in deposited street dusts in Amman, Jordan. Environ Geochem Health 29(1):1–10

Al-Taani AA, Nazzal Y, Howari FM et al (2019) Assessment of heavy metals in roadside dust along the Abu Dhabi–Al Ain National Highway, UAE. Environ Earth Sci 78(14):1–13

Apeagyei E, Bank MS, Spengler JD et al (2011) Distribution of heavy metals in road dust along an urban-rural gradient in Massachusetts. Atmos Environ 45(13):2310–2323

Bajgai Y, Hulugalle N, Kristiansen P, McHenry M et al (2013) Developments in fractionation and measurement of soil organic carbon: a review. Open J Soil Sci 2013

Baltrėnas HP, Januševičius T, Chlebnikovas A et al (2017) Research into the impact of speed bumps on particulate matter air pollution. Measurement 100:62–67

Bica S, Radoslav R, Nicolau I et al (2014) Green belts and the improving of air quality-case study: Timisoara and its surroundings. Bull Univ Agric Sci Vet Med Cluj-Napoca, Vet Med 71(1):8

Bourliva A, Christophoridis C, Papadopoulou L et al (2017) Characterization, heavy metal content and health risk assessment of urban road dusts from the historic center of the city of Thessaloniki, Greece. Environ Geochem Health 39(3):611–634

Button K, Nijkamp P (1997) Social change and sustainable transport. J Transp Geogr 5(3):215–218

Cao S, Duan X, Zhao X et al (2014) Health risks from the exposure of children to As, Se, Pb and other heavy metals near the largest coking plant in China. Sci Total Environ 15(472):1001–1009

Chan DSEA, Stachowiak GW (2004) Review of automotive brake friction materials. P I Mech Eng D J Automob Eng 218(9):953–966

Chaudhari PR, Gupta R, Gajghate DG, Wate SR et al (2012) Heavy metal pollution of ambient air in Nagpur City. Environ Monit Assess 184(4):2487–2496

Chen T, Zhang Y, Wang H et al (2014) Influence of pyrolysis temperature on characteristics and heavy metal adsorptive performance of biochar derived from municipal sewage sludge. Bioresour Technol 1(164):47–54

Chow JC, Watson JG, Houck JE et al (1994) A laboratory resuspension chamber to measure fugitive dust size distributions and chemical compositions. Atmos Environ 28:3463–3481

Chowdhury R, Apul D, Fry T et al (2010) A life cycle based environmental impacts assessment of construction materials used in road construction. Resour Conserv Recycl 54(4):250–255

Councell TB, Duckenfield KU, Landa ER et al (2004) Tire-wear particles as a source of zinc to the environment. Environ Sci Technol 38(15):4206–4214

Dahl A, Gharibi A, Swietlicki E et al (2006) Traffic-generated emissions of ultrafine particles from pavement–tire interface. Atmos Environ 40(7):1314–1323

Davis AP, Shokouhian M, Ni S et al (2001) Loading estimates of lead, copper, cadmium, and zinc in urban runoff from specific sources. Chemosphere 44(5):997–1009

Doabi SA, Karami M, Afyuni M et al (2018) Pollution and health risk assessment of heavy metals in agricultural soil, atmospheric dust and major food crops in Kermanshah province, Iran. Ecotoxicol Environ Saf 163:153–164

Dockery DW, Pope CA, Xu X et al (1993) An association between air pollution and mortality in six US cities. N Engl J Med 329(24):1753–1759

Duan J, Tan J (2013) Atmospheric heavy metals and arsenic in China: situation, sources and control policies. Atmos Environ 74:93–101

Duong TT, Lee BK (2011) Determining contamination level of heavy metals in road dust from busy traffic areas with different characteristics. J Environ Manag 92(3):554–562

Environmental Protection Agency (1998) Department of Pesticide Regulation. Environmental monitoring, Pest management branch, & California. Environmental Hazards Assessment Program

Faiz Y, Tufail M, Javed MT et al (2009) Road dust pollution of Cd, Cu, Ni, Pb and Zn along Islamabad expressway, Pakistan. Microchem J 92(2):186–192

Fauser P, Tjell JC, Bjerg PL et al (1999) Particulate air pollution with emphasis on traffic generated aerosols. Risø National Laboratory

Ferner DJ (2001) Toxicity, heavy metals. eMed J 2(5):1

Fukuzaki N, Yanaka T, Urushiyama Y et al (1986) Effects of studded tires on roadside airborne dust pollution in Niigata, Japan. Atmos Environ (1967) 20(2):377–386

Gawade A, Deshmukh P, Shivankar V et al (2016) Analysis of roadside dust for heavy metal pollutants in Navi Mumbai. Int J Eng Technol Manage Appl Sci 4:2349–4476

Goudarzi G, Alavi N, Geravandi S et al (2018) Health risk assessment on human exposed to heavy metals in the ambient air PM_{10} in Ahvaz, Southwest Iran. Int J Biometeorol 62(6):1075–1083

Gualtieri M, Andrioletti M, Vismara C et al (2005) Toxicity of tire debris leachates. Environ Int 31(5):723–730

Gunawardena J, Egodawatta P, Ayoko GA et al (2012) Role of traffic in atmospheric accumulation of heavy metals and polycyclic aromatic hydrocarbons. Atmos Environ 54:502–510

Gupta V (2020) Vehicle-generated heavy metal pollution in an urban environment and its distribution into various environmental components. In: Environmental concerns and sustainable development. InTech, Rijeka, pp 113–127

Gustafsson M, Blomqvist G, Gudmundsson A (2008) Properties and toxicological effects of particles from the interaction between tyres, road pavement and winter traction material. Sci Total Environ 393(2–3):226–240

Hare CT (1977) Light-duty diesel emission correction factors for ambient conditions. SAE Paper 770717 (No. CONF-7709217-1). Society of Automotive Engineers, Inc., Warrendale, PA

Health Effects Institute (1995) Identifying subgroups of the general population that may be susceptible to short-term increases in particulate air pollution: a time-series study in Montreal, Quebec report 97

Hjortenkrans DS, Bergbäck BG, Häggerud AV et al (2008) Transversal immission patterns and leachability of heavy metals in road side soils. J Environ Monit 10(6):739–746

Huang Y, Chen Q, Deng M et al (2018) Heavy metal pollution and health risk assessment of agricultural soils in a typical peri-urban area in Southeast China. J Environ Manag 207:159–168

Hussein T, Johansson C, Karlsson H et al (2008) Factors affecting non-tailpipe aerosol particle emissions from paved roads: on-road measurements in Stockholm, Sweden. Atmos Environ 42(4):688–702

Hwang HM, Fiala MJ, Park D et al (2016) Review of pollutants in urban road dust and stormwater runoff: part 1. Heavy metals released from vehicles. Int J Urban Sci 20(3):334–360

Institute of Environmental Conservation and Research (INECAR) (2000) Position paper against mining in Rapu-Rapu. INECAR, Ateneo de Naga University Philippines

Izhar S, Goel A, Chakraborty A et al (2016) Annual trends in occurrence of submicron particles in ambient air and health risk posed by particle bound metals. Chemosphere 146:582–590

Jeba Rajasekhar RV, Bhaskar BV, Kulandaisamy I et al (2001) SPM concentration ambient air of Madurai city. J Environ Sci 5(1):56–61

Jena S, Singh G (2016) Assessment of impacts of vehicular emission on ambient air quality along the National Highway-32 at Dhanbad

Jena S, Perwez A, Singh G et al (2019) Trace element characterization of fine particulate matter and assessment of associated health risk in mining area, transportation routes and institutional area of Dhanbad, India. Environ Geochem Health 41(6):2731–2747
Kampa M, Castanas E (2008) Human health effects of air pollution. Environ Pollut 151(2):362–367
Kannan N (1991) A study on removal of nickel by adsorption on fly ash. Indian J Environ Prot 11(7):514–518
Karlsson H, Lindbom J, Ghafouri B et al (2011) Wear particles from studded tires and granite pavement induce pro-inflammatory alterations in human monocyte-derived macrophages: a proteomic study. Chem Res Toxicol 24(1):45–53
Karlsson HL, Holgersson Å, Möller L et al (2008) Mechanisms related to the genotoxicity of particles in the subway and from other sources. Chem Res Toxicol 21(3):726–731
Kebede L, Tulu GS, Lisinge RT et al (2022) Diesel-fueled public transport vehicles and air pollution in Addis Ababa, Ethiopia: effects of vehicle size, age and kilometers travelled. Atmos Environ X 13:100144
Kemp K (2002) Trends and sources for heavy metals in urban atmosphere. Nucl Instrum Methods Phys Res, Sect B 189(1-4):227–232
Kenya National Bureau of Statistics, UNICEF, Kenya Country Office (2017) Childpoverty in Kenya: A multidimensional approach
Keshavarzi B, Tazarvi Z, Rajabzadeh MA et al (2015) Chemical speciation, human health risk assessment and pollution level of selected heavy metals in urban street dust of Shiraz, Iran. Atmos Environ 1(119):1
Kolonel LN (1976) Association of cadmium with renal cancer. Cancer 37(4):1782–1787
Kumari R, Attri AK, Panis LI et al (2013) Emission estimates of particulate matter and heavy metals from mobile sources in Delhi (India). J Environ Sci Eng 55(2):127–142
Kumari S, Jain MK, Elumalai SP et al (2021) Assessment of pollution and health risks of heavy metals in particulate matter and road dust along the road network of Dhanbad, India. J Health Pollut 11(29):210305
Li B, Ho SS, Xue Y (2017) Characterizations of volatile organic compounds (VOCs) from vehicular emissions at roadside environment: the first comprehensive study in northwestern China. Atmos Environ 161:1–2
Iijima A, Sato K, Yano K (2008) Emission factor for antimony in brake abrasion dusts as one of the major atmospheric antimony sources. Environ Sci Technol 42(8):2937–2942
Lim LP, Lau NC, Garrett-Engele P et al (2005) Microarray analysis shows that some microRNAs downregulate large numbers of target mRNAs. Nature 433(7027):769–773
Lin CC, Lin HL (2005) Remediation of soil contaminated with the heavy metal (Cd2+). J Hazard Mater 122(1–2):7–15
Liu E, Yan T, Birch G et al (2014) Pollution and health risk of potentially toxic metals in urban road dust in Nanjing, a mega-city of China. Sci Total Environ 476:522–531
Liu P, Ren H, Xu H et al (2018a) Assessment of heavy metal characteristics and health risks associated with $PM_{2.5}$ in Xi'an, the largest city in northwestern China. Air Qual Atmos Health 11(9):1037–1047
Liu Y, Xing J, Wang S et al (2018b) Source-specific speciation profiles of $PM_{2.5}$ for heavy metals and their anthropogenic emissions in China. Environ Pollut 239:544–553
Loranger S, Zayed J (1995) Environmental and occupational exposure to manganese: a multimedia assessment. Int Arch Occup Environ Health 67(2):101–110
Lowenthal DH, Zielinska B, Chow JC et al (1994) Characterization of heavy-duty diesel vehicle emissions. Atmos Environ 28(4):731–743
Luo XS, Ding J, Xu B et al (2012) Incorporating bioaccessibility into human health risk assessments of heavy metals in urban park soils. Sci Total Environ 424:88–96
Lv J, Xia Q, Yan T et al (2018) Identifying the sources, spatial distributions, and pollution status of heavy metals in soils from the southern coast of Laizhou Bay, eastern China. Hum Ecol Risk Assess Int J 25:1953

Madany IM, Akhter MS, Al Jowder OA et al (1994) The correlations between heavy metals in residential indoor dust and outdoor street dust in Bahrain. Environ Int 20(4):483–492

Maher BA, Moore C, Matzka J (2008) Spatial variation in vehicle-derived metal pollution identified by magnetic and elemental analysis of roadside tree leaves. Atmos Environ 42(2):364–373

McCluggage D (1991) Heavy metal poisoning, NCS magazine, published by the Bird Hospital, CO, USA

Meister K, Johansson C, Forsberg B et al (2012) Estimated short-term effects of coarse particles on daily mortality in Stockholm, Sweden. Environ Health Perspect 120(3):431–436

Melaku S, Morris V, Raghavan D et al (2008) Seasonal variation of heavy metals in ambient air and precipitation at a single site in Washington, DC. Environ Poll 155(1):88–98

Ministry of Petroleum and Natural Gas (MOPNG) (2014) Indian Petroleum and Natural Gas Statistics 2014–15

Ministry of Road Transport & Highways Yearbook (2021). https://morth.nic.in/sites/default/files/RTYB-2017-18-2018-19.pdf

Nagpure AS, Gurjar BR, Kumar V et al (2016) Estimation of exhaust and non-exhaust gaseous, particulate matter and air toxics emissions from on-road vehicles in Delhi. Atmos Environ 127:118–124

Najmeddin A, Keshavarzi B, Moore F et al (2018) Source apportionment and health risk assessment of potentially toxic elements in road dust from urban industrial areas of Ahvaz megacity, Iran. Environ Geochem Health 40(4):1187–1208

Namdeo AK, Colls JJ, Baker CJ et al (1999) Dispersion and re-suspension of fine and coarse particulates in an urban street canyon. Sci Total Environ 235(1–3):3–13

Ogwugbu MO, Muhanga W (2005) Investigation of Lead concentration in the blood of people in the Copperbelt Province of Zambia. J Environ 1:66–75

Peng X, Shi G, Liu G et al (2017) Source apportionment and heavy metal health risk (HMHR) quantification from sources in a southern city in China, using an ME2-HMHR model. Environ Pollut 221:335–342

Piron-Frenet M, Bureau F, Pineau A et al (1994) Lead accumulation in surface roadside soil: its relationship to traffic density and meteorological parameters. Sci Total Environ 144(1–3):297–304

Popescu V, Bace M, Nedevschi S et al (2011) Lane identification and ego-vehicle accurate global positioning in intersections. In 2011 IEEE intelligent vehicles symposium (IV), pp 870–875

Pothumsetty R, Viswam N, Thomas MR et al (2020) Health benefits of shifting from Bharat stage four to Bharat stage six. Int J Behav Healthcare Res 7(2):77–91

Pulles T, van der Gon HD, Appelman W et al (2012) Emission factors for heavy metals from diesel and petrol used in European vehicles. Atmos Environ 61:641–651

Rahman MS, Khan MDH, Jolly YN et al (2019) Assessing risk to human health for heavy metal contamination through street dust in the southeast Asian megacity: Dhaka, Bangladesh. Sci Total Environ 660:1610–1622

Roy S, Gupta SK, Prakash J et al (2019) Ecological and human health risk assessment of heavy metal contamination in road dust in the National Capital Territory (NCT) of Delhi, India. Environ Sci Pollut Res 26(29):30413–30425

Sansalone JJ, Buchberger SG, Al-Abed SR et al (1996) Fractionation of heavy metals in pavement runoff. Sci Total Environ 189:371–378

Schwab O, Bayer P, Juraske R et al (2014) Beyond the material grave: life cycle impact assessment of leaching from secondary materials in road and earth constructions. Waste Manag 34(10):1884–1896

Sharma YC, Singh B, Upadhyay SN et al (2008) Advancements in development and characterization of biodiesel: a review. Fuel 87(12):2355–2373

Shi G, Chen Z, Bi C et al (2011) A comparative study of health risk of potentially toxic metals in urban and suburban road dust in the most populated city of China. Atmos Environ 45(3):764–771

Silbergeld EK (1995) The international dimensions of lead exposure. Int J Occup Environ Health 1(4):336–348

Silbergeld EK (1996) Blood Lead levels, scientific misconduct, and the Needleman case: 5. Silbergeld responds. Am J Public Health 86(1):114–115

Silva LI, Yokoyama L, Maia LB et al (2015) Evaluation of bioaccessible heavy metal fractions in PM10 from the metropolitan region of Rio de Janeiro city, Brazil, using a simulated lung fluid. Microchem J 118:266–271

Silva SA (2020) Respiratory support for patients with COVID-19 infection. Lancet Respir Med 8(4):e18

Song F, Gao Y (2011) Size distributions of trace elements associated with ambient particular matter in the affinity of a major highway in the New Jersey–New York metropolitan area. Atmos Environ 45(37):6714–6723

Sternbeck J, Sjödin Å, Andréasson K et al (2002) Metal emissions from road traffic and the influence of resuspension—results from two tunnel studies. Atmos Environ 36(30):4735–4744

Sun J, Peng H, Chen J et al (2016) An estimation of CO_2 emission via agricultural crop residue open field burning in China from 1996 to 2013. J Clean Prod 112:2625–2631

Suryawanshi PV, Rajaram BS, Bhanarkar AD et al (2016) Determining heavy metal contamination of road dust in Delhi. Atmósfera 29(3):221–234

Suvarapu LN, Seo YK, Baek SO et al (2018) Heavy metals in the Indian atmosphere: a review. Res J Chem Environ 18(8):99–111

Tanushree B, Chakraborty S, Bhumika F et al (2011) Heavy metal concentrations in street and leaf deposited dust in Anand City, India. Res J Chem Sci 1:61–66

Tian HZ, Zhu CY, Gao JJ (2010) Quantitative assessment of atmospheric emissions of toxic heavy metals from anthropogenic sources in China: historical trend, spatial distribution, uncertainties, and control policies. Atmos Chem Phys 15(17):10127–10147

Tian HZ, Lu L, Cheng K et al (2012) Anthropogenic atmospheric nickel emissions and its distribution characteristics in China. Sci Total Environ 417:148–157

Union Europian (2002) Heavy metals in wastes. European Commission on Environment

U.S. Environmental Protection Agency (2000) Heavy-duty engine and vehicle standards and highway diesel fuel sulfur control requirements, 2000. Regulatory announcement. Washington, DC: U.S. EPA Office of Transportation and Quality, EPA420-F-00-057

US EPA (United States Environmental Protection Agency) (1989) Risk assessment guidance for superfund. Human Health Evaluation Manual (Part A). Interim Final, vol. I. Washington (DC): EPA/540/1–89/002

USEPA (1996) Compilation of air emission factors, Vol. 1: Stationary Point and Area Sources, 5th Edition, AP-42. Section 2.2 Sewage Sludge Incineration. United States Environmental Protection Agency, Office of Air Quality Planning and Standards, Research Triangle Park, North Carolina, USA

USEPA (2001). www.epa.gov/opptintr/lead/index.html

Wang N, Tang L, Pan H et al (2019) A global comparison and assessment of incentive policy on electric vehicle promotion. Sustain Cities Soc 44:597–603

Xia L, Gao Y et al (2011) Characterization of trace elements in $PM_{2.5}$ aerosols in the vicinity of highways in Northeast New Jersey in the US east coast. Atmos Pollut Res 2(1):34–44

Yang B, Shu WS, Ye ZH et al (2003) Growth and metal accumulation in vetiver and two Sesbania species on lead/zinc mine tailings. Chemosphere 52(9):1593–1600

Zhang C, Yu ZG, Zeng GM et al (2014) Effects of sediment geochemical properties on heavy metal bioavailability. Environ Int 73:270–281

Zheng J, Wang X, Men K et al (2013) Aggregation model-based optimization for electric vehicle charging strategy. IEEE Trans Smart Grid 4(2):1058–1066

Zheng N, Liu J, Wang Q et al (2010) Health risk assessment of heavy metal exposure to street dust in the zinc smelting district, northeast of China. Sci Total Environ 408(4):726–733

Life Cycle Assessment of Heavy Metal Toxicity in the Environment

10

Oliver Khonyongwa and Suman

Abstract

Heavy metals are metallic elements naturally found in the environment (Lithogenic source), possessing relatively high levels of density and atomic weight, but are also denser than water. Even though these metals are naturally occurring, because of various anthropogenic activities, industrial, agricultural, domestic, and medical, the transformation in technology has led to a major concern for heavy metal toxicity to the environment and human health. Localized contamination may have marginal effects as it's predominantly from a single source; however, nonspecific deposition or emission of heavy metals will cause significant damage. These are known as systematic toxicants based on their high potential of toxicity and will vary in their effects based on several factors, e.g., dose and exposure, and they can cause systematic organ failure. They also possess the potential for genotoxicity and are carcinogenic even after low levels of exposure. Most governments have made initiatives to monitor atmospheric deposition for particular heavy metals like lead (Pb) that usually come through traffic and burning of fossil fuels. On the other hand, pinpointing the source or origins of emission for these heavy metals is possible, but the lack of realizing the importance and their underestimation is also there. This chapter characterizes the life cycle of these heavy metals, describing every heavy metal, classifying their movement and how they evolve regarding human activities, and clearly explaining the effects of toxicity both on the environment and on public health.

Keywords

Heavy metals · Toxicity · Exposure · Carcinogenicity · Genotoxicity

O. Khonyongwa · Suman (✉)
Department of Environmental Science, School of Basic Science and Research, Sharda University Greater Noida, Noida, Uttar Pradesh, India

R. P. Singh et al. (eds.), *Heavy Metal Toxicity: Environmental Concerns, Remediation and Opportunities*, https://doi.org/10.1007/978-981-99-0397-9_10

10.1 Introduction

Heavy metals are defined broadly as "metals with a high density, electronic configuration, and include all metallic elements that are denser than water" (Wikipedia 2021). These heavy metals include elements such as lead, mercury, arsenic, cadmium, and chromium (Tchounwou et al. 2012). It has become an issue of global concern affecting both the environment and public health and has been discovered to bring about negative effects on the environment, components of the environment, and people. Recently, there have been issues associated with heavy metals, due to the high use of substances that pose high levels of these heavy metals. Although these are those elements that are present naturally in the piles of the Earth, their increasing levels of acquaintance for humans and the environment generally exist due to increasing anthropogenic activities. Sources of heavy metals may include industrial operations (production and use), agricultural activities, and smelting activities.

Other sources of literature have shown that heavy metals possess similar properties as essential ions and because of this they compete for biological binding sites, and this eventually is causing a disruption in the biomolecular structure and function, which results in a lack of mental balance. Furthermore, these trace elements are known to elicit free radicals of oxygen and nitrogen species which cause oxidative stress and an imbalance in the cellular reactions, a process that usually causes cancer.

The careless use of agricultural chemicals such as pesticides over a long period in the past years has extensively contributed to the contamination of both surface and subsurface waters and left the soil with high levels of toxicants. The building up of these elements in both the soil and the aquatic environment and the uptake by various vegetation which when finally consumed by humans may pose a risk to the health.

10.2 Heavy Metals

10.2.1 Copper

Copper is a soft, malleable, and ductile metal with extremely high thermal and electrical conductivity, denoted by the symbol Cu (Wikipedia 2022b). It is one of the crucial elements needed for bodily functions and essential in other aspects of daily activities.

10.2.1.1 Properties

Copper has a pinkish-orange look after it has been freshly cut. It is a very good conductor of heat and electricity, and because of this, it is used in the making of building materials and acts as a good constituent of different alloys like sterling silver for jewellery and others like nickel for the production of marine hardware and also money coins (Wikipedia 2022b).

Copper is one of the heavy metals of public health importance as it is one of the metals that is an essential element for some particular bodily functions such as the balance and maintenance of hormones that make nerve cells, the development of new blood vessels and connective tissue, the formation of enzymes for energy production and iron metabolism, and the regulation of the gene expression and boosting of healthy immune functions.

10.2.1.2 Exposure

Exposure to copper may occur in different ways that includes inhalation, consumption of food, or skin contact with water, soil, or substances that are containing the element. Copper can be a source in ambient air through both natural sources and anthropogenic ones. Some of the natural sources include volcanic activities, bush fires, and windblown dust. Other means are man-made sources which include the smelting of copper, iron, steel, and municipal incinerators (Barceloux and Vanadium 1999). Copper can also find its way into the environment using copper water pipes and various other plumbing materials, kitchen sets (cutlery, pots, and pans), medicines, and pharmaceuticals (birth controls, vitamins, minerals supplements, and intrauterine devices) and food (Pohl et al. 2011).

10.2.1.3 Effects

Effects of copper will greatly vary with the intake of the food options and dietary customs of individuals. Individuals can manifest copper effects because of deficiency as well as excessive uptake; thus, it is a complex matter to determine the requirements and the upper safe limit. Since there is a lack of sensitive biomarkers to flag excessive intake of copper, there are no detectable adverse effects, and it is hard to associate upper levels of copper intake with clinical disease; however, individuals may indicate with unexplained liver cirrhosis (Uauy et al. 2008).

10.2.2 Lead

It has a very vital role to play in the human body, and lead toxicity is very crucial and requires mindfulness as it brings a broad range of negative effects to components of the atmosphere and human well-being (Wani et al. 2015).

10.2.2.1 Properties

The applicability of lead dates back to ancient times when its use was very vital as the metal had several other characteristics which are as follows:

- Smoothness
- Flexibility
- Ductility
- Low conductibility
- Resistance to corrosion

Despite its lethal effects and environmental unfriendliness, it is regarded as one of the heavy metals that was globally distributed and widely used (Mahaffey 1990).

10.2.2.2 Exposure

There are several ways in which lead finds its way into the environment and the human body. A lot of leakage is observed for lead and its related compounds in industrial processes and activities in the environment. Some simple processes associated with this include the smelting of lead and the process of its combustion, combustion of leaded gasoline, smelting and recycling of batteries, and the production of arms.

Several occupations can render an individual to lead exposure, and these include boat building, painting and using of pigments with lead-based materials, using of pipes and conduits that contain lead, and also printing of books.

10.2.2.3 Effects

Lead toxicity has been shown to cause numerous health problems in the human body. One of the health problems in this category is anemia where the element lead inhibits porphobilinogen synthase and ferrochelatase causing ineffective heme synthesis eventually resulting in microcytic anemia (Cohen et al. 1981). Lead interferes with the cognition of ion channels as it acts as a calcium analog which has the potential to act as a selective blocker of voltage for the calcium channels depending on low concentration (Büsselberg et al. 1993).

Studies conducted on Uruguayan children indicated that with low-level exposure to lead, children are at risk of developing adverse effects of oxidative stress (Roy et al. 2015). Other research depicted that there is an optimistic correlation between lead experience and respiratory dysfunction among workers who have been exposed (Jurdziak et al. 2015).

10.2.3 Chromium

Chromium (Cr) is "an element with atomic number 24 that is a rigid steel-greyish, lustrous and brittle unstable metal" (Wikipedia 2022a). It is one of the most abundant metals and has several oxidation states ranging from Cr^{2+} to Cr^{6+} in the environment. Other Cr oxidation states are elements that are naturally occurring in the environment and are inactive.

10.2.3.1 Properties

Chromium (VI) is usually known as trivalent and hexavalent chromium. Wastes that are mostly produced by industrial activities and are commonly discharged into water contain chromium (Rodríguez et al. 2007). Cr is greatly utilized in various fields such as ore dressing, plating, manufacturing of dyes and stains, bronzing, timber conservation, biochemical making, and pulp and paper milling. Due to the majority of products coming out of these industries and the fast use of the products, it is safe to consider that they contribute largely to pollution and with them are accompanied adverse effects on both individual human and other ecological species (Ghani and Ghani 2011).

Chromium is a naturally occurring substance that comes from the combustion of fossil fuels such as oil, coal, and petroleum. It also has other sources such as pigments, agricultural materials such as fertilizers, the electroplating industry, the tannery industry, and the mining for oil. After all the processes have been done in different industries, effluents and other wastewater will find their way into the environment through nonpoint source pollution (Ghani and Ghani 2011). The abundance of oxygen in the atmosphere enhances chromium (III) to change state through oxidation to chromium (IV) which is very poisonous and highly decipherable in water (Cervantes et al. 2001). The International Agency for Research on Cancer has categorized chromium to be one of the elements that are highly mutagenic making Cr (VI) a group 1 human carcinogen (Dayan and Paine 2001).

10.2.3.2 Exposure

Chromium's most abundant forms in the environment are trivalent and hexavalent, both of which are toxic to animals, humans, and plants (Monalisa and Kumar 2013). Chromium (IV), an oxidized form of Cr (III), is highly soluble in water, and humans are exposed as a result of drinking polluted water (Gürkan et al. 2017).

10.2.3.3 Effects

Hexavalent chromium Cr (VI) is one of the most active forms of chromium which is capable of penetrating in the cell membrane using its anions such as $SO_4{}^{2-}$ and $HPO_4{}^{2-}$ usually taken up through phagocytosis.

The reactions between chromium (IV) and other reducing agents like thiols may end up producing other forms of reactive species of oxygen like hydroxyl radicals that may eventually lead to oxidative stress in cells denaturing proteins which will lead to DNA damage (Stohs and Bagchi 1995).

10.2.4 Mercury

Mercury, also known as quicksilver, is a chemical element with the symbol Hg and the electronic configuration 80, and it is the only metallic element that is liquid at room temperature and pressure (Wikipedia 2022c). Mercury can be found in the environment as a naturally occurring element as deposits of mercuric sulfide, a shiny silver/white odorless liquid.

Mercury has a wide range of uses from medical, e.g., thermometers, dental procedures as amalgams and as a catalyst, but also in industries to produce batteries, mercury arc lamps, fluorescent lamps, and also pulp and paper. It is also useful for the production of barometers, pyrometers, and hydrometers.

10.2.4.1 Properties

It is one of the utmost toxic trace elements due to its physicochemical properties and bioaccumulation characteristic. It is a metal of high vapor pressure and volatility and is highly mobile because of its low viscosity which allows its droplets to join easily with other metals like tin, copper, gold, and silver forming amalgams. Its atomic mass is also high up to 200.59 grams per mole with its relative gravity many times high than water (Wikipedia 2022c).

It exists in various arrangements, namely, metallic elements, inorganic salts, and organic compounds, having different bioavailability and levels of toxicity. These forms of mercury are dominant in marine environments, and once exposed to aquatic microbiota, they have transformed into methyl mercury, a substance that has a high affinity for biomagnification to other aquatic organisms (Trasande et al. 2005).

10.2.4.2 Exposure

Mercury leakages are highly present and distributed in marine environments and cause adverse harm to a wide range of living organisms. Mercury finds its way into the environment and water resources most commonly through depositions that occur due to several human activities such as agriculture, mining, and discharges of wastewater and/or material from industrial, commercial, and medical facilities (Chen et al. 2012).

10.2.4.3 Effects

Mercury is one of the heavy metals that causes acute and chronic poisoning, and several effects may result due to the inhalation of vaporized mercury causing systematic damage to the nervous, digestive, respiratory, and immune systems including the organs like the kidneys. People may experience a wide range of symptoms such as vision and hearing problems, lack of sleep, emotional distress, and cognitive skill impairment but can also affect the expecting mothers during fetal development and delay child development in childhood (Wani et al. 2015).

However, subacute exposure and chronic exposure have been known to cause more severe damage resulting in a wide range of multisystem function disturbances. Among the wide range of symptoms that occur, a lot are related to the central nervous system where individuals will face hallucinations, suicidal tendency, irritability/excitability, and extreme shyness erethism, and if the exposure is chronic, the individuals may develop tremors, muscular spasms, memory loss, and even depression (Büsselberg et al. 1993).

10.2.5 Arsenic

Arsenic is a chemical element with the symbol As and the atomic number 33. It is a metalloid that can be found in many minerals, usually in conjunction with sulfur and metals, but it can also be found as a pure elemental crystal (Wikipedia 2022). Among the several allotropes of arsenic, one gray/metallic appearance is the most vital for industrial use. Arsenic is another important potential toxic metal that is known for causing adverse effects on both the environment and individual health (Hughes et al. 1988).

10.2.5.1 Properties

It is an extremely poisonous element that is found in nature in the form of oxides or sulfides of salt, and because of its properties as an inorganic substance forming compounds of arsenite and arsenate, it possess great harm to the environment and organisms of the ecological species including individual human (Singh et al. 2007). Various types of living things such as bacteria, algae, and fungi including humans harbor inorganic forms of arsenic compounds and get them methylated turning them into intermediate products that are highly toxic and are capable of causing arsenic-induced carcinogenesis (Singh et al. 2007).

10.2.5.2 Exposure

Exposure to arsenic can occur in various ways. Although natural means are a possibility, most of the ways may be due to various anthropogenic activities that occur in industries. However, exposure to arsenic can also be done in unintended ways thus by the consumption of water that has been polluted with wastes containing arsenic materials or deposits such as arsenic pesticides (Mazumder 2008).

10.2.5.3 Effects

Arsenic is an element that targets sulfhydryl group of cells in the individual body, and this results in the distortion of various cellular functions like cell respiration, damaging the formations of cell enzymes, and disturbing the process of cell division (Gordon and Quastel 1948). Acute poisoning can occur with the consumption of arsenic in cases where children have taken the substance accidentally or in cases of suicidal attempts (Mazumder 2008; Saha et al. 1999).

10.2.6 Cadmium

Cadmium, with the symbol Cd and atomic number 48, is a soft, silvery-white metal that is chemically similar to zinc and mercury and has a lower melting point than the transition metals in groups 3 through 11(Wikipedia 2022). Cd is yet another toxic heavy metal known for negatively affecting cellular functions, causing oxidative stress and causing nutritional inadequacy in plants (Irfan et al. 2013).

10.2.6.1 Properties

Early literature dates back to the use of cadmium as a substitute for tin in the making of paint and other forms of pigment. Throughout the years, it has been used for its purpose in rechargeable batteries and as a stabilizer for plastics. It has also been used in the production of special alloys, and tobacco smoke also possesses a percentage of the element (Irfan et al. 2013).

10.2.6.2 Exposure

Cadmium is an element that has a long resident time in the atmosphere and exists in the soil as residues and living organisms can get exposed to it through contact, consumption, and inhalation. Another dominant way to get exposure is through work environments in industries. It is one of the toxic metals that will affect the food chain as it accumulates in plants and tends to proceed in such a similar manner as you go up the trophic levels of the food chain (Bernard 2008; Mutlu et al. 2012).

10.2.6.3 Effects

Cadmium, because of its properties as a heavy metal, will end up accumulating in the body which will lead to reactions with cysteine-rich protein eventually forming metallothionein which upon accumulating in the liver will cause hepatotoxicity and the kidney's nephrotoxicity. It also can bind with cysteine which causes iron deficiency (Castagnetto 2002).

10.3 Life Cycle of Heavy Metals

Anthropogenic activities have always been a major contributor and source of the production and accumulation of trace elements. The main sources where these trace elements are coming from include automobiles, industries, and the degradation of various materials. These trace elements are very helpful at low doses and are extremely hazardous and cause significant damage to the environment and living organisms particularly human beings causing a wide range of diseases (Finkelman 2004).

Because of the potential for trace elements to cause harm to the environment and living things, these concerns have allowed several agencies to guard these trace elements by monitoring and controlling the levels of how much can be discharged into the environment and consumable limits by setting up a priority list (Liu et al. 2019).

One of the most abundant ways in which heavy metals find their way into the environment is the burning and combustion of coal and the gangue with it. The study done by Tian et al. (2013) shows a significantly positive relationship between burning of the trace elements and adverse ecological environmental problems. Further evidence shows that after hot gaseous emissions that come in contact with ambient air, it eventually forms mats of concentrated minerals that get deposited in the environment

10.4 Recommended Limits for Heavy Metals

According to the UNEP, the region lacks legislation and regulations governing the import, export, transportation, use, production, emission, storage, and disposal of persistent toxic substances (PTS) like heavy metals. International conventions relevant to PTS have been ratified to some extent, and legislation within the Region has been harmonized. Table 10.1 shows the recommended and consumable limits for heavy metals.

Table 10.1 Recommended consumption and highest permissible intake limits

EU	• Max 100 g/week of big fish • Avoid consumption of other fishery products
FDA (Food and Drug Administration) and Environmental Protection Agency	• Avoid consumption of shark, swordfish, king mackerel • Up to 340 g/week of fish or canned tuna fish • Up to 170 g/week of fresh tuna fish or canned tuna fish
UK Food Standards Agency	• Avoid consumption of big fish • Recommendation of max 140 g/week of cooked tuna fish
Ireland Food Standards Agency	• Recommendation of max 2 tuna fish/ week

10.4.1 Environment-Specific Pollution

10.4.1.1 Water Pollution

A lot of water resources are heavily contaminated with pollutants based on different reasons, but the main ones are urbanization and industrialization, all of which are a factor of anthropogenic activities. Pollutants, particularly toxic metals, are carried to marine environments through runoff of water from places where they are carelessly dumped such as residential areas, commercial areas, and industries. Toxic metals, because of their nature, tend to accumulate in water bodies as sediments, and upon the use of such contaminated water, humans and other living organism may be affected directly or indirectly. If these toxic metals come into contact with the organisms, they affect the entire food chain, and particular toxicity will depend greatly on the type of the toxic metals involved, their nature, the type of organism that has been exposed, and the period it has been exposed. Human beings are usually the last organisms in this trophic level, and because of the nature of the trace elements, adverse effects of bioaccumulation and biomagnification are severe for humans.

Trace elements can both be found in industrial and domestic raw sewage, and they are found in high concentrations. These trace elements remain nondegradable in the sewage treatment process and have to be managed in the process after the treatment of wastewater or during sludge treatment (Masindi and Muedi 2018; Foster 2012; Carrondo et al. 1978; Innovative Solutions for Wastewater Treatment, Organica Water Inc. 2017; Di Bonito 2008).

Trace elements can travel in marine environments from one water body to another. This is basically due to the staste at which they are in the particular environment, where others are diluted in the solution and others are suspended. The distance traveled by toxic metals will greatly vary due to physical state such as the size of particulate materials or sediment, current, and stability. If they make their way to seas and oceans, other forces like wind and ocean currents will play a role in transporting the pollutants further. There, fish is a major part of the marine ecosystem, and if exposed to heavy metals, they may also spread further through migration (Foster 2012).

10.4.1.2 Air Pollution

Air pollution is another factor coming in due to several factors related to both natural and anthropogenic activities. Different kinds of interactions in the atmosphere play a very big role in how the pollutants behave once they have been exposed. They may find their way into different processes, and as a result of different reactions, they may be present in different forms. Due to the advancement and increasing levels of urbanization and industrialization, elements realized as particles, droplets, or gases bond and create other forms of pollutants that result in hazardous materials eventually causing adverse effects on health for both animals and individual humans after exposure to contact.

Some natural events include sandstorms, volcanic events, soil erosion, and the weathering of rock material. On the other hand, anthropogenic activities include industrial activities, burning of fossil fuels, and smelting which is known to produce particulate matter, sulfur oxides, nitrogen oxides, and heavy metals that mix and react with other environmental agents. They cause harmful effects on getting mixed with water vapor and precipitating to the ground surface.

10.4.1.3 Soil Pollution

Various ecosystems of the Earth lie either on the soil or beneath it. If the soil is polluted with various pollutants, it may lead to the development of adverse effects for the organisms and living things that reside in these ecosystems. Although at some level soil pollution is not deliberate, it can occur and contribute to pollution; however, the major contributor is deliberate where most of it happens due to human activities. These include issues of open dumping of wastes and wastewater, the use of extreme agriculture methods involving fertilizers and animal manure, and the use of wastewater for irrigation, mining, burning of fossil fuels like coal and petroleum, and spillage associated with it. The heavy metals present in these materials end up being dumped without proper treatment; as a result, it accumulates in the soil. These heavy metals stay for a longer period in the soil because of their inability for microbial and chemical degradation and their properties of not being degradable. Once they are in the soil, it is definite that they enter and disturb the food chain. This leads to heavy metal absorption in plants which are eaten by animals and humans. Furthermore, the soil loses quality and its fertility properties since these chemicals disturb the chemistry impacting the pH, color, porosity, and other characteristics.

10.5 Toxicological Processes

Toxicological processes happen particularly because of the properties of these metals and metalloids especially due to their high potential to form covalent bonds. This is most vital because this property allows for them to bind to organic groups, and the compounds or ions that are therefore formed are lipophilic, having the potential to generate adverse toxic effects after contact with cellular molecules. Here, arsenic in methylated forms is an example that due to various interactions with the living things may end up forming lyophilic compounds which are highly toxic (Masindi and Muedi 2018; Foster 2012).

Heavy metals are part of essential elements for life because other body functions rely on them; thus, their use has been very abundant in sectors like agriculture, pharmaceuticals, and industries (Gautam et al. 2016; Duffus 2002). One interesting fact is that metals are nonbiodegradable, and this simply means that when the heavy metals are ingested or inhaled into the body of living organisms, they tend to be stored for a long term in the system, a process called bioaccumulation (Lenntech 2019; Wang 2009). Organisms may detoxify these heavy metal ions usually by binding the active elements with proteins and depositing insoluble granules to be excreted by organisms in their fecal matter. However, since their resident time in the organism's body is relatively longer, it possesses a high risk as it may cause some biological and physiological complications (Duffus 2002; Lenntech 2019). The toxicological process of different heavy metals is shown in Fig. 10.1.

10.6 Carcinogenesis

Carcinogenesis, also known as oncogenesis or tumorigenesis, is the process by which normal cells are transformed into cancer cells, characterized by changes at cellular, genetic, and epigenetic levels, as well as abnormal cell division (Wikipedia 2022).

Exposure to toxic metals can cause direct or indirect production of free radicals in the human body (Jomova and Valko 2011). Oxidative stress is when cells build up free radicals like reactive oxygen or nitrogen species, and this results in a cellular redox imbalance which is usually associated with carcinogenesis (Poli et al. 2004; Valko et al. 2004; Valko et al. 2006).

Elements make their way into the body using different methods like drinking water, food air, and dermal absorption, and after any kind of contact with cellular components, they may cause several damages and disturbances to some fundamental cellular processes.

Many heavy metals have been known for being widespread in the environment as pollutants and for their capability to bioaccumulate in the individual bodies eventually causing adverse health effects such as cancer (Hemdan et al. 2007; Hou et al. 2012). The International Agency for Research on Cancer has classified a lot of heavy metals such as chromium lead, mercury, cadmium, and arsenic in a group of human carcinogens (Kim and Seo 2011; Park and Seo 2011; Park et al. 2012). These are responsible for binding with biomolecules forming toxic ions with characteristics being the same as essential ions hindering the biological binding of the essential ions and as a result distorting the normal biomolecular structure and the function which therefore leads to an imbalanced environment (Galaris and Evangelou 2002).

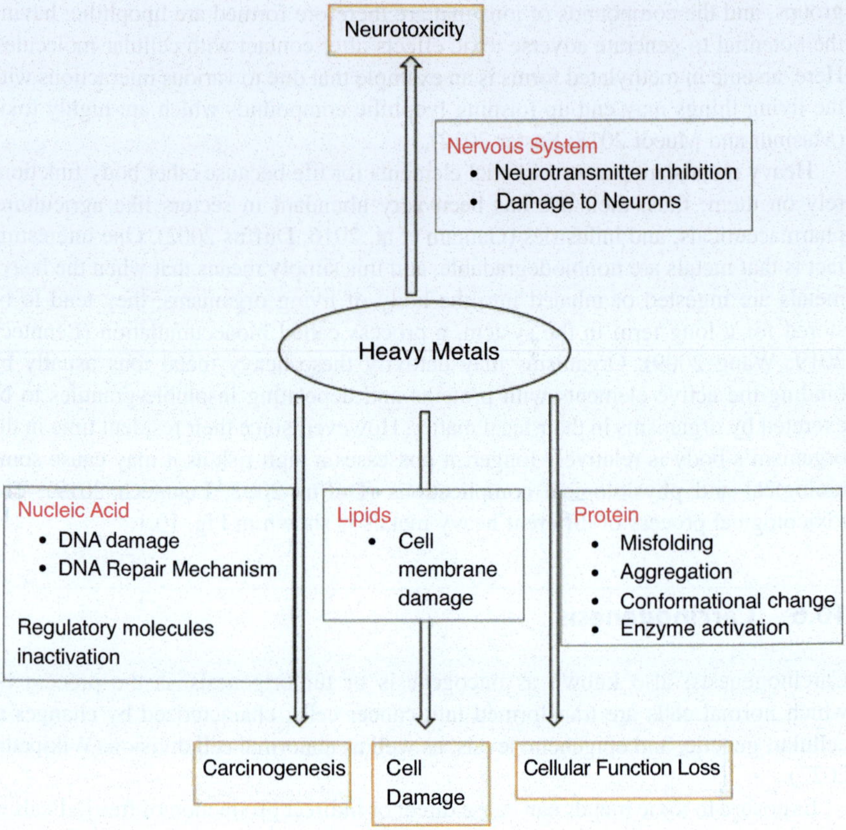

Fig. 10.1 Toxicological effects of heavy metals

10.7 Conclusion

Heavy metals have various modes of entry into the human body including drinking water, food, air, and exposure through the skin. Therefore, due to their special characteristics, they are capable of being retained and accumulating after absorption, which in most cases leads to a variety of toxic or adverse effects on the body tissues and organs.

Exposure to heavy metals can vary particularly in dose, concentration, period, and route of exposure. Regardless of exposure, metal toxicity can manifest differently depending on the individual body, and effects may be acute or chronic.

Effects of heavy metals also vary but have common characteristics that they can bind with macromolecules and they are capable of disrupting cellular functions. They are known to alter cell growth and damage gene expression and repairing process, thus leading to cell death. This, therefore, reduces the body's ability to have a strong immune system by weakening the antioxidant defense mechanism and the enzyme inactivation but also the oxidative stress. The capability of the heavy metals to affect the oxidative stress tells us that they can damage DNA and, because the DNA repairing process has been altered, then follows a high risk of cancer in the individual body; hence, toxic metals like Cr, Cd, and As are categorized as carcinogens.

This chapter has highlighted the effects of the toxic metals pointing out their sources, life cycle in the environment, and pathways to humans. Regardless of several agencies developing biomarkers for monitoring and control, there still stands a problem of the presence of these metals that is mainly due to anthropogenic activities and lifestyle. Although major achievements have been done in the field of toxicology, discovering through toxicogenomic analysis provides better knowledge of the effect-response relationship of a biosystem following exposure to toxicants, and more effort is yet to be put into practice by appropriate authorities to manage the further problem of production of toxic metals.

Agencies should develop new targets for toxic metals but also encourage the use of alternative materials for products instead of toxic metals.

References

"Arsenic," Wikipedia [Online] Available https://en.wikipedia.org/w/index.php?title=Arsenic&oldid=1076606683. Accessed 21 Mar 2022

Barceloux DG, Vanadium J (1999) Toxicol. Clin Toxicol 37(2):231–237

Bernard A (2008) Cadmium & its adverse effects on human health. Indian J Med Res 128(4):557

Büsselberg D, Evans ML, Haas HL, Carpenter DO (1993) Calcium channels by lead. Neurotoxicology 14:2–3

Cadmium, Wikipedia [Online] Available https://en.wikipedia.org/w/index.php?title=Cadmium&oldid=1077398277. Accessed 22 Mar 2022

Carcinogenesis, Wikipedia (2022) [Online] Available https://en.wikipedia.org/w/index.php?title=Carcinogenesis&oldid=1073624226. Accessed 28 Mar 2022

Carrondo MJT, Lester JN, Perry R, Loveland S (1978) Analysis of heavy metals in sewage sludge, sewages, and final effluent. Public Health and Water Resource Engineering Section, Civil Engineering Department, Imperial College, London

Castagnetto JM (2002) MDB: the metalloprotein database and browser at the Scripps Research Institute. Nucleic Acids Res 30(1):379–382

Cervantes C, Campos-García J, Devars S, Gutiérrez-Corona F, Loza-Tavera H, Torres-Guzmán JC, Moreno-Sánchez R (2001) Interactions of chromium with microorganisms and plants. FEMS Microbiol Rev 25(3):335–347

Chen CW, Chen CF, Dong CD (2012) Distribution and accumulation of mercury in sediments of Kaohsiung River Mouth, Taiwan. APCBEE Procedia 1:153–158

Cohen AR, Trotzky MS, Pincus D (1981) Reassessment of the microcytic anemia of lead poisoning. Pediatrics 67(6):904–906

Dayan AD, Paine AJ (2001) Mechanisms of chromium toxicity, carcinogenicity, and allergenicity: a review of the literature from 1985 to 2000. Hum Exp Toxicol 20(9):439–451

Di Bonito M (2008) Sewage sludge in Europe and the UK: environmental impact and improved standards for recycling and recovery to land. In: Environmental geochemistry. Elsevier, Amsterdam, pp 251–286

Duffus JH (2002) Heavy metals IUPAC technical report. Pure Appl Chem 74:793–807

Finkelman RB (2004) Potential health impacts of burning coal beds and waste banks. Int J Coal Geol 59(1–2):19–24

Foster KS (2012) Synthesis and evaluation of naturally occuring halogenated bipyrroles in a relevant marine species, fundulus heteroclitus. Electronic Theses and Dissertations 109. https://egrove.olemiss.edu/etd/109

Galaris D, Evangelou A (2002) The role of oxidative stress in mechanisms of metal-induced carcinogenesis. Crit Rev Oncol Hematol 42(1):93–103

Gautam PK, Gautam RK, Banerjee S, Chattopadhyaya MC, Pandey JD (2016) Heavy metals in the environment: fate, transport, toxicity, and remediation technologies. In: Heavy metals: sources toxicity and remediation techniques, vol 60. Nova Publishers, New York, pp 101–130

Ghani A, Ghani A (2011) Effect of chromium toxicity on growth, chlorophyll, and some mineral nutrients of brassica juncea L. Egypt Acad J Biol Sci H Botany 2(1):9–15

Gordon JJ, Quastel JH (1948) Effects of organic arsenicals on enzyme systems. Biochem J 42(3): 337–350. https://doi.org/10.1042/bj0420337

Gürkan R, Ulusoy Hİ, Akçay M (2017) Simultaneous determination of dissolved inorganic chromium species in wastewater/natural waters by surfactant sensitized catalytic kinetic spectrophotometry. Arab J Chem 10:S450–S460

Hemdan NY, Emmrich F, Faber S, Lehmann J, Sack U (2007) Alterations of TH1/TH2 reactivity by heavy metals: possible consequences include induction of autoimmune diseases. Ann N Y Acad Sci 1109(1):129–137

Hou L, Zhang X, Wang D, Baccarelli A (2012) Environmental chemical exposures and human epigenetics. Int J Epidemiol 41(1):79–105

Hughes JP, Polissar L, Van Belle G (1988) Evaluation and synthesis of health effects studies of communities surrounding arsenic-producing industries. Int J Epidemiol 17(2):407–413

Innovative Solutions for Wastewater Treatment, Organica Water Inc (2017) Primary, secondary, and tertiary wastewater treatment: how do they work [online]. Available at https://www.organicawater.com/primary-secondary-tertiary-wastewater-treatment-work

Irfan M, Hayat S, Ahmad A, Alyemeni MN (2013) Soil cadmium enrichment: allocation and plant physiological manifestations. Saudi J Biol Sci 20(1):1–10

Jomova K, Valko M (2011) Advances in metal-induced oxidative stress and human disease. Toxicology 283(2–3):65–87

Jurdziak M, Gać P, Martynowicz H, Poręba R (2015) The function of the respiratory system was evaluated using selected spirometry parameters in persons occupationally exposed to lead without evident health problems. Environ Toxicol Pharmacol 39(3):1034–1040

Kim HL, Seo YR (2011) The synergistic genotoxic effect between gene and environmental pollutant: oxidative DNA damage induced by thioredoxin reductase 1 silencing under nickel treatment. Mol Cell Toxicol 7(3):251–257

Lenntech (2019) Heavy metals - Lenntech [online]. Lenntech.com. https://www.lenntech.com/processes/heavy/heavy-metals/heavy-metals.htm

Liu C, Zhou C, Zhang N, Pan J, Cao S, Tang M, Ji W, Hu T (2019) Modes of occurrence and partitioning behavior of trace elements during coal preparation—a case study in Guizhou Province, China. Fuel 243:79–87

Mahaffey KR (1990) Environmental lead toxicity: nutrition as a component of intervention. Environ Health Perspect 89:75–78. https://doi.org/10.1289/ehp.908975

Masindi V, Muedi KL (2018) Environmental contamination by heavy metals. In: Heavy metals, vol 10. IntechOpen, London, pp 115–132

Mazumder DG (2008) Chronic arsenic toxicity & human health. Indian J Med Res 128(4):436–447

Monalisa M, Kumar PH (2013) Effect of ionic and chelate assisted hexavalent chromium on mung bean seedlings (*Vigna radiata* L. Wilczek. var k-851) during seedling growth. J Stress Physiol Biochem 9(2):232–241

Mutlu A, Lee BK, Park GH, Yu BG, Lee CH (2012) Long-term concentrations of airborne cadmium in metropolitan cities in Korea and potential health risks. Atmos Environ 47:164–173

Park JY, Seo YR (2011) The protective role of Nrf2 in cadmium-induced DNA damage. Mol Cell Toxicol 7(1):61–66

Park JY, Lee YJ, Koedrith P, Seo YR (2012) Protective role of thioredoxin reductase 1 in cadmium-induced DNA damage. Mol Cell Toxicol 8(3):289–295

Pohl HR, Roney N, Abadin HG (2011) Metal ions affect the neurological system. Met Ions Life Sci 8(247):62

Poli G, Leonarduzzi G, Biasi F, Chiarpotto E (2004) Oxidative stress and cell signaling. Curr Med Chem 11(9):1163–1182

Rodríguez MC, Barsanti L, Passarelli V, Evangelista V, Conforti V, Gualtieri P (2007) Effects of chromium on photosynthetic and photoreceptive apparatus of the alga Chlamydomonas reinhardtii. Environ Res 105(2):234–239

Roy A, Queirolo E, Peregalli F, Mañay N, Martínez G, Kordas K (2015) Association of blood lead levels with urinary F2-8α isoprostane and 8-hydroxy-2-deoxy-guanosine concentrations in first-grade Uruguayan children. Environ Res 140:127–135

Saha JC, Dikshit AK, Bandyopadhyay M, Saha KC (1999) A review of arsenic poisoning and its effects on human health. Crit Rev Environ Sci Technol 29(3):281–313

Singh N, Kumar D, Sahu AP (2007) Arsenic in the environment: effects on human health and possible prevention. J Environ Biol 28(2):359

Stohs SJ, Bagchi D (1995) Oxidative mechanisms in the toxicity of metal ions. Free Radic Biol Med 18(2):321–336

Tchounwou PB, Yedjou CG, Patlolla AK, Sutton DJ (2012) Heavy metal toxicity and the environment. Exp Suppl 101(1):133–164. https://doi.org/10.1007/978-3-7643-8340-4_6

Tian L, Wu H, Deng H, Cheng D, Bai X, Li J (2013) Ecological risks of heavy metal contamination in soils around the coal gangue dump in the coal mining area by toxicity characteristic leaching procedure. Guizhou Agricultural Sciences 41(01):123–127

Trasande L, Landrigan PJ, Schechter C (2005) Public health and economic consequences of methyl mercury toxicity to the developing brain. Environ Health Perspect 113(5):590–596

Uauy R, Maass A, Araya M (2008) Estimating risk from copper excess in human populations. Am J Clin Nutr 88(3):867S–871S

Valko M, Izakovic M, Mazur M, Rhodes CJ, Telser J (2004) Role of oxygen radicals in DNA damage and cancer incidence. Mol Cell Biochem 266(1):37–56

Valko M, Rhodes CJB, Moncol J, Izakovic MM, Mazur M (2006) Free radicals, metals, and antioxidants in oxidative stress-induced cancer. Chem Biol Interact 160(1):1–40

Wang LK (2009) Heavy metals in the environment. Crc Press, Boca Raton

Wani AL, Ara A, Usmani JA (2015) Lead toxicity: a review. Interdiscip Toxicol 8(2):55–64. https://doi.org/10.1515/intox-2015-0009

Wikipedia (2021) Heavy metals (Online). Available https://en.wikipedia.org/wiki/Heavy_metals. Accessed 18 July 2022

Wikipedia (2022a) Chromium (Online). Available https://en.wikipedia.org/wiki/Chromium. Accessed 18 July 2022

Wikipedia (2022b) Copper (Online). Available https://en.wikipedia.org/wiki/Copper. Accessed 18 July 2022

Wikipedia (2022c) Mercury (Online). Available https://en.wikipedia.org/wiki/Mercury. Accessed 18 July 2022

Metalliferous Soil Remediation Through Heavy Metal-Resistant Plant Growth-Promoting Bacteria: Prospects and Paradigms

11

Basudev Majhi, Pradeep Semwal, Shashank Kumar Mishra, Vaibhav Srivastava, Rajeev Pratap Singh, and Puneet Singh Chauhan

Abstract

Heavy metals (HMs) are natural assets on the planet, but they have become a serious threat to environmental pollution due to fast industrial enterprise, metropolitanism, and recent technological breakthroughs. The principal causes of HMs pollution in soil, water, and air are industrial waste, organic compounds, pesticides, paints (including small-medium industries), and mining activities. The land, water, and air utilized in farming are of major significance and may have an impact on the health of living organisms. Novel and robust ecotechnologies are needed to avoid HM contamination in the environment. Microbial bioremediation has long been recognized as the most well-understood biotechnological process for environmental restoration. For the treatment of HM-contaminated environmental sites, microbial bioremediation is a cost-effective option. Researchers worldwide are making strides in discovering new bacterial strains with plasmid-linked degradation/reduction ability. Genetic engineering and molecular biology aided in the development of microbes that would produce the desired results in the environment. Recent advances in microbial bioremediation techniques include biostimulation, bioaugmentation, bioaccumulation, biosorption, and the

B. Majhi · P. Semwal · S. K. Mishra · P. S. Chauhan (✉)
Microbial Technology Division, Council of Scientific and Industrial Research-National Botanical Research Institute (CSIR-NBRI), Rana Pratap Marg, Lucknow, Uttar Pradesh, India
e-mail: puneet@nbri.res.in

V. Srivastava
Sustainability Cluster, School of Engineering, University of Petroleum and Energy Studies, Dehradun, Uttarakhand, India

R. P. Singh
Institute of Environment and Sustainable Development, Banaras Hindu University, Varanasi, Uttar Pradesh, India

R. P. Singh et al. (eds.), *Heavy Metal Toxicity: Environmental Concerns, Remediation and Opportunities*, https://doi.org/10.1007/978-981-99-0397-9_11

use of biofilms. This chapter assembles data on recent developments and applications of microbe-mediated bioremediation of HM-contaminated soils.

Keywords

Heavy metal · Microbial remediation · Metal toxicity

11.1 Introduction

The environment is made up of several complicated variables such as land, air, and water. The existence of humans and other living entities such as animals, plants, and bacteria is based on their positive correlation (Arora et al. 2018). Heavy metal (HM) pollution has become a serious risk to the environment and food production as a result of the massive expansion in the global population and quick progress in modern agriculture (Selvi et al. 2015). Many studies have reported it as a global problem in countries like India, Bangladesh, Italy, Germany, Greece, Hong Kong, China, Turkey, Iran, etc. (Chikumbusko et al. 2017). The lack of understanding about safe effluent disposal and the failure to impose strong regulatory standards have contributed to environmental degradation (Khalid et al. 2017). Vast amounts of solid waste in various harmful forms have been generated as a consequence of these circumstances leading to contamination of the entire ecosystem. The wastewaters, which may exceed authorized limits specified by international regulatory bodies, will have an impact on the quality of surface water and land (EPA 1992, 2002).

Because of their nonbiodegradability, bioaccumulation, environmental stability, persistence, and biotoxicity, HMs make a major environmental risk to living creatures and environments (Khan et al. 2019). They can directly affect the physical and chemical properties of the soils, air, and water, which cause environmental pollution (Fig. 11.1) (Omwene et al. 2018). They can also disrupt the natural ecosystem and have a direct and continual impact on human health through food chains, leading to various diseases (Weber et al. 2020; Suhani et al. 2021),such as paralysis agitans, Alzheimer's, Sclerosis, cancer, hardening of the arteries, etc. (Muszyńska and Hanus-Fajerska 2015).

Many treatment approaches, such as physical, chemical, and biological, have recently been proposed to clean up HM pollution in the air, water, and soil. HM treatment processes include adsorption, heat treatment, chlorination, ion exchange, chemical extraction, bioleaching, and electrokinetics. According to reports, the majority of the aforementioned techniques are only intended to be used as single remedial methods. Despite their success, these methods have downsides such as inefficiency, cost, and failure during large-scale adoption, among other things (Volesky and Holan 1995; Selvi et al. 2019).

In this chapter, we have discussed the HM sources, their harmful effects, issues associated with the disposal and recycling of HM-containing products, and different microbial-based methods for abatement and opportunities.

Fig. 11.1 Heavy metals cause environmental pollution

11.2 Toxicity of Heavy Metals and the Environment

The biosphere is the most significant part of the environment where biotic entities, i.e., animals and plants, interact with abiotic surroundings, viz., soil, water, and air (Mahmoudi 2003). Pollution is defined as any human action that reduces the quality of the natural environment. Pollution of the environment is nothing new, yet it is still the world's most serious problem and the primary cause of diseases and loss of life. Environmental pollution is generally worse in middle- and low-income countries as compared to the developed ones, partly due to poverty, ineffective legislation, and a lack of awareness about pollution. Environmental pollution is caused by HMs through a variety of factors, including industrial growth, urbanization, population increase, exploration, mining, deforestation, bush burning, dumping of agricultural and residential wastes in water bodies, use of pesticides in aquatic animal harvesting, inappropriate disposal of technological wastes, etc. (Landrigan et al. 2018) The repercussions affect not just people but also other land and aquatic species, including microbes, which support biogeochemical cycles required for a healthy ecosystem (Ukaogo et al. 2020).

11.3 Heavy Metals

HMs are metallic chemical elements with a relatively high density, which are harmful or lethal even at low doses. Almost all the HMs are hazardous to human health above a certain concentration and pose a risk to the environment. HMs include cadmium (Cd), zinc (Zn), molybdenum (Mo), mercury (Hg), nickel (Ni), chromium (Cr), strontium (Sr), arsenic (As), vanadium (V), boron (B), cobalt (Co), copper (Cu), molybdenum (Mo), tin (Sn), lead (Pb), etc. HMs including Cu, Ni, Fe, Zn, B, and Mo are necessary for plant growth, but when their concentrations exceed the permissible limits, they can harm animals and plants. Among all the HMs, Pb, Hg, Cd, and As are not required for the growth and development of plants and animals.

11.3.1 Sources of Heavy Metal Pollution

HMs in the soil aggregate due to a variety of causes that may be natural and anthropogenic sources (Fig. 11.2) (He et al. 2012).

11.3.1.1 Natural Sources

Under different environmental conditions, HMs are naturally emitted through volatile organic compounds, forest fire, sea-salt sprays, volcanic ash, rock depletion, and dirt particles, which are the causes of HM pollution. HMs can be found in the form of oxides, silicates, sulfides, phosphates, sulfates, hydroxides, and organic molecules. Pb, Ni, Cr, Cd, As, Hg, Se, Zn, and Cu are some of the most regularly used HMs. While these HMs are found in smaller quantities in humans and other animals, they may cause severe health concerns (Ali et al. 2021).

11.3.1.2 Anthropogenic Sources

There are a lot of anthropogenic factors of HM concentrations in the environment, but the most significant are rising industrialization and urbanization in recent days. Fertilization, pesticide application, air deposition, sewage irrigation, mining, and sludge application are responsible for HM accumulation in the environment; in addition to these factors, melting activities for metallic ores, industrial wastes, combustion of fossil fuel refinement, and refinishing contribute to HM accumulation

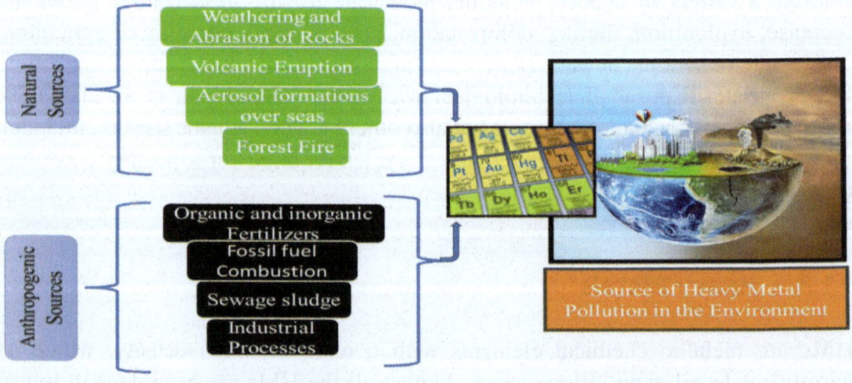

Fig. 11.2 Source of heavy metals in the environment

(Srivastava et al. 2017). The primary cause of metal pollution in the atmosphere is assumed to be coal burning (Antoniadis et al. 2017).

11.3.2 Environmental Impacts of Heavy Metals

Heavy metals have a lot of negative consequences when they are present in the environment. The hydrosphere, lithosphere, and biosphere are all affected due to these HMs (Masindi and Muedi 2018) (Fig. 11.2). They can directly impact the physical and chemical properties of the sediment, soils, and water to hinder microbial activities. They can also destabilize the natural ecosystem and have a direct and long-term impact on the human body, leading to a variety of diseases and issues in mankind (Ali et al. 2021).

Population increase is one of the major issues in the world. Due to the increasing population, the use of HMs also increased, and with this contamination, major environmental components cause serious issues worldwide (Masindi and Muedi 2018).

11.3.2.1 Effect on Soil

HMs affect the agroecosystem through both natural and artificial sources. Several studies have found that natural HM pollution sources are usually high when compared to anthropogenic activity. The parent material from which HMs are derived is the major source of HMs in soil. Human activities disrupt the nature's slow-moving geochemical cycle of HMs, resulting in accumulation in the soil (Dixit et al. 2015). Different anthropogenic activities like the refinement of fossil fuels through combustion (Muradoglu et al. 2015), smelting and extracting metals (Chen et al. 2015), municipal trash disposal (Khan et al. 2016), fertilizer application (Atafar et al. 2010), pesticide usage (Ogunlade and Agbeniyi 2011), sewage application (Sun et al. 2013; Srivastava et al. 2016), etc. cause HM pollution.

HMs in soil cause a severe problem because they accumulate in food chains, damaging the entire ecosystem. Organic pollutants are biodegradable, but their biodegradation rate is slowed by the presence of HMs in nature, which doubles the organic and HM pollution. HMs can harm animals, humans, plants, and ecosystems in a variety of ways. Direct ingestion, absorption by plants, transfer through food chains, drinking polluted water, and changes in soil color, pH, porosity, and natural chemistry all impact soil quality (Musilova et al. 2016).

11.3.2.2 Effects on Water

The kind of soil, rock, and water movement, all influence the metal composition of surface water such as rivers, lakes, and ponds. Metals on the soil's surface are carried away by the wind and end up in sewage and reservoirs (Salem et al. 2000). Rainwater becomes polluted as it travels through the atmosphere. The passage of numerous industrial wastewaters into water sources contaminates them, which contain a large number of HM leachates from landfills and liquid disposal in deep wells, and contaminates the groundwater (Oyeku and Eludoyin 2010). The metal level of water is affected by a variety of elements, including pH, life forms, ion exchange, temperature, vaporization, absorption, and others.

11.3.2.3 Effects on Air

Surface degradation and loss of colloids release HMs into the atmosphere as vapors. Mineral dust, particles of sea salt, volcanoes, and forest fires are all atmospheric sources of HMs (Colbeck 1995). HM air pollution can come from a variety of industrial activities that produce dust particles, such as metal smelters and cement factories, in addition to these natural sources. In the atmosphere, unstable metals such as gaseous pollutants particles of Sb, Se, Hg, and As are transmitted. Metals like Zn, Pb, and Cu are carried as particulates which pollutes the air (Selvi et al. 2019).

11.3.2.4 Human Exposure to Heavy Metals

Poison HMs get into the human body through a variety of mechanisms, including ingestion, inhalation, and skin absorption. People in underdeveloped countries are more exposed to hazardous metals (Eqani et al. 2016) because many people are unaware of the dangers of HM exposure and the ramifications for human health (Afrin et al. 2015). HMs may be present in the workplace and the environment. HMs are ingested by mining and industrial workers through metal particles containing dust and particulate matter. Welders exposed to welding fumes for an extended period had considerably greater blood levels of the HMs such as Cr, Ni, Cd, and Pb than the control group, as well as elevated oxidative stress (Mahmood et al. 2015). Cigarette smoking is a major source of human exposure to harmful HMs, i.e., Cd (Järup 2003) and other HMs found in tobacco leaves. Among these HMs, As is one of the most hazardous metalloids on the earth. According to the World Health Organization (WHO), the limit of As in the drinking water is 5–10 μg/L, but in many countries like Bangladesh and some parts of India, As concentration is more than 50 μg/L, causing many diseases such as skin cancers, kidney cancer, lung cancer, and bladder cancer apart from long-term exposure of inorganic arsenic hypertension, diabetes, reproductive disorders, and cardiovascular diseases (Santra et al. 2013).

The overall public is exposed to HMs through food and water. Intensification of industrial and agricultural activity has resulted from globalization, urbanization, and rapid economic development. Toxic HMs could be released into the water, air, and soil as a result of these actions. HMs bioaccumulate in human food systems, eventually reaching the human body and causing different diseases (Ali et al. 2019) (Fig. 11.3).

11.4 Environmental Heavy Metal Remediation

HMs emitted from several sites are released into the environment, either directly or indirectly, affecting humans, animals, and plants. Increased human exposure to HMs has serious health consequences and causes environmental degradation (Rzymski

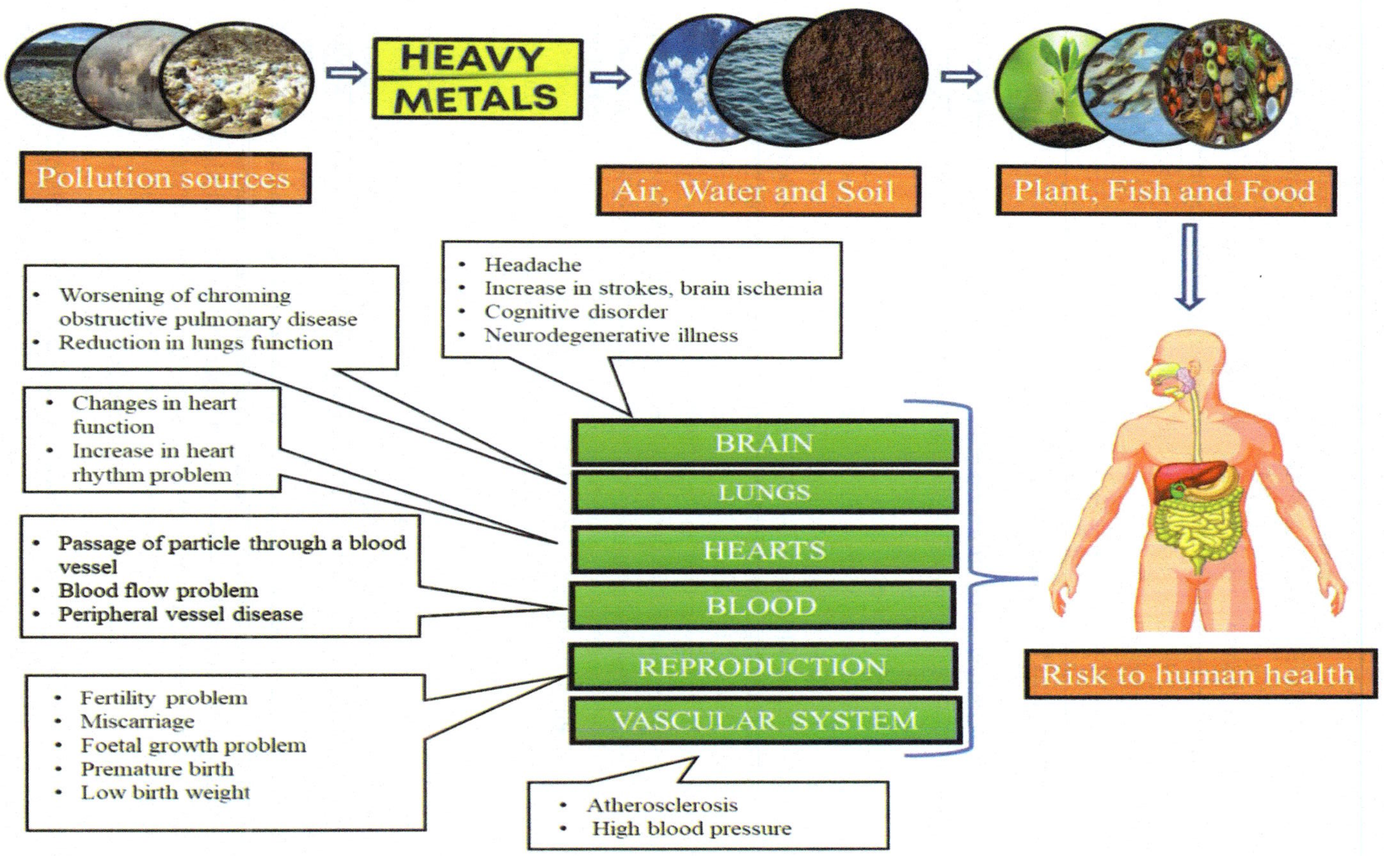

Fig. 11.3 Effect of heavy metals on human health

et al. 2014). The severity of negative health impacts varies depending on the duration of exposure, concentration, chemical form, and type of HMs. HM pollution in soil has led to ecosystem degradation, reduced food quality, and decreased soil health. HM concentrations in India's industrial zones are far greater than the WHO's allowed limit, putting humans at risk (Manivasagam 1987). The failure of respective government environmental safety agencies in developing countries to impose strict regulations and the unreliability of current individual treatment technologies in situ and a wide range of applications are all contributing to the health problems associated with metal pollution.

11.4.1 Biological Remediation

Biological remediation is the utilization of living organisms to clean up HMs in the soil. Here we mainly discussed microbial remediation and phytoremediation.

11.4.1.1 Microbial Remediation

Environmental HM contamination has seriously threatened all ecosystems (Okolo et al. 2016). According to Environmental Protection Agency, parameterization is a natural activity in which microbial processes are utilized to break down or convert dangerous substances into less harmful ones, ultimately eliminating toxins from the environment. During the microbial process, microorganisms utilize chemical pollutants as an energy source in their metabolic processes. Synthetic nutrients limit microbial development in the soil (Ahirwar et al. 2016). Microorganisms can degrade, detoxify, and decontaminate substances and even accumulate toxic organic and inorganic substances from a variety of places in the environment; some bacteria will be described biochemically, and it will be determined whether they can tolerate heavy metals like copper and zinc. In the last 20 years, bioremediation strategies have made significant advances, with the ultimate goal of efficiently restoring damaged regions in an eco-friendly and low-cost manner (Ambaye et al. 2022).

11.4.1.1.1 Bacterial Remediation Capacity of Heavy Metals

Microbial biomass contains a variety of biosorption properties that range greatly among microorganisms. However, each microbial cell's biosorption ability is determined by its pretreatment and experimental settings. Bacteria are essential bioabsorbents because of their widespread distribution, size, capacity to thrive in controlled environments, and resistance to environmental conditions (Srivastava et al. 2015). Their remarkable biosorption abilities are due to their high surface-to-volume ratios and probable active chemisorption sites on the cell wall (Mosa et al. 2016). Bacteria thrive in mixed cultures because they are more stable and survive longer (Sannasi et al. 2006). As a result, consortia are metabolically efficient for metal biosorption and more suitable for field application (Kader et al. 2007). De et al. (2008) used an *Acinetobacter* sp. bacterial consortium to decrease Cr by 78%. *B. megaterium*, *B. niger*, and *Penicillium* sp. had the greatest ability to reduce Pb (2.13–0.03 mg/L), Cr (1.38–0.08 mg/L), and Cd (0.4–0.03 mg/L), respectively.

11.4.1.1.2 Plant Growth-Promoting Endophyte-Mediated Phytoremediation

Endophytic bacteria have shown to assist host plants in adjusting to harsh soil conditions and improve phytoremediation capacity by modifying metal bioavailability in soil, increasing plant growth, decreasing metal phytotoxicity, lowering mental stress, and changing metal translocation in plants (Ma et al. 2011). These bacterial endophytes contribute to the detoxification of metal-polluted soils by improving plant metal tolerance capacity and growth and increasing uptake capacity in plants as discussed (Table 11.1).

11.4.1.1.3 Fungi Remediation Capacity of Heavy Metal

Because of their strong metal uptake and restoration capabilities, fungi are often used as biosorbents to eliminate toxic metals (Fu et al. 2012). Dead fungal biomass from *Penicillium chrysogenum*, *Saccharomyces cerevisiae*, *Aspergillus niger*, and *Rhizopus oryzae* can convert hazardous Cr (VI) to less dangerous or nontoxic Cr (Park et al. 2005). Luna et al. (2016) also reported that *Candida sphaerica* creates biosurfactants that remove Pb (79%), Zn (90%), and Fe (95%). Likewise surfactin,

Table 11.1 Microorganisms used in heavy metal remediation of contaminated sites

Class of microorganisms	Heavy metal removed	References
A. Bacteria		
Pseudomonas veronii	Cd, Zn, Cu	(Coelho et al. 2015)
Pseudomonas putida	Cr (VI)	(Balamurugan et al. 2014)
Bacillus cereus	Cr (VI)	(Coelho et al. 2015)
Bacillus cereus strain XMCr-6	Cr (VI)	(Dong et al. 2013)
Bacillus subtilis	Cr (VI)	(Balamurugan et al. 2014)
Enterobacter cloacae B2-DHA	Cr (VI)	(Rahman et al. 2015)
Kocuria flava	Cu	(Coelho et al. 2015)
Sporosarcina ginsengisoli	As (III)	(Coelho et al. 2015)
B. Fungi		
Gloeophyllum sepiarium	Cr (VI)	(Achal et al. 2011)
Rhizopus oryzae (MPRO)	Cr (VI)	(Sukumar 2010)
Aspergillus fumigatus	Pb	(Kumar Ramasamy et al. 2011)
Aspergillus versicolor	Ni, Cu	(Taştan et al. 2010)
C. Algae		
Hydrodictyon, *Oedogonium*, and *Rhizoclonium* spp.	As	(Srivastava and Dwivedi 2015)
Spirogyra spp. and *Cladophora* spp.	Pb (II), Cu (II)	(Lee and Chang 2011)
Spirogyra spp. and *Spirulina* spp.	Cr Cu, Fe, Mn, Zn	(Mane and Bhosle 2012)
D. Yeast		
Saccharomyces cerevisiae	Pb, Cd	(Lívia and Benedito 2015)

rhamnolipid, and sophorolipid were also tested for HM (Cu and Zn) removal by Mulligan et al. (2001).

11.4.1.1.4 Algae Remediation Capacity of Heavy Metal

In comparison to other microbial bioabsorbents, algae are autotrophic, meaning they consume few nutrients and produce a large amount of biomass. These bioabsorbents have also been utilized to remove HMs from the environment due to their high sorption ability (Abbas et al. 2014). Adsorption or integration of algae biomass into cells is employed for bioremediation of HM-polluted wastewater. Phycoremediation is employing several forms of algae and cyanobacteria to remove or degrade toxicants to remediate HMs (Chabukdhara et al. 2017). Algae include chemical moieties on their surface that act as metal-binding sites, including hydroxyl, carboxyl, phosphate, and amide (Abbas et al. 2014). Dead *Chlorella vulgaris* cells were utilized by Hussian and Napiórkowska-Krzebietke et al. to remove Cd^{2+}, Cu^{2+}, and Pb^{2+} ions from aqueous solutions under various pH, bioabsorbent dosage, and contact time conditions. These findings demonstrate that the biomass of *C. vulgaris* is an exceptionally efficient bioabsorbent for the removal of Cd^{2+}, Cu^{2+}, and Pb^{2+} at 95.5%, 97.7%, and 99.4%, respectively, from a mixed solution containing 50 mg/dm^3 of each metal ion (Goher et al. 2016).

11.4.1.2 Heavy Metal Removal Using Biofilm

In many experiments, biofilms were utilized to remove HMs. Biofilm is a type of bioremediation that also serves as a biological stabilizer. Biofilms have an extremely high tolerance for hazardous inorganic elements, even at deadly doses. According to research on *Rhodotorula mucilaginosa*, metal elimination efficacy varied from 4.79 to 10.25% for planktonic cells and from 91.71 to 95.39% for biofilm cells (Goher et al. 2016). Biosorbents or exopolymeric substances present in biofilms that contain molecules with a surfactant or emulsifying qualities might be used in biofilm bioremediation approaches (El-Masry et al. 2004).

11.4.1.2.1 Metal-Microbe Interaction

The bacterial cell takes up the heavy metal by different methods, either active transport, ion exchange, electrostatic interaction, complexation, or the production of extracellular polysaccharides (Srivastava et al. 2017) (Fig. 11.4).

When microbes interact with heavy metals, they accumulate in the microbial cell and can be detoxified by mechanisms such as bioadsorption, biomineralization, biodegradation, bioleaching, biotransformation, and bioaccumulation.

11.4.1.3 Methods for Heavy Metal Remediation Using Microorganisms

11.4.1.3.1 Biosorption

Although the terms bioaccumulation and biosorption are often used interchangeably, they differ in how pollutants are sequestered. Biosorption, according to Volesky, is the adsorption of chemicals from solution by biological materials via

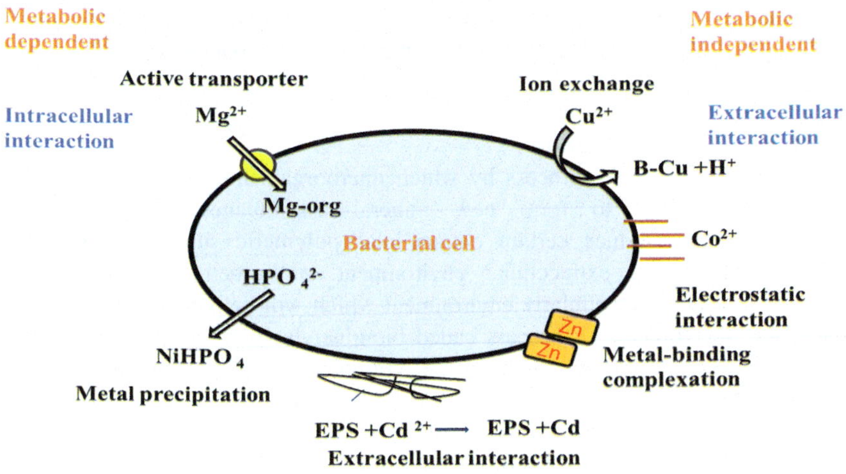

Fig. 11.4 Metal-microbe interaction

physiochemical absorption pathways such as electrostatic forces and ion/proton displacement (Volesky and Holan 1995). Biosorption has been proven to remove a wide range of HMs from aqueous solutions, including very hazardous metal ions such as Cd, Cr, Pb, Hg, and As (Saba et al. 2019). As a result, the diversity of cell wall architectures is critical to biosorption success. By using bioabsorbents like microorganisms (both live and dead), agricultural waste, and other industrial wastes, biosorption can remove pollutants and build the framework for long-term metal removal and recovery (Inoue et al. 2017). Numerous parameters such as pH, temperature, shaking speed, initial pollutant concentration, and bioabsorbent amount are considered to improve biosorption effectiveness. The chemical composition of each contaminant, biomass size, interaction between distinct metallic ions, and ionic strength influence the binding mechanism. Biosorption is also appealing because of a variety of benefits, including low operating costs due to its reversible process, no increase in chemical oxygen demand (COD), ease of desorption, and high adsorption rate. However, other factors must be considered, such as the potential toxicity of pollutants to bacterial cells if living cells are used in this procedure.

11.4.1.3.2 Bioaccumulation Process

Bioaccumulation, on the other hand, is a natural active metabolic process in which HMs accumulate and are taken up by proteins into intracellular living bacterial cells. The initial stage is the adsorption of HMs onto cells, and the metal species are then carried inside the cells, where the HMs can be sequestered by proteins, the lipid bilayer as an import system, and peptide ligands as a storage system (Mishra and Malik 2013). Metal ions were uptake by numerous substances inside the cell cytoplasm to create big ions in intracellular sequestration. Gram-negative bacteria enhanced absorption from the periplasm into the cytoplasm is dependent on the expression of inner membrane importers in extracellular sequestration (Saier 2016; Diep et al. 2018).

11.4.1.3.3 Biomineralization

Biomineralization is the process by which microorganisms mediate and catalyze inorganic reactions to form new mineral assemblages. Therefore, some microorganisms produce certain extracellular polymeric substances (EPSs) or biosurfactant in the extracellular environment, or sometimes they will also produce it in the intracellularly environment which will help to convert the heavy metal into the minerals, a process called biomineralization. Various bacteria like *Bacillus*, *Streptococcus*, etc. help in the biomineralization process (Arnold et al. 2021).

11.4.1.3.4 Bioleaching for Bioremediation

"The dissolving of metals from their mineral source using specific naturally existing microbes" or "the use of microbes to alter metal elements so that the elements can be recovered when water is filtered through it" are two definitions of bioleaching (Mishra et al. 2005). Ni, Cu, Zn, Co, Au, Pb, and As have all been dissolved with it. This process is useful for extracting valuable metal compounds from solid substrates and detoxifying HM-contaminated wastes such as ores, energy, or landfill space as a technology (Singh and Li 2015). However, an investigation has demonstrated that chemolithotrophic techniques cannot handle industrial waste materials containing substantial levels of important metals (Sajjad et al. 2019). Additionally, this is dependent on the metal compounds in the waste as vanadium, chromium, copper, and zinc may all be fully recovered (Blaise et al. 2010). *Thiobacilli* have also been able to detoxify HM-contaminated sewage sludge, soil, sediment, and water (Blaise et al. 2010). From bacteria to fungi and algae, many microorganisms have been isolated from mining and environmental bioleaching settings. *Rhodotorula* sp., *Trichosporon* sp. (yeasts), *Acidithiobacillus* sp. (Bacteria), *Eutrepia* sp. (flagellates), protozoa, and amoebas have all been isolated from a copper mine. Some thermophilic bacteria (particularly *Sulfolobus* sp.) have also been isolated and enhanced from bioleaching environments.

Researchers can examine the particular processes that sustain microbial successions and the impact of community structure on the environment in bioleaching heaps used for copper removal (Wang et al. 2020). Advances in DNA sequencing technology have made it possible to gather unprecedented amounts of information on the genomes of bioleaching bacteria, allowing for the development of metabolic potential estimation techniques and environmental level interactions.

11.4.1.4 Phytoremediation

The use of plants to clean up HMs in the soil is defined as phytoremediation. These include phytoaccumulation, phytostabilization, and phytodegradation (Fig.11.5). The hyperaccumulating plant will take up the heavy metal that will get accumulated

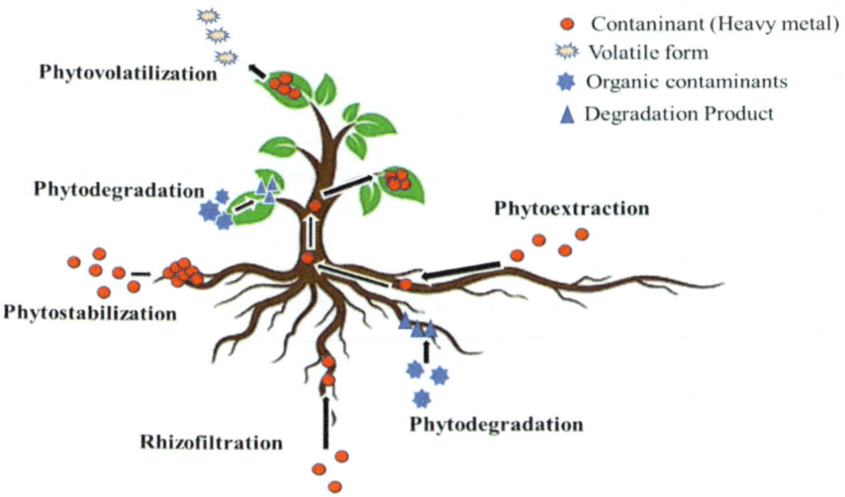

Fig. 11.5 Plants use a variety of phytoremediation techniques

Table 11.2 Microorganisms used in heavy metal remediation of contaminated sites

Family	Species	Heavy metals	References
Brassicaceae	*Arabidopsis halleri*	Cd, Zn	(Zhang et al. 2017)
Brassicaceae	*Alyssum bertolonii*	Ni	(Mengoni et al. 2012)
Brassicaceae	*Arabidopsis halleri*	Cd	(Claire-Lise and Nathalie 2012)
Brassicaceae	*Alyssum murale*	Ni	(Broadhurst and Chaney 2016)
Asteraceae	*Helianthus annuus*	Zn, Pb, Cd	(Fulekar 2016)
Asteraceae	*Berkheya coddii*	Ni	(Slatter 1998)
Caryophyllaceae	*Minuartia verna*	Pb, Zn, Cd,	(Bothe 2011)
Poaceae	*Spartina argentinensis*	Cr	(Nalla et al. 2012)
Pteridaceae	*Pteris vittata*	As	(Rathinasabapathi 2011)
Pteridaceae	*Pteris vittata*	Hg	(Su et al. 2008)
Euphorbiaceae	*Euphorbia cheiradenia*	Cu, Fe, Pb, Zn	(Nematian and Kazemeini 2013)
Fabaceae	*Astragalus racemosus*	Se	(Alford et al. 2012)
Fabaceae	*Medicago sativa*	Pb	(Chibuike and Obiora 2014)
Crassulaceae	*Sedum alfredii*	Pb	(Chen et al. 2012)
Violaceae	*Viola boashanensis*	Pb, Zn, Cd	(Zhuang et al. 2005)

within the plant (Marques et al. 2009; Muthusaravanan et al. 2018; Chaney and Baklanov 2017) (Table 11.2).

In this case, plants will take it up, and sometimes, phytodegradation or the breakdown, phytostabilization, and rhizosphere degradation, that is, degrading of metal in the rhizosphere will happen.

11.5 Conclusion

HM pollution is a serious environmental problem that occurs due to a variety of human activities and has a significant impact on humans and the environment. Because of the biotechnological potential of microbes in removing or recovering metals, our focus has shifted to eco-friendly treatments such as phytoremediation and microbial remediation, which entail HM absorption by microorganisms. Apart from their contributions, biosorbents are potentially beneficial and readily available for removing HMs and for protecting nature and the environment using a bioremediation process. Although just a few studies have been done on this subject, bacteria are one of the most significant microbiological approaches for bioremediation. As heavy-metal pollution alleviators, more research is needed to get the most out of bacterial systems and to determine the specific and unambiguous mechanisms involved in HMs removal by bacteria, fungus, and algae. For the betterment of our environment, we need environmentally friendly remediation solutions based on plants and microorganisms that are viable alternatives to physical and chemical removal methods.

Acknowledgments The authors acknowledge the Director, CSIR National Botanical Research Institute, for providing facilities and support during the study. This work is supported by the CSIR-Network (MLP0048 and MLP0049) and an in-house project (OLP109) funded by the Council of Scientific and Industrial Research, New Delhi, India.

References

Abbas SH, Ismail IM, Mostafa TM, Sulaymon AH (2014) Biosorption of heavy metals. A review. J Chem Sci Technol 3:74–102

Achal V, Kumari D, Pan X (2011) Bioremediation of chromium contaminated soil by a brown-rot fungus *Gloeophyllum sepiarium*. Res J Microbiol 6:166

Afrin R, Mia M, Ahsan M, Akbor A (2015) Concentration of heavy metals in available fish species (Bain, Mastacembelus armatus; Taki, Channa punctatus and Bele, *Glossogobius giuris*) in the Turag River, Bangladesh. Biol Sci - PJSIR 58:104–110

Ahirwar NK, Gupta G, Singh R, Singh V (2016) Isolation, identification and characterization of heavy metal resistant bacteria from industrial affected soil in Central India. Int J Pure Appl Biosci 4:88–93

Alford ÉR, Pilon Smits EA, Fakra SC, Paschke MW (2012) Selenium hyperaccumulation by *Astragalus* (Fabaceae) does not inhibit root nodule symbiosis. Am J Bot 99:1930–1941

Ali H, Khan E, Ilahi I (2019) Environmental chemistry and ecotoxicology of hazardous heavy metals. Environmental Persistence, Toxicity, and Bioaccumulation. J Chem:1–14

Ali MM, Hossain D, Khan MS, Begum M, Hasan Osman MH, Nazal MK, Zhao H (2021) Environmental pollution with heavy metals: a public health concern. IntechOpen

Ambaye TG, Chebbi A, Formicola F, Prasad S, Gomez FH, Franzetti A, Vaccari M (2022) Remediation of soil polluted with petroleum hydrocarbons, and their reuse for agriculture. Recent progress, challenges, and perspectives. Chemosphere:133572

Antoniadis V, Levizou E, Shaheen S, Ok Y, Sebastian A, Baum C, Prasad M, Wenzel W, Rinklebe J (2017) Trace elements in the soil-plant interface. Phytoavailability, translocation, and phytoremediation. A review Earth Sci Rev 171:621–645

Arnold A, Dennison E, Kovacs CS, Mannstadt M, Rizzoli R, Brandi ML, Clarke B, Thakke RV (2021) Hormonal regulation of biomineralization. Nat Rev Endocrinol 17:261–275
Arora NK, Fatima T, Mishra I, Verma M, Mishra J, Mishra V (2018) Environmental sustainability, challenges and viable solutions. Environ Sustain 1:309–340
Atafar Z, Mesdaghinia A, Nouri J, Homaee M, Yunesian M, Ahmadimoghaddam M, Mahvi A (2010) Effect of fertilizer application on soil heavy metal concentration. Environ Monit Assess 160:83–89
Balamurugan D, Udayasooriyan C, Kamaladev B (2014) Chromium (VI) reduction by *pseudomonas putida* and *Bacillus subtilis* isolated from contaminated soils. Int J Environ Sci 5:522
Blaise JF, Tyagi RD, Auclai JC (2010) Metals removal from sewage sludge by indigenous iron oxidizing bacteria. J Environ Sci 28:443–467
Bothe H (2011) Plants in heavy metal soils. In: Detoxification of heavy metals. Springer, Berlin. Heidelberg, pp 35–57
Broadhurst CL, Chaney RL (2016) Growth and metal accumulation of an Alyssum murale nickel hyperaccumulator ecotype co-cropped with Alyssum montanum and perennial ryegrass in serpentine soil. Front Plant Sci 7:451
Chabukdhara M, Gupta SK, Gogoi M (2017) Phycoremediation of heavy metals coupled with generation of bioenergy. In: Algal biofuels. Springer, Cham, pp 163–188
Chaney RL, Baklanov IA (2017) Phytoremediation and phytomining status and promise. Adv Bot Res 83:189–221
Chen X, Lu X, Yang G (2012) Sources identification of heavy metals in urban topsoil from inside the Xi'an Second Ringroad NW China using multivariate statistical methods. Catena 98:73–78. https://doi.org/10.1016/j.catena.2012.06.007
Chen ZF, Zhao Y, Fan LD, Xing L, Yan YJ (2015) Cadmium (Cd) localization in tissues of cotton (*Gossypium hirsutum* L.) and its phytoremediation potential for Cd-contaminated soils. Bull Environ Conta Toxico 95:784–789
Chibuike GU, Obiora SC (2014) Heavy metal polluted soils effect on plants and bioremediation methods. Appl Environ Soil Sci 2014:752708
Chikumbusko C, Ishmael B, Deliwe D, Rex M, Benard T, Morris M, Russel C, Stephen K, Samson M (2017) A review of heavy metals in soil and aquatic systems of urban and semi-urban areas in Malawi with comparisons to other selected countries. Afr J Environ Sci Technol 11:448–460
Claire-Lise M, Nathalie V (2012) The use of the model species *Arabidopsis halleri* towards phytoextraction of cadmium polluted soils. Nat Biotechnol 30:9–14
Coelho LM, Rezende HC, Coelho LM, De Sousa PA, Melo DF, Coelho NM (2015) Bioremediation of polluted waters using microorganisms. Adv Bioremediat Waste Water Pollut Soil 10:5772–60770
Colbeck I (1995) Particle emission from outdoor and indoor sources in airborne particulate matter. In: Kouimtzis T, Samara C (eds) The handbook of environ Chem, vol 4. Springer, Berlin, Heidelberg, pp 1–33
De J, Ramaia N, Vardanyan L (2008) Detoxification of toxic heavy metals by marine bacteria highly resistant to mercury. Mar Biotechnol 10:471–477
Diep P, Mahadevan R, Yakunin AF (2018) Heavy metal removal by bioaccumulation using genetically engineered microorganisms. Front Bioeng Biotechnol 6:157. https://doi.org/10.3389/fbioe.2018.00157
Dixit R, Wasiullah MD, Pandiyan K, Singh U, Sahu A, Shukla R, Singh B, Rai SP, Lade H, Paul D (2015) Bioremediation of heavy metals from soil and aquatic environment, An overview of principles and criteria of fundamental processes. Sustainability 7:2189–2212
Dong G, Wang Y, Gong L, Wang M, Wang H, He N, Zheng Y, Li Q (2013) Formation of soluble Cr (III) end-products and nano-particles during Cr (VI) reduction by *Bacillus cereus* strain XMCr-6. Biochem Eng J 70:166–172
El-Masry MH, El-Bestawy E, El-Adl NI (2004) Bioremediation of vegetable oil and grease from polluted wastewater using a sand biofilm system. World J Microbiol Biotechnol 20:551–557

EPA (1992) Ground water issue- behaviour of metals in soil. U.S. Environmental Protection Agency, EPA/540/S-92/018

EPA (2002) Supplemental guidance for developing soil screening levels for superfund sites. US EPA, Office of Solid Waste and Emergency Response, Washington, DC

Eqani S, Khalid R, Bostan N, Saqib Z, Mohmand J, Rehan M, Ali N, Katsoyiannis I, Shen H (2016) Human lead (Pb) exposure via dust from different land use settings of Pakistan a case study from two urban mountainous cities. Chemosphere 155:259

Fu YQ, Li S, Zhu HY, Jiang R, Yin LF (2012) Biosorption of copper (II) from aqueous solution by mycelial pellets of *Rhizopus oryzae*. Afr J Biotechnol 11:1403–1411

Fulekar MH (2016) Phytoremediation of heavy metals by Helianthus annuus in aquatic and soil environment. Int J Curr Microbiol App Sci 5:392–404

Goher ME, AM AE, Abdel-Satar AM, Ali MH, Hussian AE, Napiórkowska-Krzebietke A (2016) Biosorption of some toxic metals from aqueous solution using non-living algal cells of Chlorella vulgaris. J Elem 21:3

He B, Yun Z, Sh J, Jiang G (2012) Research progress of heavy metal pollution in China sources, analytical methods, status, and toxicity. Chin Sci Bull 58:134–140

Inoue K, Parajuli D, Ghimire KN, Biswas BK, Kawakita H, Oshima T, Ohto K (2017) Biosorbents for removing hazardous metals and metalloids. Materials 10:8–857

Järup L (2003) Hazards of heavy metal contamination. Br Med Bull 68:167–182

Kader J, Sannasi P, Othman O, Ismai BS, Salmijah S (2007) Removal of Cr (VI) from aqueous solutions by growing and non-growing populations of environmental bacterial consortia. Glob J Environ Res 1:12–17

Khalid S, Shahid M, Niazi N, Murtaza B, Bib I, Dumat C (2017) A comparison of technologies for remediation of heavy metal contaminated soils. J Geochem Explor 182:247–268

Khan I, Ghani A, Rehman AU, Awan SA, Noreen A, Khalid I (2016) Comparative analysis of heavy metal profile of *Brassica campestris* (L.) and *Raphanus sativus* (L.) irrigated with municipal waste water of Sargodha city. J Clin Toxicol 6:1–4

Khan M, Dai X, Ni Q, Zhang C, Cui X, Lu M, Deng M, Yang X, He Z (2019) Toxic metal pollution and ecological risk assessment in sediments of water reservoirs in Southeast China. Soil Sediment Contaminat Int J 28:695–715

Kumar Ramasamy R, Congeevaram S, Thamaraiselvi K (2011) Evaluation of isolated fungal strain from e-waste recycling facility for effective sorption of toxic heavy metal Pb (II) ions and fungal protein molecular characterization- a mycoremediation approach. Asian J Exp Biol Sci 2: 342–347

Landrigan PJ, Fuller R, Acosta NJ, Adeyi O, Arnold R, Baldé AB, Zhong M (2018) The lancet commission on pollution and health. Lancet 391:462–512

Lee YC, Chang SP (2011) The biosorption of heavy metals from aqueous solution by spirogyra and Cladophora filamentous macroalgae. Bioresour Technol 102:5297–5304

Lívia de CFMH, Benedito C (2015) Potential application of modified Saccharomyces cerevisiae for removing lead and cadmium. J Bioremed Biodegr 6:2

Luna JM, Rufin RD, Sarubbo LA (2016) Biosurfactant from Candida sphaerica UCP0995 exhibiting heavy metal remediation properties. Process Saf Environ Prot 102:558–566

Ma Y, Guo L, Wang H, Bai B, Shi D (2011) Accumulation, distribution, and physiological contribution of oxalic acid and other solutes in an alkali-resistant forage plant, Kochia sieversiana, during adaptation to saline and alkaline conditions. J Plant Nutr Soil Sci 174: 655–663

Mahmood Q, Wang J, Pervez A, Meryem S, Wasee M, Ullah Z (2015) Health risk assessment and oxidative stress in workers exposed to welding fumes. Toxicol Environ Chem 97:634–639

Mahmoudi S (2003) Alexandre Timoshenko, environmental negotiator handbook (London, the Hague and Boston Kluwer law international 2003) xvi 541 pages. Yearb Int Environ Law 14: 844–848

Mane PC, Bhosle AB (2012) Bioremoval of some metals by living algae *spirogyra sp.* and *Spirulina sp.* from aqueous solution

Manivasagam N (1987) Industrial effluents origin characteristics effects. In: Analysis and Treatment. Coimbatore Shakti Publications, pp 79–92

Marques AP, Rangel AO, Castro PM (2009) Remediation of heavy metal contaminated soils: phytoremediation as a potentially promising clean-up technology. Crit Rev Environ Sci Technol 39:622–654

Masindi V, Muedi K (2018) Environmental contamination by heavy metals. *Heavy Metals 10*:115–132

Mengoni A, Cecchi L, Gonnelli C (2012) Nickel hyperaccumulating plants and *Alyssum bertolonii* model systems for studying biogeochemical interactions in serpentine soils. In: Bio-Geo Interactions in Metal-Contaminated Soils. Springer, Berlin, Heidelberg, pp 279–296

Mishra A, Malik A (2013) Recent advances in microbial metal bioaccumulation. Crit Rev Environ Sci Technol 43:1162–1222

Mishra D, Dong-Jin K, Jong-Gwan A, Young-Ha R (2005) Bioleaching, a microbial process of metal recovery. A Review. Met Mater Int 11(3):249–256

Mosa KA, Saadoun I, Kumar K, Helm M, Dhankher OP (2016) Potential biotechnological strategies for the cleanup of heavy metals and metalloids. Front Plant Sci 7:303

Mulligan CN, Yong RN, Gibbs BF (2001) Remediation technologies for metal-contaminated soils and groundwater an evaluation. Eng Geol 60:193–207

Muradoglu F, Gundogdu M, Ercisli S, Encu T, Balta F, Jaafar HZ, Zia-Ul-Haq M (2015) Cadmium toxicity affects chlorophyll a and b content, antioxidant enzyme activities and mineral nutrient accumulation in strawberry. Biol Res 48(1):7

Musilova J, Arvay J, Vollmannova A, Tot T, Tomas J (2016) Environmental contamination by heavy metals in region with previous mining activity. Bull Environ Contam Toxicol 97: 569–575

Muszyńska E, Hanus-Fajerska E (2015) Why are heavy metal hyperaccumulating plants so amazing. Biotechnol Acta 4:265–271

Muthusaravanan S, Sivarajasekar N, Vivek J, Paramasivan T, Naushad M, Prakashmaran J, Al-Duaij OK (2018) Phytoremediation of heavy metals, mechanisms, methods and enhancements. Environ Chem Lett 16:1339–1359

Nalla S, Hardaway CJ, Sneddon (2012) Phytoextraction of selected metals by the first and second growth seasons of Spartina alterniflora. Instrum Sci Technol 40(1):17–28

Nematian MA, Kazemeini F (2013) Accumulation of Pb, Zn, Cu and Fe in plants and hyperaccumulator choice in Galali iron mine area, Iran. Int j Agric Crop Sci 5:4–426

Ogunlade MO, Agbeniyi SO (2011) Impact of pesticides use on heavy metals pollution in cocoa soils of Cross-River state Nigeria. Afr J Agric Res 6:3725–3728

Okolo NV, Olowolafe EA, Akawu I, Okoduwa SIR (2016) Effects of industrial effluents on soil resources in Challawa industrial area Kano Nigeria. Glob Ecol Conserv 5(1):1–10

Omwene P, Öncel M, Çelen M, Kobya M (2018) Heavy metal pollution and spatial distribution in surface sediments of Mustafakemalpaşa stream located in the world's largest borate basin (Turkey). Chemosphere 208:782–792

Oyeku OT, Eludoyin AO (2010) Heavy metal contamination of groundwater resources in a Nigerian urban settlement. Afr J Environ Sci Technol 4:201–214

Park D, Yun YS, Jo JH, Park JM (2005) Mechanism of hexavalent chromium removal by dead fungal biomass of aspergillus Niger. Water Res 39:533–540

Rahman A, Nahar N, Nawani NN, Jass J, Hossain K, Saud ZA, Saha AK, Ghosh S, Olsson B, Mandal A (2015) Bioremediation of hexavalent chromium (VI) by a soil-borne bacterium *Enterobacter cloacae B2-DHA*. J Environ Sci Health Part A 50:1136–1147

Rathinasabapathi B (2011) Arsenic hyperaccumulator fern *pteris vittata* utilities for arsenic phytoremediation and plant biotechnology in working with ferns. Springer, New York, pp 261–269

Rzymski P, Niedzielski P, Poniedziałek B, Klimaszyk P (2014) Bioaccumulation of selected metals in bivalves (Unionidae) and Phragmites australis inhabiting a municipal water reservoir. Environ Monitor Asses 186:3199–3212

Saba RY, Ahmed M, Sabri AN (2019) Potential role of bacterial extracellular polymeric substances as biosorbent material for arsenic bioremediation. J bioremediat 23(2):72–81

Saier MH (2016) Transport protein evolution deduced from analysis of sequence topology and structure. Curr Opin Struct Biol 38:9–17

Sajjad W, Zheng G, Din G, Ma X, Rafiq M, Xu W (2019) Metals extraction from sulfide ores with microorganisms: the bioleaching technology and recent developments. Trans Indian Inst Metals 72(3):559–579. https://doi.org/10.1007/s12666-018-1516-4

Salem HM, Eweida EA, Farag A (2000) Heavy metals in drinking water and their environmental impact on human health. ICEHM Cairo Uni Egypt 542:556

Sannasi P, Kader J, Othman O, Salmijah S (2006) Single and multi-metal removal by an environmental mixed bacterial isolate. Modern Multidiscip Appl Microbiol Exploit Microb Interact 1: 36–141

Santra SC, Samal AC, Bhattacharya P, Banerjee S, Biswas A, Majumdar J (2013) Arsenic in foodchain and community health risk: a study in Gangetic West Bengal. Procedia Environ 18:2–13

Selvi A, Das D, Das N (2015) Potentiality of yeast *Candida sp. SMN04* for degradation of cefdinir, a cephalosporin antibiotic, kinetics, enzyme analysis and biodegradation pathway. Environ Technol 36:3112–3124

Selvi A, Rajasekar A, Theerthagiri J, Ananthaselvam A, Sathishkumar K, Madhavan J, Rahman PK (2019) Integrated remediation processes toward heavy metal removal/recovery from various environments. A Review Front Environ Sci:7

Singh N, Li JH (2015) Bio-extraction of metals as secondary resources from E-waste. Appl Mech Mater 768:602

Slatter K (1998) Nickel accumulation and tolerance in *Berkheya coddii* and its application in phytoremediation (Doctoral dissertation)

Srivastava S, Dwivedi AK (2015) Biological wastes the tool for biosorption of arsenic. J Bioremed Biodegr 7:323

Srivastava V, De Araujo ASF, Vaish B, Bartelt-Hunt S, Singh P, Singh R (2016) Biological response of using municipal solid waste compost in agriculture as fertilizer supplement. Rev Environ Sci and Bio/Technol 677:69

Srivastava V, Sarkar A, Singh S, Singh P, De Araujo AS, Sing RP (2017) Agroecological responses of heavy metal pollution with special emphasis on soil health and plant performances. Front Environ Sci 5:64

Srivastava V, Ismail SA, Singh P, Singh RP (2015) Urban solid waste management in the developing world with emphasis on India challenges and opportunities. Rev Environ Sci and Bio/Technol 317:337

Su Y, Han FX, Chen J, Sridhar BM, Monts DL (2008) Phytoextraction and accumulation of mercury in three plant species Indian mustard (*Brassica juncea*) beard grass (*Polypogon monospeliensis*) and Chinese brake fern (*Pteris vittata*). Int J Phytoremediation 10:547–560

Suhani I, Sahab S, Srivastava V, Singh RP (2021) Impact of cadmium pollution on food safety and human health. Curr Opin Toxicol 27(1):7

Sukumar M (2010) Reduction of hexavalent chromium by Rhizopus oryzae. Afr J Environ Sci Technol 4:412–418

Sun WH, Jiang YX, Li X (2013) Research of the evaluation on heavy-metal pollution in rice by sewage irrigation. Appl Mech Mater 295:1594–1599

Taştan BE, Ertuğrul S, Dönmez G (2010) Effective bioremoval of reactive dye and heavy metals by aspergillus versicolor. Bioresour Technol 101:870–876

Ukaogo P, Ewuzi U, Onwuk C (2020) Environmental pollution, causes, effects, and the remedies. Microbes Environ:419–429

Volesky B, Holan ZR (1995) Biosorption of heavy metals. Biotechnol Prog 11:235–250

Wang Y et al (2020) Stabilization of heavy metal-contaminated soils by biochar: challenges and recommendations. Sci Total Environ 729:139060. https://doi.org/10.1016/j.scitotenv.2020.139060

Weber A, Sales C, de Souza FF, Melo R, Bazzoli N, Rizzo E (2020) Effects of metal contamination on liver in two fish species from a highly impacted neotropical river: a case study of the Fundão dam, Brazil. Ecotoxicol Environ Safety 190:110165

Zhang Z, Wen X, Huang Y, Inoue C, Liang Y (2017) Higher accumulation capacity of cadmium than zinc by *Arabidopsis halleri* ssp. Germmifera in the field using different sowing strategies. Plant Soil 418:165–176

Zhuang P, Ye ZH, Lan CY, Xi ZW, Shu WS (2005) Chemically assisted phytoextraction of heavy metal contaminated soils using three plant species. Plant Soil 276:153–162

Phytoremediation of Heavy Metals: Reaction Mechanisms and Selected Efficient Technologies of Heavy Metal Contamination

12

Ahmed Haruna Shehu, Abubakar Mohammad Gumi, and Norli Ismail

Abstract

Concepts and applications of conventional phytoremediation in comparison to other phytotechnologies, that is, phytoextraction and phytostabilization, will be discussed. The prospects of phytoremediation as an eco-friendly and sustainable approach in heavy metal removal are analyzed. Lessons learned and future directions of successes and limitation of phytotechnologies at field scale application will be deliberated. Heavy metal accumulation in various environmental conditions associated with phytoremediation factors is reviewed. The appropriate phytoremediation protocol in terrestrial and aquatic environments in regard to photostabilization of metals is a way forward for captivating the significance of phytoremediation. The current state, problems, and prospects of phytoremediation of heavy metal polluted soils are analyzed along with the efficacy of phytoremediation at different states of heavy metal speciation. Advantages and disadvantages of phytoremediation as compared to other methods of remediation of heavy metal contaminated soils (naturally and anthropogenic) are areas of concern and partly discussed in this chapter. The examples of successful phytoextraction and phytomining for cleaning up of contaminated soils globally are presented.

A. H. Shehu
Environmental Technology Division, School of Industrial Technology, Universiti Sains Malaysia, Penang, Malaysia

Nigerian Defence Academy (NDA), Kaduna, Nigeria

A. M. Gumi
Usmanu Danfodiyo University (UDUS), Sokoto, Nigeria

N. Ismail (✉)
Environmental Technology Division, School of Industrial Technology, Universiti Sains Malaysia, Penang, Malaysia
e-mail: norlii@usm.my

R. P. Singh et al. (eds.), *Heavy Metal Toxicity: Environmental Concerns, Remediation and Opportunities*, https://doi.org/10.1007/978-981-99-0397-9_12

Keywords

Phytoremediation · Soil remediation · Contaminated soil · Hyperaccumulators · Heavy metals

12.1 Introduction

Industrialization's global expansion has resulted in several polluted areas that have a detrimental effect on human health. These activities, including burning fossil fuels, quarrying, metal-bearing ore welding, sewage disposal, fertilizer use, and pesticide use, impacted the environment. Traditional cleanup methods waste time and resources while posing a greater risk to environmental health. Scientists worldwide collaborate to develop a new, practical, and cost-effective way of cleaning the atmosphere. Using phytoremediation, a plant-based treatment technique, various anthropogenic pollution problems can be resolved inexpensively in relative and with minimal environmental impact. Phytoremediation exhibited a promising strategy compared to other traditional remediation methods for heavy metal contaminants commonly found in contaminated soils and bioenergy plant species that provide aboveground benefits like wildlife habitat and biogas. Despite the numerous challenges, green remediation is the most effective method for removing heavy metals from the soil due to its environmental friendliness and efficacy. This review paper compiles essential information on numerous chemicals used and their effects on the soil environment, contamination characteristics, a picture of the main techniques of heavy metal removal, and the sustainability of the applied process.

12.2 Overview of Soil Pollution and Concept of Phytoremediation

Even though the soil is a nonrenewable natural resource, humans have increasingly used it as a pollutant sink since the Industrial Revolution. Increasing productivity requires maintaining soil fertility, and reducing soil pollution is a critical component of that effort. Soils store approximately 75% of terrestrial organic carbon and are vital to agriculture and climate change (Kassa et al. 2017). The first step toward resolving environmental toxicity in humans, other animals, and plants is to ensure that our knowledge of soil elemental concentrations and patterns is accurate. Mining, geochemistry of parent or underlying materials, weathering, lithology, and geopedological processes substantially affect these elements' concentrations (Kazapoe and Arhin 2021). Chemicals used in agriculture, industrial and domestic processes, municipal and livestock waste, and petroleum products have been implicated in soil pollution and chemistry (Bundschuh et al. 2012). The soil is critical to the human environment's survival owing to its resources and the emissions it absorbs from human activity. More importantly, its quality and pollution level influence agricultural production and human health (Adimalla et al. 2019). Toxic chemicals in urban soils threaten human health due to their ease of transfer to the body via suspended matter or direct soil contact (Chen et al. 2015).

Numerous reasons contribute to global pollution levels increasing, but heavy metals are a source of significant worry. The environmental mobility of heavy metals is influenced by the extraction of ores and the numerous processing procedures that release these elements into the environment. Heavy metal contamination is a severe problem that requires an integrated solution. Nonbiodegradable heavy metals accumulate in the background and threaten the environment and human health, such as soil pollution and water poisoning. Bioaccumulation is when these elements get into the tissues of living species, while biomagnification is how their concentration increases as they migrate up the food chain. The detrimental effects of heavy metals on soil microorganisms have resulted in their extinction (Khan et al. 2010). The inability of heavy metals to be destroyed biologically or physically and hence persist in the soil creates long-term environmental implications (Suman et al. 2018; Yan et al. 2020). Heavy metals are either essential or nonessential in biological systems according to their functions. Copper (Cu), iron (Fe), nickel (Ni), zinc (Zn), and manganese (Mn) are essential for plant physiology and biochemical processes such as photosynthesis and respiration (Devi and Kumar 2020). They can, however, be dangerous if ingested in large quantities. Heavy metals that are not essential to plants, such as lead (Pb), cadmium (Cd), mercury (Hg), and arsenic (As), have no known function in plants (Fasani et al. 2018). They can lower agricultural production, contaminate the environment, and interfere with the metabolic pathways of crop plants (Arif et al. 2016).

Plants are referred to as "phyto" in Greek and Latin and "medium" in Latin and Greek referring to the elimination of evil respectively when discussing phytoremediation. Plants remove pollutants from the environment in various ways, including absorbing and detoxifying them. Phytoremediation utilizes plants and bacteria to reduce the toxicity and concentration of contaminants in soil (Ghosh and Singh 2005; Deb et al. 2020). Heavy metals and organic pollutants such as biphenyls, insecticides, hydrocarbons, and radionuclides can be removed. A novel strategy is to use plants or green substitutes to mitigate the effects of heavy metals in soil (Vithanage et al. 2012; Fasani et al. 2018). This technique assists in removing soil contaminants while preserving topsoil fertility. Organic matter can be added to the soil's top layer to increase fertility (Gumi and Aisha 2012; Devi and Kumar 2020). Phytoremediation has emerged as a cost-effective and environmentally friendly method of treating sites contaminated with heavy metals. Onsite remediation of soil contamination is possible through various plant-based bioremediation techniques referred to as phytoremediation (Cameselle et al. 2013; Saxena et al. 2020). The methods by which plants absorb and translocate metal contaminants and the detoxification strategies they employ in response to heavy metal contamination are described in this chapter. The reliability of the phytoremediation in increasing the bioavailability, tolerance, and accumulation of heavy metals pollution is presented along with appropriate discussions.

12.3 Sources of Heavy Metal Pollution in the Soil

As parent materials deteriorate, trace amounts of heavy metals are released into the soil. These metals are generally considered safe and only rarely pose a threat (Shahid et al. 2016). Accumulation of heavy metals in the environment results from various factors, including natural processes such as rock weathering and a variety of human-caused actions. Composting, pesticide use, and phosphatic manure are all examples of naturally occurring processes, whereas human activities include metal mine tailings, electroplating, industrial effluent, leaded gasoline and lead-based paint, and land application of fertilizer (Wuana and Okieimen 2011; Mar and Okazaki 2012). Anthropogenic heavy metals being more mobile and bioavailable in soil are more harmful than pedogenic or lithogenic heavy metals (Farahat and Linderholm 2015; Rafique and Tariq 2016). Heavy metals are also found in insecticides and other wood treatments (Thangavel and Subbhuraam 2004). Cadmium (Cd) is used in phosphate fertilizers, electroplating, plastics, and paint stabilizers (Hamzah et al. 2016). Chromium (Cr) is most frequently used in steel, leather, cement, and tanneries (Luo et al. 2011). Excess fertilizer and insecticide use contain Cu (Luo et al. 2011). Mercury (Hg) can be found in the environment via coal combustion, mining, and medical waste (Wuana and Okieimen 2011; Rodriguez et al. 2012). Surgical instruments, steel alloys, automobile batteries, and household appliances contain anthropogenic Ni (Gumi and Sufiyanu 2013). Plumbum (Pb) is obtained by applying herbicides and pesticides, the combustion of gasoline, and manufacturing batteries (Thangavel and Subbhuraam 2004; Wuana and Okieimen 2011).

12.4 Fate and Pollution Pathway of Heavy Metals in Soil

As stated in the overview, the term "pollution" implies the occurrence of something undesirable. The source–pathway–receptor principle necessitates the presence of all three components and the company of significant harm. Polluters are pesticides, phosphatic manures, metal mine tailings, compost, electroplating materials, industrial effluent, leaded gasoline and Pb-based paints, fertilizers, and agricultural sludge and biosolids. The path denotes the route taken by a pollutant from one location to another to exert an effect on a receptor. A pollutant could impact a responsive organism, ecological system, or piece of property. They establish a 'pollutant linkage,' or a connection between the contaminant, the route, and the receptor. In some instances, you may be able to take multiple ways (for example, soil and water contamination). Another critical aspect of the pollutant connection concept is the ability to disrupt pollutant connections to halt pollution or clean up contaminated areas. Infection can be avoided by preventing the receptor from reaching the source or placing a barrier in its path.

12.4.1 Fertilizers

Historically, agriculture was the first significant human influence on the soil (Bahiru 2021). Micronutrients are equally as critical as macronutrients in the growth and development of plants (such as nitrogen (N), phosphorus (P), potassium (K), sulfur (S), calcium (Ca), and magnesium (Mg)). Adding heavy metals to the soil or foliar spraying crops with these metals is an option for some soils (Devi and Kumar 2020), where healthy plant growth requires their presence (Devi and Kumar 2020). Crops grown in Cu-deficient soils are supplemented with Cu, whereas crops grown in Mn-deficient soils can be complemented with Mn. In intensive farming systems, a large amount of fertilizer is applied regularly to ensure that crops receive an adequate supply of N, P, and K. After prolonged fertilizer application, heavy metals (such as Cd and Pb) may accumulate in the soil due to the chemicals used to enhance the nutrients (Khan et al. 2018). Cd and Pb are two physiologically inert metals. Certain phosphatic fertilizers may unintentionally introduce hazardous metals such as Cd, Hg, and Pb into the soil (Al et al. 2012; Thomas et al. 2012).

12.4.2 Pesticides

Heavy metals were found in high concentrations in insecticides previously used extensively in agriculture and horticulture. Cu-, Hg-, Mn-, Pb-, and Zn-based compounds were used in approximately 10% of insecticides and fungicides licensed in the UK. Cu-containing fungicidal sprays, such as the Bordeaux mixture ($CuSO_4$ and $Cu_2(OH)_3Cl$), are used in pesticides (Silva et al. 2019). $PbHAsO_4$ was used in fruit orchards to control parasitic insects for many years as soil concentrations in derelict sites in New Zealand and Australia are significantly higher than background levels due to the use of As-containing compounds in the management of cattle infestations and the preservation of timber using Cu, Cr, and As (CCA). Contamination may create complications, particularly if the areas are renovated for new agricultural or nonagricultural uses. In comparison to inorganic fertilizers, the materials are restricted to many locations and crop types (Wuana and Okieimen 2011).

12.4.3 Biosolids and Manures

Numerous biosolids, livestock manure, composts, and municipal sewage sludge, for instance, have the potential to accumulate heavy metals in the soil. As, Cd, Cr, and other heavy metals can be found in the soil due to the application of the biosolids (Khan et al. 2015; Pan et al. 2021). The biosolids were applied as soil improver directly to the farm fields and crops either in solids form or slurries. Using manures for biosolid abundance as soil improver gives both positive and negative impacts toward the heavy metals' pollution in which the Cu and Zn serve as growth promoters while As in the poultry manure creates health problems (Pan et al.

2021). Animals fed with these diets produce high As, Cu, and Zn levels in their manures. When regularly applied to small land areas, they can accumulate those metals significantly over time. Due to their potential use in agriculture, concerns have been raised about biosolids contaminating the soil with heavy metals (Oun et al. 2014). Previous work has reported that Pb, Ni, Cd, Cr, Cu, and Zn are the most frequently found heavy metals in biosolids (Pan et al. 2021). Hence, when biosolids are applied to soils, heavy metals can be absorbed and leach into the groundwater significantly (McLaren et al. 2005; Oun et al. 2014).

12.4.4 Industrial Wastewater and Mining Activity

Municipal and industrial wastewater and related effluents have been applied to the land for over 400 years, and it is now a widespread practice in many parts of the world (Reed et al. 1995; Pichtel 2016). Worldwide, it is estimated that wastewater irrigates 20 million hectares of arable land. According to studies, wastewater irrigation agriculture provides 50% of the vegetable supply for many Asian and African cities' metropolitan areas (Bjuhr 2007). Farmers are frequently indifferent to environmental consequences, preferring to maximize their harvests and profits. While metal concentrations in wastewater effluents are typically low, irrigation with them over time can result in a significant heavy metal deposition in the soil.

In many countries, soil contamination results from metal ore mining and other industries. When mining occurs, tailings are directly discharged into natural depressions, such as onsite wetlands, resulting in elevated contaminants in the environment (Lu et al. 2012; Sankhla et al. 2016). The widespread extraction and smelting of Pb and Zn ore have posed a threat to human and environmental health. Many of the restoration techniques used at these sites may be ineffective at reestablishing soil production due to their length of time and expense. Heavy metals in soil pose a health risk to humans due to their bioavailability. There are two ways to consume contaminated soils: through the food chain and oral bioavailability (Wuana and Okieimen 2011). Textiles, tanning, and petrochemicals from oil spills or petroleum-based products, insecticides, and pharmaceutical facilities are just a few of the numerous industries that produce various other materials, each with unique composition. They contain toxic organic compounds or heavy metals (Cr, Pb, and Zn) that are potentially harmful when applied to the soil. Another issue is a deficiency in soil conditioning abilities or plant nutrient availability (He et al. 2015).

12.4.5 Airborne Networking

Heavy metals in the atmosphere are also significantly increased by fugitive emissions, such as dust from storage areas or garbage dumps. Heavy metals are typically dispersed in the atmosphere as microscopic specks. Various other heavy metals can be released (such as As, Cd, and Pb). These heavy metals oxidize and form fine particles (He et al. 2015; Sankhla et al. 2016). In the absence of precipitation, natural air currents can disperse stack emissions over a large area before being removed from thc gas stream. Due to their proximity to the ground, fugitive emissions are typically concentrated in a much smaller geographic area.

On the other hand, pollutant concentrations in fugitive emissions are typically lower. Local conditions will determine the type and concentration of metals emitted by both sources. Fire smoke and other industrial chimney emissions carry a variety of solid particles that eventually end up on land or in the ocean. Due to heavy metals in fossil fuels, contamination has occurred on a massive scale since the start of the industrial revolution. Concentrations of Cd, Pb, and Zn, for example, have been detected in soils near smelting plants. Petrol, which contains $C_8H_{20}Pb$, is another significant source of soil pollution. The amount of Pb emitted into the atmosphere due to its combustion contributes significantly to the level of Pb in urban areas and soils adjacent to busy roads. Soils adjacent to highways can become contaminated with Zn and Cd for various reasons, including used tires and lubricating fluids (USEPA 1996; Lu et al. 2012).

12.4.6 Military Activities

For many years, toxic compounds have contaminated large areas of military facilities as a result of explosives and ammunition use, as well as the hazardous residues they leave behind (such as 2,4,6-trinitrotoluene (TNT), 1,3,5-trinitro-1,3,5-triazacyclohexane (RDX), Cu, Cd, Sn, Pb, U, Zn, Mn, As) (Oh et al. 2016; Tomic et al. 2018). Due to the difficulty in decomposing or treating these compounds, they endanger human health and the environment due to their potentially harmful effects (Fayiga 2019). Heavy metal fragments are some of the most persistent battle remnants in areas of military activity. The redox characteristics of the soil, specifically its proton (pH) and electron (e-) activity, are critical in determining the duration of these remnants in the soil. Over time, PTEs can become active, precipitating new minerals (primarily oxides) from the soil solution. Using bullets as an example, complex mineral and chemical reactions can release Pb into the soil, with some resulting in the precipitation of insoluble minerals or plant absorption. When soil conditions such as pH, moisture content, or organic matter (OM) change, or when the amount of Pb in the soil exceeds its capacity to hold it, as occurs in military-impacted zones, Pb partitioned into distinct soil fractions may initially be inert but later become reactive (Dinake et al. 2020). Along with Sb and Cr, other persistent toxic elements (PTEs) released into the soil by weapons include Hg, Ni, Zn, and Cd (Islam et al. 2016). It has been demonstrated that various plant species accumulate PTEs (specifically Cr, Cu, Ni, Pb, Sb, and Zn) across a broad range of military zones. However, the species determine the exact relationship between a PTE and its accumulation location (Busby et al. 2020).

12.5 Effects of Heavy Metals on the Environmental Health

Precautions must be taken to safeguard human health and the food chain from heavy metal toxicity. Even at low concentrations, certain metals can be toxic to humans, resulting in serious health consequences (Arora et al. 2008; Memon and Schröder 2009). Reactive oxygen species (ROS) are formed due to free radicals generated by heavy metals (Mudipalli 2008). As a result of ROS production, the cell membrane is damaged, and metals required by enzymes and pigments undergo chemical changes, impairing their normal function and eventually cell death (Sánchez-Chardi et al. 2009; Krystofova et al. 2009). The most toxic metals are Pb, Cd, Cu, Hg, zirconium (Zr), Cr, and tin (Sn) (Zhang et al. 2016). Compared to nonessential metals such as As and Hg, Zn and Cu are required (trace elements). Depending on the concentration and oxidative state, they can cause various health problems.

When As is present in cells in the form of arsenate, it inhibits ATP synthesis and oxidative phosphorylation (Tripathi et al. 2007). Additionally, carcinogenic Cr causes ulcers, difficulty breathing, skin cancer, and hair thinning (Salem et al. 2014). Cd is mutagenic, teratogenic, and carcinogenic and interferes with calcium (Ca) regulation in the body, resulting in renal failure and anemia (Williams et al. 2018). Due to the fluctuating nature of Cu levels, complications such as kidney and brain damage, liver cirrhosis, and stomach and intestinal discomfort are all possible (Wuana and Okieimen 2011; Salem et al. 2014). Hg poisoning has been associated with short-term memory loss, stomach ulcers, and kidney damage (Neustadt and Pieczenik 2007; Gulati et al. 2010). Ni has been associated with lung, nose, sinus, and throat cancer (Salem et al. 2014). Additionally, Pb poisoning in children can result in developmental delays, short-term memory loss, a decline in intelligence, coordination problems, and even renal failure (Wuana and Okieimen 2011). Zn overdose results in fatigue and dizziness because it elevates blood Zn levels above the safe level (Hess et al. 2009).

Unsurprisingly, soil pollutants can be detrimental to ecosystems. Even at low contamination levels, many potentially harmful compounds can significantly affect soil chemistry. These changes may affect the metabolism of microorganisms and arthropods in each soil environment. As a result, some primary food chains may become extinct, wreaking havoc on predator and consumer species. Because the concentration of extraterrestrial compounds increases as one moves up the food chain, exoplanetary compounds may enter the food chain even if their effects on lower life forms are negligible. Numerous studies have already established that DDT has a detrimental impact on birds, including eggshell thinning, which results in increased chick mortality and species extinction. Agricultural lands contaminated with specific types of soil are at risk. Contaminants disrupt the metabolic processes of plants, resulting in decreased agricultural yields. Because dying crops cannot prevent soil erosion, this has a secondary effect on soil conservation. Some of these chemical contaminants have a long half-life, while others are formed due to the decay of fundamental soil contaminants (Chae and An 2018).

12.6 Reaction Mechanisms of Heavy Metals in Soil

Zn and Cd are two additional contaminants that are frequently encountered. Pb, Cr, and As are the most commonly encountered heavy metals in contaminated sites due to their relative abundance, followed by Zn and Cr (USEPA 1996; Lu et al. 2012). If these heavy metals are present, bioaccumulation and biomagnification through the food chain may reduce crop yield. To determine the most effective remediation strategy for these pollutants, a basic understanding of their chemistry is required, as is an account of their effects on the environment and human health. A heavy metal's chemical form and species can significantly affect how it moves through the soil. Once heavy metals reach the soil, they are adsorbed by fast reactions (in minutes/hours) followed by slow responses (in days/years) and thus redistributed into different chemical forms with variable bioavailability, mobility, and toxicity (Shiowatana et al. 2001; Buekers et al. 2007).

12.6.1 Lead

The most frequently released forms of Pb are ionic Pb, Pb(II), PbO(s) and Pb $(OH)_2$(s), and Pb metal oxyanion complexes. The most stable Pb complexes are $[Pb_6O(OH)_6]^{4+}$ and $[Pb_6O(II)]$. Pb(II) is the most abundant and reactive form of Pb, forming mono- and polynuclear oxides and hydroxides in many conditions (Liang et al. 2017). When the pH exceeds 6, Pb compounds such as $Pb_3(PO_4)_2$, $PbCO_3$, and $Pb(OH)_2$ become insoluble. When increasing amounts of S are present in the soil matrix, PbS is the most stable solid form. It only occurs during times of extreme distress. Under anaerobic conditions, microbial alkylation can produce organolead-$C_4H_{12}Pb$ (Liang et al. 2017). There is a clear distinction between ionic and covalent Pb(II) compounds. PbO_2 and other Pb(IV) compounds are highly reactive oxidants. Consuming white paint peelings containing $Pb(OH)_{22}PbCO_3$, which was once the most widely used white paint pigment and a significant source of chronic Pb exposure for children, can cause Pb poisoning. Both Pb compounds (Pb(II)) and a small number of Pb compounds (Pb(IV)) are beneficial. PbO_2 and $PbSO_4$ are the most common byproducts of the reversible reaction when a Pb storage battery is charged and discharged.

12.6.2 Chromium

Cr is the most frequently encountered form of Cr in contaminated areas (IV). Cr can also be in the +3-oxidation state, depending on the pH and redox conditions. Cr (VI) is the most prevalent form of Cr in shallow aquifers with aerobic conditions. Anaerobic conditions, such as those found in deeper groundwater, soil organic matter, and S^{2-} and Fe^{2+} ions, can result in the conversion of Cr(VI) to Cr(III). It is worth noting that two major species of Cr(VI) precipitate rapidly in the presence of metal cations (particularly Ba^{2+}, Pb^{2+}, and Ag^{+}). These are CrO_4^{2-} and $Cr_2O_7^{2-}$.

Additionally, some Fe_2O_3 and Al_2O_3 adsorb on soil surface(s). Cr(III) is the predominant form of Cr at pH. 4. The heavy metal is combined with organic molecules such as H_3PO_4, O_3, H_2SO_4, F, CN, SO_4^{2-}, and NH_3 to form Cr^{3+} solutions. Cr(VI) is a less toxic and more readily available form of Cr than Cr(III). At low pH values, clays and oxide minerals adsorb Cr(III), whereas, at high pH values, the formation of $Cr(OH)_3(s)$ limits the solubility of Cr(III) (Chrostowski et al. 1991). For instance, soil sorption characteristics such as clay content, Fe_2O_3 content, and organic matter affect Cr mobility. Runoff from the surface can carry insoluble or precipitated forms of Cr into surface waters. Both soluble and nonadsorbed Cr complexes can contaminate groundwater, and the leachability of Cr decreases as soil pH increases. While Cr is deposited in sediment, the vast majority of Cr discharged into natural rivers is in the form of particles (Rue et al. 1997).

12.6.3 Arsenic

In aerobic conditions, As(V) predominates as AsO_4^{3-} in various protonation states, including H_3AsO_4, H_2AsO^{4-}, $HAsO_4^{2-}$, and AsO_4^{3-}. In the presence of metal cations, AsO_4^{3-} and another anionic As forms act as chelates and can precipitate (El-Sorogy et al. 2020). Metal arsenate complexes can be maintained under certain conditions for an extended period. Additionally, As(V) can coprecipitate or adsorb on $Fe(OH)_3$, is acidic and mildly reducing requirements(s). Co-precipitates become more mobile as the pH increases (Rue et al. 1997). AsO_3^{3-}, H_3AsO_3, and $HAsO_3^{2-}$ are protonated forms of As under reducing conditions. AsO_3^{3-} has a high affinity for other S compounds and can coexist with metal sulfides via adsorption or precipitation. As and arsine, AsH_3, can be found in significantly reducing conditions. $HAs(CH_3)_2$ and $As(CH_3)_3$ biotransformation into extremely volatile arsine methylated derivatives (via methylation). Simple anion complexes, such as Cl^- or SO_4^{2-}, do not form when As is present in anionic form. The As speciation includes organometallic species $(CH_3)AsO_2H_2$ and $(CH_3)_2AsO_2H$. Compounds can only travel a short distance in groundwater and surface water due to their strong soil binding.

12.6.4 Zinc

Zn concentrations in soil solution and availability to plants are regulated by adsorption-desorption reactions between the key and solid phases, which are influenced by pH, organic matter, soil minerals, and coexisting ions, as well as the distribution of Zn into various fractions (Catlett et al. 2002; Antoniadis et al. 2008). Chemical properties of soil Zn fractions, such as exchangeable, organic matter-bound, carbonate-bound, Fe-Mn oxides and residual Zn, can be used to identify them. Exchangeable Zn is the most prone to disintegration, making it the optimal Zn binding type for plants (Khoshgoftarmanesh et al. 2018). As a result, plants gain access to Zn that is bound to organic matter, which creates exchangeable Zn sites in the solid soil matrix and increases the soil's capacity for cation exchange

(Khoshgoftarmanesh et al. 2018). Soil carbonates and Fe-Me oxides reduce Zn bioavailability, whereas when Zn is linked to Zn, plant species' tolerance to Zn stress increases.

12.6.5 Cadmium

As with all cationic metals, soil pH significantly affects Cd chemistry. According to the Environmental Protection Agency of the United States, soil colloids, hydrous oxides, and organic matter have a negligible effect on Cd adsorption in soil (1999). Cd levels in the soil are significantly reduced when the pH exceeds six units, as it is absorbed or precipitated by the soil's solid phase. Cd can be liganded with inorganic and organic ligands, most notably chloride (Cl) ions. The formation of these complexes increases Cd's mobility in the soil (USEPA 1999). According to Linger et al. (2005), Cl may inhibit Cd^{2+} adsorption to soil particles by forming a soluble complex with Cd^{2+} known as $CdCl^{+}$. Organic matter found in kaolinite has been shown to enhance its ability to absorb Cd^{2+} ions compared to inorganic ligand ions (Linger et al. 2005). Both toxicity and bioavailability of Cd are affected by soil conditions (Mwamburi 2015).

12.6.6 Copper

Cu has been discovered in soil (as well as in soil solution) in the form of sulfides, oxides, crystal lattices of primary and secondary minerals, and soil organic matter and living organisms (Selbig et al. 2013; Vlček and Pohanka 2018). Cu ions are found in trace amounts in soil solutions, bound to colloids such as clay minerals, soil organic matter, aluminum (Al) and iron oxides, and natural zeolites. The type of soil affects the absorption and desorption of heavy metals. Cu has a significantly greater affinity for organic compounds than other heavy metals (Vlček and Pohanka 2018). As a result, bonding in the phenolic and carboxyl functional groups may accumulate in this type of medium. Meima et al. (1999) researched to understand better the effects of Cu leaching from ash following combustion. Numerous samples contained molecules with molecular weights greater than 10 kilodaltons (kDa). Cu mobility was reduced by two to three in the absence of DOC. Humic acid's carboxyl groups are critical in this process (Wu et al. 2010). According to current thinking, Cu's high affinity for organic matter is explained by its high sorption capacity and chelation proclivity (Vlček and Pohanka 2018).

This compound can bind between one-third and half of the Cu found in soil solutions (Ruan et al. 2018). Adsorption of Cu on oxides of iron (Fe), Al, and Mn (Diagboya et al. 2015). While Cu affinity decreases in the following order: Mn oxides > organic matter > Fe oxides > clay minerals, some studies have discovered that organic colloids or clay minerals have a significant effect on Cu affinity (Ruan et al. 2018). They have reduced soil response, resulting in Cu desorption from the soil. Cu desorbed concentrations are dependent on the type of soil and its pH (Ali et al. 2016). The size of the particles has a significant effect on their adsorption capacity, with metals concentrated in the coarse (sand) and fine (gravel) fractions (clay). The presence of light minerals such as silica in greater concentrations in the silt fraction than in other particles has been correlated with the silt fraction's low concentrations (Jing et al. 2007). Various factors, including adsorption, influence heavy metal bioavailability in the soil. This is accurate, as it affects the concentration of metal ions and complexes in soil solution.

12.7 Phytoremediation Technologies for Heavy Metal

Numerous new technologies developed over the last two decades to restore contaminated soils typically focus on a single contaminant or a limited set of conditions. Phytoremediation is the safest and most cost-effective treatment option in this case (Cameselle et al. 2013). Phytoremediation is an excellent technique for low-cost passive in situ decontamination and restoration because it maintains the soil's biological activity, physical structure, and fertility (Lone et al. 2008; Vithanage et al. 2012). As a result, phytoremediation is considered more environmentally friendly than traditional mechanical methods of hazardous waste disposal, which can be challenging to implement on large sites with external contamination. Additionally, the natural aesthetics of planted areas make phytoremediation more appealing than alternative methods of cleanup (Farraji et al. 2016; Ekta and Modi 2018; Hauptvogl et al. 2020; Devi and Kumar 2020). Phytoaccumulation, rhizofiltration, phytostabilization, degradation, and rhizodegradation are the most frequently used phytotechnologies for heavy metal remediation from contaminated soil, as illustrated in Fig. 12.1.

12.7.1 Phytoaccumulation

Phytoabsorption and phytoextraction are other terms for phytoaccumulation, which aid in the uptake of soil contaminants and water through their roots and accumulate biomass in top plant parts via nutrient translocation (Rafati et al. 2011; Rashid et al. 2014). Phytoextraction relies on biochemical activity to transfer heavy metals from soil to plant roots but harvesting in root biomass is impossible (Zacchini et al. 2009).

Fig. 12.1 Pollution pathway of heavy metals

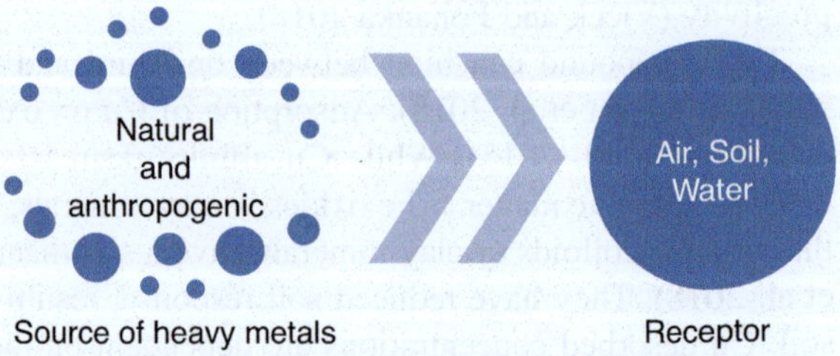

Heavy metal tolerance and phytoaccumulation physiology and molecular mechanisms in plants, on the other hand, have made significant advances. A multidisciplinary approach is required to make phytoextraction a viable commercial solution for remediating metal-polluted soils. According to Nascimento and Xing (2006), phytoextraction may one day be commercially viable. The results show that in the presence of vegetation, plant uptake of Cd in exchangeable form and subsequent nutrient intake partially eliminated Cd (Yang et al. 2017).

12.7.2 Phytofiltration

Plants are used to remove pollutants from polluted wastewater and surface water (Mukhopadhyay and Maiti 2010). Caulofiltration and rhizofiltration are respectively phytofiltration techniques that utilize the shoots and roots of plants. Rhizofiltration is an effective method for removing uranium (U) and other heavy metals from soil and groundwater (Prajapati et al. 2012; Yang et al. 2017). Seedlings are used in blastofiltration as a form of phytoremediation (Ashraf et al. 2019). Sunflower, Indian mustards, tobacco, rye, spinach, and corn were investigated for their ability to remove Pb from water (Sharma et al. 2019). Heavy metals such as Pb, Cd, Cu, Ni, Zr, and Cr retained mainly in the roots can be removed using rhizofiltration (USEPA 2000). Phytofiltration employs absorption and adsorption principles to minimize water movement.

12.7.3 Phytostabilization

Plants stabilize contaminants in contaminated soils, known as phytostabilization or phytoimmobilization. Contaminant bioavailability and mobility are reduced, which aids in preventing pollution from entering the food chain and limiting groundwater movement. Plants use a variety of mechanisms to prevent heavy metals from leaching into the soil including precipitation, heavy metal valency, rhizosphere reduction, and root sorption (Wuana and Okieimen 2011; Khalid et al. 2017). Their valency influences many heavy metals' toxicity. Plants excrete various enzymes that help reduce heavy metal stress by converting hazardous heavy metals into more minor toxic heavy metals. As an example, hexavalent Cr can be reduced to trivalent Cr, which is less harmful (Wu et al. 2010). Because it only slows their movement and inactivates the pollutants in the soil, this is not a long-term remediation (Azubuike et al. 2016).

12.7.4 Phytovolatilization

Pollutants extracted from the soil enter the environment via stomata in the leaves of plants growing in contaminated soil. These pollutants are converted to volatiles and released back into the environment (Liang et al. 2017). Some of these contaminants

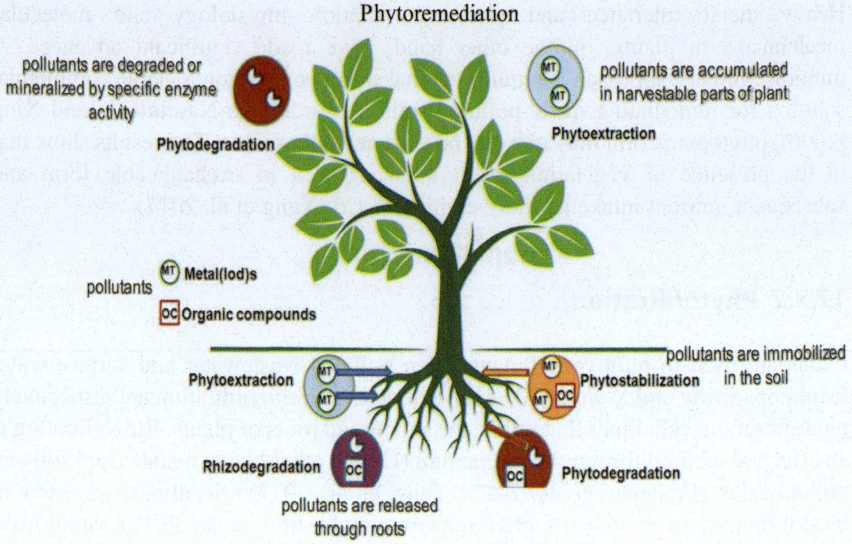

Fig. 12.2 Modified schematic phytoremediation technologies from the soil to aerial parts of plants (Gajic et al. 2018)

are emitted into the atmosphere by plant leaves. Toxins are not removed from the soil; instead, they are transferred to the air and can be redeposited in the soil via a process called adsorption (Devi and Kumar 2020). No other remediation method can prevent the spread of toxins after they have been eliminated via volatilization. Certain microorganisms can evaporate Hg (Fig. 12.2).

12.7.5 Phytodegradation

Through their metabolic processes, plants degrade organic pollutants that enter their systems via phytotransformation, also known as phytodegradation. Additionally, the plant may secrete enzymes that aid in the decomposition of contaminants (Saxena et al. 2020). Heavy metals can be removed without using microorganisms using an enzyme-based process that breaks down organic pollutants using oxygenase and dehalogenase (Yadav et al. 2018). Organic contaminants are absorbed by plants via their metabolic processes and then eliminated. In halophytes, photoconversion has been shown to convert harmful Cr(VI) to less toxic Cr(III) (Li et al. 2020). Certain plants can purify the soil, sludge, sediment, and ground and surface water by producing enzymes. This method's organic substances are herbicides, insecticides, chlorinated solvents, and inorganic pollutants. Due to the inability of heavy metals to biodegrade, their presence inhibits phytodegradation. For instance, biodegradation studies on transgenic poplars have been conducted to determine the effects of genetic engineering on plants.

12.7.6 Rhizodegradation

Numerous bacteria help decompose organic pollutants in the soil around the rhizosphere (Mukhopadhyay and Maiti 2010). This approach influences the rhizosphere around the root zone by 1 millimeter and the plants (Ali et al. 2013). Increased metabolic activity results from an increase in the number of microorganisms in the rhizosphere, which is how this technique minimizes pollutant buildup. Carbon, flavonoids, and amino acid exudates are released into the rhizosphere, resulting in a 10–100-fold increase in microbial activity! Exudation from plant roots creates a nutrient-dense habitat for bacteria, boosting their activity. Organic exudates and enzymes that degrade pollutants in the soil are secreted by plants in the rhizosphere (Yadav et al. 2018).

12.7.7 Phytoextraction

Inorganic pollutants, such as heavy metals, are either transformed into harmless molecules such as $X_2O_3(s)$ or $MHPO_3(s)$, or they accumulate in plant tissue (a process referred to as phytoaccumulation; Chirakkara and Reddy 2015a, 2015b). This is because metals and plant nutrients are chemically equivalent; hence, plants use the same transport system to absorb and distribute heavy metals for soil nutrients. Indeed, several heavy metals are essential to plant nutrients in their native state. Due to the chelating chemicals produced by plant roots, which modify the soil pH and oxidation-reduction potential (Eh), an inorganic contaminant might become dissolved in the soil around the rhizosphere (Gumi and Aisha 2012). After root absorption, plants create co- and antimembrane transporter proteins and transport channels. Phytoextraction is commercially utilized worldwide as the best phytoremediation technique for removing heavy metals from contaminated soils (Gumi and Aisha 2012; Wang et al. 2017; Milić et al. 2021). Bioavailability, soil conditions, heavy metal speciation, and plant species are the limiting factors of the phytoextraction. To be suitable for phytoextraction, plants must exhibit the following properties. The rate of growth of the plant should be rapid. Above-ground biomass output should be high, and the root system should be well-developed (Ojuederie and Babalola 2017; Selvi et al. 2019). To remove accumulated heavy metals, plant roots should be transferred to shoots. Plants should be able to endure the effects of climate change, must be disease and pest resistant, and must be more resilient due to heavy metals' possible adverse effects.

Initial concentration of metals in shoots and the biomass of shoots revealed to be the two most important determinants of a plant's phytoextraction potential (Gumi and Aisha 2012; Sharma et al. 2019). There are two approaches to phytoextraction: one uses hyperaccumulators to produce little biomass aboveground but accumulates a lot of heavy metal target, and the other uses Indian mustard to have a lot of biomasses aboveground earn a lot of heavy metal target (Rascio and Navari-Izzo 2011). High accumulation and hypertolerance are more critical in phytoremediation than biomass production (Wuana and Okieimen 2011; Gumi and Aisha 2012). Plants that have multiple harvesting cuts in a single growth period, such as *Trifolium* spp., are better for phytoextraction (Ali et al. 2013). Grasses are superior to trees and shrubs for heavy metal phytoextraction because they are more adaptable to stressors, produce more biomass, and grow faster. In recent research, maize and barley have been found helpful in the phytoextraction of heavy metals. The risk of food chain contamination is one disadvantage of using plants and crops for phytoextraction. Field crops used for heavy metal extraction should not be fed to animals or consumed directly by humans.

12.8 Hyperaccumulators of Heavy Metals

Phytoremediation of soils contaminated with heavy metals entails using plant species capable of absorbing and depositing contaminants in plant tissues, especially the aerial section or shoots, rather than merely the roots. Remediation requires hyperaccumulators or plants that can take enormous amounts of heavy metals with little influence on their growth and development. In comparison to non-hyperaccumulators, hyperaccumulators are plants that can take up the heavy metals in their shoots 100 times faster than non-hyperaccumulators (Barceló and Poschenrieder 2011). Special hyperaccumulators are plants that can uptake high quantities of heavy metals in their aerial parts at levels comparable to those found in polluted soil (Sarwar et al. 2017). Heavy metals accumulated in hyperaccumulators found to be in the shoot portions of heavy metal-tolerant plants (Rascio and Navari-Izzo 2011). Hyperaccumulators commonly employed in the phytoremediation of hazardous heavy metals', such as Pd and gold (Au), phytomining. Certain plants have natural hyperaccumulation sites for heavy metals, which are capable of absorbing and storing heavy metals from contaminated soils. The capability of phytoremediation for heavy metal pollution is shown in Table 12.1.

12.9 Drawbacks of Heavy Metal Phytoremediation

Any impairment, failure, or unknown reaction could result in the extinction of plant species and raise the dangers associated with phytoremediation. The primary disadvantages of phytoremediation, according to the literature, are plant selection and stress tolerance, as well as pollution concentration, toxicity, and bioavailability. Although this is the most successful approach for removing heavy metals from damaged soil, it has several disadvantages. While a few technologies have been helpful, they also have substantial disadvantages. Chemicals are required for some processes, such as soil cleaning, stabilization and solidification, and in situ flushing; others are so intense that they affect the soil's texture and physicochemical properties (e.g., pH or OM) (e.g., stabilization and solidification, vitrification, and electrokinetic remediation). As a result, many of the solutions outlined above are prohibitively expensive to implement. Due to the slow development and low

Table 12.1 The phytoremediation potential of different plant species for heavy metal contaminants

Plant species	Phytoremediation potential	Reference
Solanum nigrum (black nightshade)	Cd	Wei et al. 2010
Linum usitatissimum (flax)	Cd	Bjelkova et al. 2011
Populus spp. (poplar trees)	Cd	Robinson et al. 2000
Albizia Amara (oil cake tree)	Cr	Shanker et al. 2005
Tectona grandis (teak)		
Casuarina equisetifolia (common ironwood)		
Leucaena leucocephala (coffee bush)		
Spirodela polyrhiza (duckweed)	Ni	Appenroth et al. 2010
Allium fistulosum (green onion)	Pb	
Pteris cretica (moonlight fern)		
Avena sativa (oat)	Zn	Ebbs and Kochian 1998
Pinus sylvestris (pine)	Cd, Pb	Ostrowska et al. 2006
Ricinus communis (ricinus)	Cd, Pb	Niu et al. 2007
Pennisetum americanum (pearl millet)	Cd, Zn	Zhang et al. 2010
Stylosanthes guianensis (stylo)		
Paspalum atratum (crown grasses)		
Silphium perfoliatum (cup plant)		
Lolium perenne (Rye grass)	Cu, Zn	
Helianthus annuus (sunflower)	Zn, Pb	
Brassica rapa (field mustard)	Cd, Cu, Zn	
Lolium perenne (Rye grass)	Cu, Cd, As	O'Connor et al. 2003
Phragmites australis (common reed)	Cu, Hg, Pb	Weis and Weis 2004
Spartina alterniflora (smooth cordgrass)		

(continued)

Table 12.1 (continued)

Plant species	Phytoremediation potential	Reference
Amorpha fruticose (false indigo bush)	Cu, Pb, Zn	Shi et al. 2011
Styrax tonkinensis (Pierre)		
Broussonetia papyrifera		
Vitex trifolia (simple leaf chastetree)		
Glochidion puberum (Needle bush)		
Medicago sativa (alfalfa)	Cd, Cr, Ni, Zn	Peralta-Videa et al. 2002
Species from *the Brassica* genus	Heavy metals	Palmer et al. 2001
Vetiveria zizanioides (vetiver grass)	Pb, Cu, Zn, Cd, Mn	Andra et al. 2009
Eichhornia crassipes (water hyacinth)	Cd, Cu, Ni, Pb, Zn	Liao and Chang 2004
Brassica napus (canola)	Cd, Cr, Cu, Ni, Pb, Zn	
Salix spp. (willow)	Zn, Cd, Ni, Cr, Pb, Cu	Pulford and Watson 2002
Jatropha curcas (Barbados/physic nut)	Zn, Cu, Ni, Pb, Cr, Se	Gumi and Aisha 2012
Brassica juncea (Indian mustard)	Cd, Cr, Cu, Ni, Pb, U, Zn	
Thlaspi caerulescens (alpine penny grass)	Cr, Cd, Co, Cu, Mo, Ni, Pb, Zn, Mn	Robinson et al. 1998

biomass output of some hyperaccumulators and the immobility of some strongly bound metal ions, the number of heavy metals taken from the soil is reduced. A concern to consider is the risk of mismanagement and a lack of sufficient care in the food chain.

12.10 Future Prospects in Heavy Metal Phytoremediation

Heavy metal extraction from polluted soil is a relatively recent notion in the research community. Only a few field studies have been undertaken to evaluate phytoremediation's efficacy in the real world. Currently, the most prevalent way is to excavate and remove waste from the site before adequately disposing of it. In the United States, toxic substances and their environmental implications are a significant source of public concern. As a result, it is in a developer's best interest to repair the site to raise its market value. Excavation would frequently be limited due to the small number of areas requiring remediation under any future regulatory framework and any local concerns. As a result, conducting this experiment outdoors is the most dangerous alternative, as contamination happens in settings not found in a laboratory or greenhouse. The climate, moisture, nutrients, pests, soil, and plant diseases all contribute to the success of phytoremediation (Vangronsveld et al. 2009). The Environmental Agency will insist on a comprehensive evaluation of all available options before selecting one as the preferable solution for any cleanup site. This would depend on the laws and accompanying alternatives. Still, it might assess the effectiveness of various strategies for lowering dosage rates, their associated costs and disruption, and their more significant social and economic effects. Efforts to find genes involved in heavy metal hyperaccumulation in plants continue. Before producing superbug plants for phytoremediation, it is necessary to find and change genes in existing plants suited for heavy metal phytoremediation. When new phytoremediation technologies are developed, they can be improved through technical improvements by being cultivated in contaminated areas both in situ and ex situ. A new field of use is predicted to emerge when more knowledge on metal-contaminated sites and appropriate disposal technology becomes accessible.

12.11 Conclusion

Phytoremediation is an environmentally friendly, cost-effective, and efficient method of reducing heavy metal pollution in soil. Identifying plants capable of phytoremediation of specific combinations of contaminants in soils is a critical first step toward understanding the phytoremediation of soils contaminated with mixed metals. Globally increasing levels of heavy metal contamination and toxicity demand the application of proper remediation procedures to mitigate the environmental impacts. Physical and chemical remediation methods have limits due to high prices, changes in soil properties, the death of soil microbes, and the development of secondary contaminants. On the other hand, phytoremediation is the most environmentally beneficial method of resolving this issue. It is environmentally friendly, economically viable, and practically possible and has achieved widespread popularity globally. Phytoremediation requires knowledge of soil chemistry, ecology, plant biology, microbiology, and environmental engineering. Due to the sensitivity of plant development and pollutant uptake or removal to soil geochemistry and contaminant interactions, phytoremediation at a contaminated site may require contaminant- and site-specific investigations. By being aware of molecular breakthroughs and accomplishments, understanding the mechanism improves the efficacy of phytoremediation.

References

Adimalla N, Qian H, Wang H (2019) Assessment of heavy metal (HM) contamination in agricultural soil lands in northern Telangana, India: an approach of spatial distribution and multivariate statistical analysis. Environ Monit Assess 191(4):1–15

Al A, Al-Oudat L, Shamali M, Abdul K, Ghany B, Kanakri S (2012) Case study: heavy metals and fluoride contents in the materials of Syrian phosphate industry and in the vicinity of phosphogypsum piles. Environ Technol 33(2):143–152

Ali H, Khan E, Sajad MA (2013) Phytoremediation of heavy metals-concepts and applications. Chemosphere 91(7):869–881

Ali S, Rizwan M, Ullah N, Bharwana SA, Waseem M, Farooq MA, Abbasi GH, Farid M (2016) Physiological and biochemical mechanisms of silicon-induced copper stress tolerance in cotton (Gossypium hirsutum L.). Acta Physiol Plant 38(11):1–11

Andra SS, Datta R, Sarkar D, Saminathan SKM, Mullens CP, Batch SBH (2009) Analysis of phytochelatin complexes in the lead tolerant vetiver grass [vetiveria zizanioides (l.)] using liquid chromatography and mass spectrometry. J Environ Qual 38(3):868–877

Antoniadis V, Robinson JS, Alloway BJ (2008) Effects of short-term pH fluctuations on cadmium, nickel, lead, and zinc availability to ryegrass in a sewage sludge-amended field. Chemosphere 71(4):759–764

Appenroth KJ, Krech K, Keresztes A, Fischer W, Koloczek H (2010) Effects of nickel on the chloroplasts of the duckweeds spirodela polyrhiza and lemna minor and their possible use in biomonitoring and phytoremediation. Chemosphere 78(3):216–223

Arif N, Yadav V, Singh S, Singh S, Ahmad P, Mishra RK, Sharma S, Tripathi DK, Dubey NK, Chauhan DK (2016) Influence of high and low levels of plant-beneficial heavy metal ions on plant growth and development. Front Environ Sci 4:69

Arora M, Kiran RB, Rani S, Kaur A, Mittal N (2008) Heavy metal accumulation in vegetables irrigated with water from different sources. Food Chem 111(4):811–815

Ashraf S, Ali Q, Zahir ZA, Ashraf S, Asghar HN (2019) Phytoremediation: environmentally sustainable way for reclamation of heavy metal polluted soils. Ecotoxicol Environ Saf 174: 714–727

Azubuike CC, Chikere CB, Okpokwasili GC (2016) Bioremediation techniques–classification based on site of application: principles, advantages, limitations and prospects. World J Microbiol Biotechnol *32*(11):1–18

Bahiru DB (2021) Accumulation of toxic and trace metals in agricultural soil: a review of source and chemistry in Ethiopia. Int J Environ Chem 5(2):17–22

Barceló J, Poschenrieder C (2011) Hyperaccumulation of trace elements: from uptake and tolerance mechanisms to litter decomposition; selenium as an example. Plant Soil 341(1):31–35

Bjelkova M, Gencurova V, Griga M (2011) Accumulation of cadmium by flax and linseed cultivars in field-simulated conditions: a potential for phytoremediation of Cdcontaminated soils. Ind Crop Prod 33(3):761–774

Bjuhr J (2007) Trace metals in soils irrigated with wastewater in a Periurban area downstream Hanoi City, Vietnam. In: *Seminar Paper*, Institutionen for markvetenskap, Sveriges lantbruk-suniversitet (SLU), Uppsala, Sweden

Buekers J, VanLaer L, Amery F, Van Buggenhout S, Maes A, Smolders E (2007) Role of soil constituents in fixation of soluble Zn, cu, Ni and cd added to soils. Eur J Soil Sci 58(6): 1514–1524

Bundschuh J, Litter MI, Parvez F, Román-Ross G, Nicolli HB, Jean JS, Liu CW, López D, Armienta MA, Guilherme LR, Cuevas AG (2012) One century of arsenic exposure in Latin America: a review of history and occurrence from 14 countries. Sci Total Environ 429:2–35

Busby RR, Douglas TA, LeMonte JJ, Ringelberg D, Indest K (2020) Metal accumulation capacity in indigenous Alaska vegetation growing on military training lands. Int J Phytoremediation 22: 259–266

Cameselle C, Chirakkara RA, Reddy KR (2013) Electrokinetic-enhanced phytoremediation of soils: status and opportunities. Chemosphere 93(4):626–636

Catlett KM, Heil DM, Lindsay WL, Ebinger MH (2002) Soil chemical properties controlling zinc2+ activity in 18 Colorado soils. Soil Sci Soc Am J 66(4):1182–1189

Chae Y, An YJ (2018) Current research trends on plastic pollution and ecological impacts on the soil ecosystem: a review. Environ Pollut 240:387–395

Chen H, Teng Y, Lu S, Wang Y, Wang J (2015) Contamination features and health risk of soil heavy metals in China. Sci Total Environ 512:143–153

Chirakkara RA, Reddy KR (2015a) Phytoremediation of mixed contaminated soils: enhancement with biochar and compost amendments. In: IFCEE 2015, pp 2687–2696

Chirakkara RA, Reddy KR (2015b) Plant species identification for phytoremediation of mixed contaminated soils. J Hazard Toxic Radioact Waste 19(4):04015004

Chrostowski PC, Durda JL, Edelmann KG (1991) The use of natural processes for the control of chromium migration. Remediat J 1(3):341–351

Deb P, Furceri D, Ostry J, Tawk N (2020) The effects of containment measures on the COVID-19 pandemic. Covid Economics: Vetted and Real-Time Papers 19:53–86

Devi P, Kumar P (2020) Concept and application of phytoremediation in the fight of heavy metal toxicity. J Pharm Sci Res 12(6):795–804

Diagboya PN, Olu-Owolabi BI, Adebowale KO (2015) Effects of time, soil organic matter, and iron oxides on the relative retention and redistribution of lead, cadmium, and copper on soils. Environ Sci Pollut Res 22(13):10331–10339

Dinake P, Kelebemang R, Sehube N, Kereeditse T, Motswetla O (2020) Dynamic Risk Assessment of Lead Pollution of Shooting Range Soil by Applying the Delayed Geochemical Hazard Model-A Case Study in Botswana. Soil Sediment Contam 29:503–515

Ebbs SD, Kochian LV (1998) Phytoextraction of zinc by oat (Avena sativa), barley (Hordeum vulgare), and Indian mustard (Brassica juncea). Environ Sci Technol 32(6):802–806

El-Sorogy AS, Youssef M, Al-Kahtany K, Saleh MM (2020) Distribution, source, contamination, and ecological risk status of heavy metals in the Red Sea-Gulf of Aqaba coastal sediments, Saudi Arabia. Marine Pollut Bull 158:111411

Ekta P, Modi NR (2018) A review of phytoremediation. J Pharmacogn Phytochem 7(4):1485–1489

Fasani E, Manara A, Martini F, Furini A, DalCorso G (2018) The potential of genetic engineering of plants for the remediation of soils contaminated with heavy metals. Plant Cell Environ 41:1201–1232

Fayiga AO (2019) Remediation of inorganic and organic contaminants in military ranges. Environ Chem 16:81–91

Farahat E, Linderholm HW (2015) The effect of long-term wastewater irrigation on accumulation and transfer of heavy metals in *Cupressus sempervirens* leaves and adjacent soils. Sci Total Environ 51:1–7

Farraji H, Zaman NQ, Tajuddin R, Faraji H (2016) Advantages and disadvantages of phytoremediation: a concise review. Int J Env Tech Sci 2:69–75

Gajic G, Djurdjevic L, Kostic O, Jaric S, Mitrovic M, Pavlovic P (2018) Ecological potential of plants for phytoremediation and Ecorestoration of Fly ash deposits and mine wastes. Front Environ Sci 6:124

Ghosh M, Singh SP (2005) A review on phytoremediation of heavy metals and utilization of it's by products. Asian J Energy Environ 6(4):18

Gumi AM, Sufiyanu S (2013) Effect of nickel on growth, pigmentation and ion homeostasis of *Capsicum annum* L. and *Capsicum chinense* L. Adv Agric Sci Eng Res 3(12):1430–1436

Gumi AM, Aisha LM (2012) Phytoextraction of heavy metals by *Jatropha curcas* L. grown on heavy metal contaminated soil amended with cow dung and chicken droppings. Int J Phytofuels Other Allied Sci 1:97–100

Gulati K, Banerjee B, Lall SB, Ray A (2010) Effects of diesel exhaust, heavy metals and pesticides on various organ systems: possible mechanisms and strategies for prevention and treatment. Indian J Exp Biol 48(7):710–721

Hamzah A, Hapsari RI, Wisnubroto EI (2016) Phytoremediation of cadmium-contaminated agricultural land using indigenous plants. Int J Environ Agric Res 2(1):8–14

Hauptvogl M, Kotrla M, Prčík M, Pauková Ž, Kováčik M, Lošák T (2020) Phytoremediation potential of fast-growing energy plants: challenges and perspectives–a review. Pol J Environ Stud 29(1):505–516

He Z, Shentu J, Yang X, Baligar VC, Zhang T, Stoffella PJ (2015) Heavy metal contamination of soils: sources. Ind Assess J Environ Ind 9:17–18

Hess SY, Lönnerdal B, Hotz C, Rivera JA, Brown KH (2009) Recent advances in knowledge of zinc nutrition and human health. Food Nutr Bull 30(1_suppl1):S5–S11

Islam MN, Nguyen XP, Jung HY, Park JH (2016) Chemical speciation and quantitative evaluation of heavy metal pollution hazards in two army shooting range backstop soils. Bull Environ Contam Toxicol 96(2):179–185

Jing YD, He ZL, Yang XE (2007) Role of soil rhizobacteria in phytoremediation of heavy metal contaminated soils. J Zhejiang Univ Sci B 8(3):192–207

Kassa H, Dondeyne S, Poesen J, Frankl A, Nyssen J (2017) Impact of deforestation on soil fertility, soil carbon and nitrogen stocks: the case of the Gacheb catchment in the White Nile Basin, Ethiopia. Agric Ecosyst Environ 247:273–282

Kazapoe R, Arhin E (2021) Determination of local background and baseline values of elements within the soils of the Birimian terrain of the Wassa area of Southwest Ghana. Geol Ecol Landsc 5(3):199–208

Khalid S, Shahid M, Niazi NK, Murtaza B, Bibi I, Dumat C (2017) A comparison of technologies for remediation of heavy metal contaminated soils. J Geochem Explor 182:247–268

Khan A, Khan S, Khan MA, Qamar Z, Waqas M (2015) The uptake and bioaccumulation of heavy metals by food plants, their effects on plants nutrients, and associated health risk: a review. Environ Sci Pollut Res 22(18):13772–13799

Khan S, Rehman S, Khan AZ, Khan MA, Shah MT (2010) Soil and vegetables enrichment with heavy metals from geological sources in Gilgit, northern Pakistan. Ecotoxicol Environ Saf 73(7):1820–1827

Khan MN, Mobin M, Abbas ZK, Alamri SA (2018) Fertilizers and their contaminants in soils, surface and groundwater. Encyclop Anthropoc 5:225–240

Khoshgoftarmanesh AH, Afyuni M, Norouzi M, Ghiasi S, Schulin R (2018) Fractionation and bioavailability of zinc (Zn) in the rhizosphere of two wheat cultivars with different Zn deficiency tolerance. Geoderma 309:1–6

Krystofova O, Shestivska V, Galiova M, Novotny K, Kaiser J, Zehnalek J, Babula P, Opatrilova R, Adam V, Kizek R (2009) Sunflower plants as bioindicators of environmental pollution with lead (II) ions. Sensors 9(7):5040–5058

Li X, He X, Wang H, Liu Y (2020) Characteristics and long-term effects of stabilized nanoscale ferrous sulfide immobilized hexavalent chromium in soil. J Hazard Mater 389:122089

Liang SX, Jin Y, Liu W, Li X, Shen SG, Ding L (2017) Feasibility of Pb phytoextraction using nano-materials assisted ryegrass: results of a one-year field-scale experiment. J Environ Manag 190:170–175

Liao SW, Chang WL (2004) Heavy metal phytoremediation by water hyacinth at constructed wetlands in Taiwan. J Aquat Plant Manage 42:60–68

Linger P, Ostwald A, Haensler J (2005) Cannabis sativa L. growing on heavy metal contaminated soil: growth, cadmium uptake and photosynthesis. Biol Plant 49(4):567–576

Lone MI, He ZL, Stoffella PJ, Yang XE (2008) Phytoremediation of heavy metal polluted soils and water: progresses and perspectives. J Zhejiang Univ Sci B 9(3):210–220

Lu Q, He ZL, Stoffella PJ (2012) Land application of biosolids in the USA: a review. Appl Environ Soil Sci 2012:1

Luo C, Liu C, Wang Y, Liu X, Li F, Zhang G, Li X (2011) Heavy metal contamination in soils and vegetables near an e-waste processing site, South China. J Hazard Mater 186(1):481–490

Mar SS, Okazaki M (2012) Investigation of cd contents in several phosphate rocks used for the production of fertilizer. Microchem J 104:17–21

McLaren RG, Clucas LM, Taylor MD (2005) Leaching of macronutrients and metals from undisturbed soils treated with metal-spiked sewage sludge. 3. Distribution of residual metals. Aust J Soil Res 43(2):159–170

Meima JA, van Zomeren A, Comans RN (1999) Complexation of cu with dissolved organic carbon in municipal solid waste incinerator bottom ash leachates. Environ Sci Technol 33(9): 1424–1429
Memon AR, Schröder P (2009) Implications of metal accumulation mechanisms to phytoremediation. Environ Sci Pollut Res 16(2):162–175
Milić D, Bubanja N, Ninkov J, Milić S, Vasin J, Luković J (2021) Phytoremediation potential of the naturally occurring wetland species in protected Long Beach in Ulcinj, Montenegro. Sci Total Environ 797:148995
Mudipalli A (2008) Metals (micronutrients or toxicants) and Global Health. Indian J Med Res 128(4):331
Mukhopadhyay S, Maiti SK (2010) Phytoremediation of metal mine waste. Appl Ecol Environ Res 8(3):207–222
Mwamburi J (2015) Comparative evaluation of the concentrations of lead, cadmium and zinc in surficial sediments from two shallow tectonic freshwater lake basins, Kenya. Afr J Environ Sci Technol 9(6):531–544
Nascimento CWAD, Xing B (2006) Phytoextraction: a review on enhanced metal availability and plant accumulation. Sci Agric 63(3):299–311
Neustadt J, Pieczenik S (2007) Research review: heavy-metal toxicity-with emphasis on mercury. Integrat Med-Innov Commun 6(2):26
Niu ZX, Sun LN, Sun TH, Li YS, Wang H (2007) Evaluation of phytoextracting cadmium and lead by sunflower, ricinus, alfalfa and mustard in hydroponic culture. J Environ Sci 19(8):961–967
O'Connor CS, Lepp NW, Edwards R, Sunderland G (2003) The combined use of electrokinetic remediation and phytoremediation to decontaminate metal polluted soils: a laboratory scale feasibility study. Environ Monit Assess 84:141–158
Oh SY, Yoon HS, Jeong TY, Jeong TY, Kim SD (2016) Evaluation of remediation processes for explosive-contaminated soils: Kinetics and Microtoxfi bioassay. J Chem Technol Biotechnol, 91, 928–937
Ojuederie OB, Babalola OO (2017) Microbial and plant-assisted bioremediation of heavy metal polluted environments: a review. Int J Environ Res Public Health 14(12):1504
Ostrowska A, Porebsca G, Szczubialka Z (2006) Limitation of Pb and cd uptake by pine. Environ Eng Sci 23(4):596–602
Oun A, Kumar, Harrigan T, Angelakis A, Xagoraraki I (2014) Effects of biosolids and manure application on microbial water quality in rural areas in the US. Water 6(12):3701–3723
Palmer CE, Warwick S, Keller W (2001) Brassicaceae (cruciferae) family, plant biotechnology, and phytoremediation. Int J Phytoremediation 3(3):245–287
Pan M, Yau PC, Lee KC, Zhang H, Lee V, Lai CY, Fan HJ (2021) Nutrient accumulation and environmental risks of biosolids and different fertilizers on horticultural plants. Water Air Soil Pollut 232(12):1–16
Peralta-Videa JR, Gardea-Torresdey JL, Gomez E, Tiemann KJ, Parsons JG, Carrillo G (2002) Effect of mixed cadmium, copper, nickel and zinc at different pHs upon alfalfa growth and heavy metal uptake. Environ Pollut 119(3):291–301
Pichtel J (2016) Distribution and fate of military explosives and propellants in soil: a review. Appl Environ Soil Sci 2012:617236
Prajapati SK, Meravi N, Singh S (2012) Phytoremediation of chromium and cobalt using Pistia stratiotes: a sustainable approach. Proc Int Acad Ecol Environ Sci 2(2):136
Pulford ID, Watson C (2002) Phytoremediation of heavy metal-contaminated land by treesa review. Environ Int 29:529–540
Rafati M, Khorasani N, Moattar F, Shirvany A, Moraghebi F, Hosseinzadeh S (2011) Phytoremediation potential of Populus alba and Morus alba for cadmium, chromuim and nickel absorption from polluted soil. Int J Environ Res 5(4):961–970
Rafique N, Tariq SR (2016) Distribution and source apportionment studies of heavy metals in soil of cotton/wheat fields. Environ Monit Assess 188:309
Rascio N, Navari-Izzo F (2011) Heavy metal hyperaccumulating plants: how and why do they do it? And what makes them so interesting? Plant Sci 180(2):169–181

Rashid A, Mahmood T, Mehmood F, Khalid A, Saba B, Batool A, Riaz A (2014) Phytoaccumulation, competitive adsorption and evaluation of chelators-metal interaction in lettuce plant. Enviro Eng Manag J (EEMJ) 13(10):2583

Reed SC, Crites R, Middlebrooks EJ (1995) Natural Systems for Waste Management and Treatment, 2nd edn. McGraw-Hill, New York, NY

Robinson BH, Leblanc M, Petit D, Brooks RR, Kirkman JH, Gregg PEH (1998) The potential of Thlaspi caerulescens for phytoremediation of contaminated soils. Plant Soil 203:47–56

Robinson BH, Mills TM, Petit D, Fung LE, Green SR, Clothier BE (2000) Natural and induced cadmium-accumulation in poplar and willow: implications for phytoremediation. Plant Soil 227:301–306

Rodriguez JH, Wannaz ED, Salazar MJ, Pignata ML, Fangmeier A, Franzaring J (2012) Accumulation of polycyclic aromatic hydrocarbons and heavy metals in the tree foliage of Eucalyptus rostrata, Pinus radiata and Populus hybridus in the vicinity of a large aluminium smelter in Argentina. Atmos Environ 55:35–42

Ruan X, Liu Y, Wang G, Frost RL, Qian G, Tsang DC (2018) Transformation of functional groups and environmentally persistent free radicals in hydrothermal carbonisation of lignin. Bioresour Technol 270:223–229

Rue EL, Smith GJ, Cutter GA, Bruland KW (1997) The response of trace element redox couples to suboxic conditions in the water column. Deep-Sea Res I Oceanogr Res Pap 44(1):113–134

Salem ZB, Capelli N, Laffray X, Elise G, Ayadi H, Aleya L (2014) Seasonal variation of heavy metals in water, sediment and roach tissues in a landfill draining system pond (Etueffont, France). Ecol Eng 69:25–37

Sankhla MS, Kumari M, Nandan M, Kumar R, Agrawal P (2016) Heavy metal contamination in soil and their toxic effect on human health: a review study. Int J All Res Educ Sci Methods (IJARESM) 4(10):13–19

Sánchez-Chardi A, Ribeiro CAO, Nadal J (2009) Metals in liver and kidneys and the effects of chronic exposure to pyrite mine pollution in the shrew Crocidura russula inhabiting the protected wetland of Doñana. Chemosphere 76(3):387–394

Sarwar N, Imran M, Shaheen MR, Ishaque W, Kamran MA, Matloob A, Rehim A, Hussain S (2017) Phytoremediation strategies for soils contaminated with heavy metals: modifications and future perspectives. Chemosphere 171:710–721

Saxena AK, Kumar M, Chakdar H, Anuroopa N, Bagyaraj DJ (2020) Bacillus species in soil as a natural resource for plant health and nutrition. J Appl Microbiol 128(6):1583–1594

Selbig WR, Bannerman R, Corsi SR (2013) From streets to streams: assessing the toxicity potential of urban sediment by particle size. Sci Total Environ 444:381–391

Selvi A, Rajasekar A, Theerthagiri J, Ananthaselvam A, Sathishkumar K, Madhavan J, Rahman PK (2019) Integrated remediation processes toward heavy metal removal/recovery from various environments-a review. Front Environ Sci 7:66

Shahid M, Dumat C, Khalid S, Niazi NK, Antunes PM (2016) Cadmium bioavailability, uptake, toxicity and detoxification in soil-plant system. Rev Environ Contam Toxicol 241:73–137

Shanker AK, Ravichandran V, Pathmanabhan G (2005) Phytoaccumulation of chromium by some multipurpose-tree seedlings. Agrofor Syst 64(1):83–87

Sharma RK, Khatri S, Shridhar V (2019) Rapid screening of wheat cultivars suitable for cultivation in cd contaminated soils under mountain ecosystem. Int J Plant Environ 5(02):129–132

Shi X, Zhang X, Chen G, Chen Y, Wang L, Shan X (2011) Seedling growth and metal accumulation of selected woody species in copper and lead/zinc mine tailings. J Environ Sci 23(2): 266–274

Shiowatana J, Tantidanai N, Nookabkaew S, Nacapricha D (2001) A novel continuous-flow sequential extraction procedure for metal speciation in solids. J Environ Qual 30(4):1195–1205

Silva V, Mol HG, Zomer P, Tienstra M, Ritsema CJ, Geissen V (2019) Pesticide residues in European agricultural soils–a hidden reality unfolded. Sci Total Environ 653:1532–1545

Suman J, Uhlik O, Viktorova J, Macek T (2018) Phytoextraction of heavy metals: a promising tool for clean-up of polluted environment? Front Plant Sci 9:1476

Thangavel P, Subbhuraam CV (2004) Phytoextraction: role of hyperaccumulator in metal contaminated soils. Proc Indian Natl Sci Acad 70(1):109–130

Thomas EY, Omueti JAI, Ogundayomi O (2012) The effect of phosphate fertilizer on heavy metal in soils and Amaranthus caudatus. Agric Biol JN Am 3(4):145–149
Tomic NT, Smiljanic S, Jovic M, Gligoric M, Povrenovic D, Dosic A (2018) Examining the effects of the destroying ammunition, mines, and explosive devices on the presence of heavy metals in soil of open detonation pit: part 1-pseudo-total concentration. Water Air Soil Pollut 229:301
Tripathi RD, Srivastava S, Mishra S, Singh N, Tuli R, Gupta DK, Maathuis FJ (2007) Arsenic hazards: strategies for tolerance and remediation by plants. Trends Biotechnol 25(4):158–165
USEPA. (1996) Sources of lead in soil. Report number EPA 747-R-98-001b
USEPA (1999) Introduction to phytoremediation, EPA/600/R 99/107. Office of Research and Development, Washington, DC
USEPA. (2000) Methods for measuring the toxicity and bioaccumulation of sediment-associated contaminants with freshwater invertebrates, second ed. EPA 600/R-99/064
Vangronsveld J, Herzig R, Weyens N, Boulet J, Adriaensen K, Ruttens A, Thewys T, Vassilev A, Meers E, Nehnevajova E, van der Lelie D (2009) Phytoremediation of contaminated soils and groundwater: lessons from the field. Environ Sci Pollut Res 16(7):765–794
Vithanage M, Dabrowska BB, Mukherjee AB, Sandhi A, Bhattacharya P (2012) Arsenic uptake by plants and possible phytoremediation applications: a brief overview. Environ Chem Lett 10(3): 217–224
Vlček V, Pohanka M (2018) Adsorption of copper in soil and its dependence on physical and chemical properties. Acta Universitatis Agriculturae et Silviculturae Mendelianae Brunensis 66(1):219–224
Wang L, Ji B, Hu Y, Liu R, Sun W (2017) A review on in situ phytoremediation of mine tailings. Chemosphere 184:594–600
Wei S, Li Y, Zhou Q, Srivastava M, Chiu S, Zhan J, Wu Z, Sun T (2010) Effect of fertilizer amendments on phytoremediation of Cd-contaminated soil by a newly discovered hyperaccumulator Solanum nigrum L. J Hazard Mater 176(1–3):269–273
Weis JS, Weis P (2004) Metal uptake, transport and release by wetland plants: implications for phytoremediation and restoration. Environ Int 30:685–700
Williams ET, Nachana'a T, Shinggu DY (2018) Concentration of heavy metals in soil and plants along Mubi-Gombi, roadside, Adamawa state, Nigeria, Adamawa state university. J Sci Res 6(2):241–248
Wu G, Kang H, Zhang X, Shao H, Chu L, Ruan C (2010) A critical review on the bio-removal of hazardous heavy metals from contaminated soils: issues, progress, eco-environmental concerns and opportunities. J Hazard Mater 174:1–8
Wuana RA, Okieimen FE (2011) Heavy metals in contaminated soils: a review of sources, chemistry, risks and best available strategies for remediation. ISRN Ecol 2011:402647
Yan A, Wang Y, Tan SN, Mohd Yusof ML, Ghosh S, Chen Z (2020) Phytoremediation: a promising approach for revegetation of heavy metal-polluted land. Front Plant Sci 11:359
Yang J, Zheng G, Yang J, Wan X, Song B, Cai W, Guo J (2017) Phytoaccumulation of heavy metals (Pb, Zn, and cd) by 10 wetland plant species under different hydrological regimes. Ecol Eng 107:56–64
Yadav KK, Gupta N, Kumar A, Reece LM, Singh N, Rezania S, Khan SA (2018) Mechanistic understanding and holistic approach of phytoremediation: a review on application and future prospects. Ecol Eng 120:274–298
Zacchini M, Pietrini F, Mugnozza GS, Iori V, Pietrosanti L, Massacci A (2009) Metal tolerance, accumulation and translocation in poplar and willow clones treated with cadmium in hydroponics. Water Air Soil Pollut 197(1):23–34
Zhang H, Dang Z, Zheng LC, Yi XY (2010) Remediation of soil co-contaminated with pyrene and cadmium by growing maize (Zea mays L.). Int J Environ Sci Technol 6(2):249–258
Zhang C, Nie S, Liang J, Zeng G, Wu H, Hua S, Liu J, Yuan Y, Xiao H, Deng L, Xiang H (2016) Effects of heavy metals and soil physicochemical properties on wetland soil microbial biomass and bacterial community structure. Sci Total Environ 557:785–790

13 Industrial Wastewater Treatment Strategies

Harlina Ahmad, Norli Ismail, Rekah Nadarajah,
Nishalini Ratha Pukallenthy, Mohd Sohrab Hossein,
and Amer Talebi

Abstract

The sustainable approach in industrial wastewater treatments required ensuring the environment's conservation for the well-being of people, other living organisms, and the ecosystem which will be a prime topic of discussion in this chapter. The wastewater released from the industry generates various kinds of pollutants that harm human and aquatic life if not treated appropriately. The excessive concentration of toxic constituents in industrial effluents will seep into the environmental compartments and thus disturbing the ecosystem. Furthermore, groundwater systems also can be pretentious by all the toxicants due to leaching through the soil. This kind of toxicity may cause mutagenic and carcinogenic impacts not even to the aquatic organism but can result to severe damage to humans. Hence, in treating industrial wastewater, the goal is not only to comply with the stipulated regulations, but the practice should be of adopting a more sustainable approach that protects the environment.

Keywords

Industrial wastewater · Reuse · Recovery · Bioaugmentation · Energy conversion · Waste reduction

H. Ahmad (✉) · N. Ismail · R. Nadarajah · N. R. Pukallenthy · M. S. Hossein · A. Talebi
School of Industrial Technology, Universiti Sains Malaysia, Penang, Malaysia
e-mail: harlinaa@usm.my

R. P. Singh et al. (eds.), *Heavy Metal Toxicity: Environmental Concerns, Remediation and Opportunities*, https://doi.org/10.1007/978-981-99-0397-9_13

13.1 Introduction

Industrial wastewater, often known as effluent, has long been a concern. A variety of pollutants have been introduced into the environment by manufacturing-related effluents since the industrial revolution began in the middle of the eighteenth century. The issue did not appear to be serious at first. Nevertheless, "nature" is still able to absorb or tolerate exogenous substances. As a result, these pollutants were either transformed into less hazardous molecules or substances or lowered to safe levels.

Unfortunately, as time went on, the environment's ability to "self-cleanse" decreased. More hazardous, nondegradable, stable, and persistent compounds were introduced to the environment simultaneously as a result of rapid development in the industrial and chemical sectors. Massive problems suddenly started to emerge. We began to observe how the adulteration of industrial effluents was affecting the ecosystem, like species threatened with extinction, aquatic organisms behaving abnormally, a deterioration in the quality of human life, and more.

The development of the sewage treatment infrastructure sparked the growth of wastewater treatment. The Mesopotamian, Indus, Egyptian, and Greek early civilizations are known to have contributed to the evolution of sewage treatment and management. After the fall of the Roman Empire, there was a period of significant sanitary neglect, especially concerning sewage management, which resulted in the Black Death pandemic. In the eighteenth century, Britain, Germany, France, Italy, and the United States—where the industrial revolution took place—saw the beginning of the evolution of wastewater treatment and management. The nineteenth century saw the first thoughtful consideration of engineering challenges in wastewater treatment methods (Lofrano and Brown 2010).

The management and treatment of wastewater have advanced significantly throughout the years. Credit should be given to the development of analytical chemistry in particular, which assisted in establishing the benchmarks for the observed and measured parameters of the industrial wastewater effluent characteristics and undoubtedly fosters the rapid advancement of wastewater treatment technology.

13.2 Industrial Wastewater Characteristics

Industrial wastewaters or effluents are the by-products of processes like washing, heating, extracting, synthesizing, reacting, and packing. These effluents might contain solids, nutrients, pathogens, heavy metals, pigments, toxic chemicals, organic and inorganic materials, oil and grease, and heavy metals. How changeable the qualities and quantities depend on the type of industry. The main issue with industrial wastewater is the variety and volume of synthetic materials that are released into the environment from it, both of which are constantly increasing (Cisneros 2011).

Table 13.1 Industries and their major contaminants

Industry type	Contaminants	Pollutant parameters	References
Food industries (e.g., dairy, seafood, confectionery, drinks, palm oil, sugar refineries)	Carbohydrates, oil and grease, proteins, solids, additives, food colorings, cellulose, lignin, tannin, cleaning chemicals	BOD, COD O & G, SS Colors, pH, nutrients	Dhanke et al. (2018) Zajda and Aleksander-Kwaterczak (2019) Kongnoo et al. (2012)
Pulp and paper	Lignin, cellulose, solids, chlorides, phenol, tannins, acid resins, phenols, toxic metals, bleaching chemicals	COD, BOD, TDS, pH, colors, alkalinity, specific heavy metals	Kumar et al. (2021), Sharma et al. (2022)
Pharmaceutical	Antibiotics, solvent, synthetic chemicals, colorings, acid or basic solutions, heavy metals	BOD, COD, SS, TDS, pH, color, specific heavy metals, alkalinity	Vanerkar et al. (2013), (Ramola and Singh 2013)
Petrochemical	Hydrocarbons, heavy metals, sulfides, fluorides	Specific hydrocarbons (e.g., benzene, aldehydes, and others), BOD, COD, O & G, pH, specific inorganics	Capello et al. (2009)
Leather and textile	Organics, heavy metals, solvent, pigments, bleaching agents, chloride, ammonia	BOD, COD, SS, TDS, O & G, total phosphorus, alkalinity, specific heavy metal species (e.g., chromium, lead, and others), color, pH	Yurtsever et al. (2020) Sawalha et al. (2019)

Along with agriculture, mining, untreated urban runoff, and wastewater, industry accounts for 22% of worldwide water use and can be regarded as a major source of water pollution. These industrial pollutants affect groundwater, surface water, and human health, making them crucial environmental issues (Sivasubramaniam and Franks 2016). Untreated industrial wastewaters are harmful to varying degrees. For instance, pH values are frequently outside of the neutral ranges, various toxic chemicals like heavy metals and carcinogens are present, and there is concern over an excessive concentration of solids, organics, oil, and grease that, if not effectively managed, will inevitably have adverse effects on the environment.

The characteristics of the five main industries—listed in Table 13.1—that produce enormous amounts of highly contaminated wastewater containing organics, inorganics, heavy metals, colors, and extreme pH levels are discussed here. These characteristics present challenges to the traditional and current practice of wastewater treatment schemes. Organic compounds from carbohydrates, protein, oil and fat, lignin, and cellulose are frequently found in the effluents produced by the food processing sectors of dairy, seafood, confectionery, and drinks. Food coloring, cleaning products, and additives or preservatives are all sources of synthetic chemicals.

Such an example is the palm mill effluent (POME), which has unusually elevated levels of suspended particles, oil and grease, BOD, and COD. At heated temperatures between 80 and 100 °C, fresh raw POME from a mill is a viscous brownish colloidal slurry. Approximately 2.5 m^3 to 3.0 m^3 of POME are produced per ton of crude palm oil (CPO) manufactured (Hasanudin et al. 2015). According to Kongnoo et al. (2012), the POME appears blackish-brown due to the presence of humic acid, a substance that is similar to fulvic acid, and lignin. The pungent odor that POME emits is likewise unwelcome and noticeable. In POME, the majority of the constituents are water (95–96%), total solids (4–5%), suspended solids (2%), and oil (0.6–0.7%) (Abdullah and Sulaiman 2013).

The sixth most polluting industry in the world is pulp and paper, in terms of the amount of toxic effluent it generates (Ugurlu et al. 2007). The industrial effluents from the pulp and paper industry also include a variety of organic substances, including biocides, lignosulfonic acid, chlorinated phenols, and chlorinated hydrocarbons. Almost 20–100 cubic meters of wastewater produced per metric ton of the product, which contain these hazardous substances, are released into rivers and lakes, resulting in a more poisoned aquatic environment (Lindholm-Lehto et al. 2015).

Pharmaceutical compounds are usually generated in different operations in the pharmaceutical industry, where a significant volume of water is needed to wash and extract solid cake or equipment (Gadipelly et al. 2014). The existence of antibiotics and nonsteroidal anti-inflammatory drugs (NSAIDs), antidepressants, lipid regulators, β-blockers, and hormones contributes to water contamination and threatens aquatic life and human health (Kaur et al. 2016).

The cracking processes of hydrocarbon in petrochemical processing and oil refining all result in significant wastewater generation in the petrochemical industry (Asatekin and Mayes 2009). Because of complex raw materials, extensive processes, and unforeseen side effects, the wastewater produced typically contains a substantial number of potentially harmful compounds that are categorized as inorganic and organic contaminants (Tian et al. 2020). Hydrocarbons such as benzenes, aldehydes, and phenols are a few of the specific organic contaminants found in petrochemical effluent streams which are known for their high toxicity (Capello et al. 2009; Kumar et al. 2013).

The deleterious pollution that the textile industry produces involves dyes mixed with other contaminants at varying concentrations, making it one of the main contributors to environmental pollution (Yaseen and Scholz 2019). Additionally, dyes, degradable organics, detergents, stabilizing agents, desizers, inorganic salts, and heavy metals are present in the industry's effluent (Siddique et al. 2017). The ecosystem and aquatic life are seriously threatened when polluted effluents containing dyes, salts, and other organic compounds are released into the soil and water bodies (Chaturvedi et al. 2021).

Numerous chemicals, including sodium chloride and chromium sulfate, are used extensively in the leather business. As a result, its numerous operations release a large amount of wastewater that contains elevated levels of chromium, chloride, ammonia, total solids (TS), total dissolved solids (TDS), total suspended solids (TSS), and other contaminants (de Aquim et al. 2019; Tamersit et al. 2018). Table 13.1 summarized the possible contaminants that can be the sources of pollutants in the selected type of industries.

13.3 Strategies in Industrial Wastewater Treatment

Polluted wastewater must be treated to reduce the concentration of contaminants to a safe level before being reused or discharged into the environment. To increase the effectiveness of industrial wastewater treatment plants, different techniques have been developed. Untreated wastewater is one of the main contributors to water contamination; hence, a significant effort must be made to develop systems for effective and sustainable wastewater management and treatment.

The fundamental goal of wastewater treatment is to remove soluble or nonsoluble pollutants from effluents. The wastewater treatment regime includes preliminary, primary, secondary, tertiary, advanced, and sludge treatments, which can be used individually or in combination. The selection of treatment methods may differ from industry to industry, depending on the effluent characteristics, the discharge effluent limit, process flexibility, energy and chemical needs, challenges with sludge disposal, and most importantly, the treatment cost.

In general, industrial wastewater treatment is challenging because of the complex nature, strength, and production rates of the effluent. The efficiency of standard treatment methods may also be hampered by the presence of dangerous and difficult-to-degrade metals and synthetic chemicals, an extreme pH, elevated levels of organics and solids, and fat. Riffat and Husnain (2022) mentioned a few challenges in implementing wastewater treatment. Among the issues are the increasing costs of running treatment plants, environmentally friendly ways to produce biosolids or sludge, the presence of harmful compounds, and strict regulations imposed on the deteriorating water body quality. To resolve the problems outlined earlier, industrial wastewater treatment must adopt a clean and more sustainable strategy alongside concentrating on adhering to the stipulated regulatory limit as follows:

1. Reuse or recovery approach
2. Energy conversion approach
3. Bioaugmentation application
4. Waste reduction/zero waste approach
5. Integrated approach

13.3.1 Reuse and Recovery

(a) Since many of the waste products produced can be recycled on- or off-site in other plants, reusing and recovering wastewater are preferable options. The reuse, recycling, and by-product recovery strategy have frequently offered the industry an enticing payback period and paid for operating costs. For instance, a recovery method can lower the amount of textile dyes and printing pastes released with wastewater during the production process of clothing manufacturing. Recycled printing ingredients can be used in processes where a lesser quality is accepted. In this, recovered and used organic solvents from pharmaceutical production operations are reused.

The recovery of by-products from various wastewater treatment plants is also permitted when there is a need for them on- or off-site. For instance, wastewater treatment facilities can use methane recovered from anaerobic digesters to become self-sufficient or less reliant on electrical power. Heavy metal recovery from wastewater treatment facilities can be accomplished using physical-chemical, biological, or other approaches. Here are a few instances of successful reuse and recovery. Physical methods such as screening, sedimentation, flotation, filtration (including microfiltration, nanofiltration, ultrafiltration, and reverse osmosis), mixing, and adsorption can be used to accomplish these activities.

1. Screening: Screening is required to eliminate big suspended and floating materials out of wastewater to prevent blockage and breakage of pumps, pipelines, and valves. The screen might be coarse, medium, or fine based on the precise area between the bars. Fine screens might be of the rotary drum, tangential, or vibratory kind.
2. Sedimentation and flotation: Simple sedimentation is used to remove finely suspended particulate matter, but chemical-assisted sedimentation is required to remove colloidal solids. Impurities that do not settle in time because they are lighter than water are removed by flotation. Sedimentation is used in every industry that produces significant amounts of suspended organic and inorganic particles. Sedimentation and flotation contribute equally to reducing particle load in downstream processing plants.
3. Filtration: Instead of being utilized alone, industrial wastewater filtration is frequently employed in conjunction with other pretreatment processes. For example, the reuse application filters are used for discharging recycled water onto land for cultivation, groundwater injection, grass spraying, or recreational areas involving physical contact.
4. Mixing: Continuous mixing is accomplished by pumping the tank's contents and reusing part of the pumped liquid. Bubbles are injected into a liquid using compressed bubbles. If the wastewater contains greasy or oily compounds, compressed air can assist in pushing some of it to the top and scooping it up.
5. Adsorption: Organic pollutants in wastewater that are nonbiodegradable and exist in a dissolved form are effectively eliminated by the adsorbents. This

procedure is used as a purification step to improve the quality of effluent that has already been treated to remove most of the pollutants.

(b) Chemical methods: pH adjustment, coagulation, softening, oxidation, reduction, and disinfection are all examples.

1. pH correction: The pH value is vital in pH correction since it is a utilized unit process necessary to make a wastewater stream appropriate for further treatment. This procedure necessitates a proper mixing of the waste stream and chemical neutralization. PH correction and equalization in the supplied equalization tank may be coupled easily.
2. Coagulation: It is used to help settle suspended solids from wastewater. Occasionally, organic polymers are employed as coagulants.
3. Oxidation and reduction: Contaminants in industrial wastewater are infrequently removed via oxidation and reduction. For instance, chlorine oxidation is utilized to reduce BOD, regulate odors, and help in grease removal. However, the quality of industrial wastewater varies, so it is crucial to determine the appropriate amount of oxidant. Biological methods use aerobic, facultative, and anaerobic microorganisms to break down organic matter and minimize oxygen demand in wastewater. Aerobic and anaerobic mechanisms work together to transform and stabilize nonprecipitable organic material into a certain amount of dissolved organic matter. Microbes are utilized to filter wastewater, including readily biodegradable content, at a low cost. Microorganisms need the right amount of carbon, nitrogen, phosphorus, and other elements to perform their metabolic functions. For instance, anaerobic organisms are inherently sluggish and require much less nitrogen and phosphorus than aerobic organisms.

The proper units are selected and sized through (a) flow measurement, sample collection, and characterization of the wastewater flow and (b) subjecting the wastewater samples to treatability studies using laboratory-scale models that can be run on a batch feed, semi-continuous feed, or continuous feed basis.

13.3.2 Energy Conversion Approach

Sludge stabilization comes after the primary goal of normal wastewater treatment, which is the eradication of contaminants to meet discharge requirements. The continued depletion of fossil fuels, the deterioration of the environment brought on by pollution, and the scarcity of water and resources make enormous efforts necessary to meet the present sustainability standards. To reduce disposal and offer a different source of energy, wastewater can be converted into energy.

13.3.2.1 Microbial Fuel Cells (MFCs)

Microbial fuel cells (MFCs) are a renewable technology that might generate electricity while removing organic pollutants. MFCs are bio-electrochemical systems

Table 13.2 The comparison of microalgae cultivation between an open pond and a photobioreactor pond (Arun et al. 2021)

Cultivation mode	Mode of agitation (mixing)	Advantages	Limitations
Open pond	Paddlewheel	• Large yield • Low energy input • Easy setup and maintenance	• Large land needed to cultivate • Higher chance of contamination • Evaporation of medium occurs • Only selected strains are cultivated • Less mass transfer and biomass productivity
Photo bioreactor	Airlift and recirculation	• Low energy input, operating cost, and land requirement • Low contamination • Narrowed light path • Can be cultivated in outdoor and indoor conditions • Greater photosynthetic efficiency • Less hydrodynamic stress • Ease of use due to handy technique	• High-cost, energy for large mass production • Difficult to maintain cleanliness and temperature • Build with the help of sophisticated higher-end material • Photoinhibition occurs at a longer duration

that produce energy through the one-step conversion of organic molecules by microorganisms in an anaerobic condition. By oxidizing organic material, anodic compartment bacteria generate protons and electrons, releasing CO^2 and biomass as waste (Kumar et al. 2016).

In a recent study, Behera et al. (2010) created a microbial fuel cell to remove contaminants from rice mill effluents using rice-mill effluent substrates. This study demonstrated the potential properties of clay pot MFC in terms of energy production while reducing organic matter. The specific benefits and drawbacks of microalgae growing in open ponds and photobioreactors are discussed in Table 13.1. Microalgae farming in open ponds will be preferable to photobioreactors for biofuels since it can generate enormous amounts of biomass with less upkeep (Verma et al. 2020). When using carbon dioxide from industrial wastewater as a growth supplement, photobioreactors are preferable to open pond systems (Xu et al. 2020). The propagation of microalgae in an open pond and a pond with a photobioreactor is illustrated in Table 13.2.

13.3.2.2 Hydrogen and Methane Production

A high-energy, reusable, and ecologically friendly fuel is hydrogen, almost 2.75 times more power is generated by the ignition of hydrogen than by the burning of hydrocarbon fuel, and the only by-product is water (Demirbas et al. 2011).

Enzymatic processes and an acid boosted hydrogen production (Ramprakash and Muthukumar 2015). Nevertheless, in a similar investigation, a mutated version of *E. aerogenes* produced more bio-hydrogen than the natural strain. In addition, they discovered the utilization of parboiled rice effluent, acceptable hydrogen synthesis, and moderate methane production.

13.3.3 Bioaugmentation

Even though wastewater-activated sludge contains a naturally occurring microbe that biodegrades a wide range of pollutants, as previously indicated, some substances are resistant to biodegradation (Nzila et al. 2016). This resistance is brought on by several factors, including potential toxicity, poor water solubility, poor bioavailability, high stability, and low biocompatibility. Additionally, some substances might not function well as substrates for microbial metabolic enzymes. Some contaminants may have chemical structures that make it necessary for groups of microbial species to degrade them biologically, or the environment may never contain all the necessary microorganisms at once. The ability to target treatment to a particular contaminant that is common in the environment is one of the main advantages of bioaugmentation, which can overcome these difficulties. Therefore, this method is intriguing for addressing both pollutants that are already existing in significant quantities and the growing number of emerging contaminants. In contaminated environments, bioaugmentation refers to the inclusion of microorganisms that can dissolve naturally resistant compounds. This approach is more cost-effective and environmentally beneficial when compared to physicochemical methods.

A preadapted pure bacterial strain, the addition of pre-adapted consortia, the introduction of genetically modified bacteria, and the incorporation of biodegradation-relevant genes contained in a vector and conjugated into microbes already present in the biosystem are some of the far more well-known bioaugmentation techniques. In the latter case, the strategy benefits by not being reliant on the development or survival of the donor strain(s) (Herrero and Stuckey 2015). However, several of these options have only been studied in the lab.

13.3.4 Waste Reduction or Zero Waste Approach

Industrial wastewater management is a main obstacle for the circular economy (CE) since there are many different businesses having insufficient access to clean water to reduce costs. The European Union (EU) implements methodical modifications in the water-based waste disposal methods, which are significant water management components, to minimize environmental pollution caused by inadequately treated wastewater released into natural receivers (Smol et al. 2020). However, due to rising urbanization or climate instability, the water shortage has pushed people to respond by reducing water usage to relieve the strain on water

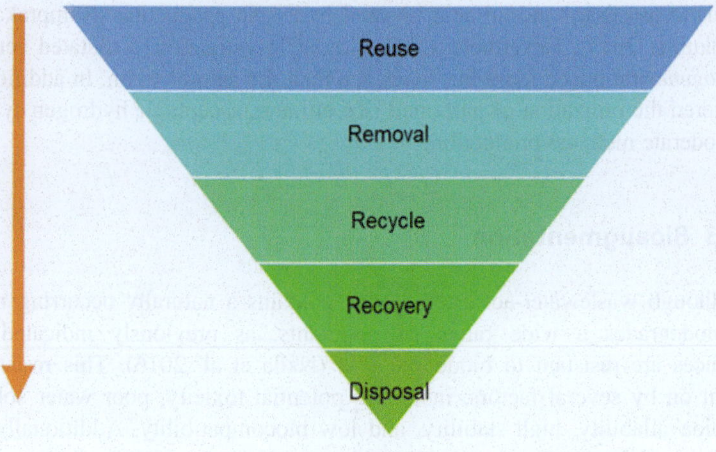

Fig. 13.1 The hierarchy of wastewater management in practicing waste reduction

supplies. In any case, this approach is insufficient to preserve the standard of living, necessitating the use of high-quality water supplies. Therefore, more initiatives that would enable reducing the demand for water resources while reducing the incidence of wastewater should be considered. Figure 13.1 depicts the industrial wastewater management hierarchy suggested in Waste Framework Directive 2008/98/EC.

The five components of the hierarchy of wastewater management are explained by Smol et al. (2020) in the excerpt that follows.

13.3.4.1 Reduce

The first suggestion in the proposal is to reduce, which is also the best course of action for protecting the environment. It implies that the first stage in lowering wastewater production would involve considering the reductions in water use and the elimination of pollution at the source through improved planning and management. The use of less water has a few benefits, including a decrease in personal expenses and time spent obtaining water, a reduction in demand for costly wastewater treatment plant (WWTP) facilities, and an increase in the quantity of water available for multiple uses (Smol et al. 2020). The idea that people should carry over their environmental attitudes and behaviors from their homes to their workplaces should be put forth. The adoption of strategies to reduce water use in industries, which is essential in processing sites that depend on water usage, is made possible by behavioral change.

13.3.4.2 Reclamation (Removal)

Reclamation, which is associated with extremely effective techniques for removing contaminants from wastewater, is the second option in the hierarchy. Because the discharge of untreated wastewater into water bodies can result in eutrophication, health issues, and greenhouse gas (GHG) emissions in the form of nitrous oxide and methane (Smol et al. 2020), treatment is required by EU regulations and national laws. To reuse the treated industrial wastewater, additional contaminants such as hazardous organic compounds, particles, nutrients, microbial pathogens, and other things must be degraded or eliminated. There are several techniques since different wastewaters (industrial and municipal) require different treatments. To meet specific water quality goals, a variety of approaches, including biological, physical, chemical, and combinations, are used in WWTPs.

The initial physicochemical processes are the focus of primary treatment, while tertiary treatment turns wastewater that has undergone primary or secondary treatment into high-quality water that can be used for both potable (drinking) and nonpotable applications (industrial). In the third stage, the largest part of impurities is removed, leaving the water of a safe quality for specific use. When water is scarce, treated wastewater provides a different source of water, which is essential. The availability/demand gap for agriculture, industry, and potable supply may also be closed by wastewater that has been treated to the required quality.

13.3.4.3 Reuse

In the model framework suggested, reuse is the third possibility. Nonpotable water can be recycled and used as an alternative source of fresh water for industrial processes, irrigation of urban landscapes, and other urban applications. It is widely accepted and incorporated into international, EU, and national initiatives to use treated wastewater as a substitute source of water supply. Reusing water allows for qualitative pressure relief from discharges of urban WWTPs to vulnerable areas as well as quantitative improvement of environmental status. When compared to other water sources like desalination or water transfer, water reuse is frequently less expensive to invest in and requires less energy (Macedonio and Drioli 2022).

Additionally, it is possible to reuse treated wastewater in agriculture, ensuring a steady supply of water for irrigation. It can also lessen the chance of crop failure and income loss. The use of treated wastewater with a controlled amount of nutrients is another crucial factor. This can help farmers save money and the environment by lowering the demand for extra fertilizers while also reducing the requirement for wastewater treatment. Water reuse in the wastewater sector is demonstrated by the utilization of gray water that enters drains from toilets and sinks for irrigation purposes (Smol et al. 2020). Following a standard of wastewater treatment, the remaining water also referred to as "black water" can also be used for irrigation. It is possible to recycle entire wastewater using this method of treatment.

13.3.4.4 Recycling

Water recycling is the process of recovering water from wastewater that can be used for potable (drinking) purposes so that it can be put straight into the water supply. It should be noted that wastewater collection and recycling have substantial financial and environmental effects, even though it helps to conserve resources and reduce waste. Recycling hence ought to be reserved for effluent that cannot be reduced or used. High-efficiency methods for the removal of contaminants from wastewater must be used in water recycling, such as membrane techniques like nanofiltration (NF), reverse osmosis (RO), or forward osmosis (FO) (Smol et al. 2020). For instance, water recovery from the wastewater of palm oil mill effluents (POME) has been made using membrane technology. This strategy provides water recycling in addition to a good possibility of resolving environmental concerns.

13.3.4.5 Recovery

Material recovery is the fifth option in the suggested model. Wastewater has a tremendous potential to support circular methods due to the number of potentially recoverable resources it contains, such as inorganic nutrients like nitrogen and phosphorus, biogas, and heavy metals. Phosphorus is one of the crucial components that can be recovered from wastewater. P-recovery potential can be found in WWTPs in the form of sedimentary liquid (leachate), dehydrated sewage sludge, and ash from sewage sludge (Smol et al. 2020).

The recovery of P in the form of struvite from the liquid phase of wastewater is one of the most promising techniques. Struvite precipitation can have various advantages, including better treatment process performance and a lessened negative environmental impact on receiving water bodies due to effluent discharge. Another option for regaining nutrients is composting sewage sludge. For instance, horticulture, gardening, urban agriculture, and organic farming might all make use of it. However, it should be mentioned that the composition of sewage sludge could be complex given that this waste stream is rich in micro- and macroelements and may also contain toxic substances and harmful organisms (Smol et al. 2020).

13.3.5 Integrated Approach

As discussed in the previous section, many biological, chemical, and physical techniques can be used to treat wastewater. Among them, biological methods are thought to be practical and cost-effective for removing these pollutants from wastewater. However, there does not seem to be a single solution that can be applied to all types of wastewaters from various industries due to the limitations associated with different techniques and the fluctuating nutritional load in wastewater. Furthermore, conventional wastewater management techniques revealed some shortcomings, such as high requirement, demanding power consumption, and stipulated standards, which increase the overall costs of treating wastewater.

To enhance resource recovery from wastewater, alternative wastewater management techniques must be investigated. The microalgae-based wastewater treatment approach is believed to be more effective and may be able to lower atmospheric carbon dioxide (CO_2) as a result, among other things. Microalgae grow by consuming CO_2 and readily available pollutants from wastewater, whereas algal biomass can be used to make biofuel.

Microalgae are promising prospects for the treatment of wastewater and the production of biofuels because they grow rapidly, produce minimal waste, and can be cultivated in nonagricultural settings. In light of this, integrating wastewater treatment with microalgae cultivation may represent a green technology for the production of biofuels and energy in the future. Both social and economic sustainability may be achieved by combining wastewater treatment with microalgae culture for the production of biofuel. Wastewater is the ideal resource for the growth of microalgae because it is:

1. A less expensive organic and inorganic carbon-rich media for growth
2. Can sustain large-scale cultivation
3. Can offer enough trace amounts
4. Can replace current wastewater treatment facilities

References

Abdullah N, Sulaiman F (2013) The oil palm wastes in Malaysia. Biomass Now-Sustainable Growth Use 1(3):75–93

de Aquim PM, Hansen É, Gutterres M (2019) Water reuse: an alternative to minimize the environmental impact on the leather industry. J Environ Manag 230:456–463

Arun J, Gopinath KP, Sivaramakrishnan R, SundarRajan P, Malolan R, Pugazhendhi A (2021) Technical insights into the production of green fuel from CO2 sequestered algal biomass: a conceptual review on green energy. Sci Total Environ 755:142636

Asatekin A, Mayes AM (2009) Oil industry wastewater treatment with fouling resistant membranes containing amphiphilic comb copolymers. Environ Sci Technol 43(12):4487–4492

Behera M, Jana PS, More TT, Ghangrekar MM (2010) Rice mill wastewater treatment in microbial fuel cells fabricated using proton exchange membrane and earthen pot at different pH. Bioelectrochemistry 79:228–233

Capello C, Wernet G, Sutter J, Hellweg S, Hungerbühler K (2009) A comprehensive environmental assessment of petrochemical solvent production. Int J Life Cycle Assess 14(5):467–479

Chaturvedi A, Rai BN, Singh RS, Jaiswal RP (2021) A comprehensive review on the integration of advanced oxidation processes with biodegradation for the treatment of textile wastewater containing azo dyes. Rev Chem Eng

Cisneros BJ (2011) Safe sanitation in low economic development areas. In: Treatise on Water Science. Elsevier, p 147

Demirbas A, Balat M, Balat H (2011) Biowastes-to-biofuels. Energy Convers Manag 52:1815–1828

Dhanke P, Wagh S, Kanse N (2018) Degradation of fish processing industry wastewater in hydrocavitation reactor. Materials Today: Proceedings 5(2):3699–3703

Gadipelly C, Pérez-González A, Yadav GD, Ortiz I, Ibáñez R, Rathod VK, Marathe KV (2014) Pharmaceutical industry wastewater: review of the technologies for water treatment and reuse. Ind Eng Chem Res 53(29):11571–11592

Hasanudin U, Sugiharto R, Haryanto A, Setiadi T, Fujie K (2015) Palm oil mill effluent treatment and utilization to ensure the sustainability of palm oil industries. Water Sci Technol 72(7):1089–1095

Herrero M, Stuckey DC (2015) Bioaugmentation and its application in wastewater treatment: a review. Chemosphere 140:119–128

Kaur A, Umar A, Kansal SK (2016) Heterogeneous photocatalytic studies of analgesic and non-steroidal anti-inflammatory drugs. Appl Catal A Gen 510:134–155

Kongnoo A, Suksaroj T, Intharapat P, Promtong T, Suksaroj C (2012) Decolorization and organic removal from palm oil mill effluent by Fenton's process. Environ Eng Sci 29(9):855–859

Kumar A, Priyadarshinee R, Roy A, Dasgupta D, Mandal T (2016) Current techniques in rice mill effluent treatment: emerging opportunities for waste reuse and waste-to-energy conversion. Chemosphere 164:404–412

Kumar A, Srivastava NK, Gera P (2021) Removal of color from pulp and paper mill wastewater-methods and techniques-a review. J Environ Manag 298:113527

Kumar N, Bauddh K, Kumar S, Dwivedi N, Singh DP, Barman SC (2013) Extractability and phytotoxicity of heavy metals present in petrochemical industry sludge. Clean Techn Environ Policy 15(6):1033–1039

Lindholm-Lehto PC, Knuutinen JS, Ahkola HS, Herve SH (2015) Refractory organic pollutants and toxicity in pulp and paper mill wastewaters. Environ Sci Pollut Res 22(9):6473–6499

Lofrano G, Brown J (2010) Wastewater management through the ages: a history of mankind. Sci Total Environ 408(22):5254–5264

Macedonio F, Drioli E (2022) Circular economy in selected wastewater treatment techniques. In: Membrane engineering in the circular economy. Elsevier, pp 101–122

Nzila A, Razzak SA, Zhu J (2016) Bioaugmentation: an emerging strategy of industrial wastewater treatment for reuse and discharge. Int J Environ Res Public Health 13(9). https://doi.org/10.3390/ijerph13090846

Ramola B, Singh A (2013) Heavy metal concentrations in pharmaceutical effluents of industrial area of Dehradun (Uttarakhand), India. J Environ Anal Toxicol 3(173):2161–0525

Ramprakash B, Muthukumar K (2015) Biohydrogen production from rice mill wastewater using mutated Enterobacter aerogenes. Eng Agri Environ Food 9(1):109–115

Riffat R, Husnain T (2022) Fundamentals of wastewater treatment and engineering. CRC Press

Sawalha H, Alsharabaty R, Sarsour S, Al-Jabari M (2019) Wastewater from leather tanning and processing in Palestine: characterization and management aspects. J Environ Manag 251: 109596

Sharma P, Iqbal HMN, Chandra R (2022) Evaluation of pollution parameters and toxic elements in wastewater of pulp and paper industries in India: a case study. Case Stud Chem Environ Eng 5: 100163

Siddique K, Rizwan M, Shahid MJ, Ali S, Ahmad R, Rizvi H (2017) Textile wastewater treatment options: a critical review. In: Enhancing clean-up of environmental pollutants, pp 183–207

Sivasubramaniam D, Franks AE (2016) Bioengineering microbial communities: their potential to help, hinder, and disgust. Bioengineered 7:137–144

Smol M, Adam C, Preisner M (2020) Circular economy model framework in the European water and wastewater sector. J Mater Cycles Waste Manag 22:682–697

Tamersit S, Bouhidel KE, Zidani Z (2018) Investigation of electrodialysis anti-fouling configuration for desalting and treating tannery unhairing wastewater: feasibility of by-products recovery and water recycling. J Environ Manag 207:334–340

Tian X, Song Y, Shen Z, Zhou Y, Wang K, Jin X et al (2020) A comprehensive review on toxic petrochemical wastewater pretreatment and advanced treatment. J Clean Prod 245:118692

Ugurlu M, Gurses A, Dogar C, Yalcin M (2007) The removal of lignin and phenol from paper mill effluents by electrocoagulation. J Environ Manag 87:420–428

Vanerkar AP, Satyanarayan S, Dharmadhikari DM (2013) Full-scale treatment of herbal pharmaceutical industry wastewater. Int J Chem Phys Sci 2:52–62

Verma P, Dwivedi G, Behura AK, Patel DK, Verma TN, Pugazhendhi A (2020) Experimental investigation of diesel engine fuelled with different alkyl esters of Karanja oil. Fuel 275:117920

Xu Y, Hu Y, Peng Y, Yao L, Dong Y, Yang B, Song R (2020) Catalytic pyrolysis and liquefaction behavior of microalgae for bio-oil production. Bioresour Technol 300:122665

Yaseen DA, Scholz M (2019) Impact of pH on the treatment of artificial textile wastewater containing azo dyes using pond systems. Int J Environ Res 13(2):367–385

Yurtsever A, Sahinkaya E, Çınar Ö (2020) Performance and foulant characteristics of an anaerobic membrane bioreactor treating real textile wastewater. J Water Proc Eng 33:101088

Zajda M, Aleksander-Kwaterczak U (2019) Wastewater treatment methods for effluents from the confectionery industry–an overview. J Ecol Eng 20(9):293

Brassica Juncea L.: A Potential Crop for Phytoremediation of Various Heavy Metals

14

Preeti Rani, Pawan Kumar Rose, Mohd. Kashif Kidwai, and Meenakshi

Abstract

Brassica juncea L. (Indian mustard) is an economically important oilseed crop of Brassicaceae family. Diverse plant species of Brassicaceae family are identified as hyperaccumulators of different heavy metals with various applications in phytoremediation. Polluted soil and water negatively impact the quality of food and nutrients of human and animal biota, which are mainly polluted by untreated effluent discharges from industries, which are broadly classified into metallic and nonmetallic pollutant-bearing effluents. Among all types of pollutants, heavy metals are one of the most toxic pollutants adversely affecting the environment. Several conventional physicochemical methods are used to clean up the heavy metal-contaminated environment, but most of them have a high cost of running along with poor efficiency. *Brassica juncea* L. has been widely investigated and applied for the phytoremediation of different toxic heavy metals such as arsenic (As), copper (Cu), nickel (Ni), cadmium (Cd), mercury (Hg), and lead (Pb). Phytoremediation is a sustainable, cost-effective, and environmentally friendly technology that offers clear advantages over traditional methods for site cleanup and detoxication. This green technology is applied to remediate the polluted soils without altering the properties of soil. *Brassica juncea* L. has the capacity to extract, sequester, and detoxify the heavy metals present in the different terrestrial environments. Phytoremediation by Indian mustard is widely investigated through different mechanisms, viz., phytoextraction, phytostabilization, etc.,

P. Rani
Department of Botany, Chaudhary Devi Lal University, Sirsa, India

P. K. Rose · M. K. Kidwai (✉)
Department of Energy and Environmental Sciences, Chaudhary Devi Lal University, Sirsa, India

Meenakshi
Department of Botany, Kurukshetra University Kurukshetra, Thanesar, Haryana, India

R. P. Singh et al. (eds.), *Heavy Metal Toxicity: Environmental Concerns, Remediation and Opportunities*, https://doi.org/10.1007/978-981-99-0397-9_14

and the relationship of *Brassica juncea* L. plants with different microorganisms for the management of heavy metal-induced stress and desired growth. The current chapter provides comprehensive information on the heavy metal accumulation and antioxidative defense potential of *Brassica juncea* L. (Indian mustard) in different phytoremediation strategies.

Keywords

Brassica juncea L. · Heavy metals · Antioxidative · Phytoextraction · Phytostabilization

14.1 Introduction

Heavy metal pollution is a significant environmental concern in all forms of ecosystems on a global scale, offering substantial risks and detrimental impacts on various components of diverse ecosystems, including humans (Chen et al. 2020; Kidwai and Dhull 2021). The properties of heavy metals (HMs), such as high density, atomic mass (greater than 20), conductivity, cation stability, etc., are often documented. HMs and metalloids such as arsenic (As), cadmium (Cd), lead (Pb), mercury (Hg), copper (Cu), nickel (Ni), chromium (Cr), etc. are hazardous to the biotic components of several ecosystems (Cristaldi et al. 2017; Ali et al. 2020; Guerra Sierra et al. 2021; Rathika et al. 2021; Kidwai et al. 2022; Kafle et al. 2022). HMs and metalloids are nonbiodegradable but can be transformed into a less harmful state and persist in different environmental spheres (Zhang et al. 2018). However, some HMs are micronutrients, such as molybdenum (Mo), manganese (Mn), zinc (Zn), Ni, Cr, Cu, iron (Fe), etc., and plants require only a minuscule quantity, but when present in high concentration, they become hazardous (Rose et al. 2022a). However, HMs and metalloids such as As, Cd, Pb, and Hg are nonessential that are not necessary for plant development and metabolic activity and induce toxicities even in trace amounts in various components of the food chain (Kanwar et al. 2015). Plants grown in HM-contaminated soil undergo morphological and physiological changes, resulting in reduced productivity. Furthermore, some HMs actively engage in several functional and biochemical activities, such as biological oxidation; DNA, glucose, and protein biosynthesis; cell wall metabolism; etc. (Li et al. 2018; Muthusaravanan et al. 2018). According to several reports, elevated HM concentration in soil may damage various plants from different families, resulting in decreased plant yield (Sterckeman et al. 2019; Ozyigit et al. 2020). HM-induced phytotoxicity affects physiological and biochemical processes such as respiration and photosynthesis, which result in lower plant yield. Prior studies have labeled Cd and Pb as "primary risks" to plants since they are not necessary to plant metabolism or growth (Tangahu et al. 2011; Raza et al. 2020). According to Chen et al. (2020), HM-induced soil pollution is a major worldwide concern, with around 16% of agricultural soil in China polluted by various HMs, followed by other countries.

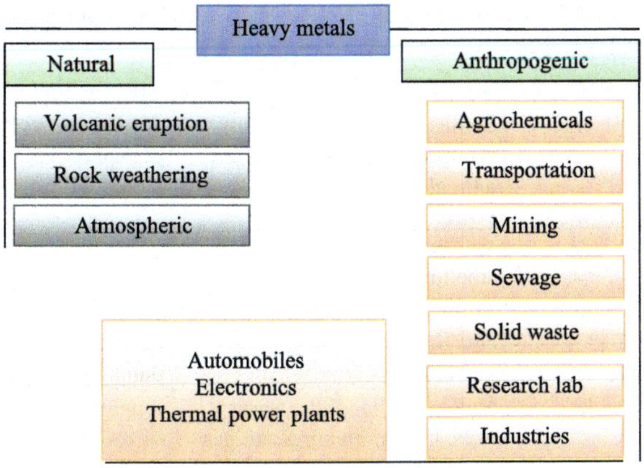

Fig. 14.1 Sources of heavy metals in the environment

HMs and metalloids are found primarily in the earth's crust, and their existence in different capacities in the soil results from various ongoing ecological, geological, and anthropogenic processes (Kidwai and Dhull 2021). As shown in Fig. 14.1, HMs are both natural and anthropogenic in origin. These metallic pollutants are toxic at the cellular level, causing various oxidative stresses that disrupt living organism's biological and physiological processes, including plants. As shown in Table 14.1, various HMs and metalloids pose various adverse effects on humans. Unfortunately, HMs and metalloids are toxic and nonbiodegradable, making their eradication from different ecosystems exceedingly challenging. The HMs-induced toxicity in plants has been widely reported (Kanwar et al. 2015; Diarra et al. 2021). Metal-tolerant plant species are called hyperaccumulators due to their propensity to accumulate significant concentrations of HMs in their tissues and protect themselves from excessive toxicity caused by natural and anthropogenic sources. The length and degree of exposure, as well as other environmental conditions, substantially influence the toxicity of HMs.

HMs pose risks to all living organisms including humans in the food chain of different ecosystems. The Agency for Toxic Substances and Disease Registry (ATSDR) considers Pb, Cd, Hg, and As to be the most lethal HMs and metalloids because of their frequency, toxicity, and possible exposure to humans (Ekta and Modi 2018; Guerra Sierra et al. 2021). Due to unsustainable anthropogenic activities, including mining, agriculture, industry, urbanization, transportation systems, civil construction, improper waste disposal, etc., there has been an upsurge in HMs pollution in natural ecosystems, including terrestrial and water-based ecosystems and man-made systems such as agroecosystems. Natural processes such as volcanic eruption and geological weathering also contribute to HMs in different concentrations in diverse ecosystems. Several human-induced operations, such as mining and other manufacturing processes in industrial units, release enormous quantities of heavy metals, resulting in pollution in various spheres across the world, as represented in Fig. 14.1.

Table 14.1 Toxic effects of some heavy metals on human health

Heavy metal	Sources	Effect on human	References
Arsenic	Semiconductors, petroleum refining, wood preservatives, animal feed additives, coal power plants, herbicides, volcanoes, mining, and smelting	Alteration in oxidative phosphorylation and ATP synthesis	Wang et al. 2022
Copper	Mining, electroplating, smelting and refining, biosolids	Liver cirrhosis, anemia, digestive organs irritation, damage of brain and kidney	Ansari et al. 2015
Nickel	Volcanoes, landfills, forest fires, bubble bursting and gas exchange in the ocean, soil and geological material weathering	Allergic skin diseases, cancer of respiratory organs, affect the immune, nervous, and reproductive systems, hair loss	Ansari et al. 2015
Cadmium	Geogenic sources, smelting, and refining of metal, fossil fuel, phosphate-based fertilizers, municipal sludge	Carcinogenic, mutagenic, damage lung and bones, affect calcium regulation in biological systems	Moosavi and Seghatoleslami 2013
Mercury	Volcanic eruptions, forest fire, caustic soda industry, coal, peat, and wood burning	Autoimmune diseases, depression, fatigue, loss of hair, insomnia, loss of memory, damage to the brain, lung, and kidney	Yadav et al. 2018
Lead	Mining and smelting of metalliferous ores, lead-containing gasoline, municipal sewage, industrial wastes, and paints	Unbalanced development, diminished IQ, short-term memory loss, learning impairments, and cardiovascular disease	Niazi et al. 2017

14.2 Bioremediation

Among all remediation strategies, including physical, chemical, and biological methods, bioremediation is an economically feasible, solar-driven, living organism-based approach incorporating bacteria, algae, fungus, plants, etc. Sustainable remediation of contaminated land and water ecosystems employs naturally occurring or genetically engineered microorganisms (Parmar and Singh 2015; Bortoloti and Baron 2022; Rose et al. 2022b; Kafle et al. 2022). Phytoremediation is another term for plant-based bioremediation.

14.3 Phytoremediation

Phytoremediation is an exclusive in situ detoxifying ecosystem service in which plants' metabolic activities in diverse ecosystems remediate various types of organic and inorganic contaminants, including metal(loid)s (Belimov et al. 2005; Kafle et al. 2022). Phytoremediation is crucial in various restoration initiatives for contaminated ecosystems due to its efficiency and outputs; several factors, as exhibited in Fig. 14.4, adversely affect the uptake of HMs and metalloids in plants and reduce the potential for environmental cleanup activities like phytoremediation. Diverse plant species from different families exhibited a high potential for the phytoremediation of many HMs and metalloids in soil and water. Some of the plants are as follows:

Agropyron smithii, *Alyssum heldreichii*, *Alyssum murale*, *Allenrolfea occidentalis*, *Amaranthus cruentus* L., *Amaranthus hybridus* L., *Amaranthus spinosus* L., *Arabis paniculata*, *Arundo donax* L., *Baccharis trimera*, *Bidens pilosa* L., *Berkheya coddii*, *Beta vulgaris* L., *Brassica juncea* L., *Brassica oleracea*, *Cannabis sativa*, *Carex pendula*, *Chenopodium album* L., *Croton bonplandianus*, *Cynodon dactylon*, *Cyperus rotundus*, *Desmostachya bipinnata*, *Dicoma niccolifera*, *Erato polymnioides*, *Helianthus annuus* L., *Hordeum vulgare*, *Ipomoea aquatica*, *Jatropha curcas*, *Lactuca sativa*, *Lolium italicum*, *Ludwigia stolonifera*, *Medicago sativa*, *Myriophyllum spicatum*, *Myriophyllum aquaticum*, *Nelumbo nucifera*, *Nicotiana tabacum*, *Oryza sativa* L., *Phaseolus vulgaris* L., *Sedum alfredii*, *Salvinia natans*, *Sorghum halepense*, *Pelargonium hortorum*, *Pteris vittate*, *Polygonum punctatum*, *Phragmites australis*, *Phyla nodiflora*, *Pistia stratiotes*, *Pennisetum purpureum*, *Phyllostachys pubescens*, *Populus cathayana*, *Ricinus communis*, *Ranunculus sceleratus*, *Rumex dentatus*, *Sesbania drummondii*, *Solanum americanum*, *Solanum tuberosum*, *Thlaspi caerulescens*, *Trifolium repens*, *Triticum aestivum*, *Typha latifolia*, *Vetiveria zizanioides*, *Vigna unguiculata*, *Vigna radiata* L., *Zea mays*, *Eichhornia crassipes* L., *Egeria densa*, *Solanum nigrum*, etc. (Kumar et al. 2018; Muthusaravanan et al. 2018; Awan et al. 2020; Sharma et al. 2020; Kidwai and Dhull 2021; Manori et al. 2021; Mustafa and Hayder 2021; Kafle et al. 2022)

The Brassicaceae or Cruciferae family has a collection of more than 350 genera which includes 3500 species and some of the genera are *Sinapis*, *Brassica*, *Thlaspi*, etc. Many plants of Brassicaceae are reported to have high nutritional and economic value (Favela-González et al. 2020). Various plants of Brassicaceae family are consumed as vegetables and edible oil by the human population in different parts of the globe as these plants are reported to supply dietary fiber, anthocyanins, dietary flavonols, vitamins, glucosinolates, etc. (Raiola et al. 2017). Apart from food and medicinal applications, more than 100 plant species of Brassicaceae are among the overall hyperaccumulator species studied for the accumulation of HMs and metalloid. A broad range of HMs and metalloids accumulated in the tissues of Brassicaceae family members, including the excluders and non-accumulators, due to the existence of a wide range of biological mechanisms in response to toxic levels. In case of HM-induced stress, several plants produce ligands such as phytochelatins,

Table 14.2 Phytoremediation of HMs and metalloids by different plant species of *Brassica* genus

Species	Heavy metal	Phytoremediation strategies	References
Brassica juncea L.	Arsenic, cadmium, chromium, copper, lead, selenium	1, 2, 3, 4, and 5	Salido et al. 2003; Belimov et al. 2005; Kumar et al. 2008; Ma et al. 2009; Srivastava et al. 2013; Bassegio et al. 2020a, b; Hasanuzzaman et al. 2020; Monei et al. 2021; Du et al. 2020
Brassica oleracea	Lead, selenium, zinc	1 and 2	Taghizadeh et al. 2018; Natasha et al. 2018; Zeremski et al. 2021
Brassica rapa	Cadmium, iron, zinc	3 and 4	Kohli et al. 2018; Bortoloti and Baron 2022
Brassica napus	Cadmium, chromium, copper, zinc	1, 2 and 5	Marchiol et al. 2004; Zhang et al. 2018; Ontañon et al. 2014; Guo et al. 2020; Bortoloti and Baron 2022
Brassica campestris L.	Chromium, copper	5	Zahoor et al. 2017
Arabidopsis thaliana	Arsenic, cadmium, lead, selenium	1	Cahoon et al. 2015; Zhang et al. 2018; Hasanuzzaman et al. 2020; Thakur et al. 2020

1, phytoextraction; 2, phytovolatilization; 3, phytostabilization; 4, rhizofiltration; 5, rhizomediation

metallothioneins, siderophores, etc. for sequestration and translocation of HMs at the cellular level along with the use of antioxidant machinery for enhancing the tolerance in plants (Bortoloti and Baron 2022) and may be one of the probable reasons associated with various species of Brassicaceae which attracts intensive research on the multiple areas of phytoremediation. *Brassica carinata* A. Braun, *Brassica oleracea* L., *Brassica juncea* L., *Brassica nigra* L. K. Koch, *Brassica rapa* L., *Brassica napus* L., and *Brassica campestris* L. are some of the most common studied species of *genus Brassica* (Zahoor et al. 2017; Raiola et al. 2017; Zeremski et al. 2021) as exhibited in Table 14.2. The mobility of HMs, morphological features, and crop management practices are some of the common factors influencing the metal's sequestration ability of plants with phytoremediation potential. The addition of organic matter, intercropping, and legumes to the cropping system, as well as the stimulation of plant growth and soil metal dissolution, are examples of crop management components that promote phytoextraction. Phytoremediation by Indian mustard (*Brassica juncea* L.) is widely investigated and it performs phytoremediation through different mechanisms such as phytoextraction, phytostabilization, phytostabilization, rhizofiltration, rhizomediation, etc. (Ma et al. 2009, Diwan et al. 2008). Because of its rapid growth rate, shorter life cycle, high biomass production, diversified photosynthetic capabilities, and excellent resistance to various HMs and metalloids, *Brassica juncea* L. is an excellent hyperaccumulator plant for HMs and metalloids (Mahmud et al. 2018). The accumulation of high metal concentrations in mustard shoots compared to surrounding soil indicates that mustard has a high natural phytoextraction potential. Mustards are reported to modify the membrane lipid composition, making it suitable for the phytoextraction of HMs such as Pb, Ni, Cd, Hg, and Se. *Brassica juncea* L. cultivated on HM-polluted soils exhibited enhanced phytoextraction efficiency and phytoextraction rate (Hall 2002; Ansari et al. 2015).

14.4 Morphology and Growth of *Brassica Juncea* L.

Brassica juncea L. is an annual herbaceous plant cultivated in the winter (rabi) season. *Brassica juncea* L. is a low-cost oilseed crop in India that belongs to the family Brassicaceae (Cruciferae) of angiosperm (Shakeel et al. 2019). Rapeseed-mustard (RM) is a common name for oleiferous *Brassica* species cultivated in Africa, Australia, Asia, and America (Nanjundan et al. 2022). *Brassica juncea* L., *Brassica carinata* (karanrai), *Brassica napus* (black mustard), *Brassica rapa* (yellow sarson), *Sinapis alba* (white sarson), and *Brassica oleracea* (cabbage) species are among the 338 genera comprised of 3700 species in the Brassicaceae family (Wanasundara et al. 2016). The Brassicaceae family includes over a hundred plants identified for their phytoremediation potential for various pollutants, including toxic metals and metalloids (Jiang et al. 2015; Yang et al. 2016; Zeremski et al. 2021).

Brassica juncea L. is a well-studied *Brassica* species with remarkable phytoremediation potential. *Brassica juncea* L. was developed by crossing *Brassica nigra* and *Brassica rapa* species and is extensively cultivated as an oilseed seed crop in dry and semi-arid regions (Srivastava et al. 2013; Zeremski et al. 2021). *Brassica juncea* L. has ecological and economic importance in the Indian subcontinent as it is used as a source of vegetables, oilseed, forage, green manure, condiments, etc., serving the oil-based demands of almost 50% of the population in all northern states (Shivran et al. 2019). *Brassica juncea* L. is a rich source of edible oil (nearly 40%, dry weight of seed), but erucic acid (a white waxy solid, about 48% of the total fatty acid composition) and glucosinolate presence in oil make it unappealing for human use in different parts of the world. India's primary source of vegetable oils is nine oil seeds cultivated in a 26 million hectare region under rain-fed conditions. Some of the oil-yielding crops cultivated in different states of India, including Haryana, are groundnut (*Arachis hypogaea*), soybean (*Glycine max*), sesame (*Sesamum indicum*), castor (*Ricinus communis*), sunflower (*Helianthus annuus*), and mustard (*Brassica juncea* L.) (Dwivedi and Sharma 2014). Among several *Brassica* species, Indian mustard (*Brassica juncea* L Czern., AABB $2n = 36$) is the most economically important and commonly cultivated species. It possesses attributes such as better adaptability and high tolerance to different abiotic and biotic factors, enabling it to be cultivated in diverse agroclimatic conditions (Nanjundan et al. 2022).

14.5 Phytoremediation Mechanisms

Phytoremediation processes use different physiological systems to remediate metals and reduce adverse toxic effects (Shah and Daverey 2020). Phytoremediation is a low-cost, ecologically sustainable remediation solution as it requires less effort without incurring extra expenses. It is an environmentally friendly approach because it maintains the original ecotype by avoiding excavation of contaminated areas. Furthermore, the aesthetic features of phytoremediation are accepted remediation strategies in or nearby human settlements (Cristaldi et al. 2017; Derakhshan et al. 2018; Kafle et al. 2022). Phytoremediation techniques include phytoextraction, phytodegradation, phytostabilization, phytofiltration or rhizofiltration, and phytovolatilization. Figure 14.2 depicts the strategy of *Brassica juncea* L. for remediating heavy metal-contaminated soils.

14.5.1 Phytovolatilization

Phytovolatilization is the unique transformation process for HMs and metalloids into different volatile states and subsequently release them into the atmospheric environment through the process of transpiration (Kafle et al. 2022). Hazardous HMs and metalloids such as Hg, Se, and As are converted into volatile and less hazardous forms before being released into the environment (Laghlimi et al. 2015). Hg and Se are reported to be remediated pollutants by phytovolatilization. Plants like canola and Indian mustard contribute to the phytovolatilization of Se. According to Moreno et al. (2008), *Brassica juncea* L. accumulates varying Hg concentrations under hydroponic conditions, accumulates Hg in its root system, and finally releases it in vapor form (HgO) into the external environment (around the root). *Brassica chinensis* also has a high concentration of Se in its roots, which is finally released into the rhizosphere (Wang et al. 2022). Both Hg and Se undergo oxidation-reduction processes in the rhizosphere, where less toxic volatile forms are produced (Hg^{2+} to HgO) as a function of a communal interaction between several bacteria resistant to these pollutants (occurring in both soil and water) (Moreno et al. 2008). One of the most notable benefits of phytovolatilization is that after plantation, low maintenance is required. Other benefits include reduced soil erosion, less soil disturbance, unrequited harvesting, and the elimination of plant biomass (Salido et al. 2003; Belimov et al. 2005; Kumar et al. 2008; Ma et al. 2009; Srivastava et al. 2013; Bassegio et al. 2020a, b; Monei et al. 2021; Du et al. 2020).

14.5.2 Phytoextraction

Phytoextraction (phytosequestration, phytoaccumulation) is a process that occurs in the roots of plants, enabling them to absorb pollutants from soil or water through their roots, followed by the translocation of the contaminants in above-ground

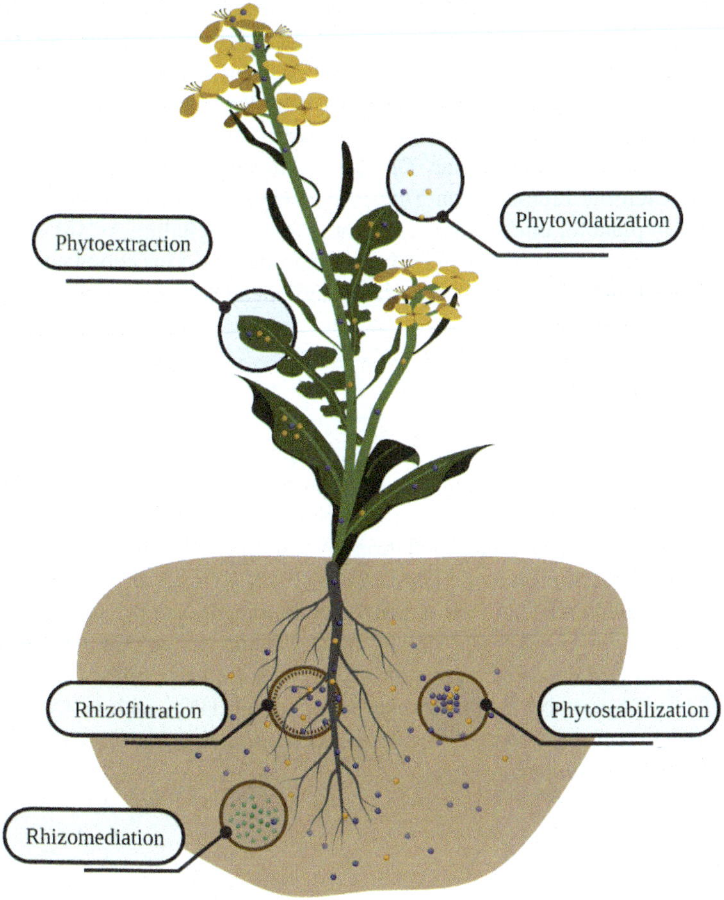

Fig. 14.2 Phytoremediation mechanisms involved for HM contamination in *Brassica juncea* L

biomass such as shoots and leaves. The plants used for phytoextraction generate huge biomass and accumulate contaminants such as HMs and metalloids. Phytoextraction is of two types, chelate-assisted phytoextraction or induced phytoextraction and continuous phytoextraction (Suman et al. 2018; Sterckeman et al. 2019). It is an efficient, low-cost, environmentally acceptable, and potentially cost-effective method of remediating metal-contaminated soils. Plants used for phytoextraction must grow rapidly, produce considerable biomass, have a long root system and high root-shoot transfer, and are resistant to high HMs and metalloids concentrations in plant tissues (Tong et al. 2004). HMs such as Cu, Cd, Cr, Pb, Ni, Se, etc. have been recorded in high concentrations in plants from the *Amaranthaceae*, *Asteraceae*, Brassicaceae, *Caryophyllaceae*, *Characeae*, *Euphorbiaceae*, *Fabaceae*, *Flacourtiaceae*, *Malvaceae*, *Poaceae*, *Urticaceae*,

Verbenaceae, *Violaceae*, etc. families (Parmar and Singh 2015; Hasanuzzaman et al. 2020; Guerra Sierra et al. 2021). Diverse Brassicaceae species are potential candidates for the phytoextraction of several HMs. As, Cd, Zn, and Ni are some of the metals scavenged by different plants of Brassicaceae species as exhibited in Table 14.2. According to Srivastava et al. (2013), *Brassica juncea* L. accumulates one-third of Zn in its tissues than *Thlaspi caerulescens* because *Brassica juncea* L. generates tenfold biomass than *Thlaspi caerulescens*. *Brassica juncea* L. is the most efficient plant for eliminating HMs from the soil, including As, Cd, Cr, Cu, Ni, Pb, Se, Hg, Zn, etc. *Brassica juncea* L. has the potential to accumulate significant levels of metal and tolerance to excessive heavy metal concentrations.

14.5.3 Phytostabilization

Phytostabilization is a specialized process of phytoremediation applied by plant species that immobilize pollutant like HMs in the soil by absorbing via the roots and precipitating them within the rhizosphere. The phytostabilization method prevents soil and groundwater pollution by limiting additional percolation and mobilization of metal pollutants (Garcia et al. 2018; Kafle et al. 2022). Since HMs are just inactivated and stabilized rather than eliminated from the soil or water, this procedure is only a management strategy for HMs such as As, Cd, Cr, Cu, Pb, etc. (Ali et al. 2013). Pérez-Esteban et al. (2014) reported the potential of *Brassica juncea* L. for phytostabilization of mine soils having HMs such as Cu and Zn in their biomass, especially roots to low bioconcentration and translocation factor and sequestration in vacuoles.

14.5.4 Rhizofiltration

Rhizofiltration is the method of employing a plant to absorb or adsorb pollutants, limiting the pollutants mobility in subsurface water; it has been reported to be applied as both ex situ and in situ bioremediation. Roots have an essential part in rhizofiltration. HMs get absorbed, adsorbed, and precipitated on the root's surface, resulting in changes in rhizospheric pH and root exudates. Once the plant has absorbed all the pollutants, it may be harvested and disposed of. Plants used in rhizofiltration must meet specific criteria, including the ability to generate an extensive root system, the accumulation of large amounts of HMs, the ease of handling, and the requirement for little maintenance. Rhizofiltration is applicable to both aquatic and terrestrial plants with long fibrous root systems. Rhizofiltration can efficiently remediate Cd, Pb, Cr, Ni, Zn, and Cu pollution. Among various terrestrial plants, Indian mustard (*Brassica juncea* L.), sunflower (*Helianthus annus* L.), etc., are reported to remove HMs from the contaminated environment. Indian mustard has been shown to be efficient in eliminating Cd, Cr, Cu, Ni, Pb, and Zn via its roots. Sunflower removes Pb, U, 137Cs, and 90Sr from hydroponic solutions. Indian mustard has a Pb bioaccumulation value of 563 and can eliminate Pb from water ranging from 4 mg/L to 500 mg/L (US Environmental Protection Agency 2000). Because various substances flow out of the roots, the pH of the rhizosphere changes, causing metals to accumulate on the root surfaces. The whole plant or the roots are removed for further processing after the saturation of roots surfaces with these toxicants (Zhu et al. 1999). Blastofiltration is a method for extracting HMs from water by employing young seedlings, an advanced process for cleaning contaminated water (Laghlimi et al. 2015). The genetically transformed *Brassica juncea* L. plants with *Agrobacterium rhizogenes*, the hairy roots of *Brassica juncea* L. plants, effectively removed uranium from the solution up to 5000 μM by the process of rhizofiltration (Eapen et al. 2003).

14.5.5 Rhizomediation

Rhizoremediation, also known as rhizodegradation, phytostimulation, and phytobial process, is one of the most widely applied, on-site, and cost-effective microbial-assisted plant-based remediation techniques. Rhizoremediation employs rhizospheric bacteria in collaboration with plants to remove organic and inorganic contaminants from polluted sites (Kumar et al. 2008; Zhang et al. 2018; Srivastava et al. 2013; Ontañon et al. 2014; Yaashikaa et al. 2020). Rhizoremediation comprises specific plants and their associated rhizosphere microbes to eliminate target contaminants. Plants contribute to this process indirectly by boosting the catalytic activities of microorganisms through root exudates. Rhizoremediation occurs in plants because their roots provide essential nutrients for microbial growth, and in return, microorganisms aid in the phytoremediation process of several types of contaminants; however, it is a slow process in comparison to phytoextraction process (Ontañon et al. 2014; Hoang et al. 2021). Microorganism-based secondary metabolites, such as biosurfactants which include glycolipids, lipopeptides, phospholipids, fatty acids, lipoproteins, polymeric compounds, etc., aid the plants by enhancing the absorption, sequestration, tolerance, etc. against different HMs (Bortoloti and Baron 2022). Some rhizospheric bacteria with plants sequester, absorb, and facilitate the bioavailability of various HMs through the production of different organic acids such as malic acid, humic acid, citric acid, etc. (Ma et al. 2009; Bortoloti and Baron 2022). According to Kumar et al. (2008), the application of *Enterobacter* sp. isolated from fly ash enhanced the phytoextraction of Ni and Cr in *Brassica juncea* L. Srivastava et al. (2013) employed *Staphylococcus arlettae* species to increase As accumulation in *Brassica juncea* L.

14.5.6 Role of *Brassica juncea* L. in Phytoremediation

Phytoremediation utilizes plant inherent capacity to eliminate contaminants from the environment and transform them into nontoxic or less toxic compounds. Indian mustard (*Brassica juncea* L.) accumulates considerable amounts of HMs in its tissues when cultivated in contaminated soil. As a metal hyperaccumulating plant, it efficiently removed HMs as Pb from the soil. Extensive research has been done to explore the ability of members of *Brassica* genus such as *Brassica nigra*, *Brassica oleracea*, *Brassica campestris*, *Brassica carinata*, *Brassica juncea* L., *Brassica napus*, etc. to effectively accumulate HMs such as Ni, Cu, Zn, Cd, and Cr (Mourato et al. 2015; Farahani et al. 2015; Napoli et al. 2018; Chen et al. 2020). One of the most studied Brassica species for its phytoremediation potential is *Brassica juncea* L. (Anjum et al. 2012; Zeremski et al. 2021). As demonstrated in Table 14.2, various *Brassica* species have diverse defensive mechanisms to deal with the harmful outcome of HM stress. The biomass of *Brassica juncea* L. and *Brassica rapa* L. was decreased in response to the elevated bioavailability of HMs when cultivated under ideal conditions. The biomass of *Brassica juncea* L. was recorded to be reduced by 40% compared to the biomass of *Brassica rapa* (Diarra et al. 2021). Mainly, there are two processes by which plants tolerate higher substrate concentrations of HMs: first is to inhibit metal intake by chelating and sequestering metals in vacuoles, while the second is to activate antioxidant mechanisms as a defense response. The cellular exclusion strategy involves actively or passively generated root exudates that attach HMs ions on the root surface, thereby hindering their adsorption (Nazir et al. 2020).

The generation of reactive oxygen species (ROS) such as O_2, H_2O_2 (hydrogen peroxide), $^1O_2$ (singlet oxygen), HO_2, OH (hydroxyl radicals), ROOH, ROO, RO, etc. is a result of several oxidative stresses such as organic pollutants, HMs and metalloids, salinity, etc. Oxidative stress is one of the prime causes of less agricultural yield worldwide as they negatively affect the alteration in membranes, redox balance, nucleic acids, proteins, carbohydrates, lipids, DNA, etc. (Gill and Tuteja 2010). Among all types of pollutants, HMs also induce ROS formation in plants, changing the redox equilibrium and ultimately causing oxidative damage and cell death. Plants evolve various ROS-detoxifying mechanisms based on an efficient enzymatic and nonenzymatic antioxidant defense system (Gonzaga et al. 2022). According to Rathika et al. (2021), ROS are natural byproducts of diverse biochemical processes in a plant. As exhibited in Fig. 14.3, some of the prime enzymatic antioxidants include SOD (superoxide dismutase), APX (ascorbate peroxidase), GPX (guaiacol peroxidase), GR (glutathione reductase), CAT (catalase), MDHAR (monodehydroascorbate reductase), and DHAR (dehydroascorbate reductase). The nonenzymatic antioxidants are carotenoids, phenolic acid, tocopherol, terpenoids, alkaloids, ascorbic acid, anthocyanin, etc. (Gill and Tuteja 2010; Kidwai and Dhull 2021). Both nonenzymatic and enzymatic antioxidants reduce the damage caused by reactive oxygen species, allowing plants such as *Brassica juncea* L. to tolerate high HM concentrations in soil-based ecosystems. *Brassica juncea* L. seedlings with high

Fig. 14.3 Antioxidants for scavenging of reactive oxygen species (ROS) in *Brassica juncea* L

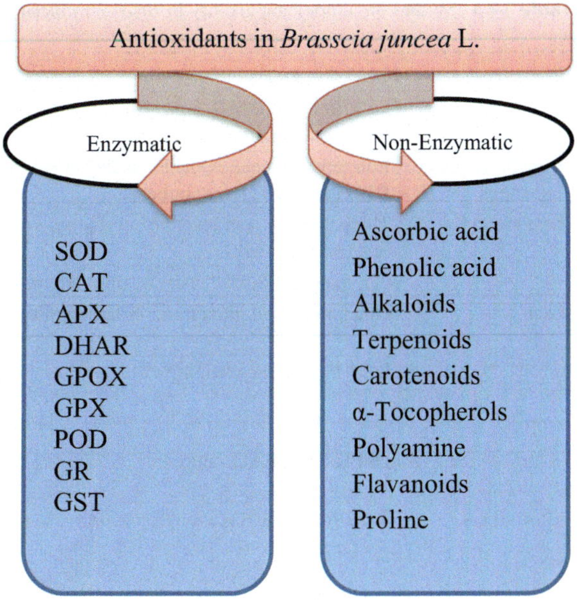

ROS activity of the antioxidant enzymes SOD, POD, and CAT as a protective strategy against lead-induced oxidative stress (Rathika et al. 2021; Zeremski et al. 2021). According to Niazi et al. (2017), for *Brassica napus* under the Pb stress, the content of H_2O_2 increased, as did the activity of the SOD, POD, CAT, and APX enzymes. Higher concentrations of Pb, on the other hand, reduced enzymatic activity (Yahaghi et al. 2018).

Brassica juncea L. seedlings raised proline levels three- to sixfold, an osmoprotectant and antioxidant that allows cellular components to self-protect under Cd stress (Alam et al. 2013; Guo et al. 2020). Another research observed an increase in proline concentration between 30 and 90 days of Cd exposure (Ahmad et al. 2021). As a defensive strategy, *Brassica juncea* L. seedlings enhanced total carbohydrate, anthocyanins, flavonoids, and polyphenol production exposed to Cd (Nazir et al. 2020). The ability of *Brassica* plants to elicit antioxidant defense responses through enzymatic and nonenzymatic mechanisms is linked to their tolerance to the adverse effects of specific metals. Mobin and Khan (2007) reported the intervarietal response of *Brassica juncea* L. cultivars to Cd-induced stress. Table 14.3 summarizes antioxidant-based research on various parts of mustard plants. *Brassica* species undertake HMs phytoremediation by physiological processes such as phytovolatilization, phytostabilization, and phytoextraction (Fig. 14.4).

Table 14.3 Studies on antioxidant activity in *Brassica juncea* L

Brassica juncea L.	Detection indicator	References
Dried leaf	DPPH (1,1-diphenyl-2-picrylhydrazyl) radical scavenging and Fe^{2+}chelating	Huang et al. 2017
Kimchi leaves and stems	Iron reduction antioxidant capacity (FRAP), scavenge DPPH and ABTS free radicals	Thiyam et al. 2006
Leaf kimchi	Thiobarbituric acid reactive substances value and free fatty acids content	Lee et al. 2010
	pH, microbiological analysis, thiobarbituric acid value	Lee et al. 2010
Leaf	Glutathione and glutathione S-transferase	Thakur et al. 2020
	DPPH radical scavenging	Thiyam et al. 2006
		Hur et al. 2014
	DPPH and TEAC value	Harbaum et al. 2008
Mustard cake	Total free radical scavenging activity, hydroperoxide	Thiyam et al. 2006

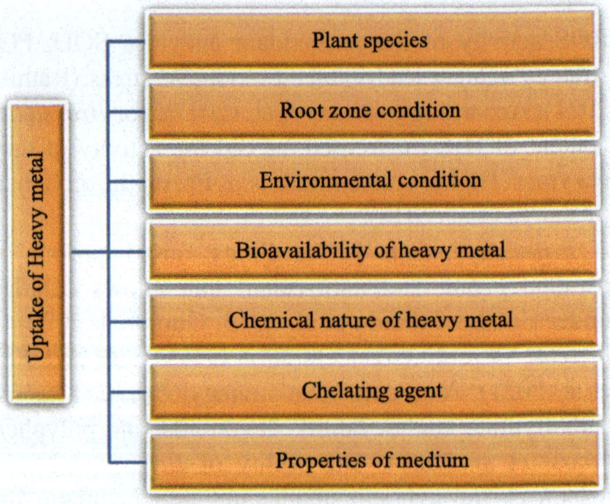

Fig. 14.4 Factors affecting the uptake mechanisms of HMs

14.5.7 Studies on *Brassica juncea* L. as a Phytoremediator

Several phytoremediation studies have revealed that *Brassica juncea* L. has the potential to remove various HMs from soil-based ecosystems, as indicated in Table 14.4.

Brassica juncea L. is an efficient plant for extracting Hg from polluted soil (Tangahu et al. 2011). However, a risk evaluation of Hg levels in the edible section of the *Brassica juncea* L. plant is necessary before the general public may consume

Table 14.4 Relevant studies on phytoremediation capacity of *Brassica juncea* L. against various HMs and metalloids

S. No	Metal	Reference
1	Arsenic	Salido et al. 2003; Ko et al. 2008; Gupta et al. 2009; Srivastava et al. 2013; Kanwar et al. 2015; Niazi et al. 2017; Rahman et al. 2018
2	Cadmium	Belimov et al. 2005; Cao et al. 2008; Lai et al. 2008; Iqbal et al. 2010; Bauddh and Singh 2012; Shanmugaraj et al. 2013; Guerra Sierra et al. 2021; Zunaidi et al. 2021
3	Copper	Ariyakanon and Winaipanich 2006; Johnson et al. 2009; Ma et al. 2009; Gonzaga et al. 2022
4	Chromium	Shahandeh and Hossner 2000; Bluskov et al. 2005; Kumar et al. 2008; Saraswat and Rai 2009
5	Lead	Salido et al. 2003; Clemente et al. 2005; Lai et al. 2008; Meyers et al. 2008; Abbaspour et al. 2012; Koptsik 2014; Zunaidi et al. 2021; Kafle et al. 2022
6	Mercury	Su et al. 2008; Raj et al. 2020
7	Nickle	Zaidi et al. 2006; Cao et al. 2008; Kumar et al. 2008; Saraswat and Rai 2009; Ansari et al. 2015; Muthusaravanan et al. 2018; Zunaidi et al. 2021
8	Selenium	Natasha et al. 2018; Hasanuzzaman et al. 2020
9	Zinc	Sharma et al. 2007

it. Pb is a common inorganic pollutant discovered in soil, with high persistence in soils even at low concentrations, and poses hazardous effects to a broad range of species. Brazilian soils have high Pb concentrations, which concern the environment and human health (Kohli et al. 2018). In recent research (Bassegio et al. 2020a), the potential and tolerance mechanism of *Brassica juncea* L. was investigated in controlled conditions with varied Pb concentrations (24, 80, 136, 362, 1150 mg/kg). The results indicated that when the soil Pb content increased, so did the Pb content in the plant aerial and root parts. Cu poisoning caused lipid peroxidation due to hydrogen peroxide and electrolyte leakage. *Brassica juncea* L. can accumulate significant amounts of Cu from various soil-based ecosystems. Cd hyperaccumulators may withstand Cd concentrations of up to 100 g/g of dry leaf weight, whereas most plants are sensitive to Cd concentrations of 5–10 g/g dry leaf weight. The heavy metal-contaminated site considerably influences plant development in terms of biomass (Chen et al. 2020). Anjum et al. (2014)) discussed that Cd treatment (100 > 50 mg/kg soil) led to a substantial reduction in plant dry mass in *Brassica campestris* and *V. radiate*.

According to Anjum et al. (2014), despite significant Cd-induced stress in *Brassica campestris* roots and *V. radiata* shoots, effective antioxidant activities of GR, GST, and GPX facilitated the plants to survive under Cd-induced stress with some abnormalities at the plant level similar observations recorded in case of *Brassica juncea* L. (Mohamed et al. 2012). The physiological and biochemical responses of *Brassica campestris* and *V. radiate* in Cd-induced stress demonstrate

Table 14.5 Heavy metal accumulation in roots (mg/kg) and shoots (mg/kg) of *Brassica juncea* L. (g/plant dry weight)

Heavy metal	Concentration in roots (mg/kg)	Concentration in shoots (mg/kg)	Dry biomass (g/plant)	Experiment type	Reference
As	70	100	1.38	Pot	Picchi et al. 2021
Cd	2.73	2.48	2.56	Pot	Guo et al. 2020
	1.123	1.974	–	Pot	Khator et al. 2021
	0.359	0.453	1.512	Pot	Goswami and Das 2015
	18	22	–	Hydroponic	Zhu et al. 1999
	14.87	11.85	–	Field	Dhanwal et al. 2017
Cu	3184	3771	3586	Pot	Ariyakanon and Winaipanich 2006
	12.0	1.74	–	Greenhouse	Gonzaga et al. 2022
Hg	358.56	436.36	586.12	Pot	Tangahu et al. 2011
Mn	0.017			Field	Oguntade et al. 2019
Ni	0.254	0.178	–	Hydroponic	Ansari et al. 2015
Pb	783.4	94.0	12.9	Greenhouse	Bassegio et al. 2020a
	80.94	67.16	3.27	Field	Yahaghi et al. 2018
	783.4	94.0	12.9	Greenhouse	Chaiane et al. 2020
	3.678	7.82	4.56	Pot	Ali et al. 2022
U	0.562	0.236	–	Field	Chen et al. 2020
	4.44	1.50	10.85	Field	Wang et al. 2022

the adaptability to Cd-contaminated conditions, which might be valuable in developing novel plant-based remediation strategies for metal HMs and metalloid contaminated ecosystems (Goswami and Das 2015). Khator et al. (2021) reported the relationship between Cd-induced oxidative stress and growth factors, photosynthetic machinery, and morphological and physiological aspects of *Brassica juncea* L. Table 14.5 demonstrates the HMs accumulated in roots and shoots parts of dried *Brassica juncea* L. (Ali et al. 2020).

14.6 Enhancement of Phytoremediation Process

As given in Fig. 14.5, some of the focus areas are identified for the next level of research in enhancing the sustainability of ecofriendly process such as phytoremediation via *Brassica* sp. The potential for phytoremediation may be enhanced using chemical or biological methods, and some of these methods are discussed.

14.6.1 Enhanced Heavy Metal Phytoextraction with Chemicals

Chelators enhance HM mobility in soils, which increases the risk of percolation into deeper layers. Chelators from the APCA (aminopolycarboxylic acids) group, i.e., EDTA (ethylene diamine tetraacetic acid) and EDDS ([S, S]-isomer of ethylenediamine disuccinate), and natural low molecular weight organic acid group (NLMWOA), i.e., gluconic (GA), oxalic (OA), and malic (MA), have been extensively explored in phytoextraction methods (Bouquet et al. 2017; Guo et al. 2019; Diarra et al. 2021; Rathika et al. 2021; Zeremski et al. 2021). EDTA has a significant risk of HM leakage due to low biodegradability issues, and EDTA–metal

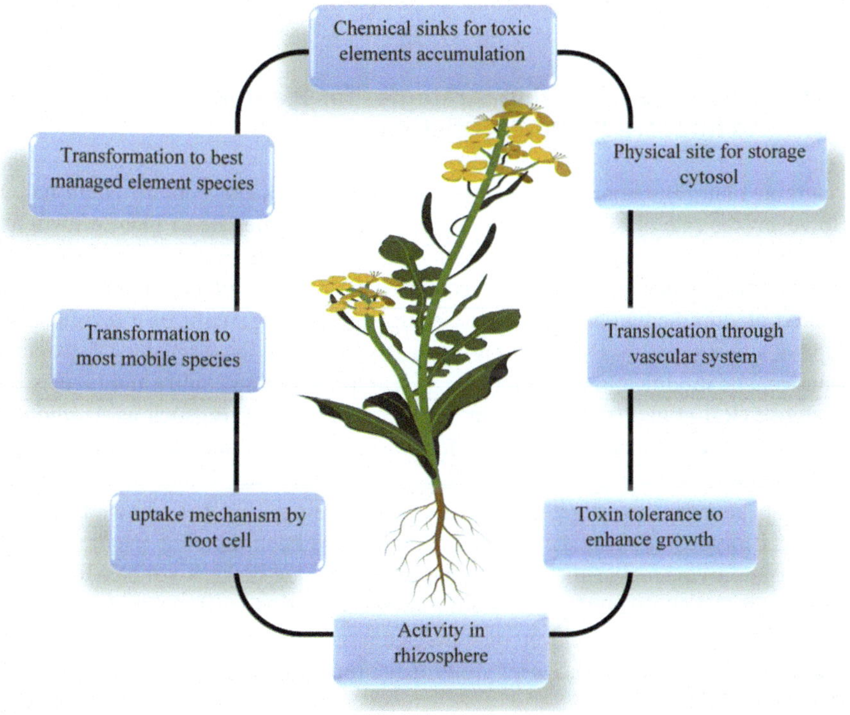

Fig. 14.5 Some of the focus areas for phytoremediation

Table 14.6 The influence of different chelators on phytoremediation in members of plants of genus *Brassica*

Plant	Metal	Chelator	Reference
Brassica juncea L.	Cd, Cr, Cu, Ni, Pb,	EDDS, EDTA, CA, GLDA	Bouquet et al. 2017; Guo et al. 2019; Diarra et al. 2021; Rathika et al. 2021; Zeremski et al. 2021
Brassica napus	Cu	EDDS, EDTA	Zeremski-Škorić et al. 2010
Brassica rapa	Cd, Cr, Cu, Ni, Pb, Zn	EDDS, CA, GLDA	Zeremski et al. 2021
Brassica oleracea	Pb, Zn	EDTA	Chatuverdi et al. 2019

complex biodegrades in the soil in a time period of 1 month after application of EDTA application (Wu et al. 2004). Citric acid has one of the highest biodegradability rates among regularly used chelators, with a half-life of 2–6 days and a cumulative degradation rate of 80% within 14 days (Brynhildsen and Rosswall 1997). Several studies have investigated the effect of various metal-chelating agents on *Brassica* sp. metal absorption and translocation. Increased metal bioavailability with decreased plant biomass production is one of the most often seen adverse effects. Table 14.6 shows the effect of several chelators on the enhancement of phytoremediation potential of various members of plants of Brassicaceae.

14.6.2 Genetic Engineering Strategies for Plant Modification to Improve Phytoremediation

Plants are naturally capable of removing hazardous pollutants from the environment. However, the rate of bioremediation is proportional to the rate of plant growth and overall biomass, rendering the process passive. Genetic engineering is a robust approach that has the potential to accelerate the creation of new plant lines with desired phytoremediation properties. Genetic engineering might be a revolutionary approach using cisgenesis and intragenesis in phytoremediation plants. These novel plant genetic modification technologies have improved crop quality and biotic stress tolerance (Rai et al. 2020). Plants of the Brassicaceae family, such as *Arabidopsis* and *Thlaspi*, are attractive candidates for genetic modifications due to the scientific understanding of their genetics and close evolutionary ties. Plants of Brassicaceae family have been investigated extensively for their ability to tolerate HMs, and various ecotypes of hyperaccumulator species have been reported (Nedjimi 2021). Recent improvements in the genetic mapping of maize make it simpler to develop plants with Cd (cadmium) accumulation in leaf tissues (Ojuederie et al. 2022). But these novel techniques can overcome transgenesis issues and accelerate research on genetic alterations to improve phytoremediation efficacy (Srivastava et al. 2021; Raza et al. 2020).

14.6.3 Enhancement in the Phytoremediation Potential of *Brassica juncea* L. by Brassinosteroids

Brassinosteroids (BRs) are plant-based polyhydroxy steroidal hormones that protect against the harmful effects produced by reactive oxygen species via enhancing the synthesis of antioxidant enzymes like CAT, SOD, and POD as a defensive mechanism in *Brassica juncea* L. plants subjected to Zn-induced stress (Sharma et al. 2007; Rajewska et al. 2016). Various research studies have shown that BRs have stress-protective capabilities in plants when subjected to several biotic and abiotic stressors such as HMs and metalloids, drought, salt, excessive temperature, diseases, etc. (Sharma et al. 2007). Plants are said to produce phytochelatins when exposed to BRs. BRs are of several varieties and belong to the C28 group, i.e., 6-oxo and 7-oxolactone types identified from leaves of *Brassica juncea* L. grown under As and Ni stress (Kanwar et al. 2015).

14.6.4 Nanotechnological Approaches to Enhance Phytoremediation

Nano-phytoremediation is a novel method for increasing plant ability to thrive in a contaminated environment and accumulate contaminants in plant tissues. Nanoparticles have shown promise in the management of contaminated agricultural lands as well as in the stimulation of plant growth and development (Chen et al. 2009). Nanoparticles (NPs) improved plant stress tolerance and aid in phytoremediation and toxicity reduction. Nanostructured silicon dioxide has shown feasibility in increasing phytoremediation potential and obtaining the desired effects (Bao-Shan et al. 2004). Salicylic acid-based NPs exhibited better As remediation in *Isatis cappadocica* (Souri et al. 2017), whereas nano-Zn particles increased As stability in *Helianthus annuus* (Vítková et al. 2018). Nanotitanium (Ti) composites such as Zr-TiO_2 and TiO_2-Fe_2O_3Ce-Ti oxide are commonly employed to remediate As-contaminated water (Ashraf et al. 2019). TiO_2, Si-NPs, and Au-NPs have been identified to mitigate the detrimental impacts of some metals on *Zea mays* (Lian et al. 2019), *Glycine max* (Li et al. 2020), and *Oryza sativa* (Jiang et al. 2021), respectively. The usage of fullerene nanoparticles might enhance the phytoavailability of soil contaminants (Song et al. 2019).

14.7 Conclusion

HM contamination threatens existing environmental security and food safety worldwide. Various physicochemical approaches have been developed and applied to remediate HM contamination. Soil and water contamination from HMs is a matter of concern since it is associated with various toxicities causing adverse consequences pertaining to human health. Phytoremediation is a long-term plant-based technique for solving pollution-related problems in various ecosystems. This chapter is an attempt to review the biochemical processes and physiological approaches such as phytostabilization, phytovolatilization, and phytoextract mechanisms used by Indian mustard (*Brassica juncea* L.) to remediate toxic effects of HMs and metalloids. Several research studies have concluded that *Brassica juncea* L. is an effective phytoremediator owing to physiological processes and an efficient enzymatic and nonenzymatic defense mechanism in response to oxidative damage caused by excessive generation of reactive oxygen species due to HMs and metalloids. The available literature also reveals various phytoremediation assistive synergistic methods such as biochar, DTPA, EDTA, brassinosteroids, bioinoculants, agronomic practices, etc. that are employed in different *Brassica* species, including *Brassica juncea* L. to enhance the uptake of HMs and metalloids and can be a promising strategy for improving phytoextraction efficiency, but such modifications may have certain disadvantages too. Intensive research must be conducted to recognize the molecular mechanisms of plant–bacteria interactions involved in plant growth, solubilization, absorption, translocation of HMs, etc. Novel molecular and nanotechnological techniques must be applied to develop transgenic hyper-accumulator plants that can increase the efficiency of phytoremediation processes for the sustainable management of degraded soil ecosystems. However, besides the phytoremediation potential, Indian mustard (*Brassica juncea* L.) has diverse food and medicinal applications; thus, mustard plants used in phytoremediation require vigil and should be discouraged from being used in different food chains and ethnobotanical applications.

References

Abbaspour A, Arocena JM, Kalbasi M (2012) Uptake of phosphorus and lead by *Brassica juncea* and *Medicago sativa* from chloropyromorphite. Int J Phytoremediation 14(6):531–542

Ahmad Z, Khan SM, Page S (2021) Politics of the natural vegetation to balance the hazardous level of elements in marble polluted ecosystem through phytoremediation and physiological responses. J Hazard Mater 414:125451

Alam MN, Bristi NJ, Rafifiquzzaman M (2013) Review on in vivo and in vitro methods evaluation of antioxidant activity. Saudi Pharm J 21(2):143–152

Ali H, Khan E, Sajad MA (2013) Phytoremediation of heavy metals—concepts and applications. Chemosphere 91(7):869–881

Ali S, Abbas Z, Rizwan M, Zaheer IE, Yavaş İ, Ünay A, Kalderis D (2020) Application of floating aquatic plants in phytoremediation of heavy metals polluted water: a review. Sustainability 12(5)

Ali I, Khan MJ, Shah A, Deeba F, Hussain H, Yazdan F, Khan MD (2022) Screening of various *brassica species* for phytoremediation of heavy metals-contaminated soil of Lakki Marwat. Pakistan Environ Sci Pollut Res 29(25):37765–37776

Anjum NA, Umar S, Iqbal M (2014) Assessment of cadmium accumulation, toxicity, and tolerance in Brassicaceae and *Fabaceae* plants—implications for phytoremediation. Environ Sci Pollut Res 21(17):10286–10293

Anjum NA, Gill SS, Ahmad I, Pacheco M, Duarte AC, Umar S, Khan NA, Pereira ME (2012) The plant family Brassicaceae: an introduction. 1–33. In: Hasanuzzaman M (ed) The plant family Brassicaceae. Springer, Singapore

Ansari MKA, Ahmad A, Umar S, Zia MH, Iqbal M, Owens G (2015) Genotypic variation in phytoremediation potential of Indian mustard exposed to nickel stress: a hydroponic study. Int J Phytoremediation 17(2):135–144

Ariyakanon N, Winaipanich B (2006) Phytoremediation of copper contaminated soil by *Brassica juncea* (L.) Czern and Bidens alba (L.) DC. var radiata J Sci Res Chula Univ 31(1):49–56

Ashraf S, Siddiqa A, Shahida S, Qaisar S (2019) Titanium-based nanocomposite materials for arsenic removal from water: a review. Heliyon 5:e01577

Awan B, Sabeen M, Shaheen S, Mahmood Q, Ebadi, Toughani M (2020) Phytoremediation of zinc contaminated water by marigold (*Tagetes Minuta* L). Cent Asian J Environ Sci Technol Innov 1(3):150–158

Bao-Shan L, Shao-Qi D, Chun-Hui L, Li-Jun F, Shu-Chun Q, Min Y (2004) Effect of TMS (nanostructured silicon dioxide) on growth of Changbai larch seedlings. J For Res 15(2): 138–140

Bassegio C, Campagnolo MA, Schwantes D, Gonçalves Junior AC, Manfrin J, Schiller ADP, Bassegio D (2020a) Growth and accumulation of Pb by roots and shoots of *Brassica juncea* L. Int J Phytoremediation 22(2):134–139

Bassegio C, Santos RF, Bassegio D, de Souza SNM (2020b) Genotypic variation in growth and lead accumulation among *Brassica juncea* accessions. Int J Phytoremediation 22(12):1249–1258

Bauddh K, Singh RP (2012) Cadmium tolerance and its phytoremediation by two oil yielding plants *Ricinus communis* (L.) and *Brassica juncea* (L.) from the contaminated soil. Int J Phytoremediation 14(8):772–785

Belimov AA, Hontzeas N, Safronova VI, Demchinskaya SV, Piluzza G, Bullitta S, Glick BR (2005) Cadmium-tolerant plant growth-promoting bacteria associated with the roots of Indian mustard (*Brassica juncea* L. Czern.). Soil Biol Biochem 37(2):241–250

Bluskov S, Arocena JM, Omotoso OO, Young JP (2005) Uptake, distribution, and speciation of chromium in *Brassica juncea*. Int J Phytoremediation 7(2):153–165

Bouquet D, Braud A, Lebeau T (2017) *Brassica juncea* tested on urban soils moderately contaminated by lead: origin of contamination and effect of chelates. Int J Phytoremediation 19(5):425–430

Bortoloti GA, Baron D (2022) Phytoremediation of toxic heavy metals by brassica plants: a biochemical and physiological approach. Environ Adv 8:100204

Brynhildsen L, Rosswall T (1997) Effects of metals on the microbial mineralization of organic acids. Water Air Soil Pollut 94:45–57

Cahoon RE, Lutke WK, Cameron JC, Chen S, Lee SG, Rivard RS, Jez JM (2015) Adaptive engineering of phytochelatin-based heavy metal tolerance. J Biol Chem 290(28):17321–17330

Cao L, Jiang M, Zeng Z, Du A, Tan H, Liu Y (2008) *Trichoderma atroviride* F6 improves phytoextraction efficiency of mustard (*Brassica juncea* (L.) Coss. Var. *foliosa bailey*) in cd, Ni contaminated soils. Chemosphere 71(9):1769–1773

Chaiane B, Marcelo AC, Daniel S, Affonso CGJ, Jéssica M, Andreia DPS, Doglas B (2020) Growth and accumulation of Pb by roots and shoots of *Brassica juncea* L. Int J Phytoremediation 22(2): 134–139

Chatuverdi R, Favas P, Pratas J, Varun M, Paul MS (2019) EDTA-assisted metal uptake in *Raphanus sativus* L. and *Brassica oleracea* L: assessment of toxicity and food safety. Bull Environ Contam Toxicol 103(3):490–495

Chen L, Long C, Wang D, Yang J (2020) Phytoremediation of cadmium (Cd) and uranium (U) contaminated soils by *Brassica juncea* L. enhanced with exogenous application of plant growth regulators. Chemosphere 242:125112

Chen S, Chen L, Ma Y, Huang Y (2009) Can phosphate compounds be used to reduce the plant uptake of Pb and resist the Pb stress in Pb-contaminated soils? J Environ Sci 21(3):360–365

Clemente R, Walker DJ, Bernal MP (2005) Uptake of heavy metals and as by *Brassica juncea* grown in a contaminated soil in Aznalcóllar (Spain): the effect of soil amendments. Environ Pollut 138(1):46–58

Cristaldi A, Conti GO, Jho EH, Zuccarello P, Grasso A, Copat C, Ferrante M (2017) Phytoremediation of contaminated soils by heavy metals and PAHs. A brief review. Environ Technol Innov 8:309–326

Derakhshan NZ, Jung MC, Kim KH (2018) Remediation of soils contaminated with heavy metals with an emphasis on immobilization technology. Environ Geochem Health 40(3):927–953

Dhanwal P, Kumar A, Dudeja S, Chhokar V, Beniwal V (2017) Recent advances in phytoremediation technology. 227-241. In: Kumar R, Sharma A, Ahluwalia S (eds) Advances in environmental biotechnology. Springer, Singapore

Diarra I, Kotra KK, Prasad S (2021) Assessment of biodegradable chelating agents in the phytoextraction of heavy metals from multi–metal contaminated soil. Chemosphere 273:128483

Diwan H, Ahmad A, Iqbal M (2008) Genotypic variation in the phytoremediation potential of Indian mustard for chromium. Environ Manag 41(5):734–741

Du J, Guo Z, Li R, Ali A, Guo D, Lahori AH, Wang P, Liu X, Wang X, Zhang Z (2020) Screening of Chinese mustard (*Brassica juncea* L.) cultivars for the phytoremediation of cd and Zn based on the plant physiological mechanisms. Environ Pollut 261:114213

Dwivedi G, Sharma MP (2014) Potential and limitation of straight vegetable oils as engine fuel–an Indian perspective. Renew Sust Energ Rev 33:316–322

Eapen S, Suseelan KN, Tivarekar S, Kotwal SA, Mitra R (2003) Potential for rhizofiltration of uranium using hairy root cultures of Brassica juncea and Chenopodium amaranticolor. Environ Res 91(2):127–133

Ekta P, Modi NR (2018) A review of phytoremediation. J Pharmacogn Phytochem 7(4):1485–1489

Farahani M, Naderi R, Mazhari M (2015) Phytoremediation of cd contaminated soils by ornamental cabbage (*Brassica oleracea*) species. J Biodivers Environ Sci 7(2):150–155

Favela-González KM, Hernández-Almanza AY, De la Fuente-Salcido NM (2020) The value of bioactive compounds of cruciferous vegetables (Brassica) as antimicrobials and antioxidants: a review. J Food Biochem 44(10):e13414

Garcia S, Kosnar M, Mercl Z, Aranda F, Tlustos P (2018) A comparative study to evaluate natural attenuation, mycoaugmentation, phytoremediation, and microbialassisted phytoremediation strategies for the bioremediation of an aged PAH-polluted soil. Ecotoxicol Environ Saf 147: 165–174

Gill SS, Tuteja N (2010) Reactive oxygen species and antioxidant machinery in abiotic stress tolerance in crop plants. Plant Physiol Biochem 48(12):909–930

Gonzaga MI, de Jesus Santos JC, Ganassali Junior LF, Fontes PT, Araujo JD, Gonzaga TA (2022) Copper uptake, physiological response, and phytoremediation potential of Brassica juncea under biochar application. Int J Phytoremediation 24(5):474–482

Goswami S, Das S (2015) A study on cadmium phytoremediation potential of Indian mustard, *Brassica juncea*. Int J Phytoremediation 17(6):583–588

Guerra Sierra BE, Muñoz Guerrero J, Sokolski S (2021) Phytoremediation of heavy metals in tropical soils an overview. Sustainability 13(5):2574

Guo D, Ali A, Ren C, Du J, Li R, Lahori AH, Xiao R, Zhang Z, Zhang Z (2019) EDTA and organic acids assisted phytoextraction of cd and Zn from a smelter contaminated soil by potherb mustard (*Brassica juncea* Coss.) and evaluation of its bioindicators. Ecotoxicol Environ Safe 167:396–403

Guo H, Jiang J, Gao J, Zhang J, Zeng L, Cai M, Zhang J (2020) Evaluation of cadmium hyperaccumulation and tolerance potential of *Myriophyllum aquaticum*. Ecotoxicol Environ Safe 195:110502

Gupta M, Sharma P, Sarin NB, Sinha AK (2009) Differential response of arsenic stress in two varieties of *Brassica juncea* L. Chemosphere 74(9):1201–1208

Hall JL (2002) Cellular mechanisms for heavy metal detoxification and tolerance. J Exp Bot 53:111

Harbaum B, Hubbermann EM, Zhu Z, Schwarz K (2008) Impact of fermentation on phenolic compounds in leaves of pak choi (Brassica campestris L. ssp. chinensis var. communis) and Chinese leaf mustard (Brassica juncea Coss). J Agric Food Chem 56(1):148–157

Hasanuzzaman M, Bhuyan MHM, Raza A, Hawrylak-Nowak B, Matraszek-Gawron R, Nahar K, Fujita M (2020) Selenium toxicity in plants and environment: biogeochemistry and remediation possibilities. Plan Theory 9(12):1711

Hoang SA, Lamb D, Seshadri B, Sarkar B, Choppala G, Kirkham MB, Bolan NS (2021) Rhizoremediation as a green technology for the remediation of petroleum hydrocarbon-contaminated soils. J Hazard Mater 401:123282

Huang D, Gong X, Liu Y, Zeng G, Lai C, Bashir H, Zhou L, Wang D, Xu P, Cheng M (2017) Effects of calcium at toxic concentrations of cadmium in plants. Planta 245(5):863–873

Hur SJ, Lee SY, Kim YC, Choi I, Kim GB (2014) Effect of fermentation on the antioxidant activity in plant-based foods. Food Chem 160:346–356

Iqbal N, Masood A, Nazar R, Syeed S, Khan NA (2010) Photosynthesis, growth and antioxidant metabolism in mustard (*Brassica juncea* L.) cultivars differing in cadmium tolerance. Agric Sci China 9(4):519–527

Jiang J, Wang Y, Zhu B, Fang T, Fang Y, Wang Y (2015) Digital gene expression analysis of gene expression differences within brassica diploids and allopolyploids. BMC Plant Biol 15:22

Jiang M, Dai S, Wang B, Xie Z, Li J, Wang L, Li S, Tan Y, Tian B, Shu Q (2021) Gold nanoparticles synthesized using melatonin suppress cadmium uptake and alleviate its toxicity in rice. Environ Sci Nano 8:1042–1056

Johnson A, Gunawardana B, Singhal N (2009) Amendments for enhancing copper uptake by *Brassica juncea* and *Lolium perenne* from solution. Int J Phytoremediation 11(3):215–234

Kafle A, Timilsina A, Gautam A, Adhikari K, Bhattarai A, Aryal N (2022) Phytoremediation: mechanisms, plant selection and enhancement by natural and synthetic agents. Environ Adv 8: 100203

Kanwar MK, Poonam, Bhardwaj R (2015) Arsenic induced modulation of antioxidative defense system and brassinosteroids in *Brassica juncea* L. Ecotoxicol Environ Saf 115:119–125

Khator K, Saxena I, Shekhawat GS (2021) Nitric oxide induced cd tolerance and phytoremediation potential of *B. juncea* by the modulation of antioxidant defense system and ROS detoxification. Biometals 34(1):15–32

Kidwai MK, Dhull SB (2021) Heavy metal induced stress and metabolic responses in fenugreek (*Trigonella foenum-graceum* L.) Plants. In: Naeem M, Aftab T, Khan MMA (eds) Fenugreek, Biology and applications. Springer, Singapore, pp 327–354

Kidwai MK, Malik A, Dhull SB, Rose PK, Garg VK (2022) Bioremediation potential of Trichoderma species for metal(loid)s. In: Malik A, Kidwai MK, Garg VK (eds) Bioremediation of toxic metal(loid)s. CRC Press, pp 137–152

Ko BG, Anderson CW, Bolan NS, Huh KY, Vogeler I (2008) Potential for the phytoremediation of arsenic-contaminated mine tailings in Fiji. Soil Res 46(7):493–501

Kohli SK, Handa N, Sharma A, Gautam V, Arora S, Bhardwaj R, Wijaya L, Alyemeni MN, Ahmad P (2018) Interaction of 24-epibrassinolide and salicylic acid regulates pigment contents, antioxidative defense responses, and gene expression in Brassica juncea L. seedlings under Pb stress. Environ Sci Pollut Res 25:15159–15173

Koptsik GN (2014) Problems and prospects concerning the phytoremediation of heavy metal polluted soils: a review. Eurasian Soil Sci 47(9):923–939

Kumar KV, Singh N, Behl HM, Srivastava S (2008) Influence of plant growth promoting bacteria and its mutant on heavy metal toxicity in *Brassica juncea* grown in fly ash amended soil. Chemosphere 72(4):678–683

Kumar Y, Gupta K, Kumar N, Reece A, Singh LM, Rezania S, Ahmad KS (2018) Mechanistic understanding and holistic approach of phytoremediation: a review on application and future prospects. Ecol Eng 120:274–298

Laghlimi M, Baghdad B, El Hadi H, Bouabdli A (2015) Phytoremediation mechanisms of heavy metal contaminated soils: a review. Open J Ecol 5(08):375

Lai HY, Chen SW, Chen ZS (2008) Pot experiment to study the uptake of cd and Pb by three Indian mustards (*Brassica juncea*) grown in artificially contaminated soils. Int J Phytoremediation 10(2):91–105

Lee SY, Choi JH, Choi YS, Han DJ, Kim HY, Shim SY, Chung HK, Kim CJ (2010) The antioxidative properties of mustard leaf (*Brassica juncea*) kimchi extracts on refrigerated raw ground pork meat against lipid oxidation. Meat Sci 84(3):498–504

Li JT, Gurajala HK, Wu LH, van der Ent A, Qiu RL, Baker AJM, Tang YT, Yang XE, Shu WS (2018) Hyperaccumulator plants from China: a synthesis of the current state of knowledge. Environ Sci Technol 52:11980–11994

Li Y, Zhu N, Liang X, Bai X, Zheng L, Zhao J, Li YF, Zhang Z, Gao Y (2020) Silica nanoparticles alleviate mercury toxicity via immobilization and inactivation of hg(ii) in soybean (Glycine max). Environ Sci Nano 7:1807–1817

Lian J, Zhao L, Wu J, Xiong H, Bao Y, Zeb A, Tang J, Liu W (2019) Foliar spray of TiO2 nanoparticles prevails over root application in reducing cd accumulation and mitigating cd-induced phytotoxicity in maize (*Zea mays* L.). Chemosphere 239:124794

Ma Y, Rajkumar M, Freitas H (2009) Inoculation of plant growth promoting bacterium *Achromobacter xylosoxidans* strain Ax10 for the improvement of copper phytoextraction by *Brassica juncea*. J Environ Manag 90(2):831–837

Monei NL, Puthiya Veetil SK, Gao J, Hitch M (2021) Selective removal of selenium by phytoremediation from post/mining coal wastes: practicality and implications. Int J Min Reclam Environ 35(1):69–77

Mahmud JA, Hasanuzzaman M, Nahar K, Borhannuddin Bhuyan MHM, Fujita M (2018) Insights into citric acid-induced cadmium tolerance and phytoremediation in *Brassica juncea* L.: coordinated functions of metal chelation, antioxidant defense and glyoxalase systems. Ecotoxicol Environ Saf 147:990–1001

Manori S, Shah V, Soni V, Dutta K, Daverey A (2021) Phytoremediation of cadmium-contaminated soil by *Bidens pilosa* L.: impact of pine needle biochar amendment. Environ Sci Pollut Res 28(42):58872–58884

Marchiol L, Assolari S, Sacco P, Zerbi G (2004) Phytoextraction of heavy metals by canola (*Brassica napus*) and radish (*Raphanus sativus*) grown on multicontaminated soil. Environ Pollut 132(1):21–27

Meyers DE, Auchterlonie GJ, Webb RI, Wood B (2008) Uptake and localisation of lead in the root system of *Brassica juncea*. Environ Pollut 153(2):323–332

Mobin M, Khan NA (2007) Photosynthetic activity, pigment composition and antioxidative response of two mustard (*Brassica juncea*) cultivars differing in photosynthetic capacity subjected to cadmium stress. J Plant Physiol 164(5):601–610

Mohamed AA, Castagna A, Ranieri A, di Toppi LS (2012) Cadmium tolerance in Brassica juncea roots and shoots is affected by antioxidant status and phytochelatin biosynthesis. Plant Physiol Biochem 57:15–22

Moosavi SG, Seghatoleslami MJ (2013) Phytoremediation: a review. Adv Agri Biol 1(1):5–11

Moreno FN, Anderson CWN, Stewart RB, Robinson BH (2008) Phytofiltration of mercury-contaminated water: volatilisation and plant-accumulation aspects. Environ Exp Bot 62:78–85

Mourato MP, Moreira IN, Leitão I, Pinto FR, Sales JR, Martins LL (2015) Effect of heavy metals in plants of the genus brassica. Int J Mol Sci 16:17975–17998

Mustafa HM, Hayder G (2021) Recent studies on applications of aquatic weed plants in phytoremediation of wastewater: a review article. Ain Shams Eng J 12(1):355–365

Muthusaravanan S, Sivarajasekar N, Vivek JS, Paramasivan T, Naushad M, Prakashmaran J, Al-Duaij OK (2018) Phytoremediation of heavy metals: mechanisms, methods and enhancements. Environ Chem Lett 16(4):1339–1359

Nanjundan J, Aravind J, Radhamani J, Singh KH, Kumar A, Thakur AK, Singh D (2022) Development of Indian mustard [*Brassica juncea* (L.) Czern.] core collection based on agro-morphological traits. Genet Resour Crop Evol 69(1):145–162

Napoli M, Cecchi S, Grassi C, Baldi A, Zanchi CA, Orlandini S (2018) Phytoextraction of copper from a contaminated soil using arable and vegetable crops. Chemosphere 219:122–129

Natasha S, Niazi M, Khalid NK, Murtaza S, Bibi I, Rashid MI (2018) A critical review of selenium biogeochemical behavior in soil-plant system with an inference to human health. Environ Pollut 234:915–934

Nazir MM, Ulhassan Z, Zeeshan M, Ali S, Gill MB (2020) Toxic metals/metalloids accumulation, tolerance, and homeostasis in brassica oilseed species. 379–408. In: Hasanuzzaman M (ed) The plant family Brassicaceae. Springer, Singapore

Nedjimi B (2021) Phytoremediation: a sustainable environmental technology for heavy metals decontamination. SN Appl Sci 3(3):1–19

Niazi NK, Bibi I, Fatimah A, Shahid M, Javed MT, Wang H, Ok YS, Bashir S, Murtaza B, Saqib ZA, Shakoor MB (2017) Phosphate-assisted phytoremediation of arsenic by Brassica napus and Brassica juncea: morphological and physiological response. Int J Phytoremediation 19(7):670–678

Oguntade OA, Olagbenro TS, Odusanya OA, Olagunju SO, Adewusi KM, Adegoke AT (2019) Assessment of composted kitchen waste and poultry manure amendments on growth, yield and heavy metal uptake by jute mallow Corchorus olitorius Linn. Int J Recycl Org Waste Agric 8: 187–195

Ojuederie OB, Amoo AE, Owonubi SJ, Ayangbenro AS (2022) Nanoparticles-assisted phytoremediation: advances and applications. In: Vimal Pandey V (ed) Assisted phytoremediation. Elsevier, pp 155–178

Ontañon OM, González PS, Ambrosio LF, Paisio CE, Agostini E (2014) Rhizoremediation of phenol and chromium by the synergistic combination of a native bacterial strain and *Brassica napus* hairy roots. Int Biodeterior Biodegradation 88:192–198

Ozyigit II, Can H, Dogan I (2020) Phytoremediation using genetically engineered plants to remove metals: a review. Environ Chem Lett 19:669–698

Parmar S, Singh V (2015) Phytoremediation approaches for heavy metal pollution: a review. J Plant Sci Res 2:1–8

Pérez-Esteban J, Escolástico C, Moliner A, Masaguer A, Ruiz-Fernández J (2014) Phytostabilization of metals in mine soils using Brassica juncea in combination with organic amendments. Plant Soil 377:97–109

Picchi C, Giorgetti L, Morelli E, Landi M, Rosellini I, Grifoni M, Barbafieri M (2021) *Cannabis sativa* L. and *Brassica juncea* L. grown on arsenic-contaminated industrial soil: potentiality and limitation for phytoremediation. Environ Sci Pollut Res 29:15983–15998

Rahman M, Jakariya M, Haq IMA (2018) Prospect of phytoaccumulation of arsenic by Brassica juncea (L.) in Bangladesh. Int J Phytoremediation 20(10):1025–1032

Rai PK, Kim KH, Lee SS, Lee JH (2020) Molecular mechanisms in phytoremediation of environmental contaminants and prospects of engineered transgenic plants/microbes. Sci Total Environ 705:135858

Raiola A, Errico A, Petruk G, Monti DM, Barone A, Rigano MM (2017) Bioactive compounds in Brassicaceae vegetables with a role in the prevention of chronic diseases. Molecules 23(1):15

Raj D, Kumar A, Maiti SK (2020) *Brassica juncea* (L.) Czern (Indian mustard): a putative plant species to facilitate the phytoremediation of mercury contaminated soils. Int J Phytoremediation 22(7):733–744

Rajewska I, Talarek M, Bajguz A (2016) Brassinosteroids and response of plants to heavy metals action. Front Plant Sci 7:629

Rathika R, Srinivasan P, Alkahtani J, Al-Humai LA, Alwahibi MS, Mythili R, Selvankumar T (2021) Influence of biochar and EDTA on enhanced phytoremediation of lead contaminated soil by *Brassica juncea*. Chemosphere 271:129513

Raza A, Habib M, Kakavand SN, Zahid Z, Zahra N, Sharif R, Hasanuzzama M (2020) Phytoremediation of cadmium: physiological, biochemical, and molecular mechanisms. Biology 9(7):177

Rose PK, Dhull SB, Kidwai MK (2022a) Utilization of lignocellulosic biomass as substrate for the cultivation of wild mushroom. In: Dhull SB, Bains A, Chawla P, Sadh PK (eds) Wild mushrooms: characteristics, nutrition, and processing. CRC Press

Rose PK, Kidwai MK, Dhull SB (2022b) Food industry waste: potential pollutants and their bioremediation strategies. In: Dhull SB, Singh A, Kumar P (eds) Food processing waste and utilization. CRC Press, pp 343–359

Salido AL, Hasty KL, Lim JM, Butcher DJ (2003) Phytoremediation of arsenic and lead in contaminated soil using Chinese brake ferns (*Pteris vittata*) and Indian mustard (*Brassica juncea*). Int J Phytoremediation 5(2):89–103

Saraswat S, Rai JPN (2009) Phytoextraction potential of six plant species grown in multimetal contaminated soil. Chem Ecol 25(1):1–11

Shah V, Daverey A (2020) Phytoremediation: a multidisciplinary approach to clean up heavy metal contaminated soil. Environ Technol Innov 18:100774

Shahandeh H, Hossner LR (2000) Enhancement of Cr (lll) phytoaccumulation. Int J Phytoremediation 2(3):269–286

Shakeel A, Khan AA, Ahmad G (2019) The potential of thermal power plant fly ash to promote the growth of Indian mustard (*Brassica juncea*) in agricultural soils. SN Appl Sci 1(4):1–5

Shanmugaraj BM, Chandra HM, Srinivasan B, Ramalingam S (2013) Cadmium induced physio-biochemical and molecular response in *Brassica juncea*. Int J Phytoremediation 15(3):206–218

Sharma P, Bhardwaj R, Arora N, Arora HK (2007) Effect of 28-homobrassinolide on growth, zinc metal uptake and antioxidative enzyme activities in *Brassica juncea* L. seedlings. Braz J Plant Physiol 19:203–210

Sharma P, Tripathi S, Chandra R (2020) Phytoremediation potential of heavy metal accumulator plants for waste management in the pulp and paper industry. Heliyon 6(7):e04559

Shivran A, Patel BJ, Gora M (2019) Effect of irrigation schedule and bioregulators on yield attributes and yield of mustard (*B. juncea* (L.) Czern & Coss) crop. Intl J Chem Studies 7: 1874–1877

Song B, Xu P, Chen M, Tang W, Zeng G, Gong J, Zhang P, Ye S (2019) Using nanomaterials to facilitate the phytoremediation of contaminated soil. Crit Rev Environ Sci Technol 49:791–824

Souri Z, Karimi N, Sarmadi M, Rostami E (2017) Salicylic acid nanoparticles (SANPs) improve growth and phytoremediation efficiency of Isatis cappadocica Desv., under as stress. IET Nanobiotechnol 11:650–655

Srivastava S, Shukla A, Rajput VD, Kumar K, Minkina T, Mandzhieva S, Suprasanna P (2021) Arsenic remediation through sustainable phytoremediation approaches. Fortschr Mineral 11(9): 936

Srivastava S, Verma PC, Chaudhry V, Singh N, Abhilash PC, Kumar KV, Singh N (2013) Influence of inoculation of arsenic-resistant *staphylococcus* on growth and arsenic uptake in *Brassica juncea* (L.) Czern. Var. R-46. J Hazard Mater 262:1039–1047

Sterckeman T, Gossiaux L, Guimont S, Sirguey C (2019) How could phytoextraction reduce cd content in soils under annual crops? Simulations in the French context. Sci Total Environ 654: 751–762

Su Y, Han FX, Chen J, Sridhar BM, Monts DL (2008) Phytoextraction and accumulation of mercury in three plant species: Indian mustard (*Brassica juncea*), beard grass (*Polypogon monospeliensis*), and Chinese brake fern (*Pteris vittata*). Int J Phytoremediation 10(6):547–560

Suman J, Uhlik O, Viktorova J, Macek T (2018) Phytoextraction of heavy metals: a promising tool for clean-up of polluted environment? Front Plant Sci 9:1476

Taghizadeh M, Solgi M, Karimi M, Sanati MH, Khoshbin S (2018) Heavy metals effects on *Brassica oleracea* and elements accumulation by salicylic acid. Arch Hyg Sci 7(1):1–11

Tangahu BV, Sheikh ASR, Basri H, Idris M, Anuar N, Mukhlisin M (2011) A review on heavy metals (As, Pb, and Hg) uptake by plants through phytoremediation. Int J Chem Eng 939161:1

Thakur S, Choudhary S, Majeed A, Singh A, Bhardwaj P (2020) Insights into the molecular mechanism of arsenic phytoremediation. J Plant Growth Regul 39(2):532–543

Thiyam U, Stöckmann H, Zum Felde T, Schwarz K (2006) Antioxidative effect of the main sinapic acid derivatives from rapeseed and mustard oil by-products. Eur J Lipid Sci Technol 108 (3):239–248

Tong YP, Kneer R, Zhu YG (2004) Vacuolar compartmentalization: a second–generation approach to engineering plants for phytoremediation. Trends Plant Sci 9:8–9

US Environmental Protection Agency (2000) National Risk Management Research Laboratory (US). Introduction to phytoremediation. National Risk Management Research Laboratory, Office of Research and Development

Vítková M, Puschenreiter M, Komárek M (2018) Effect of nano zero-valent iron application on as, cd, Pb, and Zn availability in the rhizosphere of metal(loid) contaminated soils. Chemosphere 200:217–226

Wanasundara JPD, Mcintosh TC, Perera SP, Withana-Gamage TS, Mitra P (2016) Canola/rapeseed protein-functionality and nutrition. OCL 23(4):D407

Wang C, Yue L, Cheng B, Chen F, Zhao X, Wang Z, Xing B (2022) Mechanisms of growth-promotion and se-enrichment in *Brassica chinensis* L. by selenium nanomaterials: beneficial rhizosphere microorganisms, nutrient availability, and photosynthesis. Environ Sci Nano 9: 302–312

Wu LH, Lou YM, Xing XR, Christie P (2004) EDTA-enhanced phytoremediation of heavy metal contaminated soil with Indian mustard and associated potential leaching risk. Agric Ecosyst Environ 102:307–318

Yadav KK, Gupta N, Kumar A, Reece LM, Singh N, Rezania S, Khan SA (2018) Mechanistic understanding and holistic approach of phytoremediation: a review on application and future prospects. Ecol Eng 120:274–298

Yahaghi Z, Shirvani M, Nourbakhsh F, De La Pena TC, Pueyo JJ, Talebi M (2018) Isolation and characterization of Pb-solubilizing bacteria and their effects on Pb uptake by *Brassica juncea*: implications for microbe-assisted phytoremediation. J Microbiology Biotechnol 28(7): 1156–1167

Yang J, Liu D, Wang X, Ji C, Cheng F, Liu B, Hu Z, Che S, Pental Ju Y, Yao P, Li X, Xie K, Zhang J, Wang J, Liu F, Ma W, Shopan J, Zheng H, Mackenzie SA, Zhang M (2016) The genome sequence of allopolyploid *Brassica juncea* and analysis of differential homolog gene expression influencing selection. Nat Genet 48:1225–1232

Yaashikaa PR, Senthil Kumar P, Varjani S, Saravanan A (2020) Rhizoremediation of cu (II) ions from contaminated soil using plant growth promoting bacteria: an outlook on pyrolysis conditions on plant residues for methylene orange dye biosorption. Bioengineered 11(1): 175–187

Zahoor M, Irshad M, Rahman H, Qasim M, Afridi SG, Qadir M, Hussain A (2017) Alleviation of heavy metal toxicity and phytostimulation of Brassica campestris L. by endophytic Mucor sp. MHR-7. Ecotoxicol Environ Saf 142:139–149

Zeremski T, Randelović D, Jakovljević K, Marjanović JA, Milić S (2021) *Brassica* species in Phytoextractions: real potentials and challenges. Plan Theory 10:2340

Zeremski-Škorić T, Sekulić P, Maksimović I, Šeremešić S, Ninkov J, Milić S, Vasin J (2010) Chelate-assisted phytoextraction: effect of EDTA and EDDS on copper uptake by *Brassica napus* L. J Serbian Chem Soc 75:1279–1289

Zhang X, Li M, Yang H, Li CZ (2018) Physiological responses of Suaeda glauca and Arabidopsis thaliana in phytoremediation of heavy metals. J Environ Manag 223:132–139

Zhu Y, Pilon Smits EAH, Tarun A, Weber SU, Juanin L, Terry N (1999) Cadmium tolerance and accumulation in Indian mustard is enhanced by overexpressing glutamyl cysteine synthetase. Plant Physiol 121:1169–1177

Zaidi S, Usmani S, Singh BR, Musarrat J (2006) Significance of Bacillus subtilis strain SJ-101 as a bioinoculant for concurrent plant growth promotion and nickel accumulation in *Brassica juncea*. Chemosphere 64(6):991–997

Zunaidi AA, Lim LH, Metali F (2021) Transfer of heavy metals from soils to curly mustard (*Brassica juncea* (L.) Czern.) grown in an agricultural farm in Brunei Darussalam. Heliyon 7(9):e07945

Phytoremediation of Heavy Metals 15

Manu Pant, Sonam Dolma, Monika Gahlot, Aditi Sharma, and Shreya Mundepi

Abstract

Heavy metals are nonbiodegradable metals (like Cd, Hg, Zn, Cu, Ni), nonmetals (like Se), and metalloids (like As) that are highly toxic to the environment in both their elemental and soluble salt forms. These heavy metals pose a huge threat to the environment, calling for appropriate treatment solutions. Phytoremediation is an emerging technology that employs plants for heavy metal detoxification with the advantages of being less expensive, time-saving, eco-friendly, and having a long application. Phytoremediation entails the cultivation of plants at contaminated sites and subsequent harvesting, to permanently eliminate accumulated heavy metals from the contaminated sites refortifying the vegetative potential of the area. *Hordeum vulgare*, *Brassica juncea*, *Euphorbia* spp., and other common plants have been examined for this purpose. Furthermore, many medicinal plants such as *Centella asiatica*, *Orthosiphon stamineus*, *Hypericum perforatum*, etc. have shown a higher capacity of accumulating heavy metals in cleaning up the environmental pollutants. The information provided in this chapter covers the concept and applications of phytoremediation, underlying mechanism of phytoaccumulation, phytoextraction, phytostabilization, phytodegradation, and plants used for phytoremediation, factors affecting phytoremediation potential, and analogous strategies involved in improving the phytoremediation potential of plants.

M. Pant (✉) · S. Dolma · M. Gahlot · S. Mundepi
Department of Life Sciences, Graphic Era Deemed to be University, Dehradun, Uttarakhand, India
e-mail: manupant@geu.ac.in

A. Sharma
Genetics and Tree Improvement Division, Forest Research Institute, Dehradun, Uttarakhand, India

R. P. Singh et al. (eds.), *Heavy Metal Toxicity: Environmental Concerns, Remediation and Opportunities*, https://doi.org/10.1007/978-981-99-0397-9_15

Keywords

Phytoremediation · Heavy metals · Phytoremediation mechanisms · Medicinal plants · Phytoremediation potential

15.1 Introduction

As a result of increased manufacturing and globalization, environmental degradation around the world is of common occurrence. Heavy metal contamination is abundant due to numerous natural (geological) and anthropogenic processes, which not only damage soil structure and productivity but also enter the food chain via crops and concentrates inside the body of the humans via "biomagnification." Heavy metals (HMs) are metallic elements having atomic numbers more than 20 and correspondingly high densities. They are nonbiodegradable, survive in the soil over long periods of time, and give a fatal influence on mortal organisms. As a result, soil, which serves as a disposal location for the majority of heavy metals, needs to be treated for HM removal. Traditional methods like soil washing, excavation, leaching, solidification, electric field application, etc. demand a large amount of energy and expensive machinery, besides being time-consuming, labor-intensive, and frequently leading to secondary contamination and the destruction of natural habitats. As a result, plant-based remediation can be used as a viable alternative remediation approach because it is nondestructive to the environment and economical (Salt et al. 1995). Phytoremediation entails the use of specific plants in extracting and eliminating HM contaminants out of the environment, as well as reducing the bioavailability within the soil (Berti and Cunningham 2000). Many plants like *Brassica juncea*, *Alyssum murale*, *Thlaspi caerulescens*, *Pteris vittata*, etc. are identified as HM hyperaccumulators and are being employed as phytoremediators. Furthermore, several therapeutic plants (coriander, mint, aloe vera, lavender, etc.) are more resistant to HMs than other crops. Phytoremediation has made significant progress in the successful revegetation of HM-contaminated locations. The strategy involves contaminant removal through phytoaccumulation, phytoextraction, phytostabilization, phytovolatilization, rhizofiltration, and rhizodegradation. Which suitable strategy has to be used is contingent to the type of pollutants, their bioavailability, plant species, soil conditions, and climate (Laghlimi et al. 2015). With recent advances in the biotechnological applications, transgenic plants are being developed via genetic engineering with an improved ability to decontaminate metal-polluted sites.

The present chapter gives an insight into the concept, mechanism, and application of phytoremediation as a green solution to HM contamination in the environment and discusses recent advances and the scope of future studies in this area.

15.2 Heavy Metals and Their Sources

HMs are natural components found in the earth's crust possessing high density, atomic weight, and toxicity in comparison to other naturally occurring elements (Tchounwou et al. 2012). With their persistent nature and potential to cause a high degree of injury to the organisms, HM toxicity has been realized as a serious abiotic environmental stress (Jadia and Fulekar 2009; Igiri et al. 2018). The typically hazardous HM pollutants are arsenic (As), cadmium (Cd), chromium (Cr), copper (Cu), lead (Pb), mercury (Hg), and nickel (Ni). Of these, Pb, Cd, and Hg are toxic and biologically nonessential HMs (Ali et al. 2019), while others are micronutrients vital for plant growth and development. However, if their concentration exceeds the optimal range, it leads to serious impacts on the functioning of cellular organelles and components in biological systems, eventually killing the organism. Furthermore, some HMs, such as Cd, are harmful even in lower quantities (Verma and Dwivedi 2013) suggesting their potential threat conditions.

The entrance of HMs into the environment is attributed to natural and anthropogenic conditions. Weathering of rocks, seepage, forest fire, erosion, and volcanic eruptions are examples of natural sources of HMs. Anthropogenic activities account for exceptionally high levels of HMs in the environment (Garrett 2000). These include petroleum production, vehicle exhaust, waste engine oil release, smelting, oil spills, unscientific use of fertilizers and pesticides, and industrial wastewater effluents which release HMs contaminating the mainland by dry and wet deposition and into the water bodies (Ali et al. 2019). Long-term irrigation of agricultural lands with contaminated wastewater contaminates the agricultural vegetation leading to subsequent health risks to humans and animals through their consumption (Jadia and Fulekar 2009; Waheed et al. 2019).

15.3 Traditional Strategies for Removing Heavy Metals

The common methods for HM removal from the polluted site include physical, chemical, biological methods.

15.3.1 Excavation

This is the oldest method involving physically removing contaminants out of the soil (Dhaliwal et al. 2019). The method involves any of the three options: (1) replacement of contaminated soil with fresh soil (appropriate for the small contaminated area), (2) mixing of fresh soil with old contaminated soil to reduce the effect of toxic HMs, and (3) natural degradation of metals by the deep excavation of contaminated sediments.

15.3.2 Capping

A cost-effective physical method of remediation to restrict HM mobilization, solubilization and translocation in the sediment. Subaqueous form is applied to the contaminated region; apatite and sandy material are tiered in appropriate proportion and put like a cap onto the contaminated sediment (Vandenbossche et al. 2014).

15.3.3 Immobilization

A chemical method to eliminate the HM contaminants out of the sediment, this remediation task involves inorganic and organic reagents to decline the bioavailability of toxic compounds in the contaminated area. The main concern of this method is to lower the bioavailability phase of metal into a more geochemical biostable phase by using reagents in immobilized form (Lwin et al. 2018). Mechanisms like absorption, complexation, and precipitation are used to achieve a good result. Reagents like silico-calcium, phosphate ions, and mineral-based amendments are commonly used for this purpose (Vandenbossche et al. 2014).

15.3.4 Vertification

This process involves high thermal power or energy for the neutralization of volatile toxic compounds in the contaminated soil. During the procedure, the contaminated compound is mixed with glass-forming matters at a high temperature until a liquid solution is prepared. The process is time-consuming, expensive, and challenging to perform (Lwin et al. 2018).

15.3.5 Electrokinetic

An in situ chemical remediation method, which implicates the use of a low-voltage direct electric field current on a wet soil matrix for mobilizing metal ions toward the anode and cathode. The efficiency of this method relies on the type of soil, soil age, and composition of sediments. This technique shows its highest efficiency in fine-grained clayey soil; in this soil, HMs persist in soluble form as a high and strong electric field is applied. Chelating agents like EDTA and nitriloacetic acid are used to strengthen the effect of this process (Lwin et al. 2018; Peng et al. 2018).

15.3.6 Biological Methods

These methods involve the removal of contamination by biological means mainly microflora and more lately plants. An appropriate environment is imperative for the growth of organisms (Sylvia 2005). Bioremediation is the eco-friendly, green

Table 15.1 Commonly studied microorganisms for bioremediation

S. No.	Name of microorganism	Microorganism type	Target heavy metal	Reference
1.	*Alcaligenes eutrophus*	Bacteria (Gram-negative)	Cd, Co, Zn	Nies et al. 1989
2.	*Aspergillus* spp.	Fungi	Ni	Taştan et al. 2010
3.	*Bacillus firmus*	Bacteria (Gram-positive)	Zn, Cr	Salehizadeh and Shojaosadati 2003
4.	*Methylobacterium organophilum*	Bacteria (Gram-negative)	Pb, Hg	Salehizadeh and Shojaosadati 2003
5.	*Pseudomonas aeruginosa*	Bacteria (Gram-negative)	Cd, Zn, Hg	Pardo et al. 2003
6.	*Pseudomonas putida*	Bacteria (Gram-negative)	Pb, Zn, Cd	Lee et al. 2001

solution to remove the hazardous contaminants from the habitat using microorganisms like and plant growth-promoting rhizobacteria (PGPR) and arbuscular mycorrhizal fungi (AMF) (Lucy et al. 2004).

Microorganisms have the potential to accumulate and degrade HMs alone or in association with the plants. An example is *Pseudomonas* spp. which has great potential to accumulate a high level of organic and inorganic compounds and shows high resistance to HMs, antibiotics, and organic solvents (Pardo et al. 2003). Some prominent microorganisms used for bioremediation are listed in Table 15.1.

15.4 Phytoremediation Strategies

Using plants for contaminant removal from the environment is termed "phytoremediation." The main mechanisms of phytoremediation are described as under.

15.4.1 Phytoextraction (PE)

PE involves the utilization of plants in absorbing contaminants either via water or soil, translocation, and accumulation of these contaminants in their aerial biomass (Jacob et al. 2018). It has been suggested as one of the most important and permanent solutions for eliminating HMs and metalloids out of the contaminated land (Sarwar et al. 2017). The procedure involves (1) mobilization and permeability of HMs in the rhizosphere, (2) uptake of HMs through the roots of the plant, (3) HM ions' translocation via roots to the aboveground plant parts, and (4) compartmentalization of HM ions within the plant tissues. The success of PE can be influenced by the factors like selection of appropriate plant, plant performance, soil, rhizosphere properties, and heavy metal bioavailability. The selected plant species must exhibit

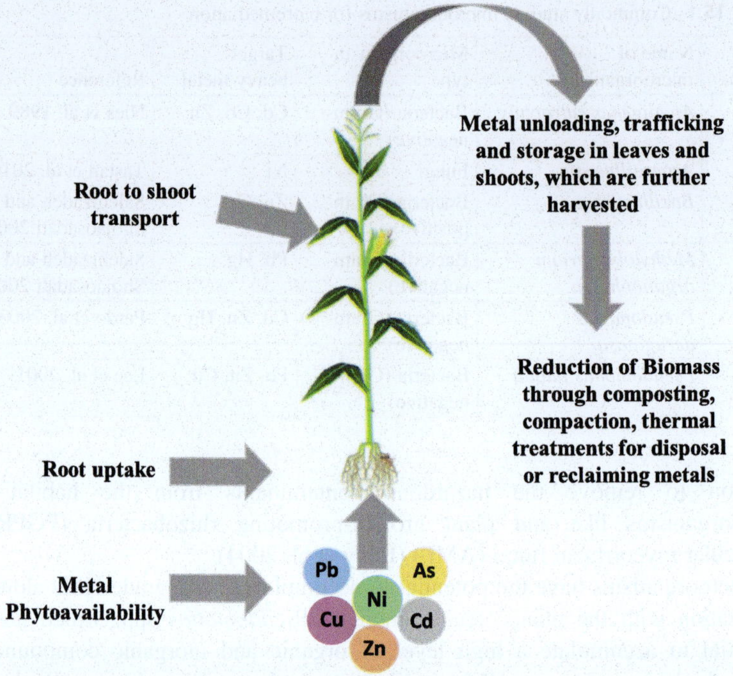

Fig. 15.1 Diagrammatic representation of mechanism of hyperaccumulation (PE) (source: Kukreja and Goutam 2012)

(1) increased tolerance against lethal outcomes of HMs, (2) increased ability of extraction along with accumulating elevated amount of HMs in plant parts which are present above the ground, (3) rapidly growing along with increased production of biomass, (4) profuse shoots and substantial root system, (5) quality modification for inducing environment, indestructible capacity to propagate in deficient soils, and stress-free cultivation and harvest, and (6) high pest resistance (Seth 2012). Out of these characteristics, the ability to accumulate metals and aboveground biomass acts as a major part in defining the PE potential of a plant species. Metal accumulation occurs by a process that involves multiple steps including mobilization through the soil following the soil solution, uptake by the roots, loading through xylem, transportation to the shoots, sequestration, and storage in the aerial tissue (Nascimento and Xing 2006) (Fig. 15.1).

Metal hyperaccumulators possess the exceptional capacity of accumulating high concentrations of HMs in the aboveground biomass, making PE an economically feasible substitute for the otherwise expensive remediation methods. Two different strategies can be employed for hyperaccumulator plant selection: (1) utilization of hyperaccumulator plants, for the accumulation of HMs in aboveground regions to a grander magnitude, and (2) utilization of plants having increased production of aboveground biomass, with the lesser capacity of metal accumulation; however, this is analogous knowingly to hyperaccumulators (Salt et al. 1998). On the basis of this conventional standard, a hyperaccumulator must be able to tolerate and accumulate in the shoot tissue above 10 mg kg^{-1} of Hg; 10,000 mg kg^{-1} of Mn and Zn; 1000 mg kg^{-1} of Co, Cr, Pb, Cu, and Ni; and 100 mg kg^{-1} of Cd. Over 400 plant species from 80 families are reported to act as hyperaccumulators for various types of metals (Baker et al. 2000). *Sebertia acuminata* can be considered an extreme example of a hyperaccumulator as it shows the accumulation elevated to 250,000 mg kg^{-1} of Ni in dry leaf sap (Jaffré et al. 1976). In some species, hyperaccumulation can be induced by applying chelators in the soil agents. According to a report, the numbers of plant species identified to accumulate one or more than one metals at a concentration of 1000 mg kg^{-1} or more are 1 for Cd; 2 for Tl; 4 for As; 14 for Pb; 20 for Se; 34 each for Co and Cu; and > 320 for Ni (Blaylock et al. 1997; Huang et al. 1998; Anderson et al. 1999).

15.4.2 Phytostabilization (PS)

PS involves the utilization of metal-tolerant plant species immobilizing the HMs which are present below the ground and decreasing their bioavailability, thus avoiding the migration within the ecosystem contamination of food chain (Marques et al. 2009). PS can be observed by precipitating HMs or reducing the metal valence in the rhizosphere, adsorption against root cell walls, or absorption sequestration inside the root tissues (Gerhardt et al. 2017). Figure 15.2 depicts the process of PS which involves reducing the mobility and bioavailability of contaminants through:

- Uptake and repossession of contaminants within the root system
- Modifications in the soil features influencing speciation and contaminant immobilization (pH, redox levels, organic matter)
- Root exudates controlling the contaminant immobilization and precipitation
- Establishing vegetation barricading which decreases the possibility of physical contact through the soil
- Mechanically stabilizing the location to reduce erosion caused due to water and wind

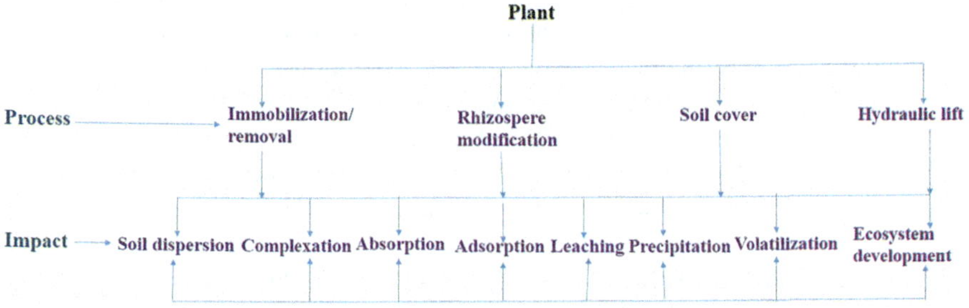

Fig. 15.2 Flowchart depicting the process of phytostabilization (source: Bolan et al. 2011)

- Improving evapotranspiration, thus decreasing the leaching of contaminants (Vangronsveld et al. 2009; Bolan et al. 2011)

15.4.2.1 Factors Affecting Phytostabilization

15.4.2.1.1 Edaphic Factors

Soil acts as a major component that controls the bioavailability and immobilization of the contaminants within the environment, hence influencing PS at contaminated sites. Principal features of soil involve organic matter, pH, and soil type and texture that includes clay content and cation and anion exchange capacities (AECs; available charged sites on soil surfaces). Soil pH is a major parameter that influences the sorption of ionizable organic and inorganic contaminants along with virtually controlling all the features of contaminant and biogeochemical procedures in soils including solubility, sorption, speciation, precipitation, and microbial activity. Additionally, there are several other environmental components including the presence of inorganic and organic ligands which show their explicit binding with the colloidal surface of the soil and also affect the ion exchange features of the soils. On the other hand, the net surface charge (i.e., CEC) of soils has been observed to be increased, thus subsequently increasing the binding ability of soils to cationic metals, with increasing sorption of cations and the net positive charge, by the precise sorption of anions on variable charge constituents, and finally resuming improved anion retention (Bolan et al. 1999). Besides, organic [bark and wood pieces, biosolids, composts, manures, sewage sludge, sawdust, and wood ash] and inorganic compounds [liming materials, e.g., $Ca(OH)_2$, $CaCO_3$, $CaMg(CO_3)_2$, CaO, gypsum (CaSO4), and phosphorus amendments (synthetic and natural apatites, phosphoric acid, hydroxyl apatites, rock phosphate (PR), diammonium phosphate (DAP) and phosphate-based salts)] can be mixed with the soil to boost the efficiency of PS as these soil amendments can lead to the alteration of metal speciation along with reducing HM solubility and bioavailability through change in soil pH and its redox status (Epelde et al. 2009; Alvarenga et al. 2009; Burges et al. 2018).

15.4.2.1.2 Plant Factors

The plants suitable for PS must exhibit certain essential characteristic features like tolerance to HM conditions and dense rooting systems (for HM immobilization) (Berti and Cunningham 2000; Marques et al. 2009). It is also important that the plants have the capacity for rapid propagation, an enormous quantity of biomass production, and easy maintenance under field conditions. These factors assist in revegetating heavy metal(loid)-contaminated areas to meet the objectives of PS.

These elements help to achieve PS's objective of revegetating HM contaminated areas. It is also suggested that the plant which is selected must have the ability to tolerate any nutrient imbalances in the substrate. It is very risky to use an exotic species in this case. However, the use of exotic species for the purpose is a tricky choice as they are less likely to be targeted by inherent herbivores, hence improving growth and decreasing the number of contaminants entering the food chain.

However, they can be established as weeds, and competing with weeds becomes frequently more challenging than soil contaminants (Dickinson et al. 2009).

15.4.2.1.3 Contaminant Concentration

Contaminants affect plant stabilization by altering microbial communities related to plant growth. In a case study on PS in mine tailings, excessive concentrations of As, Cd, Co, Mn, Pb, and Zn inhibited the application of PS-based remediation (Mendez and Maier 2008). The germination of *Brassica juncea* seeds has been observed to be inhibited as a result of the presence of an increased amount of As5þ in the tailings of the mine 176 (Ko et al. 2008). In such cases, soil amendments and pretreatments are required to decline the bioavailability of phytotoxic metals (metalloids).

15.4.2.1.4 Environmental Factors

Environmental factors such as rainfall and temperature influence PS by affecting soil erosion plant growth and contaminant reactions. In many contaminated areas, the lack of irrigation leads to high dependence on vegetation/leaching of soil elements and sediments of rainfall. Temperature conditions have an impact on the plant growth and on the features of the soil surface like cracking and formation of crust. Although cracking enhances the leaching of contaminants, slack, bare, and dry soil is prone to wind erosion through dispersion.

15.4.2.1.5 Plant-Microbial Interactions

Rhizobacteria and mycorrhizal associations of plants have also been reported to assist in PS by improving the efficiency of HM immobilization, increasing plant root surface area and depth, producing chelators, stimulating precipitation processes, and helping to prevent heavy metal ion transfer via roots following the shoots by acting as a filtration barrier (Göhre and Paszkowski 2006; Ma et al. 2011). A comparative assessment of PS and PE shows that while PE holds the better potential for HM removal, PS has the advantage of nonessential disposal of hazardous biomass.

15.4.3 Phytovolatilization

Phytovolatilization (PV) entails the utilization of the plants to absorb pollutants out of the soil and transform the contaminated components to less harmful volatile forms, thereby releasing these components into the atmosphere by the mechanism of plant transpiration. The method is utilized in detoxifying the organic contaminants along with HMs like selenium, mercury, and arsenic (Mahar et al. 2016). Brassicaceae family members, for instance, *Brassica juncea*, operate as noble volatilizers of Se. The organic selenoamino acids selenocysteine (SeCys) and selenomethionine are formed from the inorganic Se (SeMet). In comparison to inorganic Se, SeMet biomethylates to generate a volatile dimethyl selenide (DMSe) and so can be distributed in the air being less toxic (de Souza et al. 2000). A study with mercury (Hg) found that after being absorbed through roots or leaves,

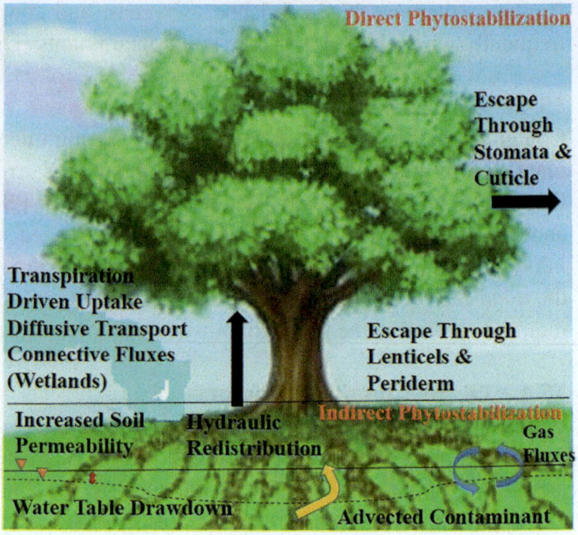

Fig. 15.3 Direct and indirect phytostabilization process (source: Limmer and Burken 2016)

methyl-Hg converts to ionic form of Hg, which then changes into a less hazardous form further volatilizing in the atmosphere (Marques et al. 2009).

PV can occur as direct or indirect phytovolatilization (Fig. 15.3). Direct PV is a more natural and well-studied form of PV that occurs due to uptake by the plants and contaminants' translocation, which results in volatilizing compound either out of the stem/trunk or the leaves. Because this channel resembles the water transpiration vascular pathway, this process is commonly referred to as "phytovolatilization." Multiple phytovolatilized chemicals are discretely hydrophobic and hence can diffuse over hydrophobic barriers like cutin or suberin, making direct PV channel commonly different from transpiration. Volatilization of substances transformed or produced by the plant is not regarded to be immediately phytovolatilized due to the requirement of absorption, translocation, and volatilization of the compound by the plant.

The compounds that are not directly phytovolatilized are released by plants after phytotransformation, e.g., transformation of selenite to dimethyl selenide that is more volatile (Kesselmeier and Staudt 1999). Indirect PV involves the increment in volatile contaminant flux from the subsurface as a result of activities in the plant root system. Plants have the ability to move vast amounts of water while simultaneously investigating large areas of soil. These activities lead to transport and reflective modifications in subsurface chemical fate. Plant roots may upsurge the flux of volatile pollutants from the subsurface by using particular strategies (Jasechko et al. 2013), such as:

- Dropping of the water table
- Advection with gas fluxes initiated by variabilities in the diel water table
- Increase in permeability of soil
- Hydraulic redistribution-based chemical transportation
- Managing rainfall pattern which might interfere with concentration of pollutants

When compared with other phytoremediation strategies, PV shows advantageous features as it can remove heavy metal (metalloid) contaminants of the site and is distributed in the form of gaseous compounds without requiring plant harvesting and disposal. PV, on the other hand, is not capable of completely eliminating contaminants as a curative strategy. It can only transport contaminants from the soil to the atmosphere, where they contaminate the ambient air with harmful volatile chemicals. They can also be redeposited in the soil by precipitation, necessitating a proper risk assessment prior to field application (Vangronsveld et al. 2009).

15.4.4 Rhizofiltration

Rhizofiltration (RF) has been identified as a better long-term solution for the cleanup of aqueous contaminated sites, as well as being resource-efficient. RF involves the use of aquatic plants for the absorption, concentration, and precipitation of metal contaminants through the roots, along with organic and inorganic contaminants from wastewater. This also lowers pollutants' mobility and prevents them from migrating to groundwater, lowering their bioavailability for entrance into the food chain. In addition, terrestrial species have been used in the treatment of ponds by cultivating them hydroponically or on floating platforms (Salt et al. 1995). Plants having a vast surface area capable of rapidly producing root biomass and the ability to translocate metals can be employed for RF. Land plants are a preferred choice owing to their long and fibrous root systems, providing a large surface area for action (Straczek et al. 2010).

As per a study by Rawat et al. (2012), the following points need to be contemplated while planning an RF strategy for a contaminated site:

- Depth of contamination
- Varying heavy metal forms present and level of contamination
- Type of vegetation to be employed (aquatic/emergent/submergent)
- Hydraulic detention time and sorption by the plant roots

Sunflower, rye, Indian mustard, spinach, corn, and tobacco plants have been assessed for removing possibly toxic metals Cu, Cd, Cr, Pb, Ni, and Zn from aqueous solutions, with sunflower showing a higher RF potential against uranium and lead (Dushenkov et al. 1995). A bioaccumulation coefficient of 563 for Pb has been reported in Indian mustard, which has also shown the efficiency to work against a wide concentration range of Pb (4–500 mg/L). *Pistia*, duckweed, and water hyacinth (*Eichornia crassipes*) have been reported to hold potential for remediation of aquatic environments polluted by coal ash comprising heavy metals (Karkhanis et al. 2005). *Plectranthus amboinicus* has also been shown to remove Pb from wastewater through RF, wherein metal accumulation occurred in the roots limiting its translocation to the aerial plant parts (Ignatius et al. 2014).

The technique is best with some limitations that can be summarized as under periodic harvesting and plant disposal:

- Continuous adjustment of the pHs to attain optimal uptake of metals
- Deciphering chemical speciation and interaction among all species (within the influent) for appropriate usage
- Need to have a well-designed system to regulate the influent concentration along with its flow rate
- Development of plants in a greenhouse or nursery before placing them in a rhizofiltration system
- Inability of plants to replicate in vitro results of RF in greenhouse/field conditions

15.4.5 Rhizodegradation

Rhizodegradation (RD) is the phytoremediation process that completely depends on the symbiotic association of plants and microorganisms present in the rhizosphere (Kennen and Kirkwood 2015). Several contaminants are utilized by fungi, yeast, and bacteria as energy sources, and by the degradation, metabolization, and/or mineralization of these contaminants, they are reduced to harmless or less harmful forms (Mani and Kumar 2014; Cristaldi et al. 2017). Plants play a primary role in an RD system by releasing compounds, or exudates, within the rhizosphere which helps these microorganisms grow and flourish. Plant root exudates show a variation based on species, hence attracting different microorganisms that might be more suitable for degrading different kinds of contaminants highlighting the need to consider plant species' root exudates and related microorganisms while executing a RD system (Mani and Kumar 2014; Kennen and Kirkwood 2015). Organic contaminants like chlorinated solvents, petroleum hydrocarbons, polychlorinated biphenyls (PCBs), PAHs, PCP, pesticides, and surfactants are potential candidates for RD. (Jordahl et al. (1997)) reported that a higher population of benzene-, o-xylene-, and toluene-degrading bacteria were observed in soil from the rhizosphere of poplar trees in comparison to the non-rhizosphere soil. Varying levels of RD effect on the total petroleum hydrocarbon (TPH) by some plant species have also been reported (Schwab and Banks 1999). Pesticides (like parathion and diazinon) and herbicides (like atrazine, metolachlor, and trifluralin) in the soil are known to be degraded by plant species like *Phaseolus* and *Kochia*, respectively (Anderson et al. 1994).

The benefits and limitations of RD are briefly discussed below.

15.4.5.1 Advantages

1. Less likelihood of contaminant translocation to the plant or atmosphere: In situ contaminant destruction and complete mineralization of organic contaminants are the characteristics of RD that limit contaminant movement.
2. No need to harvest the plant for contaminant removal: RD causes contaminant degradation at the source of contamination, checking the translocation of HMs to the plant body.
3. Remediation of a large surface area of soil: It is expected as enhanced root penetration ensures a large part of the soil can be contacted for phytoremediation.

15.4.5.2 Challenges

1. Although the RD method appears to be overall advantageous over other methods of phytoremediation, it is best with some limitations (Molina et al. 1995; Olson and Fletcher 2000).
2. Depth of root penetration: the most important hindrance to the success of RD. Many plants have shallow root zones, and soil moisture levels or soil structures that are inaccessible to roots, such as hardpans or claypans, can limit root penetration depth. Roots, on the other hand, can go rather deep into the soil (e.g., 110 cm) with high pollutant concentrations in some cases.
3. Time-consuming: Due to the time required to develop root into the new regions of soil after root dieback and disintegration of an older root system.
4. The actual soil area available for RD is restricted: Because merely a minor fraction of the entire soil volume lives through the active roots at any given time, this is the case. In addition, unfavorable soil conditions or locations with high pollutant concentrations might reduce root penetration, resulting in some areas of the soil never being reached by roots.
5. The inefficiency of rhizospheric microbiome stimulation practices: Unlike expected, stimulation of rhizospheric organisms may not necessarily result in enhanced contaminant breakdown since nondegrading microorganism populations may grow at the expense of degraders. The degree of biodegradation can also be influenced by competition between plants and microbes. Furthermore, organic matter produced by the plant can be employed as a source of carbon rather than the pollutant, limiting the amount of contaminant biodegradation.

15.5 Future Prospects of Medicinal and Aromatic Plants for Phytoremediation

Medicinal plants have been shown to have the capacity of extracting increased levels of certain HMs from the contaminated areas and translocating them to various components of the plant body. Table 15.2 provides important medicinal plants that can be utilized for phytoremediation purposes. Many of these plants are hyperaccumulators, facilitating maximal removal of HM contamination from the substratum. It is the biologically active components and specific secondary metabolites of the medicinal plants which contribute to medicinal plants' increased tolerance and remediation of HMs from the environment (Mafakheri and Kordrostami 2021). These include the action of secondary metabolites and transcription factors that detoxify the HMs (Pirzadah et al. 2019). An interesting study has been conducted concerning metallothioneins (MTs). Under the stress condition and in the HM-contaminated soil, medicinal plants produce MTs (cysteine-rich proteins) to maintain their metabolism. These MTs show higher remediation potential against HMs like copper and zinc and enhance the phytoremediation potential of medicinal plants.

Another prospective class of phytoremediation-suitable plants can be aromatic plants. Aromatic grasses have enormous potential to meet the ever-increasing

Table 15.2 Plant species established to have phytoremediation potential

S. No.	Category	Plant species	Common name	Contaminants	References
A1.	Aromatic plants	*Cymbopogon citratus*	Lemon grass	Pb(II), Cd(II), Zn(II)	Hassan 2016
2.		*Cymbopogon flexuosus*	East Indian lemon grass	Cr	Patra et al. 2015
3.		*Cymbopogon winterianus*	Lemongrass	Cr	Sinha et al. 2013
4.		*Mentha species*	Mint	Al, Cr, Ni, Cd	Manikandan et al. 2015
5.		*Ocimum gratissimum*	African basil	Ar	Mafakheri and Kordrostami 2021
6.		*Ocimum tenuiflorum*	Holy basil	As	Siddiqui et al. 2013
7.		*Pelargonium sps.*	Geranium	Pb	Abdullah and Sarem 2010
B1	Crop plants	*Alyssum murale*	Yellow tuft	Ni	Bani et al. 2010
2.		*Amaranthus* spp.	Pigweed	Ni, Fe	Shevyakova et al. 2011
3.		*Brassica juncea*	Indian mustard	Pb, Ni, Cd, Hg, U	Raj et al. 2020
4.		*Helianthus annuus*	Sunflower	Pb, Zn, U, Cd	de Andrade et al. 2008
5.		*Hordeum vulgare*	Barley	Ni, Cd, Pb, As, Hg	Belimov et al. 2004
6.		*Nicotiana tabacum*	Tobacco	Pb, Cd	Gorinova et al. 2007
7.		*Oryza sativa*	Rice	Cd, Cr, Pb, As, Hg	Liu et al. 2007
8.		*Pteris vittata*	Brake Fern	As, Cr, Cu	Kalve et al. 2011
9.		*Raphanus sativus*	White radish	Pb, Cd, As, Zn, phenols	Cobb et al. 2000
10.		*Salix viminalis*	Willow	Cu, Cd, Hg, Zn	Wang et al. 2005
11.		*Solanum lycopersicum*	Tomato	Ethidium bromide	Uera et al. 2007
12.		*Solanum tuberosum*	Potato	Pb, phenols, As, Cr, Ag, Sb	Baghour et al. 2001
13.		*Sorghum bicolor*	Sorghum	Zn, Cd	Soudek et al. 2012
14.		*Spinacia oleracea*	Spinach	Pb, Cr, Cu, Cd, Ni	Giordani et al. 2005
15.		*Thlaspi caerulescens*	Alpine pennycress	Ni, Cd, Zn	Baker and Walker 1989
16.		*Triticum aestivum*	Wheat	Cu, Pb, Cr, Ni, Zn, Cd, Fe, Mn	Chandra et al. 2009
17.		*Zea mays*	Maize	Pb, Cd	Mojiri 2011

C1.	Medicinal plants	*Allium sativum*	*Garlic*	Cd	Jiang et al. 2001
2.		*Alyssum pintodasilvae*	Sweet Alison	Ni	Varennes et al. 1996
3.		*Cannabis sativa*	Marijuana	Cd, Ni, Cr, Pb, Cu	Linger et al. 2002
4.		*Carum carvi*	Caraway	Cd, Pb	Jeliazkova et al. 2003
5.		*Catharanthus roseus*	Madagascar periwinkle	Pb, Cr, Cd, Ni	Ahmad and Misra 2014
6.		*Centella asiatica*	Gotu kola	Zn, Cu, Pb	Salim et al. 2013
7.		*Chrysopogon zizanioides*	Vetiver grass	Hg	Mangkoedihardjo and Triastuti 2011
8.		*Coriandrum sativum*	Coriander	Cu, Zn, Ni, Cr	Zheljazkov et al. 2008
9.		*Euphorbia hirta*	Asthma weed	Cd	Hamzah et al. 2016
10.		*Hypericum perforatum*	St. John's wort	Cd	Schneider and Marquard 1996
11.		*Matricaria chamomilla*	Chamomile	Pb, Cd, Zn	Stancheva et al. 2014
12.		*Matricaria recutita*	Chamomile	Cd, Zn, Pb	Stancheva et al. 2014
13.		*Mentha arvensis*	Corn mint	Cu, Zn	Malinowska and Jankowski 2017
14.		*Orthosiphon stamineus*	Java tea	Zn, Cu, Pb	Abdu et al. 2011
15.		*Rosmarinus officinalis*	Rosemary	Cd, Pb	Ramazanpour 2015
16.		*Senecio coronatus*	Woolly grassland Senecio	Ni	Przybylowics et al. 1995
17.		*Solanum nigrum*	Black nightshade	Hg, U, Cd, As	Raj et al. 2020
18.		*Vetiveria zizanioides*	Vetiver/Khus	Cd, Pb, Zn, Cu	Vo et al. 2011; Ng et al. 2017
D1.	Ornamental plants	*Althaea rosea*	Hollyhock	Cd	Liu et al. 2008
2.		*Alyssum maritima*	Sweet alyssum	Cr, Co	Budak et al. 2011
3.		*Amaranthus caudatus*	Love-lies-bleeding	Ni, Pb	Appel and Ma 2002
4.		*Calendula officinalis*	Marigold	Cu, Cd	Wang 2005
5.		*Chlorophytum comosum*	Spider plant	Cd	Wang et al. 2017
6.		*Chrysanthemum indicum*	Indian chrysanthemum	Pb	Brickell et al. 2009
7.		*Crassula portulacea*	Jade plant	Benzene	Ugrekhelidze et al. 1997

(continued)

Table 15.2 (continued)

S. No.	Category	Plant species	Common name	Contaminants	References
8.		*Dracaena sanderiana*	Lucky bamboo	Hg, Cd	Fischer et al. 2011
9.		*Euphorbia milii*	Crown of thorns	Formaldehyde, Cr	Chhon 2013
10.		*Nerium oleander*	Oleander	Pb, Zn, Cd	Naira et al. 2019
E1.	Tree species	*Agrostis capillaris*	Colonial bent or Brown top	Pb, Zn	Cotter-Howells and Caporn 1996
2.		*Betula pendula*	White birch	TCE- trichloroethylene, Zn, Pb	Lewis et al. 2015
3.		*Eucalyptus globulus*	Tasmanian blue gum	Cd (major) Hg	Luo et al. 2016
4.		*Hymenaea courbaril*	Jatoba	Cd, Ni	Dos Santos et al. 2018
5.		*Mimosa caesalpiniifolia*	Mimosa thorn	Pb	de Souza et al. 2012
6.		*Paulownia tomentosa*	Empress tree	Zn, Ni, Pb, Cr, Cd, Cu	Doumett et al. 2008
7.		*Populus nigra*	Black poplar	Mn, Zn, Cd	Saba et al. 2005
8.		*Salix* spp.	Willows	Cu, Zn	Marmiroli et al. 2011
9.		*Thuja orientalis*	Chinese arborvitae	Fe	Kacálková et al. 2015

demand for essential oils because they may be planted in contaminated areas to restore soil health as well as oil production and eco-tourism (Verma et al. 2014). These are nonedible and help in confiscation of pollutants into the plant body, restricting their entry into soil system and essential oils and resisting any change in their chemical configuration due to extraction processes (Pandey and Singh 2015). Because HMs remain in the extracted plant, essential oils produced using steam distillation have been shown to be least affected by HM contamination resulting in a marketable product (Scora and Chang 1997). Prominent families include Poaceae, Lamiaceae, Asteraceae, and Geraniaceae.

15.6 Biotechnology-Based Strategies to Enhance Phytoremediation Potential of Plants

For effective phytoremediation, a plant should have numerous characteristics like intensive growth, extensive root system, increased biomass production, increased metal tolerance, and accumulation. However, a single plant is incapable to show all the required characteristics. With the advancement of molecular biology, scientists may utilize genetic engineering, a strong technique for modifying plants with desirable qualities, to reduce the constraints of plant species with phytoremediation potential. The basic aim of genetic engineering is to modify traits including plant absorption, pollutant accumulation, and tolerance. It either enhances plants with extensive growth by introducing hyperaccumulation characteristics or hyperaccumulators for fast growth and biomass, or it improves plants with limited growth by introducing hyperaccumulation features (Bell et al. 2014).

15.6.1 Improving Phytoremediation by Utilizing Targeted Genes

For genetic engineering of higher plants, the selection of target genes encompasses:

- Genes accountable for translocation, metal uptake, and sequestration (Cherian and Oliveira 2005)
- Genes encoding heavy metal transporters
- Genes encoding metal chelators (Kärenlampi et al. 2000; Cherian and Oliveira 2005)
- Genes involved in antioxidant machinery
- Genes for enhanced biomass production and growth rate
- Genes responsible for detoxification

The overexpression and manipulation of these genes provide a direct way to upgrade the phytoremediation ability of plants (Table 15.3).

Table 15.3 Target genes for genetic transformation for improved phytoremediation through different mechanisms

Gene product	Gene origin	Gene host	Phytoremediation performance	Reference(s)
AtATM3	*Arabidopsis thaliana*	*Brassica juncea*	Cd and Pb (phytoextraction)	Bhuiyan et al. 2011
AtZIP1, AtMTP1	*Arabidopsis thaliana*	*Manihot esculenta*	Zn (phytostabilization)	Siemianowski et al. 2014
OsMTP1	*Oryza sativa*	*Nicotiana tabacum*	Cd (phytoextraction)	Das et al. 2016
YCF1	*Saccharomyces cerevisiae*	*Populus alba*, *P. tremula* var. glandulosa	Zn and Cd (phytoextraction) Pb (phytostabilization)	Shim et al. 2013
ZAT	*Thlaspi goesingense*	*Arabidopsis thaliana*	Zn	Ahmad 2015
BjMT2	*Brassica juncea*	*Arabidopsis thaliana*	Improved tolerance to Cu and Cd	Zhigang et al. 2006
OsMT2c	*Oryza sativa*	*A. thaliana*	Improved Cu tolerance and enhanced ROS scavenging ability	Liu et al. 2015
CUP1	Yeast	*N. tabacum*	Higher accumulation and tolerance to Cd	Krystofova et al. 2012
rgMT	*Oryza sativa*	*S. cerevisiae* *A. thaliana*	Increased biomass and improved seed germination	Jin and Daniell 2014
SaMT2	*Sedum alfredii*	*S. cerevisiae/ Nicotiana tabacum*	Improved accumulation and tolerance for Cd	Zhang et al. 2014
ThMT3	*Tamarix hispida*	*Salix matsudana* *S. cerevisiae*	Improved Cu tolerance Improved tolerance to Cu, Zn, and Cd	Yang et al. 2015 Yang et al. 2011

15.6.1.1 Overexpressing the Genes that Encode Metal Transporters and Their Role in Phytoremediation

For the uptake, sequestration, and distribution of HMs, the plant utilizes different transporters which are present in cell membrane and tonoplast. The manipulation of genes encoding HM transporters can improve HM tolerance in the plants (Yang et al. 2015). Several gene families encode protein transporters. For instance, the proteins accounting for detoxification and ion regulation procedures are encoded by ATP-binding cassette (ABC) family (Martinoia et al. 2002). ABC genes encoded transporters are restricted in the transgenic plant cells' tonoplasts, hence improving the vacuolar transport of HMs (Song et al. 2014). Overexpression of *AtABCC1*, *AtABCC2*, and *AtABCC3* can increase the tolerance and accumulation of Hg and Cd in *Arabidopsis* (Park et al. 2012). The cation diffusion facilitator (CDF) family encodes a protein responsible for the sequestration of metal ions to various cell compartments. OsMTP1 gene from rice has been shown to assist in hyperaccumulation of Cd and As in *N. tabacum* (Das et al. 2016) and can also be utilized for Cd removal through phytoextraction (Sun et al. 2018). Transporters of the zinc-iron permease (ZIP) family shows the involvement in the transport of Zn, Fe, and Cd, and it has been reported that overexpressing these genes in *Thlaspi caerulescens* and *Arabidopsis halleri* has resulted in increased hyperaccumulation. Transmission of Zn transporter *ZAT* gene from *Thlaspi goesingense* enhanced Zn accumulation in roots of *Arabidopsis thaliana* (Pence et al. 2000; Becher et al. 2004).

15.6.1.2 Overexpressing Genes Encoding Metal Chelators

Metallothioneins (MTs) and phytochelatins (PCs) are the main classes of metal chelators in plants. MTs are cysteine-rich metal-binding proteins having low molecular weight that are involved in metal detoxification, metal homeostasis, and oxidative stress protection. They are gene-encoded polypeptides and show a high affinity to Cu, Zn, and Cd. HM tolerance in plants has been reported to be improved by MT gene overexpression (Leszczyszyn et al. 2013). The thiol group present in cysteine amino acid is specific for the heavy metals and assists the plant tissues to alter the lethal properties of contaminants on plants. The N-terminal (α) domain of MTs possesses three metal binding sites that show binding with divalent ions, whereas the C-terminal (β) domain is bound with four divalent ions of heavy metals (Berta et al. 2009). MTs form an aggregate with Cu and Zn efficiently and experience minimal binding with Cd, Pb, and Hg.

PCs are small metal-chelating peptides synthesized by phytochelatin synthase (PCS) with glutathione as its substrate. PCs help in intracellular HM detoxification in plants by chelating HMs and sequestering complex materials via tonoplast transfer inside the vacuoles (Cobbett and Goldsbrough 2002).

15.6.1.3 Overexpressing Genes Involved in Oxidative Stress Mechanisms

Increasing the activity of enzymes of antioxidant machinery and ROS scavengers is another approach to the genetic transformation toward improved phytoremediation. The toxic form of HMs can be converted to a less toxic form by the enzymes of antioxidant system (Koźmińska et al. 2018). Glutathione S-transferase (GST) enzyme helps in the free detoxification of radicals (Hellou et al. 2012). The overexpression of genes like *SbSLSP* gene (from *Salicornia brachiate*), *1* -aminocyclopropane-1-carboxylic acid deaminase (ACCd), and γ-TMT have shown to provide increased resistance to metal-induced oxidative stress plants (Grichko et al. 2000; Jin and Daniell 2014).

15.6.2 Chloroplast Engineering for Enhanced Phytoremediation of Mercury

Chloroplast engineering is an emerging technique of plant transgenics wherein the chloroplast genome is modified to develop "transplastomic plants." This technique has several advantages over nuclear transgenics like multigene engineering, gene containment, elimination of gene silencing, etc. (Quesada et al. 2005; Daniell et al. 2016).

Chloroplast engineering has been shown to have a potential advantage in phytoremediation in some plant species. Studies on *Nicotiana tabacum* showed that genome transformation with native bacterial genes *merA* and *merB*, which encodes mercuric ion reductase and organomercurial lyase, resulted in increased capacity of the transformed plant to withstand the soil polluted with highly toxic organomercurial compound. This was attributed to the fact that enzymes mercuric ion reductase and organomercurial lyase can convert lethal methyl-Hg in a considerably less contaminated metallic Hg [0], which can be volatilized further (Bizily et al. 1999). However, the release of elemental Hg back into the environment is the disadvantage of this strategy. Therefore, a better strategy of phytoaccumulation of mercuric compounds in plant tissues was tried by expression of mouse metallothioneins gene (*mt1*) in chloroplasts (Ruiz et al. 2011; Zhang et al. 2013). The transgenic plants so developed were found to be resistant to up to 20 μM Hg. Moreover, they retained a high chlorophyll and biomass concentration.

Overexpressing the type-2 MT in *Arabidopsis thaliana* from Indian mustard (*BjMT2*) has improved Cu and Cd tolerance (Zhigang et al. 2006). Similar studies on chloroplast transgenics were also performed against copper and cadmium stress in plants like *Brassica rapa*, *Arabidopsis*, and *Nicotiana tabacum* (Kim et al. 2007; Krystofova et al. 2012; Jin and Daniell 2014). Chloroplast transformation may also be used for the phytoremediation of other HMs that affect chloroplast function and hence can help effectively in the revegetation of the locations that are suffering from heavy metal contamination. However, transgenics for phytoremediation have to be plant-specific depending on the comparative effect of nuclear versus chloroplast transformation for phytoremediation.

15.7 Conclusion

There is no denying that phytoremediation is extremely beneficial in the revegetation of HM-contaminated soil and offers significant benefits over other conventional and bioremediation procedures. Microorganism-based bioremediation has shown limited potential in many cases. However, microbial communities can be used to enhance specific plants' responses to the contaminants. Different crop plants, ornamentals, and tree species along with aromatic and medicinal plants have been identified as hyperaccumulators and are being used as effective phytoremediators. Opting the plant species for phytoremediation is critical and it relies on various features including target HM, rapid growth, production of a huge amount of biomass, high metal tolerance and accumulation, etc. Along with the plant species utilized, the performance of phytoremediation procedures is significantly influenced by a number of environmental conditions like water availability, soil composition, pH, rhizospheric bacteria, etc. Therefore, the establishment of a suitable environment and improved features in the target plant are crucial factors to develop a successful phytoremediation system. Transgenic plants with desirable features are being generated with the use of genetic engineering which would be a sustainable method to decontaminate polluted sites. In addition, new techniques like chloroplast engineering and genome editing are in the developing phase and expected to show significantly improved results in phytoremediation.

References

Abdu A, Aderis N, Abdul-Hamid H, Majid NM, Jusop S, Karam DS, Ahmad K (2011) Using *Orthosiphon stamineus* B. for phytoremediation of heavy metals in soils amended with sewage sludge. Am J Appl Sci 8(4):323–331

Abdullah S, Sarem SM (2010) The potential of *chrysanthemum* and *pelargonium* for phytoextraction of lead-contaminated soils. J Civ Eng 4:409–416

Ahmad P (2015) Plant metal interaction: emerging remediation techniques. Elsevier, Amsterdam

Ahmad R, Misra N (2014) Evaluation of phytoremediation potential of *Catharanthus roseus* with respect to chromium contamination. Am J Plant Sci 5:2378–2388

Ali H, Khan E, Ilahi I (2019) Environmental chemistry and ecotoxicology of hazardous heavy metals: environmental persistence, toxicity, and bioaccumulation. J Chem 2019:1. https://doi.org/10.1155/2019/6730305

Alvarenga P, Gonçalves A, Fernandes R, De Varennes A, Vallini G, Duarte E, Cunha-Queda AC (2009) Organic residues as immobilizing agents in aided phytostabilization:(I) effects on soil chemical characteristics. Chemosphere 74:1292–1300

Anderson CWN, Brooks RR, Chiarucci A, LaCoste CJ, Leblanc M, Robinson BH, Simcock R, Stewart RB (1999) Phytomining for nickel, thallium and gold. J Geochem Explor 67:407–415

Anderson TA, Kruger EL, Coats JR (1994) Enhanced degradation of a mixture of three herbicides in the rhizosphere of a herbicide-tolerant plant. Chemosphere 28:1551–1557

Appel C, Ma L (2002) Heavy metals in the environment concentration, pH and surface charge effects on Cd and Pb sorption in three tropical soils. J Environ Quality (JEQ) 21:154–155

Baghour M, Moreno DA, Hernández J, Castilla N, Romero L (2001) Influence of root temperature on phytoaccumulation of as, ag, Cr, and Sb in potato plants (*Solanum tuberosum* L. var. Spunta). J Environ Sci Heal Part A Toxic Hazard Subst Environ Eng 36:1389–1401

Baker AJM, Walker PL (1989) Ecophysiology of metal uptake by tolerant plants. In: Shaw AJ (ed) *Heavy metal tolerance in plants: evolutionary aspects*. CRC Press, Boca Raton, FL, pp 155–177

Baker AJM, Mcgrath SP, Reeves RD, Smith JAC (2000) Metal hyperaccumulator plants: a review of the ecology and physiology of a biological resource for phytoremediation of metal-polluted soils. In: Terry N, Bañuelos G (eds) Phytoremediation of Contaminated Soil And Water. Lewis, Boca Raton, FL, pp 85–108

Bani A, Pavlova D, Echevarria G, Mullaj A, Reeves RD, Morel JL, Sulce S (2010) Nickel hyperaccumulation by the species of *alyssum* and *Thlaspi* (Brassicaceae) from the ultramafic soils of the Balkans. Bot Serb 34:3–14

Becher M, Talke I, Krall L, Kramer U (2004) Cross-species microarray transcript profiling reveals high constitutive expression of metal homeostasis genes in shoots of the zinc hyperaccumulator *Arabidopsis halleri*. Plant J 37:251–268

Belimov AA, Kunakova AM, Safronova VI, Stepanok VV, Yudkin LY, Alekseev YV, Kozhemyakov AP (2004) Employment of rhizobacteria for the inoculation of barley plants cultivated in soil contaminated with lead and cadmium. Microbiology 73:99–106

Bell TH, Joly S, Pitre FE, Yergeau E (2014) Increasing phytoremediation efficiency and reliability using novel omics approaches. Trends Biotechnol 32(5):271–280

Berta M, Giovannelli A, Potenza E, Traversi ML, Racchi ML (2009) Type 3 metallothioneins respond to water deficit in leaf and in the cambial zone of white poplar (Populus alba). J Plant Physiol 166:521–530

Berti WR, Cunningham SD (2000) In: Raskin I, Ensley BD (eds) Phytostabilization of metals in phytoremediation of toxic metals: using plants to clean-up the environment. John Wiley and Sons, Inc., New York, NY, pp 71–88

Bhuiyan MSU, Min SR, Jeong WJ, Sultana S, Choi KS, Lee Y, Liu JR (2011) Overexpression of AtATM3 in *Brassica juncea* confers enhanced heavy metal tolerance and accumulation. Plant Cell Tissue Organ Cult 107(1):69–77

Bizily S, Rugh CC, Summers AO, Meagher RB (1999) Phytoremediation of methylmercury pollution: merB expression in *Arabidopsis thaliana* plants confer resistance to organomercurial. Proc Natl Acad Sci U S A 96:6808–6813

Blaylock MJ, Salt DE, Dushenkov S, Zakharova O, Gussman C, Kapulnic Y, Ensley BD, Raskin I (1997) Enhanced accumulation of Pb in Indian mustard by soil-applied chelating agents. Environ Sci Technol 31:860–865

Bolan NS, Naidu R, Syers JK, Tillman RW (1999) Surface charge and solute interactions in soils. Adv Agron 67:88–141

Bolan NS, Park JH, Robinson B, Naidu R, Huh KY (2011) Advances in agronomy. Phytostabilization 112:145–204

Brickell CD, Alexande C, David JC, Hetterscheid WLA, Leslie AC, Malecot V, Jin X, Cubey JJ (2009) International code of nomenclature for cultivated plants (ICNCP). Eigth Edition. Scri Hortic 10:1–184. [accessed 2019 Sep 5]

Budak F, Zaimoglu Z, Basci N (2011) Uptake and translocation of hexavalent chromium by selected species of ornamental plants. Pol J Environ Stud 20:857–862

Burges A, Alkorta I, Epelde L, Garbisu C (2018) From phytoremediation of soil contaminants to phytomanagement of ecosystem services in metal contaminated sites. Int J Phytoremediation 20: 384–397

Chandra R, Bharagava RN, Yadav S, Mohan D (2009) Accumulation and distribution of toxic metals in wheat (*Triticum aestivum* L.) and Indian mustard (*Brassica campestris* L.) irrigated with distillery and tannery effluents. J Hazard Mater 162(2–3):1514–1521

Cherian S, Oliveira MM (2005) Transgenic plants in phytoremediation: recent advances and new possibilities. Environ Sci Technol 39(24):9377–9390

Chhon Y (2013) Phytoremediation of trimethylamine by Euphorbia milii and agricultural residues. Master's thesis,. King Mongkut's University of Technology Thonburi, Bangkok

Cobb GP, Sands K, Waters M, Wixson BG, Dorward-King E (2000) Accumulation of heavy metals by vegetables grown in mine wastes. Environ Toxicol Chem 19(3):600–607

Cobbett C, Goldsbrough P (2002) Phytochelatins and Metallothioneins: roles in heavy metal detoxification and homeostasis. Annu Rev Plant Biol 53(1):159–182

Cotter-Howells J, Caporn S (1996) Remediation of contaminated land by formation of heavy metal phosphates. J Appl Geochem 11(1–2):335–342

Cristaldi A, Conti GO, Jho EH, Zuccarello P, Grasso A, Copat C, Ferrante M (2017) Phytoremediation of contaminated soils by heavy metals and PAHs. A brief review. Environ Technol Innov 8:309–326

Daniell H, Lin C, Yu M, Chang W (2016) Chloroplast genomes: diversity, evolution and applications in genetic engineering. Genome Biol 17:134

Das N, Bhattacharya S, Maiti MK (2016) Enhanced cadmium accumulation and tolerance in transgenic tobacco overexpressing rice metal tolerance protein gene OsMTP1 is promising for phytoremediation. Plant Physiol Biochem 105:297–309

de Andrade SAL, da Silveira APD, Jorge RA, de Abreu MF (2008) Cadmium accumulation in sunflower plants influenced by arbuscular mycorrhiza. Int J Phytoremediation 10:1–14

de Souza MP, Lytle CM, Mulholland MM, Otte ML, Terry N (2000) Selenium assimilation and volatilization from dimethylselenoniopropionate by Indian mustard. Plant Physiol 122:1281–1288 de Souza SCR, de Andrade SAL, de Souza LA, Schiavinato MA (2012) Lead tolerance and phytoremediation potential of Brazilian leguminous tree species at the seedling stage. J Environ Manag 110:299–307

Dhaliwal SS, Singh J, Taneja PK, Mandal A (2019) Remediation techniques for removal of heavy metals from the soil contaminated through different sources: a review. Environ Sci Pollut Res 27:1319–1333

Dickinson N, Baker A, Doronila A, Laidlaw S, Reeves R (2009) Phytoremediation of inorganics: realism and synergies. Int J Phytoremediation 11:97–114

Dos Santos NMC, da Costa VAM, de Araújo FV, Alencar BTB, Ribeiro VHV, Okumura F, Simeone MLF, Dos Santos JB (2018) Phytoremediation of Brazilian tree species in soils contaminated by herbicides. Environ Sci Pollut Res 25(27):27561–27568

Doumett S, Lamperi L, Checchini L, Azzarello E, Mugnai S, Mancuso S, Petruzzelli G, Del Bubba M (2008) Heavy metal distribution between contaminated soil and *Paulownia tomentosa*, in a pilot-scale assisted phytoremediation study: influence of different complexing agents. Chemosphere 72(10):1481–1490

Dushenkov V, Nanda Kumar PBA, Motto H, Raskin I (1995) Rhizofiltration-the use of plants to remove heavy metals from aqueous streams. Environ Sci Technol 29:1239–1245

Epelde L, Becerril JM, Mijangos I, Garbisu C (2009) Evaluation of the efficiency of a phytostabilization process with biological indicators of soil health. J Environ Qual 38:2041–2049

Fischer L, Falta T, Koellensperger G, Stojanovic A, Kogelnig D, Galanski M, Krachler R, Keppler BK, Hann S (2011) Ionic liquids for extraction of metals and metal containing compounds from communal and industrial waste water. Water Res 45(15):4601–4614

Garrett RG (2000) Natural sources of metals to the environment. Hum Ecol Risk Assess 6(6): 945–963

Gerhardt KE, Gerwing PD, Greenberg BM (2017) Opinion: taking phytoremediation from proven technology to accepted practice. Plant Sci 256:170–185

Giordani C, Cecchi S, Zanchi C (2005) Phytoremediation of soil polluted by nickel using agricultural crops. Environ Manag 36:675–681

Göhre V, Paszkowski U (2006) Contribution of the arbuscular mycorrhizal symbiosis to heavy metal phytoremediation. Planta 223:1115–1122

Gorinova N, Nedkovska M, Todorovska E, Simova-Stoilova L, Stoyanova Z, Georgieva K, Demirevska-Kepova K, Atanassov A, Herzig R (2007) Improved phytoaccumulation of cadmium by genetically modified tobacco plants (*Nicotiana tabacum* L.). physiological and biochemical response of the transformants to cadmium toxicity. Environ Pollut 145:161–170

Grichko VP, Filby B, Glick BR (2000) Increased ability of transgenic plants expressing the bacterial enzyme ACC deaminase to accumulate Cd, Co, Cu, Ni, Pb, and Zn. J Biotechnol 81:45–53

Hamzah A, Hapsari RI, Wisnubroto EI (2016) Phytoremediation of cadmium-contaminated agricultural land using indigenous plants. Int J Environ Agric Res (IJOEAR) 2(1):8–14

Hassan E (2016) Comparative study on the biosorption of Pb (II), cd (II) and Zn (II) using lemon grass (*Cymbopogon citratus*): kinetics, isotherms and thermodynamics. Chem Int 2(2):89–102

Hellou J, Ross NW, Moon TW (2012) Glutathione, glutathione S-transferase and glutathione conjugates, complementary markers of oxidative stress in aquatic biota. Environ Sci Pollut Res 19:2007–2023

Huang JW, Blaylock MJ, Kapulnik Y (1998) Phytoremediation of uranium-contaminated soils: role of organic acids in triggering uranium hyperaccumulation by plants. Environ Sci Technol 32: 2004–2008

Igiri BE, Okoduwa SI, Idoko GO, Akabuogu EP, Adeyi AO, Ejiogu IK (2018) Toxicity and bioremediation of heavy metals contaminated ecosystem from tannery wastewater: a review. J Toxicol 2018:1

Ignatius A, Arunbabu V, Neethu J, Ramasamy EV (2014) Rhizofiltration of lead using an aromatic medicinal plant *Plectranthus amboinicus* cultured in a hydroponic nutrient film technique (NFT) system. Environ Sci Pollut Res 21:13007–13016

Jacob JM, Karthik C, Saratale RG, Kumar SS, Prabakar D, Kadirvelu K et al (2018) Biological approaches to tackle heavy metal pollution: a survey of literature. J Environ Manag 217:56–70

Jadia CD, Fulekar MH (2009) Phytoremediation of heavy metals: recent techniques. Afr J Biotechnol 8(6)

Jaffré T, Brooks RR, Lee J, Reeves RD (1976) *Sebertia acuminata*: a nickel-accumulating plant from New Caledonia. Science 193:579–580

Jasechko S, Sharp ZD, Gibson JJ, Birks SJ, Yi Y, Fawcett PJ (2013) Terrestrial water fluxes dominated by transpiration. Nature 496(7445):347–350

Jeliazkova E, Craker EL, Xing B (2003) Seed germination of Anise, caraway and fennel in heavy metal contaminated solutions. Int J Geogr Inf Syst 10:83–93

Jiang W, Liu D, Hou W (2001) Hyperaccumulation of cadmium by roots, bulbs and shoots of garlic (*Allium sativum L.*). Bioresour Technol 76:9–13

Jin S, Daniell H (2014) Expression of gamma-tocopherol methyl-transferase in chloroplasts results in massive proliferation of the inner envelope membrane and decreases susceptibility to salt and metal-induced oxidative stresses by reducing reactive oxygen species. Plant Biotechnol J 12: 1274–1285

Jordahl JL, Foster L, Schnoor JL, Alvarez PJJ (1997) Effect of hybrid poplar trees on microbial populations important to hazardous waste bioremediation. Environ Toxicol Chem 16(6): 1318–1321

Kärenlampi S, Schat H, Vangronsveld J, Verkleij JAC, Lelie D, Mergeay M, Tervahauta AI (2000) Genetic engineering in the improvement of plants for phytoremediation of metal polluted soils. Environ Pollut 107:225–231

Kacálková L, Tlustoš P, Száková J (2015) Phytoextraction of risk elements by willow and poplar trees. Int J Phytoremediation 17(5):414–421

Kalve S, Sarangi BK, Pandey RA, Chakrabarti T (2011) Arsenic and chromium hyperaccumulation by an ecotype of *Pteris vittata*-prospective for phytoextraction from contaminated water and soil. Curr Sci India 100:888–894

Karkhanis M, Jadia CD, Fulekar MH (2005) Rhizofiltration of metals from coal ash leachate. Asian J. water. Environ Pollut 3(1):91–94

Kennen K, Kirkwood N (2015) Phyto: principles and resources for site remediation and landscape design. Routledge, New York

Kesselmeier J, Staudt M (1999) Biogenic volatile organic compounds (VOC): an overview on emission, physiology and ecology. J Atmos Chem 33(1):23–88

Kim SH, Lee HS, Song WY, Choi KS, Hur Y (2007) Chloroplast targeted BrMT1 (*Brassica rapa* Type-1 Metallothionein) enhances resistance to cadmium and ROS in transgenic Arabidopsis plants. J Plant Biol 50(1):1–7

Ko BG, Anderson CWN, Bolan NS, Huh KY, Vogeler I (2008) Potential for the phytoremediation of arsenic-contaminated mine tailings in Fiji. Aust J Soil Res 46:493–501

Koźmińska A, Wiszniewska A, Hanus-Fajerska E, Muszyńska E (2018) Recent strategies of increasing metal tolerance and phytoremediation potential using genetic transformation of plants. Plant Biotechnol Rep 12:1–14

Krystofova O, Zitka O, Krizkova S, Hynek D, Shestivska V, Adam V, Hubalek J, Mackova M, Macek T, Zehnalek J, Babula P, Havel L, Kizek R (2012) Accumulation of cadmium by transgenic tobacco plants (*Nicotiana tabacum* L.) carrying yeast metallothionein gene revealed by electrochemistry. Int J Electrochem Sci 7:886–907

Kukreja S, Goutam U (2012) Phytoremediation: a new hope for the environment. Front Recent Develop Plant Sci:149–171

Laghlimi M, Baghdad B, El Hadi H, Bouabdli A (2015) Phytoremediation mechanisms of heavy metal contaminated soils: a review. Open J Ecol 5:375–388

Lee SW, Glickmann E, Cooksey DA (2001) Chromosomal locus for cadmium resistance in *pseudomonas putida* consisting of a cadmium-transporting ATPase and a MerR family response regulator. Appl Environ Microbiol 67:1437–1444

Leszczyszyn OI, Imam HT, Blindauer CA (2013) Diversity and distribution of plant metallothioneins: a review of structure, properties and functions. Metallomics 5:1146–1169
Lewis J, Qvarfort U, Sjöström J (2015) Betula pendula: a promising candidate for phytoremediation of TCE in northern climates. Int J Phytoremediation 17(1):9–15
Limmer M, Burken J (2016) Phytovolatilization of organic contaminants. Environ Sci Technol 50(13):6632–6643
Linger P, Mussig J, Fischer H, Kobert J (2002) Industrial hemp (Cannabis sativa L.) growing on heavy metal contaminated soil: fibre quality and phytoremediation potential. *Ind Crops Prod* 16(1):33–42
Liu J, Shi X, Qiana M, Zhenga Lm Lianb C, Yan Xiaa Y, Shen Z (2015) Copper-induced hydrogen peroxide upregulation of a metallothionein gene, OsMT2c, from *Oryza sativa L.* confers copper tolerance in *Arabidopsis thaliana*. J Hazard Mater 294:99–108
Liu JN, Zho QX, Sun T, Ma LQ, Wang S (2008) Growth responses of three ornamental plants to cd and cd-Pb stress and their metal accumulation characteristics. J Hazard Mater 151:261–267
Liu WX, Shen LF, Liu JW, Wang YW, Li SR (2007) Uptake of toxic heavy metals by rice (Oryza sativa L.) cultivated in the agricultural soil near Zhengzhou city, People's Republic of China. Bull Environ Contam Toxicol 79(2):209–213
Lucy M, Reed E, Glick BR (2004) Applications of free living plant growth promoting rhizobacteria. Antonie Van Leeuwenhoek 86:1–25
Luo J, Qi S, Peng L, Wang J (2016) Phytoremediation efficiency of cd by *Eucalyptus globulus* transplanted from polluted and unpolluted sites. Int J Phytoremediation 18(4):308–314
Lwin CS, Seo B, Kim H, Owens G, Kim K (2018) Application of soil amendments to contaminated soils for heavy metal immobilization and improved soil quality: a critical review. Soil Sci Plants Nutr 64(2):156–167
Ma Y, Prasad M, Rajkumar M, Freitas H (2011) Plant growth promoting rhizobacteria and endophytes accelerate phytoremediation of metalliferous soils. Biotechnol Adv 29:248–258
Mafakheri M, Kordrostami M (2021) Recent advances toward exploiting medicinal plants as phytoremediators. In: Handbook of bioremediation. Academic Press, pp 371–383
Mahar A, Wang P, Ali A, Awasthi MK, Lahori AH, Wang Q, Li R, Zhang Z (2016) Challenges and opportunities in the phytoremediation of heavy metals contaminated soils: a review. Ecotox Environ Safe 126:111–121
Malinowska E, Jankowski K (2017) Copper and zinc concentrations of medicinal herbs and soil surrounding ponds on agricultural land. Landscape Ecol Eng 13(1):183–188
Mangkoedihardjo S, Triastuti Y (2011) Vetiver in phytoremediation of mercury polluted soil with the addition of compost. J Appl Sci Res 7(4):465–469
Mani D, Kumar C (2014) Biotechnological advances in bioremediation of heavy metals contaminated ecosystems: an overview with special reference to phytoremediation. Int J Environ Sci Technol [Online] 11(3):843–872
Manikandan R, Sahi SV, Venkatachalam P (2015) Impact assessment of mercury accumulation and biochemical and molecular response of *Mentha arvensis*: a potential hyperaccumulator plant. Sci World J 2015:1
Marmiroli M, Pietrini F, Maestri E, Zacchini M, Marmiroli N, Massacci A (2011) Growth, physiological and molecular traits in Salicaceae trees investigated for phytoremediation of heavy metals and organics. Tree Physiol 31(12):1319–1334
Marques AP, Rangel AO, Castro PM (2009) Remediation of heavy metal contaminated soils: phytoremediation as a potentially promising clean-up technology. Crit Rev Env Sci Technol 39: 622–654
Martinoia E, Klein M, Geisler M, Bovet L, Forestier C, Kolukisaoglu Ü, Schulz B (2002) Multifunctionality of plant ABC transporters—more than just detoxifiers. Planta 214(3):345–355
Mendez MO, Maier RM (2008) Phytostabilization of mine tailings in arid and semiarid environments-an emerging remediation technology. Environ Health Perspect 116:278
Mojiri A (2011) The potential of corn (*Zea mays*) for phytoremediation of soil contaminated with cadmium and lead. J Biol Environ Sci 5(13):17–22

Molina M, Araujo R, Bond JR (1995) Abstract: Dynamics of oil degradation in coastal environments: Effects of bioremediation products and some environmental parameters. In: Symposium on Bioremediation of Hazardous Wastes: Research, Development, and Field Evaluations, August 10–12, 1995, Rye Brook, NY

Naira AA, Naira ZA, Inas AZ, Ashraf AE (2019) Bioremediation of heavy metals by using some shrubs in three different locations of Alexandria city. (B) N. Oleander plant Alex Sci Exch 40: 99–114

Nascimento CWA, Xing B (2006) Phytoextraction: a review on enhanced metal availability and plant accumulation. Sci Agric 63:299–311

Ng CC, Boyce AN, Rahman M, Abas R (2017) Tolerance threshold and phyto-assessment of cadmium and lead in vetiver grass, *Vetiveria zizanioides* (Linn.) Nash. Chiang Mai J Sci 44(4): 1367–1368

Nies A, Nies DH, Silver S (1989) Cloning and expression of plasmid genes encoding resistances to chromate and cobalt in *Alcaligenes eutrophus*. J Bacteriol 171(9):5065–5070

Olson PE, Fletcher JS (2000) Ecological recovery of vegetation at a former industrial sludge basin and its implications to phytoremediation. Environ Sci Pollut Res 7:1–10

Pandey VC, Singh N (2015) Aromatic plants versus arsenic hazards in soils. J Geochem Explor 157:77–80

Pardo R, Herguedas M, Barrado E, Vega M (2003) Biosorption of cadmium, copper, lead and zinc by inactive biomass of *pseudomonas putida*. Anal Bioanal Chem 376:26–32

Park J, Song WY, Ko D, Eom Y, Hansen TH, Schiller M, Lee TG, Martinoia E, Lee Y (2012) The phytochelatin transporters AtABCC1 and AtABCC2 mediate tolerance to cadmium and mercury. J Plant 692:278–288

Patra HK, Marndi DS, Mohanty M (2015) Chromium toxicity, physiological responses and tolerance potential of lemon grass (*Cymbopogon flexuosus* Nees ex steud. Wats.). Ann Plant Sci 4(05):1080–1084

Pence NS, Larsen PB, Ebbs SD, Letham DL, Lasat MM, Garvin DF, Eide D, Kochian LV (2000) The molecular physiology of heavy metal transport in the Zn/cd hyperaccumulator Thlaspi caerulescens. Proc Natl Acad Sci U S A 97:4956–4960

Peng W, Li X, Xiao S, Fan W (2018) Review of remediation technologies for sediments contaminated by heavy metals. J Soils Sediment 18:1701–1719

Pirzadah TB, Malik B, Dar FA (2019) Phytoremediation potential of aromatic and medicinal plants: a way forward for a green economy. J Stress Physiol Biochem 15(3):62–75

Przybylowics WJ, Pineda CA, Prozesky VM, Mesjasz-przybylowicz J (1995) Investigation of Ni hyperaccumulation by true elemental imaging. Nucl Instrum Methods Phys Res B 104:176–181

Quesada T, Ruiz ON, Daniell H (2005) Characterization of heterologous multigene operons in transgenic chloroplasts: transcription, processing and translation. Plant Physiol 138:1–17

Raj D, Kumar A, Maiti SK (2020) Brassica juncea (L.) Czern. (Indian mustard): a putative plant species to facilitate the phytoremediation of mercury contaminated soils. Int J Phytoremediation 22(7):733–744

Ramazanpour H (2015) Study effect of soil and amendments on phytoremediation of cadmium (cd) and lead (Pb) from contaminated soil by rosemary (*Rosmarinus officinalis* L.). [doctoral dissertation]. University of Zabol, Zabol

Rawat K, Fulekar MH, Pathak B (2012) Rhizofiltration: a green technology for remediation of heavy metals. Int J Innov Sci 2(4):193–199

Ruiz ON, Alvarez D, Torres C, Roman L, Daniell H (2011) Metallothionein expression in chloroplasts enhances mercury accumulation and phytoremediation capability. Plant Biotechnol J 9:609–617

Saba G, Parizanganeh AH, Zamani A, Saba J (2005) Phytoremediation of heavy metals contaminated environments: screening for native plants in Zanjan-Iran. Int J Environ Res 9(1):309–3016

Salehizadeh H, Shojaosadati SA (2003) Removal of metal ions from aqueous solution by polysaccharide produced from Bacillus firmus. Water Res 37(17):4231–4235

Salim MRJ, Adenan MI, Amid A, Jauri MH, Sued AS (2013) Statistical analysis of metal chelating activity of *Centella asiatica* and *Erythroxylum cuneatum* using response surface methodology. Biotechnol Res Int

Salt DE, Blaylock M, Kumar NPBA, Dushenkov V, Ensley BD, Chet I, Raskin I (1995) Phytoremediation: a novel strategy for the removal of toxic metals from the environment using plants. Nat Biotechnol 13:468–474

Salt DE, Smith RD, Raskin I (1998) Phytoremediation. Ann Rev Plant Physiol Plant Mol Biol 49: 643–668

Sarwar N, Imran M, Shaheen MR, Ishaque W, Kamran MA, Matloob A, Rehim A, Hussain S (2017) Phytoremediation strategies for soils contaminated with heavy metals: modifications and future perspectives. Chemosphere 171:710–721

Schneider M, Marquard R (1996) Investigation on the uptake of cadmium in *Hypericum perforatum L.* (St. John's wort). Acta Hortic 426:435–442

Schwab AP, Banks MK (1999) Phytoremediation of petroleum-contaminated soils. Chapter 28. In: Adriano DC, Bollag J-M, Frankenburger WT Jr, Sims RC (eds) Bioremediation of contaminated soils. Agronomy monograph 37. American Society of Agronomy, Madison, WI. 772

Scora RW, Chang AC (1997) Essential oil quality and heavy metal concentrations of peppermint grown on a municipal sludge-amended soil. J Environ Qual 26:975–979

Seth CS (2012) A review on mechanisms of plant tolerance and role of transgenic plants in environmental clean-up. Bot Rev 78:32–62

Shevyakova N, Cheremisina A, Kuznetsov V (2011) Phytoremediation potential of *Amaranthus* hybrids: antagonism between nickel and iron and chelating role of polyamines. Russ J Plant Physiol 58(4):634–642

Shim D, Kim S, Choi YI, Song WY, Park J, Youk ES, Jeong S-C, Martinoia E, Noh E-W, Lee Y (2013) Transgenic poplar trees expressing yeast cadmium factor 1 exhibit the characteristics necessary for the phytoremediation of mine tailing soil. Chemosphere 90(4):1478–1486

Siddiqui F, Krishna SK, Tandon PK, Srivastava S (2013) Arsenic accumulation in *Ocimum* spp. and its effect on growth and oil constituents. Acta Physiol Plant 35(4):1071–1079

Siemianowski O, Barabasz A, Kendziorek M, Ruszczyńska A, Bulska E, Williams LE, Antosiewicz DM (2014) AtHMA4 expression in tobacco reduces cd accumulation due to the induction of the apoplastic barrier. J Exp Bot 65(4):1125–1139

Sinha S, Mishra RK, Sinam G, Mallick S, Gupta AK (2013) Comparative evaluation of metal phytoremediation potential of trees, grasses, and flowering plants from tannery-wastewater contaminated soil in relation with physicochemical properties. Soil Sediment Contamin 22(8): 958–983

Song WY, Park J, Eisenach C, Maeshima M, Lee Y, Martinoia E (2014) ABC transporters and heavy metals. In: Plant ABC transporters. Springer International Publishing, New York, pp 1–17

Soudek P, Petrová Š, Vaněk T (2012) Phytostabilization or accumulation of heavy metals by using of energy crop sorghum sp. In: In: 3rd international conference on biology, environment and chemistry IPCBEE. IACSIT Press, Singapore

Stancheva I, Geneva M, Boychinova M, Mitoval MY (2014) Physiological response of foliar fertilized *Matricaria recutita L.* grown in industrially polluted soil. J Plant Nutr 37(12): 1952–1964

Straczek A, Duquene L, Wegrzynek D, Chinea-Cano E, Wannijn J, Navez J, Vandenhove H (2010) Differences in U root-to-shoot translocation between plant species explained by U distribution in roots. J Environ Radioact 101(3):258–266

Sun SK, Chen Y, Che J, Konishi N, Tang Z, Miller AJ, Zhao FJ (2018) Decreasing arsenic accumulation in rice by overexpressing OsNIP1;1 and OsNIP3;3 through disrupting arsenite radial transport in roots. New Phytol 219:641–653

Sylvia DM (2005) Mycorrhizal symbioses. In: Sylvia DM, Fuhrmann JJ, Hartel PG, Zuberer DA (eds) Principles and applications of soil microbiology. Pearson, Prentice Hall, NJ, pp 263–282

Taştan BE, Ertuğrul S, Dönmez G (2010) Effective bioremoval of reactive dye and heavy metals by *aspergillus versicolor*. Bioresour Technol 101(3):870–876

Tchounwou PB, Yedjou CG, Patlolla AK, Sutton DJ (2012) Heavy metal toxicity and the environment. Mol Clinic Environ Toxicol:133–164

Uera RB, Paz-Alberto AM, Sigua GC (2007) Phytoremediation potentials of selected tropical plants for ethidium bromide. Environ Sci Pollut Res 14(7):505–509

Ugrekhelidze D, Korte F, Kvesitadze G (1997) Uptake and transformation of benzene and toluene by Plant leaves. Ecotoxicol Environ Saf 371June) 24 29:0147–6513

Vandenbossche M, Jimenez M, Casetta M, Traisnel M (2014) Remediation of heavy metals by biomolecules: a review. Crit Rev Environ Sci Technol 45(15):1644–1704

Vangronsveld J, Herzig R, Weyens N, Boulet J, Adriaensen K, Ruttens A, Thewys T, Vassilev A, Meers E, Nehnevajova E, van der Lelie D, Mench M (2009) Phytoremediation of contaminated soils and groundwater: lessons from the field. Environ Sci Pol 16:765–794

Varennes A, de Torres MO, Coutinho JF, Rocha MMGS, Neto MMPM, De-Varennes A (1996) Effects of heavy metals on the growth and mineral composition of a nickel hyperaccumulator. J Plant Nutr 19:669–676

Verma R, Dwivedi P (2013) Heavy metal water pollution-a case study. Recent res. Sci Technol 5:5

Verma SK, Singh K, Gupta AK, Pandey VC, Trevedi P, Verma SK, Patra DD (2014) Aromatic grasses for phyto management of coal fly ash hazards. Ecol Eng 73:425–428

Vo VM, Nguyen VK, Le VK (2011) Potential of using vetiver grass to remediate soil contaminated with heavy metals. VNU J Sci Earth Sci 27(3):146–150

Waheed H, Ilyas N, Iqbal Raja N, Mahmood T, Ali Z (2019) Heavy metal phytoaccumulation in leafy vegetables irrigated with municipal wastewater and human health risk repercussions. Int J Phytoremediation 21(2):170–179

Wang XF (2005) Resource potential analysis of ornamentals applied in contaminated soil remediation (in Chinese). A dissertation in Graduate School of Chinese Academy of Sciences, Beijing

Wang Y, Stauffer C, Keller C (2005) Changes in hg fractionation in soil induced by willow. Plant Soil 275:67–75

Wang Y, Zhu C, Yang H, Zhang X (2017) Phosphate fertilizer affected rhizospheric soils: speciation of cadmium and phytoremediation by *Chlorophytum comosum*. Environ Sci Pollut Res 2: 3934–3939

Yang JL, Wang YC, Liu GF, Yang CP, Li CH (2011) Tamarix hispida metallothionein-like ThMT3, a reactive oxygen species scavenger, increases tolerance against Cd2+, Zn2+, Cu2+, and NaCl in transgenic yeast. Mol Biol Rep 38:1567–1574

Yang T, Chen ML, Wang JH (2015) Genetic and chemical modification of cells for selective separation and analysis of heavy metals of biological or environmental significance. Trend Anal Chem 66:90–102

Zhang J, Zhang M, Tian S, Lu L, Shohag MJI, Yang X (2014) Metallothionein 2 (SaMT2) from *sedum alfredii* Hance confers increased cd tolerance and accumulation in yeast and tobacco. PLoS One 9(7):102750

Zhang Y, Liu J, Zhou Y, Gong T, Wang J, Ge Y (2013) Enhanced phytoremediation of mixed heavy metal (mercury)–organic pollutants (trichloroethylene) with transgenic alfalfa co-expressing glutathione S-transferase and human P450 2E1. J Hazard Mater 260:1100–1107

Zheljazkov VD, Craker LE, Xing B, Nielson NE, Wilcox A (2008) Aromatic plant production on metal contaminated soils. Sci Total Environ 395(2–3):51–62

Zhigang A, Cuijie L, Yuangang Z, Yejie D, Wachter A et al (2006) Expression of BjMT2, a metallothionein 2 from Brassica juncea, increases copper and cadmium tolerance in Escherichia coli and Arabidopsis thaliana, but inhibits root elongation in *Arabidopsis thaliana* seedlings. J Exp Bot 57:3575–3582

Bioremediation of Mining Sites: Sustainable Approach to Restore a Healthy Ecosystem 16

Shekhar Nimbrana, Poonam Ranga, and Anju Malik

Abstract

Ecological damage due to pollutants resulting from industries, mining, and other anthropogenic activities is one of the major global challenges in the rapidly developing world. Thus, removal of toxic organic and inorganic pollutants has become the need of the hour to replenish the health of the ecosystem. Conventional cleanup techniques like chemical precipitation, ion-exchange adsorption, and immobilization are very expensive and create hazardous secondary pollutants having adverse effects on biological activity, soil structure, and soil fertility. Bioremediation is an alternative less expensive strategy to combat toxic pollutants through biological agents like plants, microbes, or natural biological actions to contaminated soil and water. However, it is used very less. Phytoremediation uses plants which accumulate contaminants into roots and aboveground shoots or leaves and transform them to less toxic, more stable, and less mobile form with low economic cost. Microbial bioremediation techniques – biosparging, bioventing, and bioaugmentation – are also advantageous than conventional methods as these are able to mineralize most of the hydrocarbons and organic compounds, use intrinsic microbes, and produce non-toxic byproducts. Bioremediation techniques used for various mining sites in the world have been discussed in this chapter. Significant reduction of heavy metal contamination was found by phytoremediation via hyperaccumulator plant species, etc., microbial remediation via heavy metal-resistant bacteria and fungus species, and genoremediation via genetic engineering methods. Various

S. Nimbrana · A. Malik (✉)
Department of Energy and Environmental Sciences, Chaudhary Devi Lal University, Sirsa, Haryana, India

P. Ranga
ICAR-Central Institute for Cotton Research, Regional Station, Sirsa, Haryana, India

R. P. Singh et al. (eds.), *Heavy Metal Toxicity: Environmental Concerns, Remediation and Opportunities*, https://doi.org/10.1007/978-981-99-0397-9_16

modern biotechnological approaches such as genomics, metabolomics, and transcriptomics for bioremediation have also been discussed. Further, in situ (onsite) bioremediation is more effective than ex situ because it removes pollutants without disturbing normal activities; a complete destruction of target pollutants is possible without transferring the pollutants from one environmental medium to another. Thus, this chapter will describe the advantages of bioremediation techniques – cost reliability, their role in ecosystem restoration, and as sustainable practices to tackle pollutants.

Keywords

Bioremediation · In situ and ex situ bioremediation · Biosparging · Bioventing · Bioaugmentation · Ecosystem restoration

16.1 Introduction

Contaminants in the environment are produced as by-products of various human activities like agriculture, mining, manufacturing, and other industrial activities out of which some are inert and harmless; however, others are toxic and harmful to the ecosystem which affect soil, water, air, human, plants, animals, etc. Heavy metal contamination in the ecosystem is one of the major issues across the globe. Heavy metals are the elements having atomic number greater than 20 and atomic density greater than 5 gcm^{-3} (Nriagu and Pacyna 1988; Hawkes 1997). These are present in the environment naturally or introduced through various anthropogenic activities. Some heavy metals are essential for the certain processes and act as micronutrients (iron, manganese, copper, zinc, etc.); however, they are harmful to life above threshold values (Knox 1999). Some are toxic at very low concentrations such as arsenic, cadmium, chromium, nickel, lead, mercury, etc.

Due to technological advances and industrialization in the last few decades, increased demand of new materials and energy sources has led to exploitation of natural resources at a large scale and enhanced the release of heavy metals in the environment (Mireles et al. 2012; Wei and Yang 2010; Yaylali and Abanuz 2011). This further led to generation of large amounts of waste materials causing environmental pollution. The mining waste is categorized into four major types:

1. Mining waste which contains low-grade ore, overburden, and barren rocks excavated from surface and underground operations. Mining waste is a heterogeneous geological material and may consist of sedimentary, metamorphic, or igneous rocks, soils, and loose sediments.
2. Tailings which are processed wastes from a mill, washery, or concentrator from which the economic metals, minerals, mineral fuels, or coal resource have been removed, generally deposited in tailing dam or pond near the mine sites.
3. Dump heap leach which is a process used for metal extraction from low-grade ore.

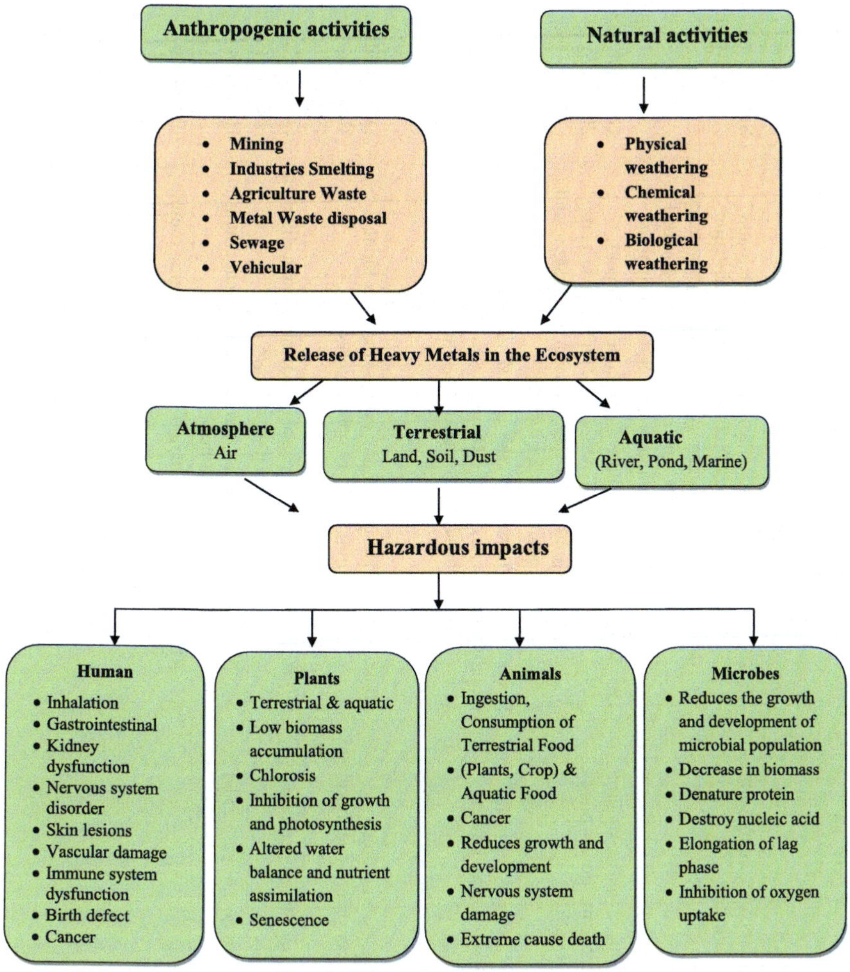

Fig. 16.1 Schematic diagram for the sources of contaminants in the ecosystem, their fate map, and hazardous impacts on human, plants, animals, and microbes

4. Acid mine water which is the water that infiltrates a mine. These mining wastes are emitted into the air, discharged into the water system, or disposed of on land.

Heavy metals are released from mining waste disposal sites, industrial and smelting activities, agricultural waste, and vehicular emissions in the soil and water through physical, chemical, and biological weathering (Fig. 16.1). Heavy metals have long-term persistence in the environment because of their ability to accumulate in the biological systems (Fig. 16.1; Table 16.1). Persistence time of various heavy metals in the environment is 100 to 1 million years (Table 16.1). This metal contamination will further directly or indirectly affect the diversity of living

Table 16.1 Anthropogenic sources of heavy metals, their persistence in the environment and health impacts

S. No.	Heavy metals	Anthropogenic activities	Persistence time (years)	Hazards caused	References
1	Chromium (Cr)	Mining, industrial coolants, chromium salts manufacturing, leather tanning	4.6–18	Occupational asthma, eye irritation and damage, eardrums, respiratory irritation, kidney damage, liver damage, pulmonary congestion and edema, upper abdominal pain, respiratory cancer, skin irritation	Schmidt and Andren 1984
2	Lead (Pb)	Lead acid batteries, paints, e-waste, smelting operations, coal-based thermal power plants, ceramics, bangle industry	150–5000	Affect children's brain development, anemia, hypertension, renal impairment, immunotoxicity, reproductive system, and neurological disorder	Gisbert et al. 2003
3	Mercury (Hg)	Chlor-alkali plants, thermal power plants, fluorescent lamps, hospital waste (broken thermometers, barometers, sphygmomanometers), electrical appliances, etc.	Several months to years	Irritation to the eyes, skin, and stomach, chest pain, difficulty in breathing, irritability, headache, exhaustion, and weight loss	Sprovieri and Pirrone 2010
4	Arsenic (As)	Geogenic/natural processes, smelting operations, thermal power plants, fuel burning	9000	Harmful to the eyes, skin, liver, kidney, lungs, and lymphatic system	Donal and Hernandez 1999
5	Copper (Cu)	Mining, electroplating, smelting operations	5.0×10^4	Vomiting, nausea, abdominal pain, diarrhea, kidney and liver damage	Goldberg et al. 1963
6	Cadmium (Cd)	Zinc smelting, waste batteries, e-waste, fuel combustion	5×10^5	Highly toxic, carcinogenic, damage lungs, kidney, and bones	Hem 1970
7	Molybdenum (Mo)	Spent catalyst	5×10^5	Headache, fatigue, loss of appetite, muscle and joint pain, repeated exposure may raise the uric acid level, damage to the liver and kidney	Crompton 2015

8	Zinc (Zn)	Smelting, electroplating	1.8×10^5	Stomach cramps, skin irritations, vomiting, anemia, damages the pancreas and disturbs the protein metabolism, arteriosclerosis	Roy 1998
9	Beryllium (Be)	Alloy (with Cu), electrical insulators in power transistors, moderator, or neutron deflectors in nuclear reactors	1.5×10^2	Chronic granulomatous lung disease, shortness of breath, unexplained coughing, fatigue, weight loss, fever, and night sweats	Beryllium 2022
10	Cobalt (Co)	Metallurgy (in super alloys), ceramics, glasses, paints	1.8×10^4	Asthma-like allergy, asthma attack, shortness of breath, wheezing, cough and chest tightness, heart attack, liver and kidney damage, thyroid problems	Matthew brennan 2021
11	Iron (Fe)	Cast iron, wrought iron, steel, alloys, construction, transportation, machine manufacturing	1.4×10^2	Hereditary hemochromatosis, liver disease, cancer, heart problem, and diabetes	Dallman et al. 1975
12	Manganese (Mn)	Production of ferromanganese steels, electrolytic manganese dioxide for use in batteries, alloys, catalysts, fungicides, antiknock agents, pigments, dryers, wood preservatives, coating welding rods	1.4×10^3	Damage to the lungs, liver, and kidney, nervous system disorder	Aschner 2008
13	Nickel (Ni)	As an alloy in the steel industry, electroplating, Ni/Cd batteries, arc welding, rods, pigments for paints and ceramics, surgical and dental protheses, molds for ceramic and glass containers, computer components, catalysts	1.8×10^4	Skin irritation, lung damage, affects the stomach and kidney, may lead to cancer	Jose et al. 2019
14	Antimony (Sb)	Type-metal alloy, electrical applications, Britannia metal, pewter, Queen's metal, Sterline, in primers and tracer cells in semiconductors, flameproof pigments and glass, medicines for parasitic diseases, nauseant, expectorant, combustion	3.5×10^5	Respiratory effect, inflammation of the lungs, chronic bronchitis, and chronic emphysema	Boreiko and Rossman 2020

(continued)

Table 16.1 (continued)

S. No.	Heavy metals	Anthropogenic activities	Persistence time (years)	Hazards caused	References
15	Titanium (Ti)	White pigments, UV-filtering agents, nucleation agent for glass ceramics, alloy in aeronautics Tl, used for alloys (with Pb, Ag, or Au) with special properties, electronics industry, infrared optical systems, catalyst, deep temperature thermometers, low melting glasses, semiconductors, supraconductors	1.6×10^2	Slight changes in the lungs, tightness and pain in the chest, coughing and difficulty in breathing, eye and skin irritation	Wojcik et al. 2019

beings, resulting in hazardous and lethal effects on humans, plants, animals, and microorganisms. They directly and indirectly create imbalance in the ecosystem by disturbing the biogeochemical cycles. This is a global concern for environmental health and safety. The sources of heavy metals and their persistence duration in the environment are summarized in Fig. 16.1 and Table 16.1.

Accumulation and persistence of heavy metals due to household and industrial waste disposal in aquatic bodies affect aquatic flora, fauna, and microbes (Fig. 16.1). This further leads to disturbance in the food chain, and ingestion of contaminated aquatic food further affects human health. These pollutants are evident to be the major contributor to the global epidemic of cancer and various degenerative diseases (Fig. 16.1, Table 16.1). For instance, exposure to lead (Pb) causes defects in brain development, anemia, immunotoxicity, and reproductive and neurological disorder; chromium (Cr) exposure leads to asthma and damage to the eardrums, kidney, and liver; cadmium (Cd) is highly carcinogenic and causes kidney, lung, and bone impairment; copper (Cu) exposure causes vomiting, nausea, abdominal pain, diarrhea, and kidney and liver damage. Further, the harmful impact of heavy metals on human health is summarized in Table 16.1.

Heavy metal's accumulation in plant tissues affects their physiological and metabolic activities like photosynthesis and respiration, which leads to premature leaf fall, delayed germination, stunted growth, crop yield reduction, and loss of enzyme activities, even leading to degeneration of main cell organelles and death of plants (Mohanty et al. 2012; Fig. 16.1).

Therefore, heavy metals disturb the whole ecosystem one way or another. Thus, these pollutants need to be remediated to reduce their toxic effects and pollution potential. Various remediation technologies are required for these nondegradable pollutants to decrease or completely remove the toxicity of waste materials in the environment. The conventional methods like chemical stabilization/immobilization, acid mine drainage, soil washing, and soil flushing are used to remediate the waste from mining sites. These methods are expensive as specific engineering costs are required and have adverse impacts on soil structure, soil fertility, and biological health. Thus, more cost-effective and safe methods are required to mitigate or eliminate the noxious impacts of pollutants and make it safe for living beings in a more sustainable manner.

16.2 Bioremediation

Bioremediation is the use of living beings (plants and microorganisms, i.e., bacteria, algae, fungi, yeast, etc. to remediate the pollutants in soil, water, etc. either by removing the toxic elements or reducing the toxicity potential of pollutants (Margesin et al. 2005; Zhao and Poh 2008; Singh et al. 2008). Depending upon the site of remediation, it is of two types: in situ or onsite and ex situ or excavation and removal of contaminated material to a remote location for treatment (Fig. 16.2). In situ bioremediation is the method of on-site remediation. Its efficiency depends upon the site conditions like composition, compaction, etc. Bioventing, biosparging,

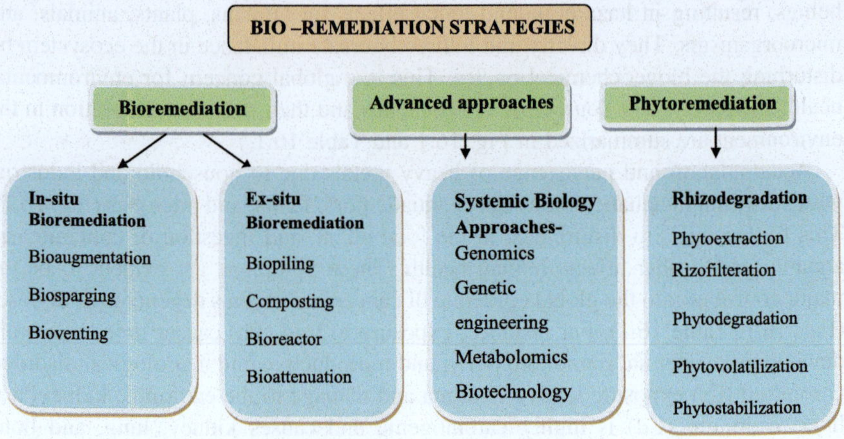

Fig. 16.2 Different strategies of bioremediation

and bioaugmentation are the three main in situ bioremediation strategies. Bioventing is the air ingression and ventilation of ground to enhance microbial action. Biosparging is high-pressure air injection into the soil or underground water table for increasing oxygen concentration to enhance microbial activity. Bioaugmentation is the implantation of exogenous microbe species to the site.

Ex situ bioremediation is the off-site treatment of contaminated material at remote locations. It involves the excavation of polluted soil first and then transfers it to the remote location. Biopiling, composting, bioreactor, and bioattenuation are the methods of ex situ bioremediation. Ex situ bioremediation is the nonpreferred method because it includes high cost of transportation and poses health risk to the personnel.

16.2.1 Methods of Bioremediation

Methods of bioremediation are categorized further based on the living organisms used for remediation. These categories are phytoremediation by plants, microbial remediation by bacteria, mycoremediation by fungi, phycoremediation by algae, and genoremediation by genetic engineering methods.

16.2.1.1 Phytoremediation

It is the remediation of contaminated sites by the plants and their associated microbes (Pilon-Smits 2005). Various mechanisms of bioremediation by plants are phytodegradation, phytovolatilization, phytostabilization, and phytosequestration.

Phytodegradation is the degradation of contaminants by plants through various internal enzymatic reactions and metabolic processes. It is also called phytotransformation. The uptake of contaminants is followed further by subsequent breakdown and metabolization so that contaminants may pass partially or fully through the rhizosphere. Phytodegradation depends on various factors, viz., concentration and composition of contaminants, soil conditions, and the plant species. The contaminants are further subjected to biological processes within the plant itself and phytoextraction can be done. Enzyme catalytic reactions help in reduction and breakdown of contaminants metabolize or mineralize partially into less toxic contaminants or completely into carbon dioxide and water, for example, oxygenases to degrade aliphatic and aromatic hydrocarbons and nitroreductases to reduce and break down the explosives trinitrotoluene (TNT), triazine (RDX), etc. (McCutcheon and Schnoor 2003; Van Aken 2009).

Phytovolatilization is the volatilization of contaminants from the plant either from the root, shoots, or leaves of the plant. The volatized product may be the product of breakdown during rhizodegradation and/or phytodegradation. These contaminants are released in the environment in altered form which is less toxic or nontoxic. This method is used only for highly volatile contaminants. For organic contaminants, the ability to volatilize is derived from Henry's constant and vapor pressure, for example, phytovolatilization of trichloroethene by poplar plants and mercury by tobacco plants and uptake of selenium and its release into the atmosphere as volatile dimethyl selenium by Brassica plants.

Phytostabilization refers to the in situ capturing of contaminated soil and sediments and immobilization of toxic contaminants by vegetation. Windblown dust from the hazardous site is prevented by a rooted vegetation zone. This method is an alternative method where other in situ methods of remediation are not possible. Phytostabilization is generally of two types:

(a) Soil/sediment stabilization: Uncontrolled water flows and blowing wind lead to soil and sediment mobilization known as erosion and leaching. The mechanism to control the erosion is the growing of rooted plants with fibrous root systems such as grasses, herbs, and wetland species to control bulk migration of soil and sediments. Halophytes and hyperaccumulators are also grown as they have phytosequestration ability to accumulate contaminants in aboveground tissues.
(b) Infiltration control: In this case, the interaction between water and waste is prevented to avoid migration of contaminants. This approach is used for minimizing the surface water recharge of groundwater plumes.

Phytosequestration is the process of reducing the mobility and migration of the contaminant to soil, water, and air by plants. There are three mechanisms of phytosequestration:

(a) Phytochemical complexation: Precipitation and immobilization of the contaminants occur in the rhizosphere or root zone, which reduces the bioavailability of contaminants.
(b) Transport protein inhibition on the root membrane: Transport proteins bind to the surface of the root membrane and inhibit the entering of pollutants in the plants through root surfaces and stabilize the contaminants on the root surfaces associated with the exterior.

(c) Vacuolar storage in the root cells: Vacuoles are the compartments in the cell and act as storage and waste receptacle for the plant. Transport proteins facilitate the transfer of contaminants in the vacuoles of root cells and inhibit their transport to the transpiration stream of the plant.

16.2.2 Factors Affecting Bioremediation

The most important parameters for bioremediation are described as follows and illustrated in Fig. 16.3.

Nutrients The nutrients like carbon, nitrogen, phosphate, and potassium are essential for biological activity of microbes. Carbon and nitrogen required for the biodegradation are in a ratio of 25:1 (Atagana et al. 2003).

Nature of pollutants Bioremediation ability also depends on the nature of pollutant which are (a) solid, semi-solid, liquid, or volatile in nature, (b) toxic or nontoxic organic and inorganic pollutants, (c) heavy metals, and (d) polycyclic aromatic hydrocarbons, pesticides, chlorinated solvents, etc.

Soil Structure Effective ingression of air, water, and nutrients is facilitated by granular and well-structured soil. The soil structure contains different textures ranging from low to high contents of sand, silt, and clay.

pH The optimum pH range to facilitate the microbial growth for the bioremediation is 5.5–8.0 (Vidali 2001).

Moisture content Dielectric constant is determined by the moisture content of soil and other media.

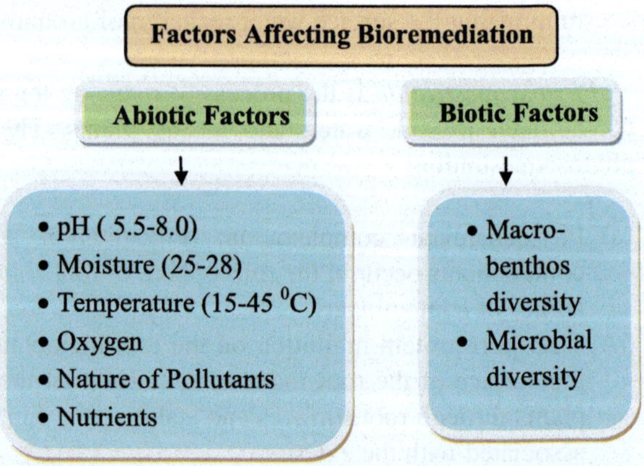

Fig. 16.3 Various factors affecting the process of bioremediation

Microbial diversity The microbial diversity of the site such as *Pseudomonas*, *Aeromonas*, *Flavobacteria*, *Chlorobacteria*, *Corynebacteria*, *Acinetobacter*, *Mycobacteria*, *Streptomyces*, *Bacilli*, *Arthrobacter*, *Aeromonas*, *Cyanobacteria*, etc.

Macrobenthos diversity Consortium of aquatic plants *E. crassipes*, *S. molesta*, and *C. demersum* with aquatic animals *A. woodiana* and *L. hoffmeisteri* have high potential to degrade turbidity, BOD, COD, ammonia, nitrite, and nitrate in domestic wastewater (Mangunwardoyo et al. 2013).

Temperature Optimum temperature ranges from 15 °C to 45 °C. Temperature affects biochemical reaction rates and the rates are double for each 10 °C rise in temperature.

Oxygen The breakdown of the hydrocarbon in the contaminated sites is initiated by the oxygen. It also determines the aerobic and anaerobic conditions of bioremediation.

16.3 Bioremediation of Mining Sites

Bioremediation is used for restoring various mining sites. Various studies for bioremediation of mining sites are summarized in Table 16.2 and also discussed here.

16.3.1 Phytoremediation

Several studies are evident for the use of phytoremediation for the restoration of various mining sites in the world. Significant reductions in the concentrations of lead (30.4%), cadmium (85.8%), and arsenic (55.3%) were found when hyperaccumulator plant species *P. vittata* and *S. alfredii* were grown in agricultural land in South-western China (Huang et al. 2006). Landfill phytocover systems were established by constructing wetland with macrophytes at the mine waste storehouse at National Aluminium Company Ltd. (NALCO)'s Angul Plant in Orissa (Mathiyazhagan and Natarajan 2011). This storehouse waste contains hazardous metals like cyanide and fluoride and is located near the aluminum smelter and coal fly ash slurry. This phytocover system has resulted in minimal infiltration, isolated waste, and controlled landfill gas (Mathiyazhagan and Natarajan 2011). Abandoned landfills are now recovered and abundant tree growth is evident from these sites. Further, significant reductions in Cr VI from mine water at South Kaliapani Chromite mine, Orissa, India, were done by *Eichhornia crassipes* (24–54%) and *Brachiaria mutica* (18–33%) (Mohanty et al. 2012).

The onsite restoration of Cu-contaminated sites was achieved by phytoextraction through Cu-tolerant plants *Helianthus annuus* and *Brassica juncea* (Ariyakanon and Winaipanich 2006; Wilson-Corral et al. 2011). Siderophore-producing bacteria have

Table 16.2 Bioremediation techniques used for treating mining sites

Sr. No.	Study	Location	Technology	Factor	Sites	References
1	Bioremediation of mined wasteland	India	Microorganisms, in situ bioremediation	pH, heavy metals (Cd, Pb, Cu, Fe, Ca, Mg, Zn, Cr, Mn, Pb, Hg)	Kashlog Mining Area, Darlaghat, Solan, Himachal Pradesh, India	Rani et al. 2018
2	Engineering bacteria for bioremediation	Brazil	Microorganisms-bacteria, archaea, fungi, yeast	Ph, EC, and heavy metals	Tailing and soil	Perpetuo et al. 2011
3	Bacterial communities in mining soils and surrounding areas under regeneration process in a former ore mine	Brazil	Bacterial communities in soil	Organic matter, pH, heavy metals (K, P, Zn, Fe, Mn, Cu, Br, Ni)	Mining	Fernandes et al. 2018
4	Bioremediation of Hg-contaminated soil by combining a novel Hg-volatilizing *Lecythophora* sp. fungus, DC-F1, with biochar: performance and the response of soil fungal community	China	Fungal community	Heavy metal-Hg	Mining and soil	Chang et al. 2019
5	Bioremediation methods for the recovery of lead-contaminated soils: a review	Italy	Phytoremediation, fungi, or bacteria	Heavy metal-lead	Mineral rock weathering, melting of metals and anthropogenic sources	Rigoletto et al. 2020
6	Assessing bioremediation of acid mine drainage in coal mining sites using a predictive neural network-	US-EDP156 mining sites	Microbial communities	Sulfuric acid, ferrous sulfate, and associated toxic metals	Mining and Acid Mine Drainage (AMD) sites	Ibeanusi et al. 2012

	based decision support system					
7	Biotechnologies for remediation and pollution control in the mining industry	Peak Gold Mine, Western NSW	Bioventing, biosparging, bioflushing, lagoon treatments	Organic matter, soil with heavy hydrocarbon, heavy metals (cyanide and sulfate reduction for metals removal)	Mining industry, Acid mine drainage	Bernoth et al. 2000
8	Bioremediation on effluents from magnesite and bauxite mines using *Thiobacillus* spp. and *Pseudomonas* spp.	Salem district, Tamil Nadu	Tolerant bacterial species (namely, *P. aeruginosa* and *T. ferrooxidans)*	pH and heavy metals (Cd, Pb, Cu, Fe, Ca, Mg, Zn, Cr, Mn, Hg)	Magnesite and bauxite mines, waste dump	Mathiyazhagan and Natarajan 2011
9	Heavy metal-polluted soils: effect on plants and bioremediation methods	Nigeria	Microorganisms/ phytoremediation	Heavy metals (Co, Cu, Fe, Mn, Mo, Ni, V, and Zn)	Mining and smelting of metals, burning of fossil fuels	Chibuike and Obiora 2014
10	Heavy metal contamination of mine soil, their risk assessment, and bioremediation	Korea	Bioremediation, microbial population	Heavy metals (As, Cd, Cr, Ni, Cu, Pb, Mn, Co, Hg, Se, Sb, Au, Mo, Ag, Ti, Sn, U, V, and Zn)	Trace element, tailings, mining sites	Chon et al. 2017
11	Bioremediation of copper-contaminated soils by bacteria	Switzerland, Belgium, France	Bioremediation microorganisms	pH, heavy metals (Cd, Ni, Zn).	Fungicide spraying and mining sites	Cornu et al. 2017
12	Bacterial communities in mining soils and surrounding areas under regeneration process in a former ore mine	Minas-Gerais, Brazil	Bacterial communities, bioremediation, microorganisms	Heavy metal (P, Ca, Mg, Al, Zn, Fe, Mn, Cu, B, and S), pH, organic matter	Mining sites	Fernandes et al. 2018
13	Bioremediation of a soil contaminated with radioactive elements	Black Sea coast, Southeastern Bulgaria	Microorganisms, bioremediation	Radioactive elements – uranium, radium, and thorium. Toxic heavy metals – copper, cadmium, and lead	Mining and mineral processing	Groudev et al. 2000

(continued)

Table 16.2 (continued)

Sr. No.	Study	Location	Technology	Factor	Sites	References
14	Bioremediation of Pb-contaminated soil by incubating with *Phanerochaete chrysosporium* and straw	Sweden, France	Bioremediation	Heavy metal	Lead (Pb)-contaminated soils	Huang et al. 2006
15	Engineering of microorganisms toward recovery of rare metal ions	Kyoto – Sakyo	Microorganisms, bioremediation	Toxic heavy metal	Anthropogenic activities, heavy metals	Kuroda and Ueda 2010
16	Metal ion resistance in fungi: molecular mechanisms and their regulated expression	Utah	Fungal species, metal toxicity	Toxic heavy metal	Mining, anthropogenic activities, heavy metals	Mehra and Dennis 1991
17	Copper accumulation and changes in soil physical-chemical properties promoted by native plants in an abandoned mine site in northeastern Brazil: implications for restoration of mine sites	Brazil	Plant species, phytoremediation and phytoextraction	Copper- contaminated soils, heavy metals	Mine sites, metallic mine waste	Perlatti et al. 2015
18	Bioremediation of Pb-contaminated soil by incubating with *Phanerochaete chrysosporium* and straw	China	Bioremediation – microorganism	pH, heavy metals, lead-contaminated site	Mining sites	Huang et al. 2006
19	Bioremediation of agricultural and forestry soils	Western Australia – Nedlands	Symbiotic microorganisms, bacteria, and fungi	pH, heavy metals	Contaminated sites, mine waste	Jasper 1994

	with symbiotic microorganisms					
20	A comprehensive overview of elements in bioremediation	India	Bioremediation, phytoremediation, biosorption	pH, heavy metals	Mining sites, oil-contaminated soils	Juwarkar et al. 2010
21	Bioremediation of heavy metals by using bacterial mixtures	South Korea	Bioremediation-bacterial strains	Pb, Cd, Cu, heavy and toxic metals	Mining sites	Kang et al. 2016
22	Bioremediation of copper mine waste: a case study from Mosaboni copper mines, Eastern India	India	Bioremediation-phytoremediation, bacterial community	pH, Mo, Ni, Au, Ag, Pt, Co heavy metals	Mosaboni Copper mining sites (Singhbhum)	Maiti 2010
23	Biotechnological advances in bioremediation of heavy metal-contaminated ecosystems: an overview with special reference to phytoremediation	India	Bioremediation-phytoremediation, bioaccumulation	Anthropogenic activities, heavy metals	Mine waste	Mani and Kumar 2014
24	Bioremediation treatment process through mercury-resistant bacteria isolated from Mithi river	India	Bacterial remediation	pH, EC, TOC, BOD, COD, heavy metals	Mithi river	Pushkar et al. 2019
25	Bioremediation of hexavalent chromium (VI) by a soil-borne bacterium, *Enterobacter cloacae* B2-DHA	India	Bioremediation – *Enterobacter cloacae*	Chromium and chromium, toxic metal	Anthropogenic activities, soil	Rahman et al. 2015
26	Soil bioremediation in heavy metal-contaminated mining areas: a microbiological/	Brazil	Bioremediation – plants and microorganisms	Heavy metals	Mining area	Vieira and Stefenon 2017

(continued)

Table 16.2 (continued)

Sr. No.	Study	Location	Technology	Factor	Sites	References
	biotechnological point of view					
27	Removal of heavy metals and arsenic from contaminated soils using bioremediation and chelant extraction techniques	Greece	Bioremediation – plants and microorganisms	Pb, Zn, As, soils, heavy metals	Mining area	Vaxevanidou et al. (2008)

the ability to stimulate plant growth; thus, they are also used along with plants for boosting Cu phytoextraction. A study from Mosaboni copper mines, Jharkhand, India, showed that plants of *Merremia emarginata* accumulated Cu and Ni, and *Amaranthus* plants accumulated Cu, Ni, and Zn. These plants are resistant to heavy metals and play an important role in the recovery of biogeochemical cycles (Maiti 2010).

16.3.2 Bacterial Remediation

Several mining sites have been remediated by using heavy metal-resistant bacteria species. For instance, *Thermobacillus ferrooxidans* and *Pseudomonas aeruginosa* – *Acidithiobacillus* bacterial species – are evident for removal of toxic and heavy metals from the various mining sites such as magnesite and bauxite mines of Salem District, Tamil Nadu, India (Mathiyazhagan and Natarajan 2011), and bioleaching of metal sulfide ores (Kelly 1988; Straube et al. 2003). These species are resistant to heavy metals and can survive at a low pH environment of metal contaminated tailings and soils. Significant bioremediation effects of *P. aeruginosa* and *T. ferrooxidans* are evident from the concentration of heavy metals (Zn, Mn, Cu, Cr, and Hg) in pre- and posttreated soils from the waste dumps of magnesite and bauxite mines, Tamil Nadu, India (Mathiyazhagan and Natarajan 2011).

Bacteria from the genus *Acidithiobacillus*, *Leptospirillum ferrooxidans*, and *Sulfobacillus thermotolerans* have been successfully used to extract Cu from metallurgical slag (Kaksonen et al. 2011). Chon et al. (2017) reviewed the bioremediation of heavy contamination sites in Korea. Indigenous bacterium isolated from Pb- and Cd-contaminated soil having a tolerance to high Pb (353 mg/kg) and Cd (3.2 mg/kg) toxicity. This bacterium was isolated and identified as *Bacillus thuringiensis* by rRNA sequence analysis. Further, leaching efficiency of *Acidithiobacillus ferrooxidans* and *A. thiooxidans* was observed at highest for Fe and As under the anaerobic conditions and 20 mM glucose supply.

Reduction in the levels of radioactive elements – uranium, radium, and thorium – and heavy metal contaminants such as Cu, Cd, and Pb in the agricultural lands in Southeastern Bulgaria was observed after in situ remediation by the indigenous soil microflora. The contaminants were treated by immobilization into lower soil horizons after their dissolution in the upper soil horizon (Groudev et al. 2000). Lead-contaminated soils were bioremediated by bacteria *Phanerochaete chrysosporium* along with straw in the laboratory. Lower concentration of active Pb, higher microbial mass, and microbial quotient were observed in treated soils as compared to control samples (Huang et al. 2006).

Further, synergistic abilities of bacterial mixtures *Viridibacillus arenosi* B-21, *Sporosarcina soli* B-22, *Enterobacter cloacae* KJ-46, and *E. cloacae* KJ-47 to bioremediate the Pb, Cd, and Cu from contaminated soils from an abandoned mine site in Korea were studied and compared with single culture method. Higher resistance to heavy metals and greater efficiency for bioremediation was observed in bacterial mixtures than single bacterium species (Kang et al. 2016). Remediation ability of *Desulfuromonas palmitatis* – iron-reducing microorganisms in combination with EDTA – was studied on heavy metal-contaminated soils (Vaxevanidou et al. 2008). The bacterial activity was found to have a pronounced positive effect on the removal of arsenic, lead, and zinc.

Fernandes et al. (2018) investigated the mining area in Iron Quadrangle, Minas Gerais, Brazil, for bacterial diversity, metabolic repertoire, and physiological behavior by metagenomics. *Proteobacteria* was found the most abundant phylum followed by *Acidobacteria*, *Verrucomicrobia*, *Planctomycetes*, and *Bacteroidetes*. Audu et al. (2020) studied the bioremediation potential of resident bacterial species in Zamfara mining site, Nigeria, and observed biosorption capacity of *Pantoea agglomerans* for heavy metals (lead, copper, and iron).

16.3.3 Mycoremediation

Several species of fungi are evident for remediation of heavy metals in the mining sites. The fungal species *Paecilomyces javanicus* and *Metarhizium anisopliae* were evident for producing organic acids and have the ability to transform lead into chloropyromorphite at lead mining area in Scotland (Rhee et al. 2012). Further, Chang and coworkers isolated multimetal-resistant fungus *Lecythophora* sp. *DC-F1* from the mining area soil. This study found the *Lecythophora* fungus evident for reducing the bioavailability of mercury by volatilization, ~86% of Hg (II) within 16 hours (Chang et al. 2019).

Therefore, microbial remediation is a very promising approach for heavy metal remediation; however, so far it has been used for remediation at laboratory scale and not or very less for on-site remediation.

16.3.4 Genoremediation

Enhanced ability of bioremediation by the genetic modification and gene introduction in bacteria and plant species is known as genoremediation. The genes encoding for enzymatic reduction, metabolic activities, bioaccumulation, and metal tolerance are introduced into wild species of organisms to produce transgenic one. Genetic modification in tobacco plants has been done to make them able to take up methylmercury (highly toxic) which is further altered chemically to less toxic elemental mercury, which phytovolatizes at low concentrations into the atmosphere. It is evident that mercury in air poses lower risk than in the soil (Lyyra et al. 2007). The enhanced metal tolerance and accumulation ability has been evident in transgenic poplar, *B. juncea*, and rice plants (Banuelos et al. 2005; Rugh et al. 1998; Heaton et al. 2003). Cadmium accumulation in roots of *Nicotiana tabacum* is observed after the gene MRP7 expression (Wojas et al. 2009). The genetic modification in membrane intrinsic proteins (MIPs) subfamilies resulted in 50% and 60% arsenic accumulation in rice stems and roots, respectively (Ma et al., 2008).

Genoremediation is an advanced and promising approach for bioremediation of mining sites, but there is lesser attention on this aspect.

16.4 Conclusion

Industrial and mining activities are essential driving factors for the economic growth of a nation, which are causing depletion of natural resources and harm to the ecosystem. The sustainable approach to create the balance between economic growth and environmental health is the major concern across the globe. Several studies described in this chapter investigated the effects of bioremediation of heavy metals in the laboratory as well as on mining sites. Various conventional methods used to remediate heavy metals are costly and create secondary pollutants and are destructive to the ecosystem. Bioremediation is the best method to remediate the heavy metal pollution because of reduced cost of remediation and exposure risk for cleanup personnel, in situ applications, ability to eliminate waste permanently, minimum equipment requirement, and noninvasive technique. Advanced technologies like transcriptomics, proteomics, and metabolomics are in use for profiling of heavy metal resistance microbes and plants. Genetic engineering is used for introduction of heavy metal-resistant genetic sequences in plants and microbes to create recombinant strains for bioremediation. Therefore, bioremediation is the best method to create balance between economic growth and ecosystem protection from heavy metal pollution and, hence, sustainable approach for ecosystem restoration.

References

Ariyakanon N, Winaipanich B (2006) Phytoremediation of copper contaminated soil by *Brassica juncea* (L.). Czern and Bidens alba (L.) DC. Var. radiata. J Sci Res Chula Univ 31:49–56

Aschner M (2008) Manganese: brain transport and emerging research needs. Environ Health Perspect 108:429–443

Atagana HI, Haynes RJ, Wallis FM (2003) Optimization of soil physical and chemical conditions for the bioremediation of creosote-contaminated soil. Biodegradation 14:297–307

Audu KE, Adeniji SE, Obidah JS (2020) Bioremediation of toxic metals in mining site of Zamfara metropolis using resident bacteria (*Pantoea agglomerans*): A optimization approach. Heliyon 6(8):e04704. https://doi.org/10.1016/j.heliyon.2020.e04704. PMID: 32904083; PMCID: PMC7452395

Banuelos G, Terry N, LeDuC DL, Pilon-Smits EA, Mackey B (2005) Field trial of transgenic Indian mustard plants shows enhanced phytoremediation of selenium-contaminated sediment. Environ Sci Technol 39:1771

Bernoth L, Firth I, McAllister P, Rhodes S (2000) Biotechnologies for remediation and pollution control in the mining industry. Biomineralizat Minerals Metallurg Proc 17(2):105–111

Beryllium (2022) Toxicological profile for beryllium. U.S Department of health and human service. Agency for toxic substance and disease Registry

Boreiko and Rossman (2020) Antimony and its compounds: health impact related to pulmonary toxicity, cancer and genotoxicity. Toxicol Appl Pharmacol 403:115156

Brennan M (2021) Cobalt strikes again: an analysis of obfuscated Malware. Huntress blog

Chang J, Duan Y, Dong J, Shen S, Si G, He F, Yang Q, Chen J (2019) Bioremediation of Hg-contaminated soil by combining a novel Hg-volatilizing Lecythophora sp. fungus, DC-F1, with biochar: Performance and the response of soil fungal community. Sci Total Environ 671: 676–684

Chibuike GU, Obiora SC (2014) Heavy metal polluted soils: effect on plants and bioremediation methods, applied and environmental soil science. Hindawi Publishing Corporation, p 12

Chon HT, Lee JU, Lee JS (2017) Heavy metals contamination of mine soil, their risk assessment, and bioremediation, assessment, restoration and reclamation of mining influenced soils. Elsevier Inc, pp 387–417

Cornu JY, Huguenot D, Lollier KJM, Lebeau T (2017) Bioremediation of copper-contaminated soils by bacteria. World J Microbiol Biotechnol 33:26

Crompton TR (2015) Determination of metals in natural waters, sediments and soils. Chemical Analysis Department Water Authority, Leeds Yorkshire

Dallman et al (1975) Brain iron: persistent deficiency following short term iron deprivation in the young rat. Br Soc Haematol 31(2):209–215

Fernandes CC, Kishia LT, Lopes EM, Omorib WP, Antonio J, Souzab MD, Lucia Maria L, Alves C, Lemos EGM (2018) Bacterial communities in mining soils and surrounding areas under regeneration process in a former ore mine. Braz J Microbiol 49:489–502

Gisbert et al (2003) A plant genetically modified that accumulates Pb is especially promising for phytoremediation. Biochem Biophys Res Commun 2:440–445

Goldberg LI, McDonald RH, Zimmerman AM (1963) Sodium diuresis Produced by Dopamine in Patients with Congestive Heart Failure. The New England J Med 269:1060

Groudev SN, Georgiev PS, Spasova K, Komnitsas (2000) Bioremediation of a soil contaminated with radioactive elements. Hydrometallurgy 59:311–318

Hawkes SJ (1997) What Is a Heavy Metal? J Chem Educ 74:1374

Heaton AC, Rugh CC, Kim T, Meagher RB (2003) Toward detoxifying mercury-polluted aquatic sediments with rice genetically engineered for mercury resistance. Environ Toxicol Chem 22: 2940–2947

Hem J D (1970) Study and interpretation of the chemical characteristics of Natural water, Second Edition, USGS science for a changing world, U.S Geological survey water-supply paper 2254

Hernandez DJ (1999) Department of health, Olympia Book, Children of Immigrants: Health, Adjustment, and Public Assistance. National Research Council, Division of Behavioral and Social Sciences and Education, Commission on Behavioral and Social Sciences and Education, Committee on the Health and Adjustment of Immigrant Children and Families, p 672

Huang DL, Zeng GM, Jiang XY, Feng CL, Yu HY, Huang GH, Liu HL (2006) Bioremediation of Pb-contaminated soil by incubating with *Phanerochaete chrysosporium* and straw. J Hazard Mater B 134:268–276

Ibeanusi VM, Jackson E, Coffen J, Jeilani Y (2012) Assessing bioremediation of acid mine drainage in coal mining sites using a predictive Neural Network-Based Decision Support System (NNDSS). J Bioremed Biodegr 3:3–5

Jasper DA (1994) Bioremediation of agricultural and forestry soils with symbiotic micro-organisms. soil biology and biochemistry. Aust J Soil Res 32:1301–1319

Jose et al (2019) Nickel induced transcriptional changes persist post exposure through epigenetic reprogramming. Epigenetics Chromatin 12:75

Juwarkar AA, Singh SK, Mudhoo A (2010) A comprehensive overview of elements in bioremediation. Rev Environ Sci Biotechnol 9:215–288

Kaksonen AH, Lavonen L, Kuusenaho MK, Kolli A, Närhi HM, Vestola EA, Puhakka JA, Tuovinen OH (2011) Bioleaching and recovery of metals from final slag waste of the copper smelting industry. Miner Eng 24:1113–1121

Kang CH, Kwon YJ, Seong JS (2016) Bioremediation of heavy metals by using bacterial mixtures. Ecol Eng 89:64–69

Kelly DP (1988) Evolution of the understanding of the microbiology and biochemistry of the mineral leaching habitat. Science and Technology Letters, Kew, Surrey, pp 3–14

Knox AS (1999) Sources and practices contributing to soil contamination, bioremediation of contaminated soils. Am Soc Agron 37:158–162

Kuroda K, Ueda M (2010) Engineering of microorganisms towards recovery of rare metal ions. Appl Microbiol Biotechnol 87:53–60

Lyyra S, Meagher RB, Kim T, Heaton A, Montello P et al (2007) Coupling two mercury resistance genes in Eastern cottonwood enhances the processing of organomercury. Plant Biotechnol J 5: 254–262

Ma JF, Yamaji N, Mitani N, Xu XY, Su YH et al (2008) Transporters of arsenite in rice and their role in arsenic accumulation in rice grain. Proc Nat Acad Sci USA 105(29):9931–9935

Maiti SK (2010) Bioremediation of copper mine waste: a case study from Mosaboni copper mines, Eastern India. Int J Environ Pollut 43:1–3

Mangunwardoyo W, Sudjarwo T, Patria MP (2013) Bioremediation of effluent wastewater treatment plant Bojongsofang Bandung Indonesia using consortium aquatic plants and animals. Int J Res Rev Appl Sci 14(1):150–160

Mani, Kumar (2014) Biotechnological advances in bioremediation of heavy metals contaminated ecosystems: an overview with special reference to phytoremediation. Int J Environ Sci Technol 11:843–872

Margesin R, Fonteyne P, Redl B (2005) Low-temperature biodegradation of high amounts of phenol by *Rhodococcus* spp. and *Basidiomycetous* yeasts. Res Microbiol 156:68–75

Mathiyazhagan N, Natarajan D (2011) Bioremediation on effluents from magnesite and bauxite mines using thiobacillus Spp and Pseudomonas Spp. J Bioremed Biodegr 2:1

McCutcheon SC, Schnoor JL (eds) (2003) Phytoremediation – transformation and control of contaminants. Wiley Interscience, p 985

Mehra RK, Dennis R, Winge (1991) Metal ion resistance in fungi: molecular mechanisms and their regulated expression. J Cell Biochem 45:30–40

Mireles F, Davila JI, Pinedo JL, Reyes E, Speakman RJ, Glascock MD (2012) Assessing urban soil pollution in the cities of Zacatecas and Guadalupe, Mexico by instrumental neutron activation analysis. Microchem J *103*:158–164

Mohanty M, Pattnaik MM, Mishra AK, Patra HK (2012) Bio-concentration of chromium- an in situ phytoremediation study at South Kaliapani chromite mining area of Orissa, India. Environ Monit Assess 184:1015–1024

Nriagu JO, Pacyna J (1988) Quantitative assessment of worldwide contamination of air, water and soil by trace metals. Nature 333:134–139

Perlatti F, Ferreira TO, Romero RE, Otero MCGC (2015) Copper accumulation and changes in soil physical–chemical properties promoted by native plants in an abandoned mine site in northeastern Brazil: Implications for restoration of mine sites. Ecol Eng 82:103–111

Perpetuo EA, Souza CB, Nascimento CAO (2011) Engineering bacteria for bioremediation. In: Progress in molecular and environmental bioengineering – from analysis and modeling to technology applications, pp 605–626

Pilon-Smits E (2005) Phytoremediation. Annu Rev Plant Biol 56:15–39

Pushkar B, Sevak P, Singh A (2019) Bioremediation treatment process through mercury-resistant bacteria isolated from Mithi river. Appl Water Sci 9:1–17

Rahman A, Nahar N, Nawanic NN, Jassb J, Hossaind K, SauddN ZA, Ananda K, Sahae GS, Olssona B, Mandala A (2015) Bioremediation of hexavalent chromium (VI) by a soilborne bacterium, Enterobacter cloacae B2-DHA. J Environ Sci Health Part A Toxic/ Hazardous Substances Environ Eng 50(11):1136–1147

Rani N, Sharma HR, Kaushik A, Sagar A (2018) Bioremediation of mined waste land. Springer International Publishing, Handbook of environmental materials management

Rhee YJ, Hillier S, Gadd GM (2012) Lead transformation to pyromorphite by fungi. Curr Biol 22: 237–241

Rigoletto M, Calza P, Gaggero E, Malandrino M, Fabbri D (2020) Bioremediation methods for the recovery of lead-contaminated soils: a review. Appl Sci 10:3528

Roy SK (1998) Impact of zinc supplementation on persistent diarrhoea in malnourished Bangladeshi children. Acta Paediatrica 87(12):1235–1239

Rugh CL, Senecoff JF, Meagher RB, Merkle SA (1998) Development of transgenic yellow poplar for mercury phytoremediation. Nat Biotechnol 16(10):925–928

Schmidt, Andren (1984) Serum levels of TPA in patients with gynaecological cancer. In: Protides of the biological fluids, pp 441–444

Singh S, Kang SH, Mulchandani A, Chen W (2008) Bioremediation: environmental clean-up through pathway engineering. Curr Opin Biotechnol 19:437–444

Sprovieri F, Pirrone N (2010) A review of worldwide atmospheric mercury measurements. Atmos Chem Phys 10:8245–8265

Straube WL, Nestlen CC, Hansen LD, Ringleberg D, Pritchard PJ et al (2003) Remediation of poly aromatic hydrocarbons (PAHs) through landforming with biostimulation and bioaugmentation. Acta Boitechnol 23:179–196

Van Aken B (2009) Transgenic plants for enhanced phytoremediation of toxic explosives. Curr Opin Biotechnol 20:231–236

Vaxevanidou K, Papassiopi N, Paspaliaris I (2008) Removal of heavy metals and arsenic from contaminated soils using bioremediation and chelant extraction techniques. Chemosphere 70: 1329–1337

Vidali M (2001) Bioremediation. An overview. Pure Appl Chem 73(7):1163–1117

Vieira JD, Stefenon VM (2017) Soil bioremediation in heavy metal contaminated mining areas: a microbiological/biotechnological point of view. J Adv Microbiol 4(1):1–10

Wei B, Yang L (2010) A review of heavy metal contaminations in urban soils, urban road dusts and agricultural soils from China. Microchem J 94(2):99–107

Wilson-Corral V, Anderson C, Rodriguez M, Lopez-Perez J (2011) Phytoextraction of gold and copper from mine tailings with Helianthus annuus L. and Kalanchoe serrata L. Miner Eng 24: 1488–1494

Wojas S, Henning J, Plaza S, Geisler M, Siemianowski O et al (2009) Ectopic expression of Arabidopsis ABC transporter MRP7 modifies cadmium root-to-shoot transport and accumulation. Environ Pollut 157(10):2781–2789

Wojcik et al (2019) Effect of Titanium dioxide Nanoparticles Exposure on human health- a Review. Biol Trace Elem Res 193:118–129

Yaylali, Abanuz G (2011) Heavy metal contamination of surface soil around Gebze industrial area, Turkey. Microchem J 99(1):82–92

Zhao B, Poh CL (2008) Insights into environmental bioremediation by microorganisms through functional genomics and proteomics. Proteomics 8:874–881

Industrial Pollution Management Approach 17

Geeta Singh and Seema Mishra

Abstract

Every industry uses water in its operations and hence produces wastewater, also called industrial effluent. Industrial effluent contains inorganic salts, heavy metals, pesticides, hazardous organic compounds, microbial contamination, and turbidity. Apart from this, industries also have their own effluents like chemical and biological agents in the wastewater. Hence, industries are a major source of water pollution. The present generation is witnessing harmful effects of such contaminants on aquatic, marine animal, and human health. Stringent quality requirements have been imposed on all the industrial plants to water pollution by effluent. The effluent treatment approach is a combination of different techniques and technologies which is used to treat wastewater so it can be released into the environment. These processes include several primary, secondary, and tertiary treatment and additional treatment steps of effluent treatment. These steps include processes such as filtration, coagulation, aerobic and anaerobic biodecomposition, active carbon adsorption, and disinfection using chemicals or ultraviolet (UV) light, etc. These steps leave a lot of industry-specific contaminants in the water. With the use of advanced technology, it is now possible to achieve minimal or zero liquid discharge (MLD/ZLD) and recovery of water and valuable resources, such as salts. In this chapter, we focused on the different procedures, applications, adaptability, and limitations of these modern techniques available to each industry in a single text. It also serves as a tool for personnel seeking information on different wastewater treatment technologies.

G. Singh (✉)
Department of Chemistry, Deen Dayal Upadhyay Gorakhpur University, Gorakhpur, India

S. Mishra
Department of Chemistry, Deen Dayal Upadhyay Gorakhpur University, Gorakhpur, India

Department of Chemistry, University of Lucknow, Lucknow, India

R. P. Singh et al. (eds.), *Heavy Metal Toxicity: Environmental Concerns, Remediation and Opportunities*, https://doi.org/10.1007/978-981-99-0397-9_17

17.1 Introduction

In India, industrialization, urbanization, and growth in population created an environmental imbalance due to resource exploitation and negligence (see Fig. 17.1). Environmental pollution is becoming a serious concern and creating environment conflicts. Different contaminants like organic matter, chemicals, heavy metal, pesticides, suspended solids, inorganic salts, radioactive waste, oil spillage, and many more when present in river, lake, and groundwater and create health problems to living organisms are called water pollution.

Water pollution by industrial waste is a matter of concern for the society, public, and industries (Sonune and Ghate 2004; Sharma 2015; Rathoure and Dhatwalia 2016). An effective effluent management approach minimizes the volume of waste, decreases the toxicity of wastewater, and recycles and reuses the waste materials. As our finance minister stated in his budget speech in 2019, only 4% of drinking water is available for the 1.5 billion population which is 18% of the world population. (https://www.worldbank.org/en/country/india). Apart from this, India also did not have a proper wastewater treatment system to manage the situation. The wastewater treatment approach needs immediate attention because this field is very emerging. Traditional treatment processes which have been used for many years are not sufficiently effective and stable for every kind of waste. This needs to be improvised for better efficiency and performance by the engineers and technologists in innovative ways to serve for vast numbers of wastes and industries. A balance between environment and development is very important. The imbalance in the environment can be maintained by the highly efficient industrial pollution management approach with waste minimization.

Figure 17.1 shows a depiction of how the industrial effluent is released directly without treatment into water bodies.

Fig. 17.1 Wastewater from industries is released directly in water bodies

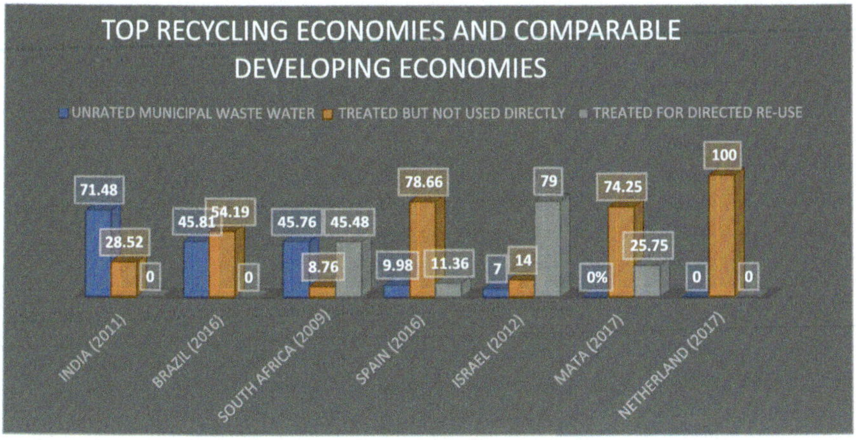

Fig. 17.2 Wastewater treatment and reuse (source: Food and Agricultural Organisation of the United Nations Data)

India has 17.7% of the world population and 2.4% of surface area, but in the case of water resources, it has less than 4%. According to the Indian NITI Aayog report of 2018, approx. 600 million Indians are facing a huge water shortage to serve their basic needs, and this water shortage problem could also affect the loss in GDP by 2050 (NITI Aayog 2019). This water shortage issue in India can be rectified by the proper waste water treatment plans, because if 80% of untreated waste water can be reused by treatment, 75% of demands can be achieved by 2025 and also will help in saving the GDP (WSP 2016) (Composite Water Management Index 2018). Approximately 33% of India urban wastewater is treated properly and can be possible to reuse (Singh 2019). The percentage of effluent treatment is higher in developed countries. The percentage of wastewater generated, treated, and reused of some developing countries are given in Fig. 17.2).

17.1.1 Industrial Effluents

Industrial effluents are all kinds of impurities like inorganic and organic harmful compounds, microbes, etc. generated by various industries. Along with industries' main products, lots of by-products are also formed, which are harmful for the environment and need special attention for their further treatment. These by-products could be cost-effective, so the industries keep them for their processing and reuse them. But every industry did not have the proper effluent treatment plans, so they also discharge the remaining waste either they are in the form of solid, liquid, or gases into the water bodies like ponds, lakes, canals, rivers, or sea and solid wastes are dumped on the land or in aquatic bodies, hence increasing the landfill (Awuchi et al. 2020a). Whatever their mode of entry is, they adversely affect the environment. Some important industries which are most polluting are caustic soda, cement,

Table 17.1 Different industrial wastes and the pollution created by them

S. No.	Industry	Produced wastes	Type of pollution
1	Caustic soda	Mercury, chlorine gas	Air, water, and land
2	Distillery	Organic waste	Land and water
3	Fertilizer	Ammonia, cyanide, oxides of nitrogen and sulfur	Air and water
4	Dye	Inorganic waste pigment	Land and water
5	Iron and steel	Smoke, gases, coal dust, fly ash,	Air, water, and land
6	Pesticides	Organic and inorganic waste	Water and land
7	Oil refineries	Smoke, toxic gases, organic waste	Air and water
8	Paper and pulp	Smoke, organic waste	Air and water
9	Sugar	Organic waste, molasses	Land and water
10	Textiles	Smoke, particulate matter	Land and water
11	Tanneries	Organic waste	Water
12	Thermal power	Fly ash, SO_2 gas	Air and water
13	Nuclear power station	Radioactive wastes	Water and land
14	Food processing	Alkalis, phenols chromates, organic wastes	Water and land

Source: https://www.environmentalpollution.in/industrial-pollution/industrial-pollution-types-effects-and-control-of-industrial-pollution/299

brewery, dyes, fertilizers, iron and steel, oil refineries, paper and pulp, food, pesticides, pharmaceuticals, textiles, thermal power plants, leather, etc. (Table 17.1).

Mostly, effluents have a major number of inorganic salts like calcium and magnesium salts, different carbonates, bicarbonates, and various metal salts, and also, they have poisonous gases, turbidity, organic matter, and microorganisms. The data of treated and untreated wastewater from Indian cities of 30 years are given in Fig. 17.3.

17.1.2 A Glance of Basic Terms Used for Impurities/in the Analysis

Turbidity: Turbidity is the dissolved particles, which reduce the clarity of the water.

pH: pH is the negative log of hydrogen ion concentration.

Alkalinity: It is hydroxyl ions present in the sample.

Volatile fatty acids: These are by-products of anaerobic degradation of organic matter from wastewater.

Total dissolved solids (TDS)/total suspended solids (TSS): These are the amounts of matter dissolved or suspended in water.

Nitrogen: The presence of organic and ammoniacal nitrogen is evidence of recent organic pollution of animal origin.

Chloride: Chloride in water may be present due to dissolution of salt deposits.

Dissolved mineral matter: Soluble inorganic salts.

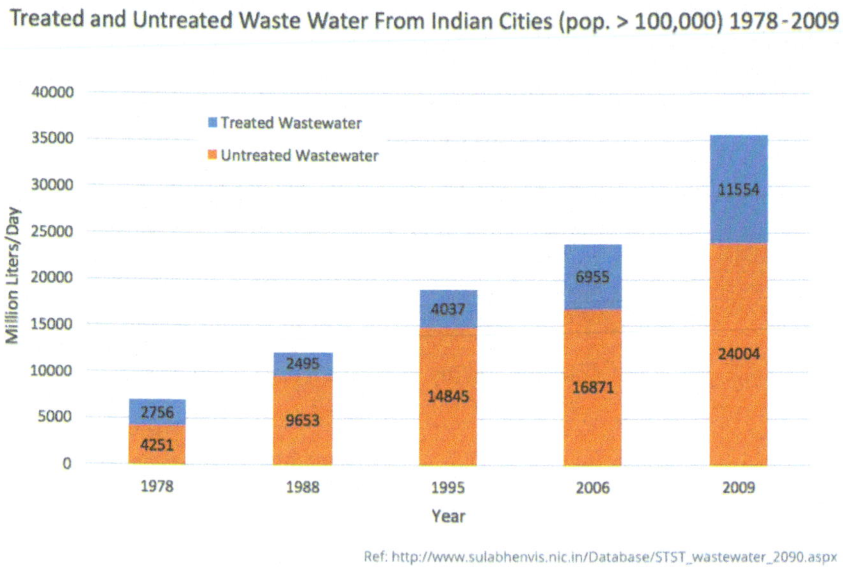

Fig. 17.3 Data from 1978 to 2009 of treated and untreated wastewater

Dissolved gases: CO_2, O_2, N_2, H_2S, and CH_4.

Microorganisms: Very small animals and plants, difficult contaminants to remove.

Chemical oxygen demand (COD): It is the amount of oxygen required to oxidize the polluting chemicals to CO_2 and H_20. The maximum acceptable COD is 90 ppm.

Biological oxygen demand (BOD): It is the amount of oxygen required by the biological microbial mass during the effluent treatment to oxidize the biologically pollutants.

Waste Characterization: A waste characterization depends upon the industrial processes that generate a waste. Some industries also characterized their waste from time to time. For this purpose, one should have proper knowledge about the specific industry and the detailed information on processes that generate wastes. This information is very important for the better management of waste. There are some useful factors for waste characterization like their physical status, volume of waste generation, composition of waste, and the constituents of waste for their effective processing. Consult with national and local regulatory agencies for determination of their specific testing requirements and select an appropriate analysis based on the above information (EPA Guide).

17.1.3 Major Sources of Industrial Wastewater

The major source of wastewater is industries like leather industry, textile industry, paper industries, pulp industries, wood industries, fertilizer industries, sugar industries, pharmaceutical industries, food processing industries, and so on which

Table 17.2 Value of wastewater produced in food processing industries

Industry	Range of ton product
Fruits and vegetables	
• Green beans	12,000–17,000
• Peaches and pears	3600–4800
• Other fruits and vegetables	960–8400
Food and beverage	
• Beer	2400–3840
• Bread	480–960
• Meat packing	3600–4800
• Milk products	2400–4800
• Whiskey	14,400–19,200

Source: Ministry of Food Processing Industries, Government of India (http://mofpi.nic.in)

are contributing to this problem (Asamudo et al. 2005). So, if we took the example of only food industries, the food production produces wastewater from many steps like washing and rinsing, sorting, peeling, juicing, blanching, cooking, and cleanup. The amount of wastewater produced from food industries is given in Table 17.2.

As described in Table 17.2, food industries create huge waste and their treatment is also expensive because the waste consists of various organic carbon, nitrogenous organics, inorganics, and suspended and dissolved solids and has high biochemical and chemical oxygen demands (https://www.watertechonline.com/wastewater/article/15550688/wastewater-treatment-challenges-in-food-processing-and-agriculture). They must be treated very efficiently so that they will not harm the receiving waters due to excessive nutrients or oxygen demand when directly discharged to sewers (European Environment Agency 2001).

Other industries like the electronic industry generated very toxic contaminants like toxic heavy metals, oil, and grease. Textile industries involve various processing of clothes like yarn manufacturing, cleaning, and fabric finishing. Different industries produce different pollutants containing effluents with various characteristics. The safety values of toxic contaminants in the industrial effluent are regulated by governmental authorities from time to time. Kulkarni et al. (2000) reported contaminants and their discharge levels in the effluent as given by the Indian Pollution Control Board (Table 17.3).

Heavy metals like lead, cadmium, chromium, arsenic, nickel, zinc, etc. are when present in the wastewater effluents adversely affect humans, animals, and plants. The summarized form of the heavy metal sources and their adverse effects on health are given in Table 17.4.

17.1.4 Risk Assessment Involved in Waste Management

As the name suggests, risk assessment is the estimation of the risk/hazard which can occur in a planned activity. So, in the case of the waste management approach, we first evaluate the risk assessment in a systematic and defined way to avoid any

Table 17.3 Toxic contaminants and their maximum discharge levels (source: Kulkarni et al. 2000)

Sl. No.	Parameters	Maximum discharge level
1	Mercury	0.01 mg/l
2	pH	5.5–9
3	Suspended solids	250 mg/l
4	Biochemical oxygen demand	150 mg/l
5	Temperature	Shall not exceed 5 °C above the ambient temperature of the receiving body
6	Free available chlorine	0.5 mg/l
7	Oil and grease	10 mg/l
8	Copper	3 mg/l
9	Iron	3 mg/l
10	Zinc	5 mg/l
11	Chromium	2 mg/l
12	Phosphate	5 mg/l
13	Sulfide	2 mg/l
14	Phenolic compounds	5 mg/l
15	Hexavalent chromium	0.1 mg/l
16	Nickel	3 mg/l
17	Cadmium	2 mg/l
18	Chloride	9000 mg/l
19	Sulfate	1000 mg/l
20	Cyanides	0.2 mg/l
21	Ammoniacal nitrogen	50 mg/l
22	Lead	0.01 mg/l
23	Total metal	10 mg/l

consequences later. This process is important to understand the basics principles by everyone like the public, industries, and all interested stakeholders before making any environmental decision. Being a good citizen, we all want our waste to be managed safely and perfectly so that we and our mother nature are safe. Risk assessment has several steps mainly risk identification, exposure check, and risk characterization. Therefore, firstly the identification source/location of risk like the waste management unit is important. Then, the pathway of risk like toxic chemicals and heavy metal should be avoided to the people for exposure to ensure their safety, and direct drainage from the industries to the water bodies should be strictly inhibited. Finally, we characterize the risk to know how much the hazard can affect us and what kind of health problems we can face if we come across the hazard. In this process, health standard information and exposure assessment must be integrated and analyzed to arrive at quantitative estimations of various health problems like cancer risks. I must say that the risk assessment procedure is an ongoing process in any study.

Table 17.4 Sources and toxic effects of heavy metals (Lohchab and Saini 2017)

Heavy metals	Source	Harmful effect
Chromium	Discharge from steel, textile manufacturing, electroplating and pulp mills; erosion of natural deposits, etc.	Gastrointestinal hemorrhage, hemolysis, acute renal failure, pulmonary fibrosis, lung cancer
Copper	Metal cleaning, plating baths, pulp and paper industry, fertilizer industry, copper/brass plating, corrosion of pipes, erosion of rocks, etc.	Gastrointestinal distress, liver or kidney damage
Cadmium	Electroplating, paint pigments, plastics, alloy preparation mining and silver-cadmium batteries; metal refinery discharge; corrosion of pipes, erosion of rocks, etc.	Pneumonitis, proteinuria, lung cancer, osteomalacia
Nickel	Processing of minerals, paints, electroplating, enameling of porcelain, etc.	Allergic sensitization, lung and nervous system damages, and dermatitis
Zinc	Printed circuit board manufacturing, metal electroplating, painting, dying, photography, etc.	Dermatitis, pneumonitis, stomach pain, nausea, lethargy, dizziness, and muscle incoordination (Bishnoi and Garima, 2005)
Lead	Batteries, smelting and alloying, paints, some types of solders, etc.	Nausea, vomiting, encephalopathy, headache, anoxia, anemia, abdominal pain, nephropathy, foot drop/wrist drop
Mercury	Old paint, industrial pollutants, leaded gasoline	Inflammation of gums and mouth, kidney disorder, neurotic disorder

17.2 Need for Industrial Wastewater Treatment

The saying "the solution to pollution is dilution" is famous. All of the mentioned industries produce their effluents continuously and this wastage is released into the environment (Awuchi and Echeta, 2019) and causes harm to aquatic organisms. The industrial pollution management approach is not well organized in most of the developing countries, thereby creating a big mess (Dao-Tuan et al. 2018; Awuchi et al. 2020b). Different industries produce different effluents and need specific way of treatment. If we talk about India, to manage the wastewater, lots of challenges have been encountered due to rapid urbanization, weak quality control systems, lack of resources, and strict regulations. With the help of new techniques, we can minimize this harm by making recycled treated wastewater. These industries can use this refined water for their personal use like boiler feed water, chemical synthesis, and other usage. Many advanced processes are established for a long time, but there is always a scope for improvement (Crini and Lichtfouse 2019). Technologists and researchers are tirelessly looking for the improved version of processes with great efficiency and performance. Sustainability is everything that is required for our well-being and it depends on our natural environment. The main application of advanced technology should be focused on nutrient removal and recovery, removal of all contaminants, water security, energy conservation, conservation of renewable energy sources, etc.

17.3 Methods of Industrial Wastewater Treatment

17.3.1 Fundamental Methods

Every industry has their responsibility to treat their waste properly and they are also benefited by the treatment of wastewater. The primary benefit of industrial waste treatment is the reuse of treated effluents by meeting compliance of regulatory bodies like CPCB, SPCB, etc. Other benefits are improved water quality, increased employment in construction and operation of wastewater treatment plants, and increased recreation uses, such as fishing, boating, and swimming, as a result of increased purity of water. Some major benefits of wastewater treatment are environmental sustainability, good public relations and improved industrial image, improved mental health of citizens, and improved conservation practices (Nemrow 2005).

The different physicochemical properties of industrial wastewater for different parameters are determined by using standard methods of analysis of wastewater (APHA 2005) as listed in Table 17.5.

The varying contaminations of wastewater also need a variety of methods to remove the contamination (Tchobanoglous et al. 2003). When we study the numerous publications and review articles, we clearly see that several industrial effluent treatment methods are proposed and mentioned by different industries like

Table 17.5 Parameters and method of analysis (APHA 2005)

S. No.	Parameters	Method of analysis
1	Colors	Visual
2	Temperature	Thermometric
3	pH	pH meter
4	EC	EC analyzer
5	Alkalinity as (CaCO3)	Titrimetric
6	Chloride	Argentometric
7	Total solids (TS)	Gravimetric
8	Total dissolved solids (TDS)	Gravimetric
9	Total suspended solids (TSS)	Gravimetric
10	Total volatile fatty acids (TVFA)	Distillation
11	Total Kjeldahl nitrogen (TKN)	Kjeldahl method
12	Total phosphate as (PO_4)	Spectrophotometer
13	BOD (biochemical oxygen demand)	Azide modification
14	COD (chemical oxygen demand)	Dichromate reflux
15	Sulfate	Spectrophotometer
16.	Sodium and potassium	Flame photometric

coagulation, precipitation, oxidation, solvent extraction, evaporation, carbon adsorption, ion exchange, membrane filtration, electrochemistry, and different biological processes (Henze 2001, Chen 2004; Forgacs et al. 2004; Anjaneyulu et al. 2005; Crini et al. 2007, Morin-Crini et al. 2017). But it's very difficult and confusing to choose the most appropriate method.

An industrial wastewater purification process consists of different levels like preliminary treatment (physical and mechanical), primary treatment (physicochemical and chemical), secondary treatment (biological processes), tertiary treatment (physical and chemical advanced level), and treatment of the sludge formed. All these steps like primary, secondary, and tertiary treatment and biological unit processes like aerobic and anaerobic and their applicability also depend on the composition of waste (Crini 2005; Cox et al. 2007). At first the pollutants are removed manually, and then primary physical operations used are screening, flocculation, sedimentation, floatation, and filtration. Afterward, pollutants are removed by converting them into nonharmful products. Chemical processes mainly change the pollutant composition by reaction of pollutants with chemicals present in the treatment system. Biological agents like microorganisms convert the organic substances in presence or absence of oxygen through enzymatic action. Predominant chemical processes are coagulation, precipitation, adsorption, and disinfection. The overall purpose of different treatment techniques is waste minimization, and the waste can be minimized by adopting techniques of volume reduction, strength reduction, neutralization, and equalization and proportioning. The main focus in industry should be on waste minimization through volume reduction, and it can be achieved by better process control, improved equipment design, use of different or better-quality raw materials, good housekeeping, and preventive maintenance (Nemerow 2007). The next objective is strength reduction of waste and it can be done by change of process, equipment modification, by-product recovery, equalization, and proportioning of waste. Then we can check the pH of waste and try to neutralize that with some neutralizing agents. Without neutralization of waste, the microbes in biological treatment systems may be killed or inactivated and the more acidic or basic waste can be harmful for the flora and fauna of the aquatic system.

17.3.2 Conditions on Which Fundamental Methods Are Chosen

As of now, we have learned about industrial effluents; their sources, composition, and constitution; the need for effluent treatment; and some fundamental practices which we consider prior to serving our purpose. All industrial effluents have their own composition and they require some specific type of treatment among primary, secondary, tertiary, or some more advanced methods. The review of literature on this topic is very vast and versatile, so before choosing the methods, we should consider some fundamental criteria of that particular wastewater stream. The major criteria that required our attention are characteristics of the wastewater stream (components, salt content, suspended solids, flow rate, variation in water quality or quantity), desired removal or purification rate (emission limits, process requirements), technical feasibility (removal rates, fouling, corrosion, scale-up experience), local economic background (energy and chemical costs), infrastructure (site condition, availability of energy, chemicals and technical personnel, resources, existing effluent treatment plan), local engineering and operation staff, investment and operating budget, and time schedule (Leonhauser et al. 2014).

17.3.3 Treatment Levels

Treatment levels are mainly preliminary, primary, secondary, tertiary, or advanced. The purpose of the preliminary level is water conditioning. Physical treatment is used for removal of suspended solids and chemicals for pH adjustment and removal of toxic metals. Secondary treatment is used to remove biodegradable organic materials (Kesalkar et al. 2012). A summarized form of different processes/methods is shown in Table 17.6 and Fig. 17.4.

Primary treatment is the first step in which suspended solids (SS) are separated from the wastewater. As a result of surface charge on small particles in the effluent, larger solid particulates remain in a suspended manner in it. This is also termed as MLSS (mixed liquor suspended solids) (Kulkarni et al. 2014). Sedimentation and screening operations usually accomplish the separation process in primary treatment. The method of removing large matters from sewage by passing through screens is termed as screening. We use a tool named screen with uniform-size openings for removing bigger suspended or floating matter in sewage. Screens are classified as (1) racks or bar screens, (2) perforated or fine screens, and (3) comminuters or cutting screens (Warren et al. 1993).

Secondary treatment is the further treatment of the effluents left from the primary level and, it is typically performed by indigenous water-borne microorganisms in a managed habitat. Basically, in secondary treatment, degradation of biological content of the sewage derived from human waste, food waste, soaps, and detergent was conducted by indigenous water-borne microorganisms in a

Table 17.6 Summarized methods of different levels of industrial effluent treatment

Primary level	Secondary level	Tertiary level
• Screening like coarse screen, medium screen, and fine screen • Flocculation • Sedimentation • Floatation • Membrane filtration • Skimmers	• Activated sludge process • Aerated lagoon • Trickling filter • Biological processes (aerobic and anaerobic)	• Precipitation, coagulation • Sand filtration • Activated carbon adsorption • Electrodialysis • Reverse osmosis • Advanced oxidation processes • Ozonation ion exchange, UV Chlorination and ozonation

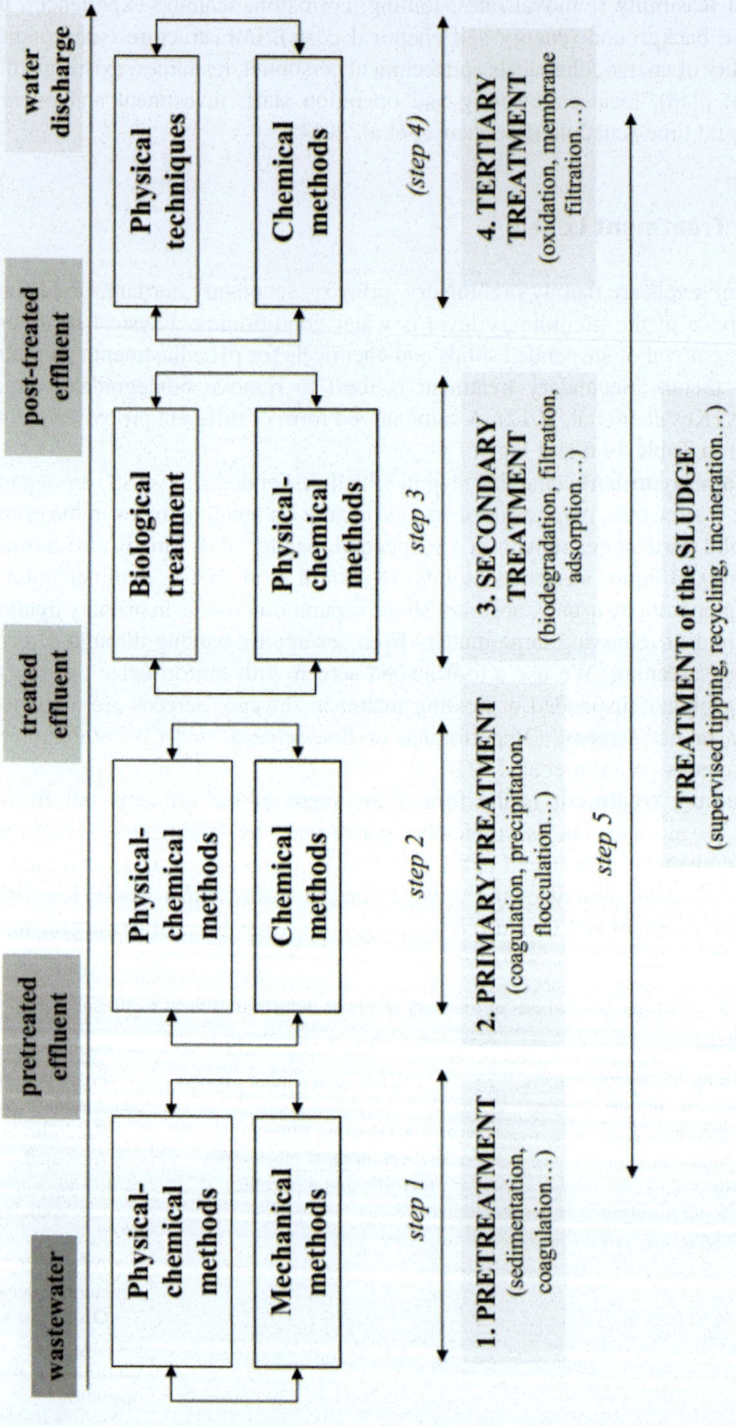

Fig. 17.4 Different levels of industrial effluent management (source: Crini and Lichtfouse 2019)

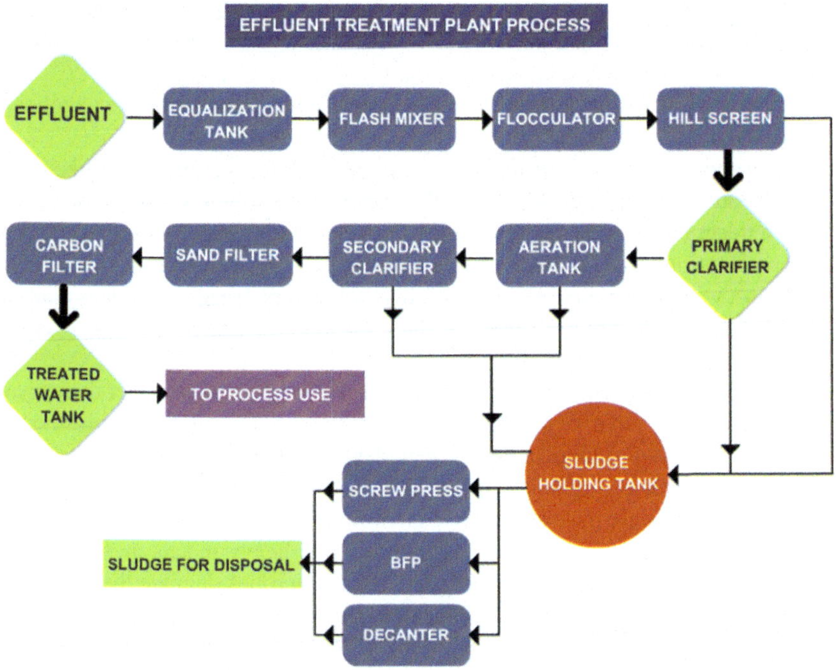

Fig. 17.5 A schematic diagram of ETP workstation (source: https://www.nrail.com/etp.html)

managed habitat. Residual suspended solids and the organic matter with high BOD are removed by different biological processes. The effluents from secondary treatment have less BOD and low suspended solids (SS) value than primary treatment. The most biologically degradable organic materials are removed during biological treatment, but some resistant material which is known as refractory organic materials still left as a residue in the effluents. These materials might be end products of biological decompositions or artificial products, such as oils, chlorophenols, nitro-compounds, synthetic detergents, pesticides, PCE, etc. (Sharma and Sanghi 2013).

Tertiary treatment is the third and the advanced level of wastewater treatment. This step removes stubborn leftover wastes that were completely removed in the secondary treatment. The unit operations and processes in tertiary treatment include precipitation, activated carbon adsorption, coagulation, filtration, electrodialysis, reverse osmosis, advanced oxidation processes like wet oxidation process, Brine treatment, ozonation, etc. and, at the same time, are often used for local treatment of industrial effluents (Crittenden et al. 2005, I.D.A.Y. 2011). These unit operations and processes are applied in treatment of specific components of waste such as refractory organics removal, heavy metal removal, etc.

The conceptual approach of the treatment includes the removal of suspended particles, dissolved organic matters and handling of sludge for disposal as shown in Fig. 17.5.

Now we discuss about some effluent treatment methods used in different industries according to their necessity:

- Removal of suspended solids is mainly performed by filtration, coagulation, sedimentation, and floatation.
- The methods employed for the removal of dissolved inorganic solids like heavy metals are also very important. So, the treatment of solids is also known as the brine treatment which includes evaporation, electrodialysis, ion exchange, membrane filtration processes like reverse osmosis, etc. The evaporation processes are the most widely used because they are more tolerant of organics, hardness salts, or hydrocarbons; in addition, they produce the highest purity effluent, even the distillate quality.
- Organic dissolved solids are result of discharge of organic wastewater in the water bodies, which result in depletion of oxygen because microorganism consumed the dissolved oxygen to oxidize the biodegradable organic matter.
- For the removal of organic matters, the biological methods are most efficient. Microorganisms are most sensitive to environmental conditions like temperature, pH, dissolved oxygen level, toxic chemicals like heavy metals, pesticides, and amount of organic matter in wastewater. These microbes degrade the biodegradable organic matter present in industrial wastewater. These methods can be aerobic or anaerobic depending upon the depth and mechanism of purification. Aerobic treatment means treatment of organic waste of wastewater in presence of oxygen, while anaerobic treatment means treatment of organic wastes of wastewater in absence of oxygen. Aerobic systems include lagoon, activated sludge processes, completely mixed system, aeration ponds, trickling filters, and aerobic composting. Anaerobic digestion takes place in four steps, i.e., hydrolysis, acidogenesis, acetogenesis, and methanogenesis.
- The removal of some persistent other organic materials like chemical solvents, pesticides, dyes, medicinal wastes, and paint which are difficult to treat with conventional methods and need some specific treatment. These methods include distillation, ozonation, advanced oxidation processing, vitrification, incineration, adsorption, and landfill disposal.

Some of the important processes which involved for the industrial effluent treatment are as follows.

17.3.3.1 Sedimentation

In sedimentation, heavy particles settle to the bottom. Sludge word used for particles that gather at the bottom of a container. This process depends on viscosity of wastewater and size and specific gravity of suspended particles. Sedimentation collects and settles down solids, which are then transported to drying beds to dry out the sludge.

17.3.3.2 Filtration

Coagulation and filtration are the conventional water purification systems. Filtration is the process of passing water through a filter or porous media that removes particulates. Filters are made of sand and charcoal layers, and the filter membranes used are in different sizes. It's a primary level process because only small particles (less than 0.1 mm can be removed in filtration). The microorganisms that are less than this size can't be treated so they go for further treatment. Filtrations are of many types like rapid sand (remove suspended solids) and flow sand filtration. Flow sand filtration was an old method, whereas rapid sand has many advancements (http://www.who.int/water_sanitation_health/dwq/S12.pdf.).

17.3.3.3 Coagulation

This technique involves adding aluminum or iron salts like aluminum sulfate, ferric sulfate, or ferric chloride to the untreated water. These compounds are known as coagulants and they have a positive charge (Government of Canada Drinking water report, 2015).

In this, tiny dirt particles stick together after mixing. These collected particles together generate larger, heavier particles, or coagulate (this process is sometimes also called flocculation). After that, these bended particles were easily removed through sedimentation and filtration. This process is the primary level of the wastewater treatment; after this, water goes for the secondary treatment because coagulation is not able to remove microorganism like virus and bacteria. Figure 17.6 clearly defines the basic function of coagulation, flocculation, and sedimentation.

17.3.3.4 Floatation

Floatation is the process in which suspended solids and colloidal substances come to the surface of water with the help of air/gas bubbles as the transport medium. The impurities higher than water will also be elevated to the surface of the floatation tank with the help of buoyant forces of air bubbles. The floatation technique has major applications in the removal of a wide variety of heavy metal ions. So, floatation contributes a major role for the purification of water and also to the effective separation of various pollutants (George and Kosta 2018).

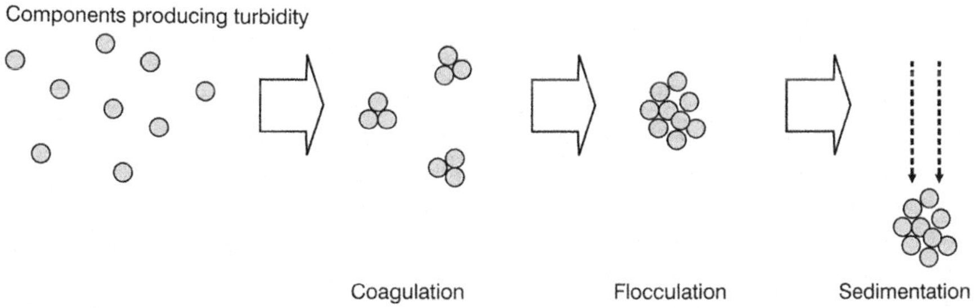

Fig. 17.6 Process of coagulation, flocculation, and sedimentation in one frame (source: https://www.safewater.org/fact-sheets-1/2017/1/23/conventional-water-treatment

17.3.3.5 Evaporation

Evaporation helps to vaporize the water and hence dissolve solids changed into residues at the bottom. It is a very basic technique like sedimentation which we often performed in the laboratories for separation of components. It is an effective application for heavy metals, removing salts and various toxic chemicals from solution. Evaporation also produces a high-quality, reusable distillate where water conservation is a priority.

17.3.3.6 Skimming

Skimming is a primary level process to remove floating matters like oil with the help of different skimming tanks.

17.3.3.7 Electrodialysis

Dialysis is the diffusion of solute from higher concentration to lower concentration with the help of selective permeable membrane. It is used to purify the solution of different chemicals. In electrodialysis, ionic components of a solution are separated through ion selective membranes. Different cations and anions move to their respective electrodes through the membrane.

17.3.3.8 Ion Exchange Method

Ion exchange method is similar to the adsorption theory. In this process we use zeolites due to their pore size and adsorbent properties. Basically, they are aluminosilicate minerals and behave as ion exchangers. In this method, atoms with higher charge have a greater affinity for ions with smaller hydrated radii that are preferentially adsorbed. It helps in the removal of nitrates and phosphates, some special toxic ions, and hardness of water by exchange of their sodium ions for calcium and magnesium ions.

17.3.3.9 Reverse Osmosis

It is a well-known process which is used in our household also for drinking water purification from tap water. In this, the basic principle is just opposite than osmosis. That is to say, water from the solution passes through a semipermeable membrane and solute remains in the solution. By applying pressure more than osmotic pressure across the membrane, solvent is made to pass from high strength wastewater to low strength wastewater. In this way dissolved solids can be removed by using reverse osmosis.

17.3.3.10 Activated Sludge Process

Activated sludge is also a biochemical method used for the wastewater treatment with the help of microorganism and air. In this method, a typical activated sludge process includes an aeration tank with oxygen supply through the surface aerator mixed into the wastewater (Subramani and Arulalan 2012). In the next step, settling tank is used to allow the waste sludge to properly settle. A portion of the waste sludge is recycled into the aeration tank and the remaining waste sludge removed for additional treatment and final disposal.

17.3.3.11 Hydrocyclone Oil Separators

As the name tells, a cyclone kind of atmosphere is created in hydrocyclone oil separator for the effluent treatment with very high centrifugal force (very high rpm). This high centrifugal force separates the water and oil droplets. Separators are designed in such a way that the both separated oil and water are discharged from the different ends of the cyclone. Finally, treated water is collected for further treatment.

17.3.3.12 Trickling Filter Process

This process is a kind of aerobic treatment that utilizes microorganisms attached to a medium to remove organic matter from wastewater. Filters are made up of a fixed bed of rocks, coke, gravel, slag, polyurethane foam, sphagnum peat moss, ceramic, or plastic media over which sewage or other wastewater flow downward. Waste treatment with trickling filters is a well-characterized treatment technology. A microbial layer is formed and it involves the adsorption of organic compounds in wastewater. Air diffusion into the slime layer gives the oxygen necessary for the biochemical oxidation of organic compounds (https://sswm.info/factsheet/trickling-filter). A typical complete trickling filter system is shown in Fig. 17.7.

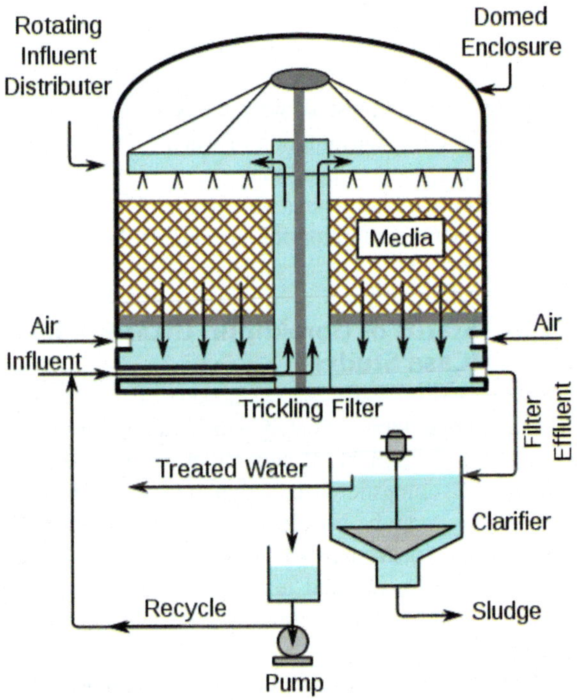

Fig. 17.7 A typical complete trickling filter system (source: https://www.wikiwand.com/en/Trickling_filter)

17.3.3.13 Use of Smart Capsules

Lead is a highly toxic metal which could cause many health problems if present in discarded industrial wastewater. Encapsulation technology is a potential advanced method for the removal system for lead ions and some other ions from polluted sources. The three types of capsules under investigation include carbon nanotubes, polymer swelling capsules, and alginate-based capsules. These capsules provide likely means for remediation of contaminated water (Tylkowski and Jastrząb 2017). This technology has many benefits over the other methods like its simplicity, its wide spectrum, selective extractants, large special interfacial area, ability for concentration of metal ions from dilute solutions, and less discharge to the environment.

17.3.3.14 Wet Air Oxidation (WAO)

WAO is a well-established technique for wastewater treatment, especially for toxic and wastewater. It is a flameless combustion procedure. Its driving force is enhanced solubility of oxygen in aqueous solutions at different conditions. WAO results in the oxidation of organic compounds to CO_2 and H_2O and into other less hazardous products like acids.

17.4 Applications of Effluent Treatment Plan

The major applications of effluent treatment plan are as follows:

- To recycle and reuse of wastewater
- To reduce the consumption of water
- To reduce water expenditure
- To meet regulatory standard
- To reduce the pollution for sustainable development

17.5 Status of Gorakhpur, Uttar Pradesh, in Effluent Treatment (Case Study)

The Gorakhpur city is situated in the plain of Saryu of mid-Gangetic valley and at the confluence of Rapti and Rohini River (en.wikipedia.org/wiki/Gorakhpur). Approximately 70% domestic water supplied is released as wastewater. Types of industries in Gorakhpur can be largely split up into food products, cotton textiles, beverages, tobacco and tobacco products, wool, silk and synthetic fiber textiles, terracotta products, jute, hemp and textiles, hosiery and garments, leather products, wood products, paper products, printing materials, chemical and chemical products, transport equipment and parts, rubber and plastic products, metal products, nonmetallic mineral products, basic metal industries, machineries, electrical machinery and apparatus, repairing and servicing industries, etc. According to the Uttar Pradesh pollution control board recent data of 2020 (UPPCB), a total of 252 industries are running except small-scale industries (https://cpcb.nic.in/Actionplan/Gorakhpur.pdf). The monitoring of pollution sources has been supervised by the UP Pollution Control Board with the help of District Ganga Committees/Zila Paryavaran Samitis. Establishment of sewage treatment plants is done by UP Jal Nigam and Nagar Nigam Gorakhpur, and regular monitoring of river water quality is done by UPPCB/District Environment Committee. A sewage treatment plant in Ramgarh Tal lake, Gorakhpur, UP, has been shown in Fig. 17.5. As per the information given by Gorakhpur development authority (GDA) office who is working tirelessly in cleaning of the lake, there are in total 24 drains falling in Ramgarh Tal of which 6 are major drains and 18 are minor drains. All these drains falling directly in Ramgarh Tal have been intercepted and diverted and are being treated in STPs 15 MLD and 30 MLD capacities each. Work on tapping and diversion of all the minor drains are under process (http://www.indiaenvironmentportal.org.in/files/file/Ramgarh-lake-Gorakhpur-pollution-report-NGT.pdf). Gorakhpur Development Authority is setting up a STP (sewage treatment plant) near Padley Ganj that is not sufficient, and according to a recent study (Kumari and Singh 2017), the value of various physico-chemical properties of the lake is way higher than limits by the general standard for India set by the Central Pollution Control Board (Fig. 17.8).

17.6 Effects of Industrial Pollution

17.6.1 On Human Health

- Irritation of eye, nose, throat, respiratory tracts, asthmatic attacks, etc.
- Increases mortality rate and morbidity rate.
- Chronic pulmonary diseases like bronchitis and asthma can happen due to high concentration of SO_2, NO_2, particulate matter, and photochemical smog.
- Heavy metals like lead, cadmium, mercury, and arsenic may enter the body through lungs and cause poisoning.
- Synthetic pesticides can remain as residues to create adverse health effects, such as respiratory problems, reproductive issues, endocrine system disruption, neurological damage, and cancers.

17.6.2 On Animal Health

- Contaminants accumulate and magnify in the environment and affect the animals.
- Poisoning of the animals may happen if they consumed the contaminated food.
- All the toxic metals (fluorine, arsenic, chromium, and lead) and pesticides are responsible for most livestock damage.

Fig. 17.8 STP in Ramgarh Tal lake, Gorakhpur, UP

17.6.3 On Plants

Environmental pollution has serious adverse effects on plants. The major pollutants affecting plants are SO_2, O_3, NO_2, NH_3, HCN, ethylene, herbicides, peroxyacetyl nitrate, etc. In the presence of pollutants, the healthy plants suffer from neurosis, chlorosis, abscission, epinasty, etc.

17.7 Control of Industrial Pollution

The utmost aim is for the control of pollution to maintain safety of man, material, and machinery. The implementation of control measures should be based on the principle of recovery or recycling of the pollutants and must be taken as an integral part of production to make waste an asset. Some important control measures are:

1. **Control at source:** It involves suitable alterations in the choice of raw materials and process in treatment of exhaust gases before finally discharged in the environment.
2. **Selection of industry site:** The industrial site should be properly examined considering the climatic and topographical characteristics before establishment.

3. **Industrial waste management:** The industrial wastes should be properly treated before their discharge. There should be strict rules and regulations for the treatment of waste.
4. **Plantation:** Intensive plantation in the region considerably reduces the dust, smoke, and other pollutants.
5. **Stringent government action:** Government should take stringent action against industries which discharge higher numbers of pollutants into the environment than the level prescribed by the Pollution Control Board.
6. **Environmental risk assessment:** Environmental risk assessment should be conducted regularly which intends to identify and evaluate the potential and harmful impacts of the industries on the natural ecosystem.
7. **Implementation of Environmental Protection Act:** Environment Protection Act should be strictly followed and the culprits of the environment should be punished strictly.

17.8 Industries Approach Toward Waste and Limitation of Industries

In 2019, the Indian economy was the world's fifth largest by nominal GDP and third largest by purchasing power parity. There are more than 680,000 + manufacturing companies in India (en.wikipedia.org/wiki/List_of_companies_of_India). If we talk about the textile industry, a recent report published by the US government says that 50 on 80 garments are found to have NPE (nonylphenol ethoxylates), which is a toxic chemical that is released during the processing of clothes. The waste created by textile industries, mainly deposits in landfill, can cause serious effects to humans and ecology (https://waste-management-world.com/a/trash-talking-textile-recycling) (Environmental Protection Agency 2017). Decomposition of such materials releases toxic greenhouse gases and is also polluting the water bodies directly and indirectly (Bertram 2016).

Welspun is the brand that sells recycled fabrics made from recycled plastic bottles (Aishwariya 2018). Two hundred polyethylene terephthalate (PET) bottles can cover up a normal-sized sofa (Charter et al. 2017). There are some more brands like Lutradur ECO that also use disposed drinking PET bottles. The reputed brands like Armani, H & M, Repreve, and Marks and Spencer also use recycled fibers and make fashionable clothing (Francoise et al. 2018). Nike has started the efforts to save the environment by reducing the size of the shoe box, monitoring the effluent discharged, recycling the cloth hangers, and recycling and using the yarns to make new apparels and footwear. Eco-friendly paper-making industry is an upcoming industry that finds old fabrics as an excellent building material. Biomass, agro-waste, and old cotton textiles serve in building the matrix in the handmade paper which is a great effort. (https://www.valmet.com/board-and-paper/recycled-fiber/). Recycled fibers are also used in automobile interiors, agro-textiles, reinforcement in geotextiles, acoustics, textiles for building construction purpose, upholstery, package textiles, and food packing materials (Zander et al. 2017). Among the other industries, the leather industry generates vast amounts of wastewater due to their tanning processes; it also produces 200 times more waste than total product output which occurs in solutions (Chojnacka et al. 2021). The non-tanned waste was used as a raw material for glue, gelatin, technical fats, protein sheaths, and even feed and fertilizers (Katarzyna et al. 2021). Agricultural and food processing wastewater treatment is also important due to the production of methane gas and biogas as a by-product by anaerobic treatment. As we know, methane gas is used in the generation of heat and electrical energy, and it also reduces the volume of waste and the most important factor, i.e., carbon footprint. Food industry is a very vast industry because it includes poultry and meat processing, dairy products, brewery, and oil production which generate high amount and a variety of wastes. So, they need specialized and costly treatment due to the contaminant loadings and the variability of the different wastes encountered in a plant. It's a tough task to summarize the recycling, reusing, and making the industrial waste useful for all the industries (Awuchi and Igwe 2017) (The 2030 WRG 2009). All the industries and regulatory authorities are doing their job to achieve the minimal or zero liquid discharge (MLD/ZLD).

But every field has some limitations. For instance, every process of wastewater treatment has some advantages and disadvantages. Some limitations are the following:

- Poor policy support of prevention by regulatory authorities and governments.
- The disposal of certain wastes, such as batteries and paper, is strongly discouraged because of their potential hazard.
- The lack of guidance for choosing the methods and resources.
- The lack of investments in waste management and other sectors such as transport and energy.
- Insufficient waste management system.
- The unavailability of adequate service infrastructure, such as emergency management, waste recycling and disposal services, sewerage services, reticulated water supply, electricity, gas, communications, and transport access.
- Wastewater discharges are difficult to manage due to pollutant volume, composition, and variability. It is tough to assess the physical and chemical characteristics of the effluent to determine a cost-effective and efficient process for industrial wastewater treatment.
- Its drainage controls to isolate potentially contaminated areas from discharge to the environment.

Environmental pollution is an unfortunate necessity of all industrialization and urbanization. Consequently, a large amount of hazardous chemicals are released into the environment and create the environment pollution (Metcalf and Eddy, 2003, Souza et al. 2004). So, protecting our environment on every level (country and worldwide) and especially the problem of water pollution should be our topmost priority. So, it's important to develop cheap, effective, and novel methods for the treatment of wastewater for the best use of that. To make pollutant-free and drinkable water from industrial wastewater, treatment is a challenging task and most important necessity in the present scenario. However, this is a difficult task (Sonune and Ghate 2004; Anjaneyulu et al. 2005; Crini 2005; Crini et al. 2007; Barakat 2011; Sharma and Sanghi 2013). Every method has its benefits and drawbacks. In this chapter, we have discussed many levels of treatment, and the process involved in that, but only a few are commonly used by the industrial sector for economic and technological reasons. Adsorption method, specially adsorption onto activated carbons, is nevertheless often cited as the procedure of choice to remove many different types of pollutants because it gives the best results in terms of efficiency and technical feasibility at the industrial scale (Crini and Lichtfouse 2019).

17.9 Conclusion

We are dealing with water scarcity, not only in India but worldwide also. Water can be saved by adhering strictly to these industrial pollution management approaches. We know about the adverse effect of these pollutants on environment and human health, so industrial waste management should ensure the safety of the environment as well as human health. The first step of waste management should be segregation of waste according to their characterization. Industrial wastewater treatment involves the processes used for wastewater treatment produced by industries as an undesirable by-product. Treatment levels of wastewater are often identified as primary, secondary, and tertiary. In these levels various methods are employed according to the waste characterization and resources available. With the use of advanced technology, it is now possible to achieve the zero liquid discharge (ZLD) and maximize the recycle/reuse of the waste. Risk assessment is also a very important step for the evaluation of potential risk. Firstly, separation of suspended/dissolved solids from the wastewater has been treated in primary level treatment, and then further treatment of the effluent was performed by secondary treatment by biological processes. Tertiary treatment is the advanced level of wastewater treatment which includes various techniques to deal with stubborn wastes. In the case of India, its cities need an effective and dedicated effort from everyone including the government, regulatory authorities, industry personnel, and even the citizens also. A complete (100%) reuse and recycling of wastewater gives a win-win situation for everyone mentioned above that even our mother nature will be thankful to us. There should be establishment of marketing of treated water to decrease the use of scarce freshwater and freshwater conservation can be encouraged. A joint effort of government agencies, nongovernment organizations, and the public is needed to overcome this problem.

Acknowledgment Geeta Singh and Seema Mishra are grateful to the University Grants Commission, New Delhi, for the research grant UGC-Start up grant (F.30- 505/2020(BSR)) and to Department of Science and Technology (SEED Division), New Delhi, for the project grant SEED/TIASN/2018/74.

References

Aishwariya S (2018) Waste Management Technologies in Textile Industry. Innov Energy Res 7(3): 211. https://doi.org/10.4172/2576-1463.1000211

Anjaneyulu Y, Sreedhara CN, Samuel SRD (2005) Decolourization of industrial effluents–available methods and emerging technologies: a review. Rev Environ Sci Biotechnol 4(4):245–273

APHA (2005) Standard methods for the examination of water and wastewater, 21st edn. American Public Health Association, Washington, DC, New York

Asamudo NU, Daba AS, Ezeronye OU (2005) Bioremediation of textile effluent using phanerochaete chrysosporium. Afr J Biotechnol 4(13):1548–1553

Awuchi CG, Igwe SV (2017) Industrial waste Management: brief survey and advice to cottage, small and medium scale Industries in Uganda. Int J Adv Acad Res 3(1):26–43. ISSN: 2488-9849

Awuchi CG, Echeta KC (2019) Current developments in sugar alcohols: chemistry, nutrition, and health concerns of sorbitol, xylitol, glycerol, arabitol, inositol, maltitol, and lactitol. Int J Adv Acad Res 5(11):1–33. ISSN: 2488–9849

Awuchi CG, Hannington T, Awuchi CG, Igwe VS (2020a) Industrial waste Management, treatment, and health Issues: wastewater, solid, and electronic wastes. Eur Acad Res 8(2):1081–1119

Awuchi CG, Gospel AC, Otuosorochi AI, Somtochukwu IV (2020b) Industrial and community waste management: global perspective. Am J Phys Sci 1(1):1–16. ISSN 2519-5549. https://www.iprjb.org/journals/index.php/AJPS/article/view/1043

Barakat MA (2011) New trends in removing heavy metals from industrial wastewater. Arab J Chem 4:361–377. https://doi.org/10.1016/j.arabjc.2010.07.019

Bertram RF, Chi T (2016) A study of companies business responses to apparel E-commerce's environmental impact. Int J Fashion Design Technol Education 11:254–264

Bishnoi NR, Garima (2005) Fungus: an-alternative for bioremediation of heavy metal containing wastewater: a review. J Sci Ind Res 64:93–100

Charter M, Polonsky MJ (2017) Greener marketing: a global perspective on greening marketing practice. Routledge

Chen G (2004) Electrochemical technologies in wastewater treatment. Sep Purif Technol 38(1): 11–41

Chojnacka K, Skrzypczak D, Mikula K, Witek-Krowiak A, Izydorczyk G, Kuligowski K, Bandrów P, Kułażyński M.(2021) Progress in sustainable technologies of leather wastes valorization as solutions for the circular economy. J Clean Prod. Volume313, 127902 DOI: https://doi.org/10.1016/j.jclepro.2021.127902 2021.

Composite Water Management Index (2018) A Tool for Water Management.

Cox M, Négré P, Yurramendi L (2007) Industrial liquid effluents. INASMET Tecnalia, San Sebastian, p 283

Crini G, Lichtfouse E (2019) Advantages and disadvantages of techniques used for wastewater treatment. Environ Chem Lett, Springer Verlag 17(1):145–155

Crini G, Montiel AJ, Badot PM (2007) Treatment and purification of polluted industrial water: membrane processes, bioadsorption and chemical oxidation. Presses Univ, Franche-Comte

Crini G (2005) Recent developments in polysaccharide-based materials used as adsorbents in wastewater treatment. Prog Polym Sci 30(1):38–70

Crittenden JC, Trussell RR, Hand DW, Howe KJ, Tchobanoglous G (2005) Water treatment principles and design. John Wiley & Sons:2

Dao-Tuan, A, Nguyen-Thi-Ngoc, A; Nguyen-Trong, K; Bui-Tuan, A; and DinhThi-Hai, V Chen, Yuanfang, Duong TQ. (2018) Optimizing vehicle routing with Path & Carbon Dioxide Emission for municipal solid wastes collection in Ha Giang, Vietnam, Industrial Networks & Intelligent Systems, Springer International Publishing, 221, pp. 212–227, doi:https://doi.org/10.1007/978-3-319-74176-5_19.

Environmental Protection Agency (2017) Textile Mills Effluent Guidelines. EPA. 06–30.

European Environment Agency (2001) Copenhagen, Denmark. "Indicator: Biochemical oxygen demand in rivers.

Forgacs E, Cserháti T, Oros G (2004) Removal of synthetic dyes from wastewaters: a review. Environ Int 30(7):953–971

Francoise R, Bertram, Chi T (2018) A study of companies' business responses to fashion e-commerce's environmental impact. Int J Fashion Design Technol Educ 11(2):254–264. https://doi.org/10.1080/17543266.2017.1406541

George ZK, Kosta AM (2018) Flotation in water and wastewater treatment. Processes 6:116 . https://doi.org/10.3390/pr6080116

Government of Canada Drinking Water report (2015). https://www.watertechonline.com/wastewater/article/15550688/wastewatertreatment-challenges-in-food-processing-and-agriculture

Henze M (ed) (2001) Wastewater treatment—biological and chemical processes. Springer, Berlin

http://www.indiaenvironmentportal.org.in/files/file/Ramgarh-lake-Gorakhpur-pollution-report-NGT.pdf.

http://www.who.int/water_sanitation_health/dwq/S12.pdf.

https://cpcb.nic.in/Actionplan/Gorakhpur.pdf.

https://waste-management-world.com/a/trash-talking-textile-recycling.

https://www.safewater.org/fact-sheets-1/2017/1/23/conventional-water-treatment

https://www.valmet.com/board-and-paper/recycled-fiber/.

https://www.watertechonline.com/wastewater/article/15550688/wastewater-treatment-challenges-in-food-processing-and-agriculture.

International Desalination Association Yearbook (2011).

Katarzyna C, Dawid S, Katarzyna M, Anna W, Krowiak GI, Ksawery K, Paulina B, Marek K (2021) Progress in sustainable technologies of leather wastes valorization as solutions for the circular economy. J Clean Prod 313:127902

Kesalkar VP, Kesalkar IP, Sudame AM (2012) Physio-chemical characteristics of wastewater from paper industry. Int J Eng Res Appl (IJERA) 2(4):137–143

Kulkarni A, Deshpande M, Pandit AB (2000) Techniques of wastewater treatment, introduction to effluent treatment and industrial methods. Resonance:56–68

Kulkarni S, Mahendrakar H, Belwal S, Rao BM (2014) Transformation of sewage water to potable water. J Eng Technol Res 2(4):63–68

Kumari M, Singh A (2017) Hematological and biochemical changes induced by water pollutants in fishes collected from Ramgarh Lake of Gorakhpur (U.P.) India. Int J Life Sci Sci Res 3(1): 792–799

Leonhauser J, Pawar J, Birkenbeul U (2014) Novel Technologies for the Elimination of pollutants and hazardous substances in the chemical and pharmaceutical industries. Chapter 5. In: Book: industrial wastewater treatment, recycling, and reuse

Lohchab RK, Saini J (2017) Industrial Pollution Management. book January Publisher. IAHRW Publications Private Limited. ISBN: 978–93–821201-7-9

Metcalf L, Eddy HP (2003) Waste water engineering. Tata McGraw Hill Pub. Co., New Delhi

Morin-Crini N, Crini G, Roy L (2017) Eaux industrielles contaminées. PUFC, Besancon 513:37 – 47

Nemerow NL (2007) Industrial waste treatment. Elsevier Inc., Oxford

Nemrow NL (2005) Industrial collaborative solutions. In: Agaedy FJ, Nemrow NL (eds) Environmental solutions. Elsevier Inc., Oxford, pp 249–295. (ISBN: 978-0-12-088441-4)

NITI Aayog (2018) Government of India, Composite Water Management Index: A Tool for Water Management, India. https://niti.gov.in/writereaddata/files/document_publication/2018-05-18-Water-Index-Report_vS8-compressed.pdf.

NITI Aayog (2019) Government of India, Composite Water Management Index, India. https://niti.gov.in/sites/default/files/2019-08/CWMI-2.0-latest.pdf.

Rathoure AK, Dhatwalia VK (eds) (2016) Toxicity and waste management using bioremediation. IGI Global, Hershey

Sharma S K, Sanghi R (2013) Waste water reuse and Management: Amazon.com 256.
Sharma SK (2015) Green chemistry for dyes removal from wastewater. LLC Wiley, Bilverely
Singh S (2019) Hidden risks and untapped opportunities: water and the Indian banking sector, India. World Wildlife Fund for Nature (WWF)India. http://www.indiaenvironmentportal.org.in/files/file/hidden_risks_and_untapped_opportunities.pdf
Sonune A, Ghate R (2004) Developments in wastewater treatment methods. Desalination 167:55–63
Souza RR, Bresolin IT, Bioni TL, Gimenes ML, Dias-Filho BP (2004) The performance of a three-phase fluidized bed reactor in treatment of wastewater with high organic load. Braz J Chem Eng 21(2):219–227
Subramani T, Arulalan K (2012) Evaluation of the efficiency of surface aerator in the activated sludge process treatment of food processing effluent. Int J Modern Eng Res 2(4):1518–1528
Tchobanoglous G, Burton LF, Stensel DH (2003) Wastewater engineering (treatment disposal reuse) / Metcalf & Eddy, Inc, 4th edn. McGraw-Hill Book Company. ISBN 0-07-041878-0.
The 2030 Water Resources Group (2030 WRG) (2009) Charting Our Water Future. McKinseyCompany. https://www.mckinsey.com/~/media/mckinsey/dotcom/client_service/sustainability/pdfs/charting%20our%20water%20future/charting_our_water_future_full_report_.ashx.
Tylkowski B and Jastrząb R (2017) Smart capsules for Lead removal from industrial wastewater. Met Ions Life Sci. 2017: 17 /books/9783110434330/9783110434330-004/9783110434330-004.xml. doi: https://doi.org/10.1515/9783110434330-004.
Warren L, Mc C, Smith JC, Harriott P (1993) Unit operations of chemical Engg. McGraw-Hill, Inc: New Delhi 931
World Bank Water and Sanitation Program (WSP) and International Water Management Institute (IWMI) (2016) Recycling and Reuse of Treated Wastewater in Urban India: A Proposed Advisory and Guidance Document, Sri Lanka. http://www.iwmi.cgiar.org/Publications/wle/rrr/resource_recovery_and_reuse-series_8.pdf.
Zander NE, Gillan M, Sweetser D (2017) Composite Fibers from Recycled Plastics Using Melt Centrifugal Spinning. Materials 10:1044

Harnessing Green Energy Along with Precious Metal Recovery from Wastewater in Bioelectrochemical Systems: A Win-Win Scenario

18

Syed Saquib, Ardiyan Harimawan, and Tjandra Setiadi

Abstract

The extraction and use of metals from the natural environment are the major events in the economic advancement and prosperity of a nation. However, increasing anthropogenic demand and resulting industrialization lead to intensive mining and processing and metal-containing wastewater. This resulted in environmental deterioration and fast depletion of natural resources. The need for sustainable technology and process emerges extensively to overcome such problems. Bioelectrochemical systems are one such sustainable technology that has the potential to solve multiple concerns including efficient metal recovery, wastewater treatment, and bioelectricity generation simultaneously at one place. This chapter reviews the possible "bioelectrochemical systems" (BES) techniques to recover precious metals from wastewater stream and other related advantages in detail.

Keywords

Bioelectrochemical systems · Precious metal recovery · Wastewater treatment · Bioenergy production

18.1 Introduction

Metals are valuable raw materials to propel the economy of a country, which is usually obtained from mining mineral ores, finitely distributed across the world. There is a need to secure them for sustainable production in various sectors such as

S. Saquib · A. Harimawan · T. Setiadi (✉)
Department of Chemical Engineering, Faculty of Industrial Technology, Institut Teknologi Bandung, Bandung, Indonesia

R. P. Singh et al. (eds.), *Heavy Metal Toxicity: Environmental Concerns, Remediation and Opportunities*, https://doi.org/10.1007/978-981-99-0397-9_18

energy, technologies, automobiles, electronics, and medical industries. Due to their continuous decline together with increased future demand, the need for metal recovery and reuse is increased from wastes such as sewage sludge, mine tailings, waste electronics scraps, etc. Rapid industrialization and anthropogenic enterprise produce a large quantity of metal-loaded wastewater. Such contaminated water possesses great health and environmental concerns, because of nonbiodegradability and bioaccumulation in flora and fauna along with disease and disorders in humans. Various physicochemical and biological methods for removal and recovery of metals with higher economic importance, such as silver, gold, copper, zinc, aluminum, etc., from wastewater, were developed in the past. However, these conventional technologies were less efficient and more capital intensive. In this chapter, we will discuss a very rapidly growing, novel technique of precious metal recovery using "bioelectrochemical systems" (BESs). Most BES bioreactors are constructed of electrodes (anode and cathode) and a membrane separator. Biodegradable materials, such as wastewater containing an organic or inorganic carbon source, are oxidized by electroactive microbes at the anode to generate electrons transferred to the cathode, whereas in the cathode compartment, the electrons can be used for direct electricity generation or used for the reduction of water or oxidized chemicals such as metal ions, CO_2, or organic chemicals (microbial electrolysis cells (MECs) or microbial electrosynthesis (MES)) (Choi and Cui 2012; Heijne et al. 2010). For metal recovery and reuse, the specific cathodic condition in BES provides a favorable situation for metal recovery/removal. This chapter aims to provide an overall perspective on the principle and mechanisms for precious metal recovery, factors influencing performances of bioelectricity-harnessing capabilities of bioelectrochemical systems (Nancharaiah et al. 2016). Such advantages over conventional methods make them a sustainable, green technology and provide us with a win-win scenario.

18.2 Bioelectrochemical Systems: Types and System Configuration

18.2.1 Bioelectrochemical Systems

Bioelectrochemistry is an applied, more practical consolidated domain between electrochemistry and microbiology with an approach of research toward environmental engineering, physical, and biochemical aspects. Systematic representation of such a technique is termed "bioelectrochemical systems" (BES) (Schröder et al. 2015). BES is associated with a variety of services such as wastewater treatment, bioelectricity generation, bioremediation of degraded soil and water bodies, and desalination (Sharma et al. 2014). BES is based on microbial electrochemistry in which electron donors and acceptors interact with electro-active bacterial biofilm for the exchange of metabolic electrons released from electron donors and transfer to acceptors through an electroconductive matrix/material, e.g., graphite (Rabaey et al.

2007; Rosenbaum and Franks 2014). Schröder put forward the classification among the type and concepts of bioelectrochemistry provided in further subheadings.

18.2.1.1 Microbial Fuel Cell

A conventional microbial fuel cell comprises two electrodes bearing chambers (anodic and cathodic) with an ion-exchange membrane separating them. Electrons and protons are generated at the anode by oxidation of organic substrate through microbial action which is then carried to the cathode. Electrons flow through the external circuit to produce electricity, whereas protons flow through the separating membrane. At the cathode, electrons and protons react with terminal electron acceptors, usually oxygen, and give rise to water by reducing it. Simply, MFCs use anode-respiring microbes as biocatalysts to change chemical energy to electrical directly via substrate oxidation (Fig. 18.1). A typical electrode reaction with a carbon source is given below:

$$\text{Anodic chamber}: CH_3COO^- + 2H_2O \underset{\rightarrow}{\text{EABs}} 2CO_2 + 7H^+ + 8e^-$$

$$\text{Cathodic chamber}: O_2 + 4e^- + 4H^+ \rightarrow 2H_2O$$

18.2.1.2 Microbial Electrolysis Cell

This configuration is mainly used to produce hydrogen and is named as *microbial electrolysis cell* (MEC). In such a system, external electrical supply is required. Under external power, electrons produced by bacterial activity in an anodic chamber along with the protons in the cathode liberated hydrogen gas. Theoretical cell voltage for hydrogen production at the cathode is evaluated to be 0.114 V, but in practicality 0.25 V is required because of overpotential. Earlier studies report that hydrogen generation by MEC has been up to 1.1 m^3 H^2/m^3/day with an overall energy generation up to 82% (Cheng and Logan 2007).

18.2.1.3 Microbial Desalination Cell

Assurance of safe potable water is the right of every person. As the name suggests, microbial desalination cell (MDC) is a sustainable approach toward water desalination along with wastewater treatment. Among various types, multichamber designs of MDCs use membranes for ion exchange (PEM) that allow salts to be separated from saline water. Other system designs include air cathode MDC, biocathode MDC, stacked MDC, recirculation MDC (rMDC), and capacitive MDC. A typical three-chamber MDC consists of anode, cathode, and desalination compartments (Al-Mamun et al. 2018). Anode and cathode exchange membrane (AEM and CEM) are used to make a desalination chamber placed adjacent to respective electrodes. MDC systems can produce a power density of 31 W/m^3 with a simultaneous salt removal of 90%.

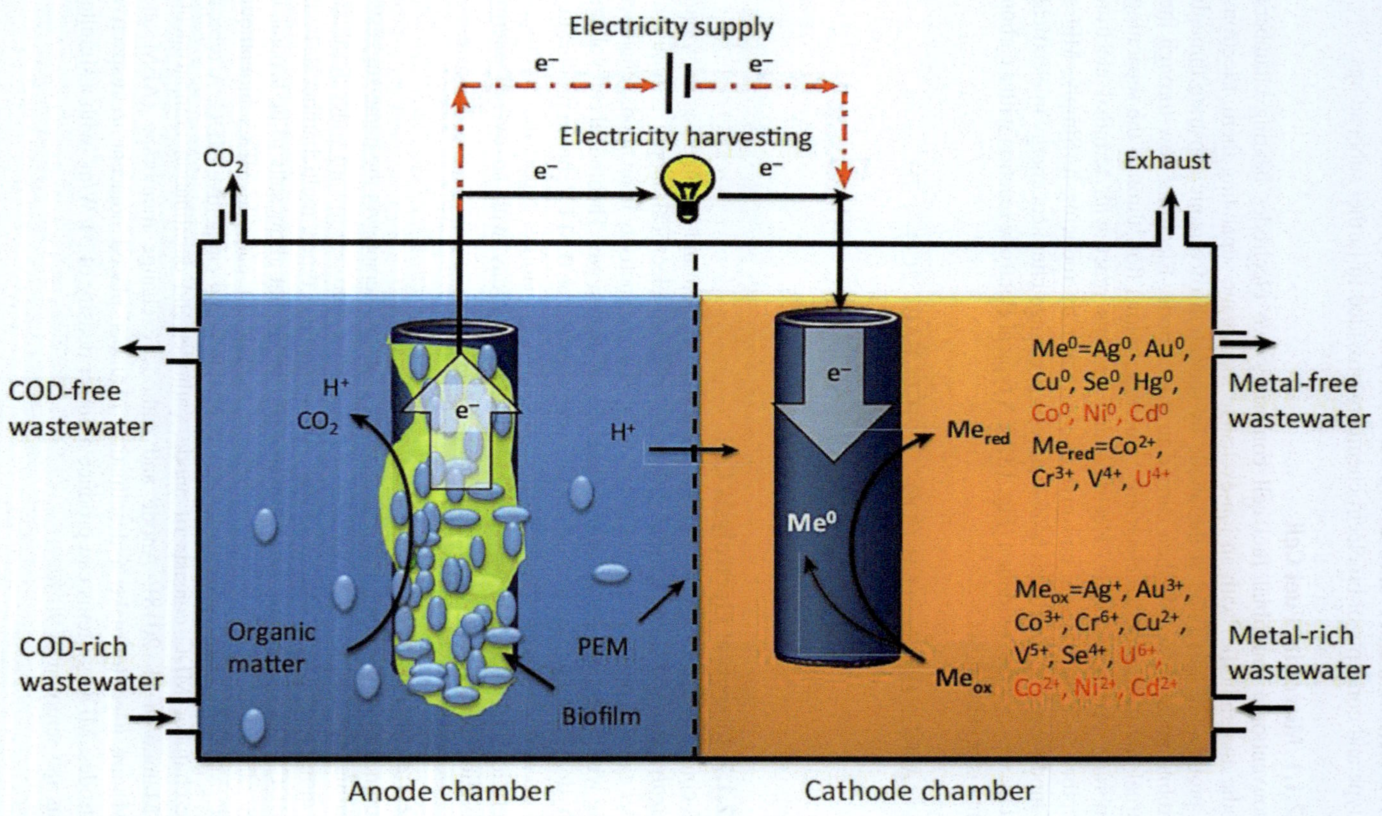

Fig. 18.1 Working principles of bioelectrochemical systems used for metal recovery. Reused from Nancharaiah et al. (2016)

18.3 Bioelectrochemical Recovery of Precious Metals

This chapter focuses to give a comprehensive overview of BES as a sustainable and effective approach for valuable metal recovery and removal from various waste and wastewater sources. In BES, electrons liberated at the anode are used for the reduction of metals at the cathode. There are mainly four categories in which metal recovery mechanisms can be divided: direct reduction of the metals, metal recovery on anode using a power supply, use of bio-cathode, and metal recovery on bio-cathode with external power supply.

18.3.1 Direct Metal Recovery on Abiotic Cathode

This process involves the recovery of metals such as Au (III), V (V), Cr (VI), Ag (I), Cu (II), Fe (III), Hg (II), etc., directly reduced on an abiotic cathode with redox potential higher than the anode. Past studies showed that valuable metals with higher redox potentials can be recovered at the cathode along with electricity production.

Cr(VI) can be directly reduced on an abiotic cathode with less toxic forms including $Cr(OH)_2^+$ and $Cr(OH)_2^+$ and also to $Cr(OH)_3$ in MFCs (Wang et al. 2008). Choi and Hu (2013) were able to recover 98.9% of gold in a two-chamber MFC setup from tetrachloroaurate wastewater. Another study demonstrates silver (Ag) recovery with 99% efficiency and 4.25 W/m^2 of power density from wastewater source (Choi and Cui 2012). Vanadium metal may occur in two states in an aqueous environment: V(V), the most toxic one, and V(IV), the less toxic form. V (V) with a reduced rate of 25.3% was achieved in an MFC supplied with sulfide and glucose for electron evolution (Zhang et al. 2009). Many studies reported high Cu (II) recovery in BES setups. The metal recovery was reported up to 99.9% under various conditions (Heijne et al. 2010; Wang et al. 2010). Catal et al. (2009) able to recover selenium Se(IV) in a single-chamber MFC system. The recovery was 99%.

18.3.2 Abiotic Cathodes Supported by the External Power Supply for Metal Recovery

Several metallic ions have lower redox potential than the anode that needed to be supplied by an external energy source for transporting electrons from the high potential anode to the lower abiotic cathode to reduce the metals. Such metals are Ni (II), Pb (II), and Cd (II). Qin et al. (2012) obtain maximum recovery of 67% in the case of nickel-Ni(II) ion under an external voltage supply of 0.5–1.1 V. BES supplied with external power could be an interesting approach toward sequential metal recovery/removal from a mixed media. Cd (II) reduction was achieved in a series-connected two double-chamber MFC setup with an initial concentration of 200 mg/l, 100 mg/l, and 50 mg/l and recovery efficiency clocks at 89.73%, 93.30%, and 93.43%, respectively (Choi et al. 2014). However, using such external input systems has their demerits including cost-benefit analysis of energy used and its audit. Another problem is hydrogen production, as cathode potential is below the hydrogen evolution potential which reduces metal recovery capabilities.

18.3.3 Metal Recovery Using Biocathode: Direct and External Supply Mediated System

Microorganisms are known to interact with metals in the natural environment. Apart from using abiotic cathode, microbes also help in metal assimilation and dissimilation. Metal dissimilation is considered as the major pathway for the metal recovery process in which microbes use metal as an external electron acceptor (Holden and Adams 2003). Various toxic metals are reduced by dissimilatory metal-reducing bacteria (DMRBs) in the bioremediation process. For instance, Cr(VI) can be reduced to less toxic Cr(III). Similarly, Se(VI) and Se(IV) can be microbially degraded to Se(0) for its precipitation. Tandukar et al. (2009) observed the biological Cr (VI) reduction to $Cr(OH)^3$ with *Trichococcus pasteurii* and *Pseudomonas aeruginosa* as primary microbes associated with the mechanism at the cathode of an MFC. For metals with lower redox potentials, the external power is responsible for a reduction on biocathode. At the biocathode, metal ions are accumulated into the biofilms on the electrodes, where microbial respiration helps in the reduction of metals. *Geobacter sulfurreducens* under externally supported cathode potential of −500 mV reduce the U(VI) to U(IV), much lower than the electrochemical reduction at −900 mV (Gregory and Lovley 2005).

18.4 Mechanism of Metal Removal and Recovery in BES

Traditionally, metal contaminant treatment and recovery are done by conventional physicochemical and biological accumulation methods. On the other hand, emerging BES technology majorly includes four prominent mechanisms, namely, direct redox reaction (DRO), indirect byproduct precipitation (IBP), ion transfer, and biological removal. Among these, the majority of past research investigates metal recovery on DRO method.

In a bioelectrochemical system, soluble metals usually precipitate as solid at the cathode. This principle is dominated in all the abovementioned mechanisms except ion migration. In the latter, migration and separation of metal ions takes first followed by precipitation. Wang and He (2020) tried to demonstrate the dominant removal mechanism among DRO and IBP. Taking Cu (II) as test metal species, they deduced that Cu accepted electrons from cathode and reduced the Cu (II), whereas in IBP hydroxide production due to O2 reduction on the cathode helps in precipitation of Cu (II). Increase in redox reaction is directly proportional to metal concentration. However, copper at low concentration ($<0.5\ g\ L^{-1}$) was more efficiently removed by IBP than that of DRO. Other differences account as follows: in DRO metal deposition mostly takes place on cathodic electrodes, while precipitates in IBP can deposit on various parts of the whole cathodic chamber (Dong et al. 2017). In comparison to

DRO where pure metallic recovery takes place, IBP produces metal hydroxide. For metal recovery via DRO from cathode, either physical method (scrapping) or chemically induced acid dissolution is done. Ali et al. (2019) used DRO method in an MFC to successfully remove 83% of silver from AgNO3 solution and were able to recover 67.8% of Ag scrapped from its cathodic electrode. In IBP, gravity-settled metal precipitates are easily collected in comparison to DRO where disassembling of the system needs to be done. Posttreatment is required in case of ion transfer method due to production of metal brine to obtain solid metal (Jiang et al. 2020). Biological removal metal recovery involves cell lysis and physical separation.

18.5 Factors Influencing System Performance

18.5.1 Initial Metal Concentration

The initial concentration of metals in BES plays a key role in recovery as well as in the overall performance of the system. As the metal concentration falls, the decrease in cathode potential is observed with increment in internal resistance. However, increasing the initial concentration will reduce the metal ion reduction capability at the cathode (Heijne et al. 2010; Cheng et al. 2013). Initial copper concentrations as high as 500–6400 mg/L resulted in the generation of $Cu_4(OH)_6SO_4$ which is non-reductive than that of the reductive species of Cu_2O and Cu at a minimal Cu (II) concentration of 200 mg/L (Tao et al. 2011a, 2011b). Thus, a moderate level of initial concentration should be maintained and controlled to obtain pure metals.

18.5.2 pH Level

Initial pH level can also tag as an influential factor in BES. In most cases, low pH of about 2–3 is required for metal reduction at the abiotic cathode. Reduction of Cr (VI) to Cr(III) was reported in the abiotic cathode in acidic medium of pH 2.0 (Wang et al. 2008). While in the case of silver, cathodic reduction depends on the type of initial substrate used. When using Ag(I) thiosulfate complex, higher pH of about 10 was preferable for its reduction, whereas ionic silver was reduced under lower pH of 4 (Tao et al. 2012). Higher Co(II) reduction was achieved in microbial electrolysis cells with up to 97% of recovery efficiency as the initial pH increased from 3.8 to 6.2. However, bioanode surrounded by a biofilm of exoelectrogens was worked at near-neutral pH for higher electricity generation from MFCs in comparison to an abiotic cathode which requires lower pH to directly recover metals from wastes with no further pH adjustments.

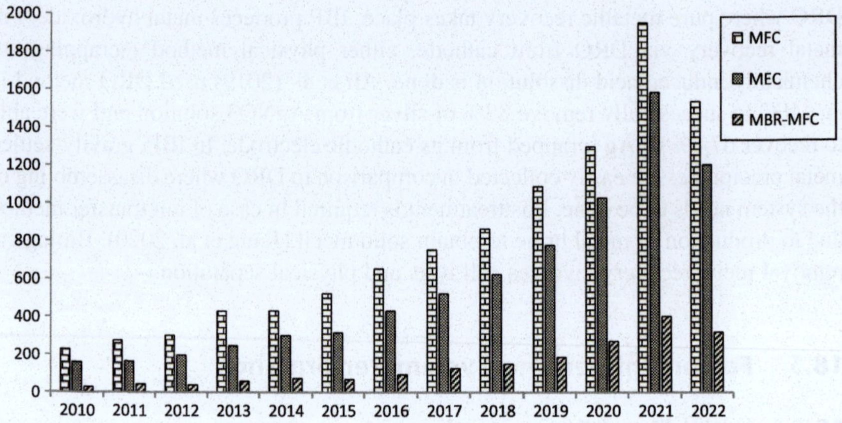

Fig. 18.2 Number of publications regarding heavy metal and precious metal recovery/removal from wastewater using various bioelectrochemical systems [years: 2010–2022 (April)] (source: SciDirect)

18.5.3 Electrode Material

Electrode material, especially cathode, is also crucial in terms of the recovery potential of valuable metals. Increasing cathodic surface area can improve electron acceptance and cathodic reactions and result in improved power production. Carbon-based air cathodes and anodes such as graphite plate and graphite mesh are generally used for the recovery of metals at high concentrations (Heijne et al. 2010; Tao et al. 2012) (Fig. 18.2).

18.6 Specific Metal Recovery

Bioelectrochemical systems are widely studied and proven to be an emerging technology for high-strength wastewater treatment along with heavy metal removal. Efficient recovery of precious metals such as Ag, Au, Cu, Zn, Pb, and others were also reported in BESs (Fig. 18.3 and 18.4).

18.6.1 Silver

Silver is one of the most important and precious metals used in various purposes including jewelry, mirror formation, and even in pharmaceuticals at a large scale. Silver can cause a disease called argyria, discoloration of the skin and eyes. Silver in the range of 0.4–1 mg/L has caused pathological damages to internal organs such as the kidneys, spleen, and liver (Tao et al. 2012). Being an electron acceptor species, silver can be recovered in a cathode code under anaerobic environment to avoid electron acceptance by oxygen. Cathode reaction is given as follows:

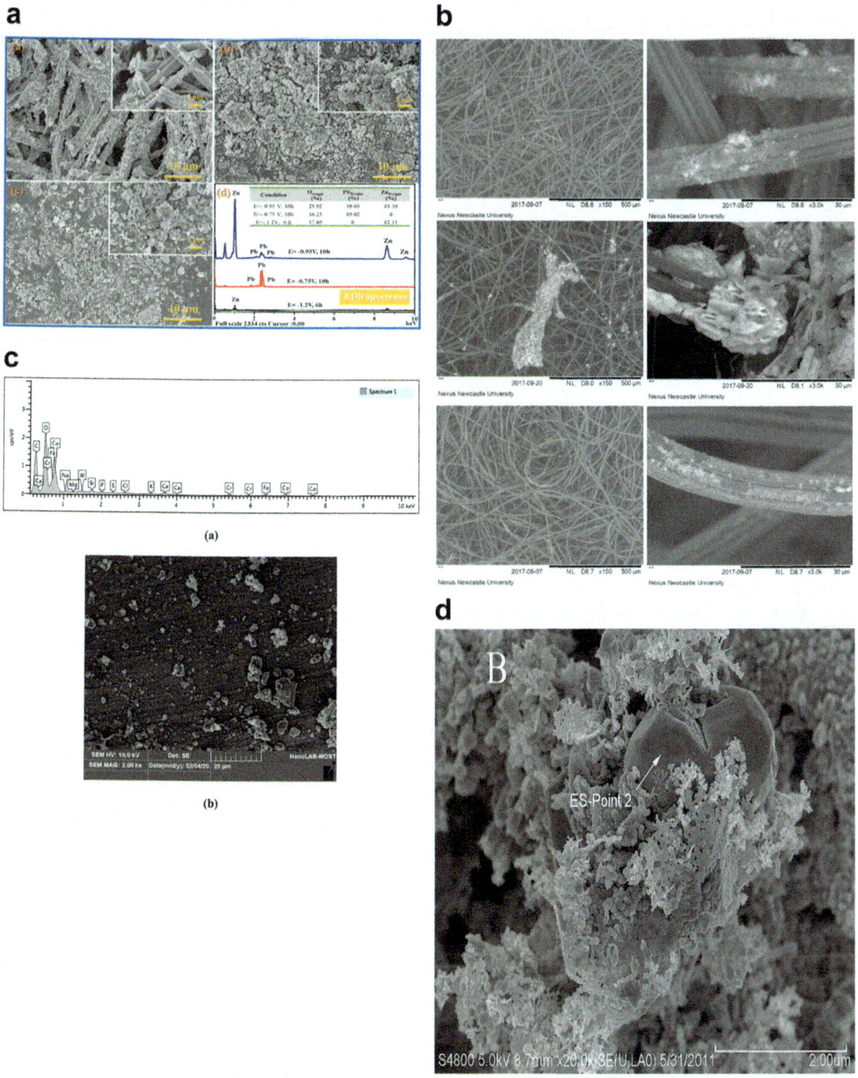

Fig. 18.3 Morphological (SEM) analysis of cathode electrode surface showing deposition of recovered metals [reused: (**a**). Pb and Zn (Zhang et al. 2020); (**b**). Zn (Lim et al. 2021); (**c**). Co (Saad et al. 2020); (**d**). Ag (Tao et al. 2012)]

$$Ag^{+} + e^{-} \rightarrow Ag \ (E^{\circ} = 0.799 \ V)$$

Tao et al. (2012)) achieved silver recovery of 99%, and another study reported silver recovery of 69 kg/ kWh energy under incremental concentration of $AgNO_3$ (Choi and Cui 2012).

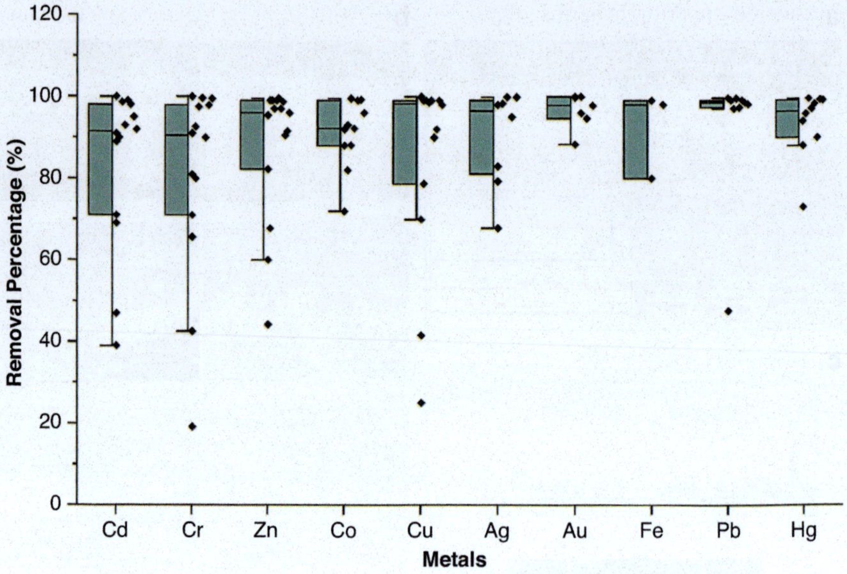

Fig. 18.4 Box plots showing specific metal recovery/removal efficiency among various studies using different BESs ($n = 14$ for Cd; $n = 14$ for Cr; $n = 15$ for Zn; $n = 11$ for Co; $n = 14$ for Cu; $n = 8$ for Ag; $n = 7$ for Au; $n = 3$ for Fe; $n = 9$ for Pb; $n = 10$ for Hg)

18.6.2 Gold

Mankind uses gold from several centuries as a major ornamental and jewel metal. It is also used in the industrial as well as electronics and aerospace sectors due to its high conductivity. This property makes it a potential recovery candidate. Bioelectrochemical recovery of gold can be done by treating tetrachloroaurate wastewater in which it acts like catholyte and acetate wastewater as anolyte. Recovery of 99.89% gold was reported at the concentration of 200 ppm with tetrachloroaurate as a potential electron accumulator (Choi and Hu 2013).

18.6.3 Copper

It is a by-product metal of copper and nickel refining. Copper ions have a high redox potential ($E° = 0.34$ V) and after reduction get recovered on the cathodes; thus, MFCs generate more electricity in comparison to others. Deposition up on cathode takes place in the form of Cu or from Cu_2O precipitates. Up to 80% of copper recovery along with power density of 2 W/m^2 was obtained (Nancharaiah et al. 2016). Various factors influence the copper recovery in bioelectrochemical systems including initial concentration, external resistance, distance between electrodes, and electrode material used (Tao et al. 2011a, 2011b; Wang et al. 2010; Rodenas Motos et al. 2015). A membraneless MFC showed a maximum power density of 314 mW/m3 and Cu2+ removal efficiency of 70%. Stainless steel mesh electrodes are considered superior in terms of recovery performances in comparison to others. 99.9% copper recovery was obtained when initial pretreatment with acid washing was done before BES recovery (Fedje et al. 2015).

18.6.4 Chromium

Chromium is a toxic metal released from many industries, such as tanning and dyeing. Cr(VI) and Cr(III) are the two major persisting species of chromium in the environment, among which highly toxic due to increased water solubility of the hexavalent one. Thus, microbial reduction of Cr(VI) to Cr(III) is considered as of water and wastewater treatment process:

$$Cr_6{}^{+} + 3e^{-} = Cr_3{}^{+}\ (E^{\circ} = 1.09\ V)$$

99.5% and 66.2% of hexavalent and total chromium removal was achieved in an MFC, respectively (Li et al. 2008). Another study reported the reduction of hexavalent chromium of about 90% and 78% in a urine-based fuel cell and a biocathode-supported dual-chamber MFC, respectively, along with power generation of 3.4 W/m^2 in former one (Wu et al. 2015; Xu et al. 2016a). Complete removal of Cr(VI) was achieved while using alumina nickel nanoparticles dispersed on carbon nanofiber electrodes. Alumina nanoparticles increase the electrical conductivity, whereas nickel acts as a catalyst for metal reduction at the electrode (Gupta et al. 2017).

18.6.5 Platinum

Platinum is one of the rare elements present in finite reserves and has extensive extraction cost. It is an expensive metal extensively used in pharmaceutical, jewelry manufacturing, and electronic items and as a chemical catalyst. These properties urge for its recovery from various wastes for sustainable approach and consumption. Previous research reported 90% of the platinum recovery in the BESs with initial metal concentration of 16.88 mg/L in wastewater (Liu et al. 2019).

18.6.6 Cobalt

Cobalt is one of the important metals extensively used in lithium-ion batteries, and recovery of it from $LiCoO_2$ cathode is essential for economic, environmental, and health point of view. Co (II) which is soluble in an MFC can be reduced from Co (III) present in $LiCoO_2$:

$$LiCoO_2 + 4H^{+} + e^{-} \rightarrow Li^{2+} + Co^{2+} + 2H_2O$$

A study with lithium cobalt particles accounts for 7 mg/L/h of copper recovery with highest power density of 5 W/m^3 on an acclimated bioanode (Huang et al. 2014). A study successfully recovered flake shaped cobalt from aqueous Co(II) in microbial electrolysis cells (MECs) while producing hydrogen. The applied voltages of 0.3–0.5 V resulted in yields of 0.81 mol Co/mol COD and 1.21 ± 0.03–1.49 ± 0.11 mol H2/mol COD (Jiang et al. 2014). Under acidic medium (pH 1.0) and mesophilic temperature of 35 °C, it tends to enhance cobalt leaching. Other factors also play an important role in cobalt recovery including solid/liquid ratio, external resistance, conductivity, etc. Figure 18.3 demonstrates various studies reported deposition of recovered metals on electrode surface using SEM.

18.7 Efficient Wastewater Treatment Using Bioelectrochemical Systems

Among the plethora of applications, efficient, cost-effective, and sustainable wastewater treatment tops the chart when we talk about bioelectrochemical systems. Various types of wastewater were treated in past studies. It also serves as a nutrient for the proper growth of microbes in the reactor. In recent years, BES emerged as a promising technology for effective wastewater treatment with high pollutant removal above 90% removal rate for chemical oxygen demand (He et al. 2012). Various parameters are considered for wastewater treatment performance of BES, primarily COD, BOD, total dissolved and suspended solids, and acidity. Some treatment performance studies are shown in Table 18.1. The use of MFCs for wastewater treatment dated back to the nineteenth century when Habermann and Pommer (1991) achieved 35% and 75% for sewage and landfill effluent, respectively, along with a maximum anodic current density of 150 mA/cm2. A higher COD removal of up to 98% was achieved in MFCs with different configurations (single- or dual-chamber reactors) and wastewater types (Yu et al. 2015). Min et al. (2005)) achieved COD and NH^{4+}-N removal efficiency of 92% and 83%, respectively, in single- and double-chamber MFCs treating swine wastewater. Another study incorporating high BOD and sugar-rich food industry wastewater is able to obtain 95% of COD removal in a double-chamber MFC setup (Oh and Logan 2005). Toxic chemical-laden wastewater especially from the textile and oil mill industry was also explored to be treated using novel bioelectrochemical systems. A study used integrated upflow MFC and UASB reactors to treat palm oil mill effluent. The removal rate was 96% and 94% for COD and nitrogen, respectively (Cheng et al. 2010). A recent study conducted with synthetic wastewater containing azo dye achieved high COD removal of 94.04% along with methyl orange removal rate of 94.22% in a novel and innovative double-chambered constructed wetland microbial fuel cell system (Mittal et al. 2021).

Table 18.1 Performance for wastewater treatment and electricity generation in Bio-electrochemical Systems

Reactor characteristics	Target study	Wastewater type	Pollutant removal efficiency	Current density (mA/cm^2)/power density	References
Two-chambered upflow MFC	Wastewater treatment and bioelectricity generation	Swine wastewater	TCOD = 71–83%	13 mW/m^2	(Ma et al. 2016)
Upflow anaerobic MFC	Real wastewater treatment	Chocolate industry wastewater	COD = 55.8–70.4% TSS = 62.6%	98 mW/m^2	(Subha et al. 2019)
Single-chamber MFC	Petrochemical wastewater treatment with acclimatized activated sludge	Petrochemical wastewater	COD = 43–82%	1500	(Sarmin et al. 2019)
Single-chamber air cathode MFC	Paper industry wastewater treatment	Diluted paper recycling wastewater	TCOD = 27–76% Cellulose = 96%	0.25	(Huang and Logan 2008)
METland	Wastewater treatment	Pretreated municipal wastewater	COD = 93% BOD = 99% NH_4-N^+ = 97% TN = 69%	N.A	(Aguirre-Sierra et al. 2016)
CW-MFC	Performance assessment of cathodic compartment of CW-MFC	Urban wastewater	COD = 61% NH4 = 60%	80 ± 56–131 ± 61 mWh/m^2. Day	(Corbella et al. 2016)
CW-MFC	Wastewater treatment	Synthetic wastewater	COD = 99% NO3 = 46% NH4 = 96%	93 mW/m^3	(Oon et al. 2016)
CW-MFC	Nitrate removal	Synthetic wastewater	COD = 57% NO3 = 80%	47.77 mA/m^2	(Wang et al. 2016)
CW-dewatered alum sludge+ MFC	Wastewater treatment	Diluted swine wastewater	COD = 81% TN = 45% NH4 = 53% TP = 88%	N.A	(Xu et al. 2016b)
Hybrid CW-MFC	Boron (B) removal and bioelectric production	Synthetic wastewater	B = 63% NO_3 = 47% NO_2 = 19%	105 mA/m^2	(Türker and Yakar 2017)
Tiered CW-MFC	Energy capture and nutrient removal	Synthetic wastewater	COD = 88% TN = 75% NH_4 = 85% PO_4 = 94%	0.5 W/m^3	(Xu et al. 2017)

(continued)

Table 18.1 (continued)

Reactor characteristics	Target study	Wastewater type	Pollutant removal efficiency	Current density (mA/cm^2)/power density	References
CW-MFC	Influence of substrate on wastewater treatment in CW-MFC	Synthetic wastewater	COD = 92% NH4 = 93% NO3 = 81% TP = 96%	15.1 mW/m^2	(Yakar et al. 2018)
CW-MFC with *H. verticillata*	Amplifying the performance of CW-MFC supported by submerged plants	Synthetic wastewater	COD = 64% NH4 = 89% NO3 = 78% TP = 94%	N.A	(Shen et al. 2018)
CW-MFC with *Typha angustifolia*	Plantation and its role in treatment and electricity generation in CW-MFC	Synthetic wastewater	COD = 88% NH4 = 98% NO3 = 63% TP = 97%	7.47 ± 13.7 mW/m^2	(Saz et al. 2018)

18.8 Electron Transfer Mechanism in BES and Bioelectricity Generation

18.8.1 Microbial Extracellular Electron Transfer [EET]

MET systems depend upon the large number of fermentative and electrogens in a synergistic consortium that can catalyze anodic and cathodic reactions on the conductive electrode material (Reimers et al. 2001). Potter (1911) was the first to report the electron transportation capability of bacteria and their interaction with the electrode.

Fermentative bacteria utilize and degrade complex organic matters into simpler forms which are further oxidized by electrogens in anaerobic conditions (Logan and Rabaey 2012). The electroactive microbes are specialized to gain energy through transferring an electron from a donor to a terminal e^- acceptor, as an extracellular conductive and insoluble form. Such interaction between microbes and conductive electrode material favors the formation of microbial biofilm on electrodes (Rabaey et al. 2007).

18.8.1.1 Electroactive Bacteria (EABs)

EABs are widely found and isolated from various natural ecosystems along with wastewater treatment plants, aquatic sediments, soil and activated sludge, etc. (Chabert et al. 2015). Majority of them are present in an anaerobic medium. Initially, Gram-negative bacteria belonging to *Proteobacteria* (*Shewanella* spp., *Geobacter* spp.) are considered to be the only species to show electron-transferring properties, but recent advances show that some Gram-positive bacteria also exhibit electrogenic properties, e.g., *Thermincola potens* strain (Wrighton et al. 2011). *Geobacter* spp. and *Shewanella* spp. are extensively well-studied "exoelectrogens" for direct ET (DET) via c-type cytochromes addressed in their outer membrane (Bond and Lovley 2003). Far-range ET has also been promisingly expedited by both microbes using conductive projection such as pilli or nanowires (Reguera et al. 2005; Malvankar et al. 2011; Pirbadian et al. 2014). Until now, there are very few Gram-positives being explored exhibiting ET mechanisms. However, direct electrode reduction was reported in *Thermincola ferriacetica* (Marshall and May 2009).

18.8.1.2 Mechanism of EET

Extracellular electron transfer is achieved by two main techniques, either by *direct extracellular electron transfer (DEET)* or by *indirect/mediated extracellular electron (MEET*; Fig. 18.5)

18.8.1.2.1 Direct Extracellular Electron Transfer (DEET)

DEET usually occurs when microbial biofilm is directly in contact with the electrode surface. Some EABs are metabolically preferential toward solid electron donors or acceptors rather than soluble ones such as glucose, acetate, O_2, and fumarate, respectively (Erable et al. 2010). Here, direct electron transfer occurs either by

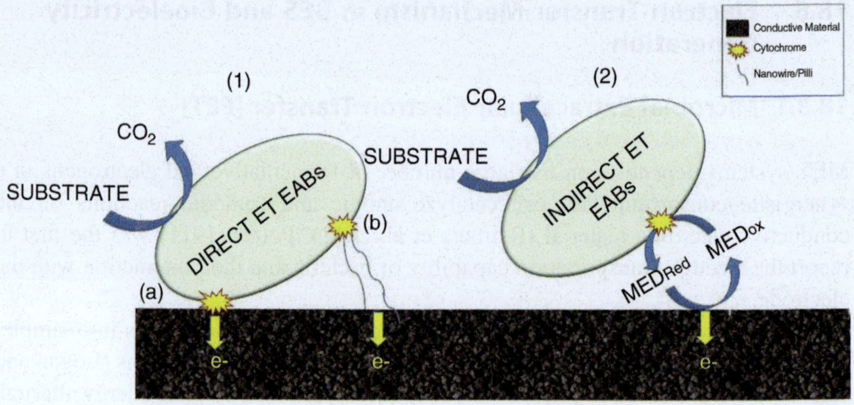

Fig. 18.5 Schematic presentation of EET mechanisms. (1) Direct extracellular electron transfer (DEET): (**a**) by cytochromes and (**b**) by electroconductive pili/nanowires. (2) Indirect or mediated extracellular electron transfer (MEET) by secondary metabolites (adapted and modified from Schröder et al. (2015)); Ramírez-Vargas et al. (2019))

external redox proteins (Cytochromes) or by electrically conductive cellular projections – nanowires/pili (Busalmen et al. 2008; Bonanni et al. 2012).

Microbial species such as *Geobacter and Shewanella* are generally involved in electron transfer through c-type outer-membrane cytochrome and redox protein (Choi et al. 2003). Nanowires (or electroconductive pili) are protruded periplasmic extensions and cells (ranges 2–3 μm in length) that help in the direct electron transmission between microbial cells and acceptors (Kracke et al. 2015; Butti et al. 2016). Some species of *Geobacter* and *Shewanella* used to develop such cellular projection to connect with distant electron acceptors (Lovley 2006; Patil et al. 2012). Malvankar et al. (2011) studied enhanced anodic performance due to the formation of multilayered biofilm of *G. sulfurreducens* with outer-membrane c-cytochrome and conductive pili.

Synergistic interaction between two bacterial species can also facilitate a direct electron transfer mechanism to the electrode (Kato et al. 2012; Zhao et al. 2015). This process of interspecial EET through direct connection can also be referred to as "direct interspecies electron transfer" (DIET).

18.8.1.2.2 Mediated Extracellular Electron Transfer (MEET)

Unlike DEET in mediated electron transfer, bacteria are in indirect connection with a conductive surface. Thus e^- is transferred via the indirect electron transfer (IET) mechanism. In such a case, EABs develop metabolic mechanisms regulated through electron shuttles which facilitate extracellular electron transfer (Arends and Verstraete 2012). Microorganisms such as *E. coli*, *Pseudomonas*, *Proteus*, and *Bacillus* can naturally manufacture and ooze out endogenous redox-active chemicals that pretend as electron conciliators comprising flavins or phenazine (Erable et al. 2010; Mao and Verwoerd 2013). MEET can either take place through the production of their secondary metabolites as in *Pseudomonas*, *Shewanella putrefaciens*, and

Geothrix fermentans or by the reduction of primary, derived from fermentation, or by anaerobic respiration.

18.8.2 Bioelectricity Generation

The most important and distinguishing feature of bioelectrochemical system-based metal removal is the utilization of organics present in wastewater and other sources to produce bioelectricity (Table 1). Although BES is still an insignificant and uneconomical tool for power production, in recent times progressive works are done in this concern. The power output of 10–50 and 250–500 mW/m^2 have been generated along with treating household wastewater and organic carbon sources such as glucose (Logan 2004). Many efforts have been made to increase the efficiency of BES in terms of current density such as electrode modifications, system designs, metal, and biocatalyst exposure at the electrodes.

The bioelectricity generation mechanism starts with the conductive biofilm formation of exoelectrogens on the anodic electrode. Several factors including optimal thickness and bacterial inoculum govern the efficient performance and higher current densities in bioelectrochemical systems. For instance, biofilm thickness usually ranges between 30 and 50 μm (Kracke et al. 2015). After biofilm attachment, EABs facilitate metabolic electron transfer from the outer membrane to the anode. This extracellular electron transfer (EET) takes place by either direct or mediated transfer mechanisms which were already discussed in earlier sections. The next step is the transport of anodic electrons to the cathode via an electrical connection. Here, reactions take place between electrons, protons, and an electron acceptor. Oxygen is the widely used and freely available electron acceptor. Reduction of oxygen takes place in presence of electrons at the cathode, leading to the production of water and generating highest open-circuit voltage (OCV) at the cathode of ~0.805 V.

In the case of metal recovery BES systems, electron transport efficiency (ETE) plays an important role. Coulombic efficiency (CE) in the anode and cathode efficiency are included in ETE. Heijne et al. (2010)) found that more than half of the studies done had CE $>$ 30% which in some cases reached up to 95%. The cathode efficiency through the direct redox reaction of three extensively studied metals, Cr (VI), Cu (II), and Cd (II), was estimated with an average value of 64%, 76%, and 32%, respectively (Wang and He 2020).

18.9 Conclusion

The BESs put forward a sustainable, eco-friendly, and cost-effective alternative to generating energy and along with wastewater treatment. Several types of wastewaters ranging from low to high strength have been used in BESs for these processes. Metal recovery from wastewater has been a key approach in resource recovery. The use of a bioelectrochemical platform has the potential to become a major driver in efficient and low-energy associated approaches. This chapter summarizes BES technique by categorizing metal recovery into four major processes, highlighting their pros and cons respectively. We also try to provide an overview of various precious metal recovery and removal from wastewater using BESs. Despite the several advantages, bioelectrochemical systems have several challenges including dissemination and sustaining laboratory techniques to real-world scenario, maintenance of large-scale setup, long-term performance monitoring, controlling physicochemical and biological parameters for effective metal recovery, cost-benefit and lifecycle assessment of systems, etc.

References

Aguirre-Sierra A, Bacchetti-De Gregoris T, Berná A, Salas JJ, Aragón C, Esteve-Núñez A (2016) Microbial electrochemical systems outperform fixed-bed biofilters in cleaning up urban wastewater. Environ Sci Water Res Technol 2(6):984–993

Al-Mamun A, Ahmad W, Baawain MS, Khadem M, Dhar BR (2018) A review of microbial desalination cell technology: configurations, optimization and applications. J Clean Prod 183: 458–480

Ali J, Wang L, Waseem H, Sharif HMA, Djellabi R, Zhang C, Pan G (2019) Bioelectrochemical recovery of silver from wastewater with sustainable power generation and its reuse for biofouling mitigation. J Clean Prod 235:1425–1437

Arends JB, Verstraete W (2012) 100 years of microbial electricity production: three concepts for the future. Microb Biotechnol 5(3):333–346

Bonanni PS, Schrott GD, Robuschi L, Busalmen JP (2012) Charge accumulation and electron transfer kinetics in *Geobacter sulfurreducens* biofilms. Energy Environ Sci 5(3):6188–6195

Bond DR, Lovley DR (2003) Electricity production by *Geobacter sulfurreducens* attached to electrodes. Appl Environ Microbiol 69(3):1548–1555

Busalmen JP, Esteve-Nuñez A, Feliu JM (2008) Whole cell electrochemistry of electricity-producing microorganisms evidences an adaptation for optimal exocellular electron transport. Environ Sci Technol 42(7):2445–2450

Butti SK, Velvizhi G, Sulonen ML, Haavisto JM, Koroglu EO, Cetinkaya AY, Singh S, Arya D, Modestra JA, Krishna KV, Verma A (2016) Microbial electrochemical technologies with the perspective of harnessing bioenergy: maneuvering towards upscaling. Renew Sust Energ Rev 53:462–476

Catal T, Bermek H, Liu H (2009) Removal of selenite from wastewater using microbial fuel cells. Biotechnol Lett 31(8):1211–1216

Chabert N, Ali OA, Achouak W (2015) All ecosystems potentially host electrogenic bacteria. Bioelectrochemistry 106:88–96

Cheng S, Logan BE (2007) Sustainable and efficient biohydrogen production via electrohydrogenesis. Proc Natl Acad Sci 104(47):18871–18873

Cheng SA, Wang BS, Wang YH (2013) Increasing efficiencies of microbial fuel cells for collaborative treatment of copper and organic wastewater by designing reactor and selecting operating parameters. Bioresour Technol 147:332–337

Choi C, Cui Y (2012) Recovery of silver from wastewater coupled with power generation using a microbial fuel cell. Bioresour Technol 107:522–525

Choi C, Hu N (2013) The modeling of gold recovery from tetrachloroaurate wastewater using a microbial fuel cell. Bioresour Technol 133:589–598

Choi C, Hu N, Lim B (2014) Cadmium recovery by coupling double microbial fuel cells. Bioresour Technol 170:361–369

Choi YJ, Kim NJ, Kim SH, Jung SH (2003) Dynamic behaviors of redox mediators within the hydrophobic layers as an important factor for effective microbial fuel cell operation. Bull Korean Chem Soc 24(4):437–440

Cheng J, Zhu X, Ni J, Borthwick A (2010) Palm oil mill effluent treatment using a two-stage microbial fuel cells system integrated with immobilized biological aerated filters. Bioresour Technol 101(8):2729–2734

Corbella C, Garfí M, Puigagut J (2016) Long-term assessment of best cathode position to maximise microbial fuel cell performance in horizontal subsurface flow constructed wetlands. Sci Total Environ 563:448–455

Dong Y, Liu J, Sui M, Qu Y, Ambuchi JJ, Wang H, Feng Y (2017) A combined microbial desalination cell and electrodialysis system for copper-containing wastewater treatment and high-salinity-water desalination. J Hazard Mater 321:307–315

Erable B, Duţeanu NM, Ghangrekar MM, Dumas C, Scott K (2010) Application of electro-active biofilms. Biofouling 26(1):57–71

Fedje KK, Modin O, Strömvall AM (2015) Copper recovery from polluted soils using acidic washing and bioelectrochemical systems. Metals 5(3):1328–1348

Gregory KB, Lovley DR (2005) Remediation and recovery of uranium from contaminated subsurface environments with electrodes. Environ Sci Technol 39(22):8943–8947

Gupta S, Yadav A, Verma N (2017) Simultaneous Cr (VI) reduction and bioelectricity generation using microbial fuel cell based on alumina-nickel nanoparticles-dispersed carbon nanofiber electrode. Chem Eng J 307:729–738

Habermann W, Pommer EH (1991) Biological fuel cells with sulphide storage capacity. Appl Microbiol Biotechnol 35:128–133

He YR, Xiao X, Li WW, Sheng GP, Yan FF, Yu HQ, Yuan H, Wu LJ (2012) Enhanced electricity production from microbial fuel cells with plasma-modified carbon paper anode. Phys Chem Chem Phys 14(28):9966–9971

Heijne AT, Liu F, Weijden RVD, Weijma J, Buisman CJ, Hamelers HV (2010) Copper recovery combined with electricity production in a microbial fuel cell. Environ Sci Technol 44(11): 4376–4381

Holden JF, Adams MW (2003) Microbe–metal interactions in marine hydrothermal environments. Curr Opin Chem Biol 7(2):160–165

Huang L, Logan BE (2008) Electricity generation and treatment of paper recycling wastewater using a microbial fuel cell. Appl Microbiol Biotechnol 80(2):349–355

Huang L, Yao B, Wu D, Quan X (2014) Complete cobalt recovery from lithium cobalt oxide in self-driven microbial fuel cell–microbial electrolysis cell systems. J Power Sources 259:54–64

Jiang L, Huang L, Sun Y (2014) Recovery of flakey cobalt from aqueous co (II) with simultaneous hydrogen production in microbial electrolysis cells. Int J Hydrog Energy 39(2):654–663

Jiang Q, Song X, Liu J, Shao Y, Feng Y (2020) In-situ cu (II) enrichment and recovery from low-strength copper-laden wastewater using a novel electrically enhanced microbial copper recovery cell (MCRC). Chem Eng J 382:122788

Kato S, Hashimoto K, Watanabe K (2012) Microbial interspecies electron transfer via electric currents through conductive minerals. Proc Natl Acad Sci 109(25):10042–10046

Kracke F, Vassilev I, Krömer JO (2015) Microbial electron transport and energy conservation–the foundation for optimizing bioelectrochemical systems. Front Microbiol 6:575

Lim SS, Fontmorin JM, Pham HT, Milner E, Abdul PM, Scott K, Head I, Yu EH (2021) Zinc removal and recovery from industrial wastewater with a microbial fuel cell: experimental investigation and theoretical prediction. Sci Total Environ 776:145934

Liu Y, Song P, Gai R, Yan C, Jiao Y, Yin D, Cai L, Zhang L (2019) Recovering platinum from wastewater by charring biofilm of microbial fuel cells (MFCs). J Saudi Chem Soc 23(3): 338–345

Li Z, Zhang X, Lei L (2008) Electricity production during the treatment of real electroplating wastewater containing Cr6+ using microbial fuel cell. Process Biochem 43(12):1352–1358

Logan BE (2004) Peer reviewed: extracting hydrogen and electricity from renewable resources. Environ Sci Technol 38(9):160A–167A

Logan BE, Rabaey K (2012) Conversion of wastes into bioelectricity and chemicals by using microbial electrochemical technologies. Science 337(6095):686–690

Lovley DR (2006) Bug juice: harvesting electricity with microorganisms. Nat Rev Microbiol 4(7): 497–508

Ma D, Jiang ZH, Lay CH, Zhou D (2016) Electricity generation from swine wastewater in microbial fuel cell: hydraulic reaction time effect. Int J Hydrog Energy 41(46):21820–21826

Malvankar NS, Vargas M, Nevin KP, Franks AE, Leang C, Kim BC, Inoue K, Mester T, Covalla SF, Johnson JP, Rotello VM (2011) Tunable metallic-like conductivity in microbial nanowire networks. Nat Nanotechnol 6(9):573–579

Mao L, Verwoerd WS (2013) Selection of organisms for systems biology study of microbial electricity generation: a review. Int J Energy Environ Eng 4(1):1–18

Marshall CW, May HD (2009) Electrochemical evidence of direct electrode reduction by a thermophilic gram-positive bacterium, *Thermincola ferriacetica*. Energy Environ Sci 2(6): 699–705

Min B, Kim J, Oh S, Regan JM, Logan BE (2005) Electricity generation from swine wastewater using microbial fuel cells. Water Res 39(20):4961–4968

Mittal Y, Dash S, Srivastava P, Mishra PM, Aminabhavi TM, Yadav AK (2021) Azo dye containing wastewater treatment in earthen membrane-based unplanted two chambered constructed wetlands-microbial fuel cells: a new design for enhanced performance. Chem Eng J 427:131856

Nancharaiah YV, Mohan SV, Lens PNL (2016) Biological and bioelectrochemical recovery of critical and scarce metals. Trends Biotechnol 34(2):137–155

Oh S, Logan BE (2005) Hydrogen and electricity production from a food processing wastewater using fermentation and microbial fuel cell technologies. Water Res 39(19):4673–4682

Oon YL, Ong SA, Ho LN, Wong YS, Dahalan FA, Oon YS, Lehl HK, Thung WE (2016) Synergistic effect of up-flow constructed wetland and microbial fuel cell for simultaneous wastewater treatment and energy recovery. Bioresour Technol 203:190–197

Patil SA, Hägerhäll C, Gorton L (2012) Electron transfer mechanisms between microorganisms and electrodes in bioelectrochemical systems. Bioanal Rev 4(2):159–192

Pirbadian S, Barchinger SE, Leung KM, Byun HS, Jangir Y, Bouhenni RA, Reed SB, Romine MF, Saffarini DA, Shi L, Gorby YA (2014) Shewanella oneidensis MR-1 nanowires are outer membrane and periplasmic extensions of the extracellular electron transport components. Proc Natl Acad Sci 111(35):12883–12888

Potter MC (1911) Electrical effects accompanying the decomposition of organic compounds. Proceedings of the royal society of London. Series b, containing papers of a biological character 84(571):260–276

Qin B, Luo H, Liu G, Zhang R, Chen S, Hou Y, Luo Y (2012) Nickel ion removal from wastewater using the microbial electrolysis cell. Bioresour Technol 121:458–461

Rabaey K, Rodríguez J, Blackall LL, Keller J, Gross P, Batstone D, Verstraete W, Nealson KH (2007) Microbial ecology meets electrochemistry: electricity-driven and driving communities. ISME J 1(1):9–18

Ramírez-Vargas CA, Arias CA, Carvalho P, Zhang L, Esteve-Núñez A, Brix H (2019) Electroactive biofilm-based constructed wetland (EABB-CW): a mesocosm-scale test of an innovative setup for wastewater treatment. Sci Total Environ 659:796–806

Reguera G, McCarthy KD, Mehta T, Nicoll JS, Tuominen MT, Lovley DR (2005) Extracellular electron transfer via microbial nanowires. Nature 435(7045):1098–1101

Reimers CE, Tender LM, Fertig S, Wang W (2001) Harvesting energy from the marine sediment–water interface. Environ Sci Technol 35(1):192–195

Rodenas Motos P, Ter Heijne A, van der Weijden R, Saakes M, Buisman CJ, Sleutels TH (2015) High rate copper and energy recovery in microbial fuel cells. Front Microbiol 6:527

Rosenbaum MA, Franks AE (2014) Microbial catalysis in bioelectrochemical technologies: status quo, challenges and perspectives. Appl Microbiol Biotechnol 98(2):509–518

Saad DR, Alismaeel ZT, Abbar AH (2020) Cobalt removal from simulated wastewaters using a novel flow-by fixed bed bio-electrochemical reactor. Chem Eng Proc-Proc Intensif 156:108097

Sarmin S, Ethiraj B, Islam MA, Ideris A, Yee CS, Khan MMR (2019) Bio-electrochemical power generation in petrochemical wastewater fed microbial fuel cell. Sci Total Environ 695:133820

Saz Ç, Türe C, Türker OC, Yakar A (2018) Effect of vegetation type on treatment performance and bioelectric production of constructed wetland modules combined with microbial fuel cell (CW-MFC) treating synthetic wastewater. Environ Sci Pollut Res 25(9):8777–8792

Schröder U, Harnisch F, Angenent LT (2015) Microbial electrochemistry and technology: terminology and classification. Energy Environ Sci 8(2):513–519

Sharma M, Bajracharya S, Gildemyn S, Patil SA, Alvarez-Gallego Y, Pant D, Rabaey K, Dominguez-Benetton X (2014) A critical revisit of the key parameters used to describe microbial electrochemical systems. Electrochim Acta 140:191–208

Shen X, Zhang J, Liu D, Hu Z, Liu H (2018) Enhance performance of microbial fuel cell coupled surface flow constructed wetland by using submerged plants and enclosed anodes. Chem Eng J 351:312–318

Subha C, Kavitha S, Abisheka S, Tamilarasan K, Arulazhagan P, Banu JR (2019) Bioelectricity generation and effect studies from organic rich chocolaterie wastewater using continuous upflow anaerobic microbial fuel cell. Fuel 251:224–232

Tandukar M, Huber SJ, Onodera T, Pavlostathis SG (2009) Biological chromium (VI) reduction in the cathode of a microbial fuel cell. Environ Sci Technol 43(21):8159–8165

Tao HC, Li W, Liang M, Xu N, Ni JR, Wu WM (2011a) A membrane-free baffled microbial fuel cell for cathodic reduction of cu (II) with electricity generation. Bioresour Technol 102(7): 4774–4778

Tao HC, Zhang LJ, Gao ZY, Wu WM (2011b) Copper reduction in a pilot-scale membrane-free bioelectrochemical reactor. Bioresour Technol 102(22):10334–10339

Tao HC, Gao ZY, Ding H, Xu N, Wu WM (2012) Recovery of silver from silver (I)-containing solutions in bioelectrochemical reactors. Bioresour Technol 111:92–97

Türker OC, Yakar A (2017) A hybrid constructed wetland combined with microbial fuel cell for boron (B) removal and bioelectric production. Ecol Eng 102:411–421

Wang G, Huang L, Zhang Y (2008) Cathodic reduction of hexavalent chromium [Cr (VI)] coupled with electricity generation in microbial fuel cells. Biotechnol Lett 30(11):1959–1966

Wang Z, Lim BS, Lu H, Fan J, Choi CS (2010) Cathodic reduction of cu 2+ and electric power generation using a microbial fuel cell. Bull Korean Chem Soc 31(7):2025–2030

Wang J, Song X, Wang Y, Abayneh B, Li Y, Yan D, Bai J (2016) Nitrate removal and bioenergy production in constructed wetland coupled with microbial fuel cell: establishment of electrochemically active bacteria community on anode. Bioresour Technol 221:358–365

Wang Z, He Z (2020) Frontier review on metal removal in bioelectrochemical systems: mechanisms, performance, and perspectives. Journal of hazardous. Mater Lett 100002:100002

Wrighton KC, Thrash JC, Melnyk RA, Bigi JP, Byrne-Bailey KG, Remis JP, Schichnes D, Auer M, Chang CJ, Coates JD (2011) Evidence for direct electron transfer by a gram-positive bacterium isolated from a microbial fuel cell. Appl Environ Microbiol 77(21):7633–7639

Wu X, Zhu X, Song T, Zhang L, Jia H, Wei P (2015) Effect of acclimatization on hexavalent chromium reduction in a biocathode microbial fuel cell. Bioresour Technol 180:185–191

Xu W, Zhang H, Li G, Wu Z (2016a) A urine/Cr (VI) fuel cell—electrical power from processing heavy metal and human urine. J Electroanal Chem 764:38–44

Xu L, Zhao Y, Doherty L, Hu Y, Hao X (2016b) The integrated processes for wastewater treatment based on the principle of microbial fuel cells: a review. Crit Rev Environ Sci Technol 46(1): 60–91

Xu L, Zhao Y, Wang T, Liu R, Gao F (2017) Energy capture and nutrients removal enhancement through a stacked constructed wetland incorporated with microbial fuel cell. Water Sci Technol 76(1):28–34

Yakar A, Türe C, Türker OC, Vymazal J, Saz Ç (2018) Impacts of various filtration media on wastewater treatment and bioelectric production in up-flow constructed wetland combined with microbial fuel cell (UCW-MFC). Ecol Eng 117:120–132

Yu YY, Guo CX, Yong YC, Li CM, Song H (2015) Nitrogen doped carbon nanoparticles enhanced extracellular electron transfer for high-performance microbial fuel cells anode. Chemosphere 140:26–33

Zhang L, Xu Z, He Z (2020) Selective recovery of lead and zinc through controlling cathodic potential in a bioelectrochemically-assisted electrodeposition system. J Hazard Mater 386: 121941

Zhao Z, Zhang Y, Wang L, Quan X (2015) Potential for direct interspecies electron transfer in an electric-anaerobic system to increase methane production from sludge digestion. Sci Rep 5(1): 1–12

Zhang B, Zhao H, Shi C, Zhou S, Ni J (2009) Simultaneous removal of sulfide and organics with vanadium (V) reduction in microbial fuel cells. J Chem Technol Biotechnol 84(12):1780–1786

Minimization of Cadmium Toxicity in Wheat by Exogenous Application of Hydroxamate Siderophore

19

Pratika Singh, Anjana, Ravinsh Kumar, Azmi Khan, Ashutosh Singh, and Amrita Srivastava

Abstract

Siderophores are considered as small molecular weight, non-ribosomal peptides with high affinity for ferric ions. Recent studies confirmed its chelation ability with metals and metalloids other than iron. Hence, they are also termed as metallophores. Owing to this property, they show dynamic properties related to bioremediation, antimicrobial effect, and virulence. In the current study, the effect of siderophore to check cadmium-induced toxicity in wheat seedlings (*Triticum aestivum*) has been analyzed. Cadmium ions adversely affected the morphological, physiological, and biochemical properties of wheat seedlings. The present study revealed positive impact of hydroxamate kind of siderophore of *Aspergillus nidulans* origin upon Cd-stressed wheat plants. The seedlings successfully recovered growth and showed improved activity of antioxidative enzymes like superoxide dismutase (SOD), peroxidase (POD), and catalase. At the same time, total soluble sugar and free amino acid content were also determined along with length of root and shoot. Restoration of chlorophyll content further established that siderophore was able to minimize the toxicity of Cd ions. Hence, siderophores can be further tested for agricultural benefits and bioremedial activities.

Keywords

Cadmium · *Triticum aestivum* · Siderophore · Bioremediation · Chelation · Antioxidative enzymes

P. Singh · Anjana · R. Kumar · A. Khan · A. Singh · A. Srivastava (✉)
Department of Life Science, Central University of South Bihar, Gaya, Bihar, India
e-mail: amritasrivastava@cub.ac.in

R. P. Singh et al. (eds.), *Heavy Metal Toxicity: Environmental Concerns, Remediation and Opportunities*, https://doi.org/10.1007/978-981-99-0397-9_19

19.1 Introduction

Heavy metals are one of the important environmental poisons. It is considered a metallic element with comparatively high density and is lethal at very low concentration (Ali and Khan 2018). It is a collection of metals possessing atomic density more than 4 g/cm^3, or five times or more, greater than water (Nagajyoti et al. 2010). Cadmium (Cd) is a notable toxic trace element which enters the environment by different sources such as industrial effluents and agriculture fields with wastewater and municipal composts and phosphatic fertilizers. It is a chief distress in plants as it collects in leaves at extremely high concentration that can be consumed by organisms through food chain. Cd in a soil occurs on an average of 0.36 mg/kg, and according to WHO guideline, permissible limit in a drinking water should not exceed beyond 3 μg/ml (Kubier et al. 2019).

Wheat (*Triticum aestivum)* is one of the most staple foods around the globe with annual production estimated to be 650 million tons (FAO 2013). Cd uptake occurs primarily through roots from where it translocates to the aerial region. Consequently, Cd uptake disturbs essential nutrient uptake, affects morphological and physiological parameters, and leads to oxidative stress (Rizwan et al. 2016). Moreover, it disturbs the water balance and causes damage to the photosynthetic apparatus including light harvesting complexes and the photosystems. The major symptoms include necrosis, chlorosis, and reduced growth and yield along with browning of root tips due to Cd toxicity that alters other nutrient levels too. Although several strategies have been employed to minimize its effect, there is a need to mitigate its entry through natural compounds. The last few years have witnessed the role of one such secondary metabolite as a bioremedial strategy called siderophore.

Siderophores are comparatively low molecular weight, non-ribosomal compounds which are specific ferric chelating agents (Khan et al. 2021; Singh et al. 2022. They have multifaceted roles like chelation, bioremediation, antimicrobial activity, virulence, molecular markers, etc. (Fig. 19.1).

They are mostly produced by bacteria, fungi, and cyanobacteria growing under low iron stress (Khan et al. 2018). While the major role of siderophore is to sequester ferric iron, they can also play a significant role in detoxifying heavy metal-contaminated surroundings as well as in effective solubilization and mobilization of different metals and metalloids (Schalk et al. 2011; Singh et al. 2020). Consequently, the siderophore-metal complexation seems a promising approach for bioremediation of heavy metal. These complexes might alleviate concentration of free metal reducing their toxic effects. Thus, siderophores might be able to stop heavy metal entry inside the cells by diffusion, effectively purifying the atmosphere for organisms (Zou and Gregory 2005). Complexation of siderophore with either essential or nonessential metals depends principally on ligand functionalities determining metal-siderophore complex stability (Schalk et al. 2011; Mehri et al. 2012; Johnstone and Nolan 2015). Siderophore from *P. azotoformans* also mobilizes arsenic from contaminated soil (Nair et al. 2007). Pyochelin and putrebactin both form complexes with vanadium; however, only vanadyl ion-pyoverdine and vanadyl ion-enterobactin complex are formed while vanadate ion complexed with

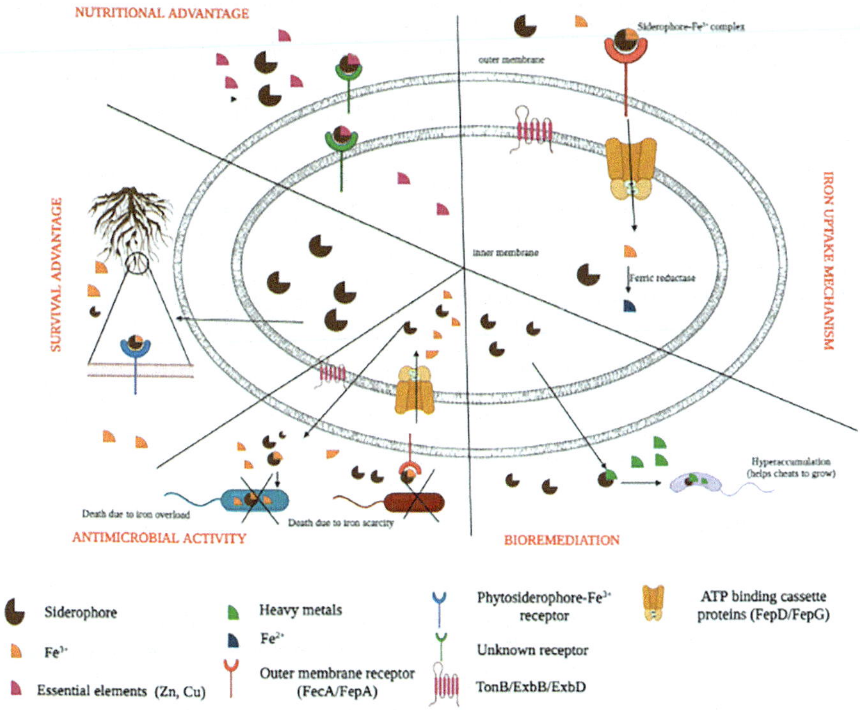

Fig. 19.1 Multifaceted role of siderophore

desferrioxamine B (Baysse et al. 2000; Pakchung et al. 2011). Siderophore producers also assist in plant growth, thereby acting as plant growth-promoting microbes (PGPM). Yu et al. (2017) reported enhanced assimilation of Pb from soil in *Brassica juncea* and further revealed that *Bacillus* sp. PZ-1 produces copious amounts of siderophores that can aid in phytoextraction of Pb from soil. Similarly, *P. cretica* can be used for phytoextraction of arsenic because of its hyperaccumulating capacity (Jeonga et al. 2014). Increased affinity of phytosiderophores toward several metals has been found in the order- Cd^{2+} > Ni^{2+-} > Pb^{2+} > Sn^{2+} > AsO_4^{-2} > AsO_2^{-1} > Mn^{2+} > Co^{2+} > Cu^{2+} > Fe^{3+}, while very less affinity was observed with Al^{3+} and Cr^{3+} (Ruggiero et al. 1999).

Aspergillus nidulans produces majorly two types of siderophores: it is responsible for excreting triacetylfusarinine C (TAFC) and also contains ferricrocin (FC) intracellularly (Eisendle et al. 2003). Previous studies confirmed that when siderophores were exogenously applied to the plants under heavy metal stress, it led to improved chlorophyll content than in the plants grown under heavy metal condition alone, even though it contains substantial amounts of bioavailable iron (Kumari et al. 2019). The act of bioremediation by microbes via siderophore production has an evolutionary aspect too. According to Brien et al. (2014), siderophore synthesis in terms of bioremediation might be an altruistic behavior (costly for individuals but benefits adjacent organisms in the niche). As a process of decontamination by "public good," siderophores potentially benefit the entire community. Preference of metal detoxification over iron scavenging by microbes may occur as the latter is species specific, whereas metal detoxification acts benefit the entire community.

Studies on siderophores as bioremedial compounds and other heavy metal chelation have enlarged in the past decade. Siderophores can be utilized in soils containing high cadmium content to nurture cereal plants like wheat which will be helpful in improving morphological, physiological, and antioxidative enzymes caused by cadmium mitigation. The present book chapter explores the possible strategies to minimize Cd toxicity in wheat seedlings through exogenously applied hydroxamate, a kind of siderophore.

19.2 Materials and Methods

19.2.1 Culture of *Aspergillus nidulans* and Isolation of Siderophore

Aspergillus nidulans was grown in YEPDA medium (yeast extract peptone dextrose agar) for siderophore isolation. Spore suspension was prepared by harvesting conidia using phosphate buffer saline (pH 7) with 0.05% tween 20. The suspension was collected in sterile falcon, and a number of conidia were counted using hemocytometer. 1×10^6/100 ml conidia were transferred in freshly prepared modified Grimm-Allen's (GA) media and incubated at 37 °C in incubator shaker at 60 rpm for 3 days. After 3 days of incubation, *Aspergillus nidulans* culture was filtered through Whatman filter paper, and filtrate was checked by CAS assay for siderophore production. CAS-positive filtrates were checked for hydroxamate-type siderophore through Atkin's assay.

19.2.2 Purification and Quantification of Hydroxamate-Type Siderophore

Hydroxamate-type siderophore was purified by adsorption chromatography using XAD-2 resins (Khan et al. 2020). Atkin's assay positive samples acidified to pH 2 ± 0.1 and loaded on XAD-2 resins and adsorbent were eluted in methanol. The siderophores were concentrated by rotary evaporator. Hydroxamate-type siderophore was quantified in siderophore units (SU). Partial purified siderophore was quantified spectrophotometrically using the following formula:

$$\%\text{Siderophore unit} = \text{Ar} - \text{As}/\text{Ar} \times 100$$

where:

Ar = Absorbance of reference

As = Absorbance of sample

Absorbance was taken at 630 nm. Blank was set using uninoculated media, while the reference contained uninoculated media with CAS dye and shuttle solution.

19.2.3 Cultivation and Treatment of Wheat Plant

K-0307 variety of wheat seed was surface sterilized in 0.1% $HgCl_2$ for 5 mins and rinsed thoroughly in distilled water and then imbibed in water for 2 h. Seeds were left for germination on moist filter paper in a petri plate under dark condition for 3 days. Seed germination frequency was calculated using formula (no. of germinated seed/ total no. of seeds) $\times$ 100. When radicals appeared, seedlings were transferred to plastic pots filled with Murashige and Skoog (MS) medium (pH -5.6). The seedlings were grown for 4–5 days at normal room temperature at light intensity of 200 $\mu M\ cm^{-2}\ s^{-1}$ (16/8 h; day/night).

19.2.4 Cadmium Treatment

Firstly, a stock solution of 10 mM cadmium nitrate was prepared. From the stock solution, different concentrations of $Cd(NO_3)2$ were prepared, i.e., 0.2 mM, 0.4 mM, 0.6 mM, 0.8 mM, 1 mM, 2 mM, 3 mM, 4 mM, and 5 mM. The 7-day-old seedlings were treated to different concentrations of cadmium salt for 3 days dissolved in MS medium. 4 mM concentration of cadmium nitrate was used for further experiment as maximum deleterious effect was observed at this concentration. Plants grown in cadmium-deficient medium served as control.

19.2.5 Siderophore Treatment

Maximum deleterious concentration of cadmium nitrate which showed 50% deleterious effect was selected for siderophore treatment. The 7-day-old seedlings were exposed to cadmium stress using cadmium nitrate, and 25 SU/ml of partial purified hydroxamate siderophore was added, and observation was taken after 3 days. Four setups were made, i.e., control (-Cd, -Sid), +cadmium, + Cd + siderophore (25 SU/ml) and + siderophore (25 SU/ml). The samples were harvested after 3 days and used for further analysis.

19.2.6 Morphological Analysis

The root and shoot length of the 7-day-old seedling was measured using a scale. The plant biomass was measured.

19.2.7 Physiological Analysis

19.2.7.1 Pigment Analysis

Chlorophyll and carotenoid were measured after harvesting the respective treated seedlings. 100 mg of the leaf samples was taken in 10 ml of DMSO and incubated at

65 °C for 3 hrs. Absorbance of filtrate was measured at 663, 640, and 470 nm. Total chlorophyll content was estimated using Arnon (1949) equation and carotenoid was estimated using Lichtenthaler and Wellburn (1983) equations.

19.2.7.2 Free Amino Acid Content Estimation

Proline amino acids were determined by FAA assay. 3% aqueous sulfosalicylic acid was used to homogenize 500 mg of plant material followed by filtration through Whatman filter paper. Reaction between 2 ml of filtrate along with 2 ml ninhydrin acid and 2 ml of glacial acetic was conducted for 1 hour at 100 °C, and the reaction was stopped by keeping the samples in water bath. 4 ml of toluene was added and mixed for 15 seconds. The absorbance was taken at 520 nm using toluene as blank. The proline concentration was detected from the standard curve and calculated as a fresh weight through the following equation: proline (μmole/g of fresh weight) = [(μg proline/ml × ml toluene)/ 115.5 μg/μmole]/5.

19.2.8 Biochemical Analysis

19.2.8.1 Antioxidative Enzyme Assay

Antioxidative enzyme assay was carried out after the 3-day treatments that were given to 7-day-old seedlings. 1 gm of sample was crushed in 10 ml of liquid nitrogen followed by addition of 4 ml phosphate buffer saline (PBS) (pH = 7.5). Sample was centrifuged at 12,000 rpm for 15 min at 4 °C. The supernatant was used for biochemical assay.

19.2.8.2 Catalase Assay

Catalase assay was measured by following the protocol of Aebi (1984). 0.05 ml of the supernatant was mixed with 1.5 ml phosphate buffer and 0.5 ml of H_2O_2 was added. Initial absorbance and final absorbance after 30 seconds were recorded at 240 nm, and catalase activity was calculated using the following formula: CAT activity (unit/min/g FW) = (3 × Abs final – Abs initial)/ (0.0028) micromole/gm/min.

19.2.8.3 Peroxidase Assay

Peroxidase assay was measured as per the protocol of Polle and Junkermann (1994). 0.05 ml supernatant was mixed with 1 ml phosphate buffer. 0.5 ml hydrogen peroxide was added followed by addition of 0.5 ml guaiacol and 0.9 ml double-distilled water. Initial and final absorbances were recorded at 470 nm. The enzyme activity was calculated as per extinction coefficient of its oxidation product, tetra guaiacol. Peroxidase activity was calculated using the following formula: POD activity (unit/min/g FW) = 11,278 × OD (micromole/tetra guaiacol/min/mg FW).

19.2.8.4 Superoxide Dismutase Assay

Superoxide dismutase assay was carried out according to the protocol of Dhindsa et al. (1981). SOD activity was measured by taking an absorbance at 560 nm and calculated in terms of unit of activity.

19.3 Results and Discussion

19.3.1 Qualitative and Quantitative Estimation of Siderophore

Siderophore production by *A. nidulans* in modified GA medium was confirmed by CAS assay. The conversion of blue to pink color confirmed the chelation of ferric ions by presence of siderophore in CAS universal assay (Fig. 19.2a), while the change of orange color in Atkin's assay confirmed that the siderophore was of hydroxamate type (Fig. 19.2b).

19.3.2 Cadmium Treatment

Wheat seedlings were grown at different concentrations of cadmium nitrate (0.2 mM, 0.4 mM, 0.6 mM, 0.8 mM, 1 mM, 2 mM, 3 mM, 4 mM, and 5 mM) in order to select optimum concentration of cadmium stress. The plant with 50% damage was selected for the siderophore treatment. Plants treated with 4 mM

(a) (b)

Fig. 19.2 Qualitative analysis of hydroxamate siderophore using CAS and Atkin's reagent. (**a**) Control and CAS positive sample, (**b**) control and Atkin's positive sample

Table 19.1 Showing extent of damage at different concentrations of Cd

Concentration	Yellowing of leaves	Leaf curling	Turgidity
Control	–	–	+++
0.2 mM	–	–	+++
0.4 mM	–	+	++
0.6 mM	+	+	++
0.8 mM	+	–	++
1 mM	++	++	+
2 mM	++	+	++
3 mM	++	++	+
4 mM	+++	++	+
5 mM	+++	+++	+

cadmium nitrate treatment showed 50% damage as observed in leaf curling, yellowing of leaves, and turgidity after 7 days of treatment (Table 19.1; Fig. 19.3).

19.3.3 Siderophore Treatment

According to our previous studies published (Kumari et al. 2019), 25 SU/ml was taken as siderophore concentration to check the effects on arsenic-treated wheat seedlings. Therefore, in the present study, 25 SU/ml was taken to check its effect. Four experimental setups were taken: control (-Cd –Siderophore), + Cd (4 mM), + Cd + $Siderophore_{25}$, + $Siderophore_{25}$. Substantial yellowing of leaves and leaf curling were observed in the case of Cd alone. The decrease in leaf yellowing and leaf curling was evident upon siderophore treatment. Better morphological performance was observed when treated with siderophore alone which confirmed the fact that siderophore acts as plant growth promoters (Table 19.2; Fig. 19.4).

19.3.4 Root and Shoot Length

Increase in the root length was observed in Cd stressed seedlings due to Cd toxicity. The length decreased as Cd-stressed sample treated with siderophore. The maximum shoot length was observed in control plants which showed a reduction when treated with cadmium (Fig. 19.5a). Similarly, maximum biomass of plants was observed in the case of control. It decreased slightly when cadmium stress was given to the seedlings (Fig. 19.5b).

19.3.5 Total Chlorophyll and Carotenoid Content

Sharp decline in chlorophyll content was observed in Cd-stressed plants. Heavy metals like As and Cd damage the chloroplast membrane as well as decrease plastid

(a)

(b)

Fig. 19.3 Morphology of plants treated with different concentrations of Cd

pigment resulting in reduction of chlorophyll content. The Cd-stressed seedlings when treated with siderophore (25SU/ml) had improved total chlorophyll content and carotenoid content decreased by 23% as compared to control (Fig. 19.6).

Table 19.2 Showing effect of siderophore on Cd-stressed wheat seedlings

Concentration	Yellowing of leaves	Leaf curling	Turgidity
Control	–	–	+++
4 mM	+++	+++	+
4 mM + 25 SU/ml	++	+	+
25 SU/ml	+	+	+

Fig. 19.4 Morphology of different experimental setups

19.3.6 Total Soluble Sugar Content

Stress conditions trigger the addition of soluble sugars. With the increase in the stress, there was an increase in the accumulation of TSS in the plants. It increased for about 56% as compared to control plants. When Cd-stressed plants were further treated with siderophore, there was reduction in TSS for about 48.73% as compared to cadmium-treated plants (Fig. 19.7).

19.3.7 Free Amino Acid (FAA) Determination

Osmoprotectants like amino acids (proline) stabilize cell membrane proteins and decrease osmotic potential, thereby preventing cellular dehydration. There was a drastic increase in the accumulation of free amino acid in the cadmium-stressed plants. Its level was lowest in case of control. Cd-stressed plants had shown an increase in FAA for about 62.10% as compared to control. When Cd-stressed plants were further treated with siderophore, there was reduction in FAA for about 48.73% as compared to cadmium-treated plants (Fig. 19.8).

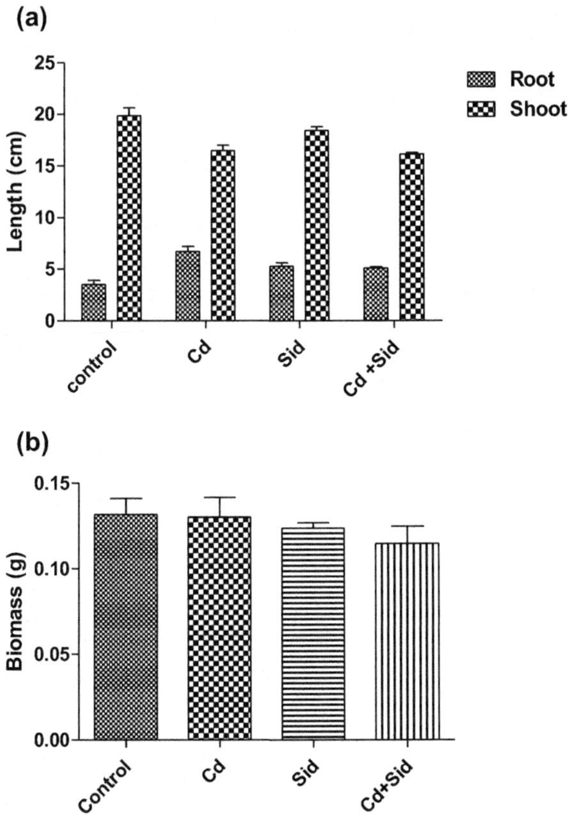

Fig. 19.5 Effect of siderophore treatment on (**a**) root and shoot length and (**b**) biomass of wheat seedlings

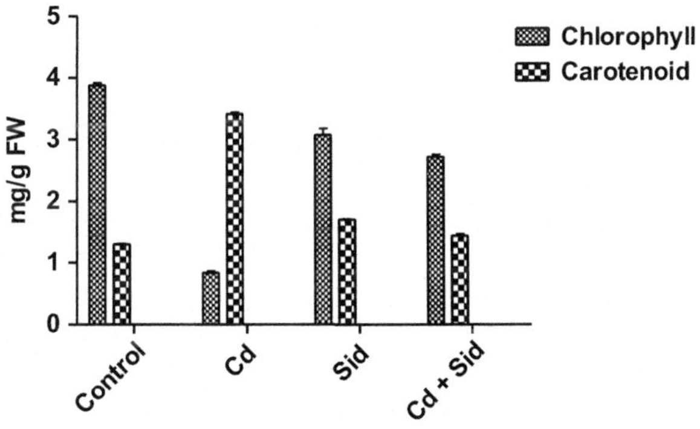

Fig. 19.6 Chlorophyll and carotenoid content of Cd-stressed wheat seedlings treated with siderophore

19.3.8 Catalase Activity

Catalase activity increased in case of combined treatment of Cd and siderophore. It was lowest in the case of cadmium-treated seedlings. When Cd-stressed seedlings were supplied exogenously with siderophore, there was improvement in catalase

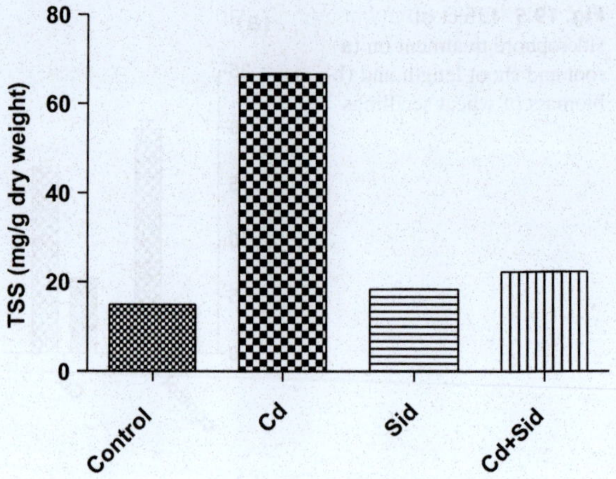

Fig. 19.7 Total soluble sugar content of Cd-stressed wheat seedlings treated with siderophore

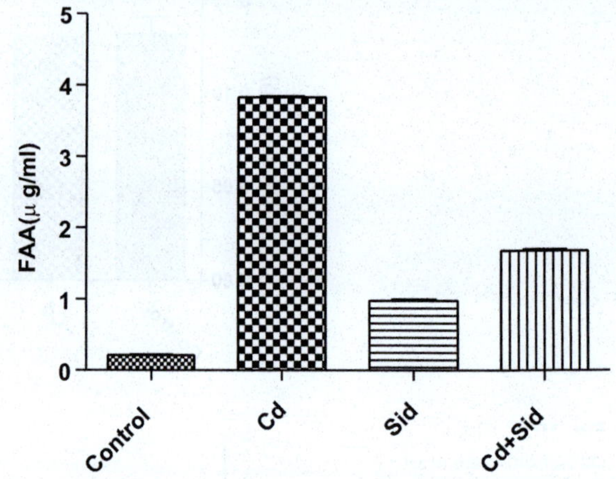

Fig. 19.8 Free amino acid content of Cd-stressed wheat seedlings treated with siderophore

activity. Increase in catalase activity was observed of about 19.23% as compared to control (Fig. 19.9).

19.3.9 POD Activity

There was a significant increase in POD activity in cadmium-stressed plants for about 53%, while in siderophore-treated plants alone, it decreased. When Cd-stressed seedlings were treated with siderophore, POD activity was reduced about 37%. Control plants had the lowest level of POD activity (Fig. 19.10).

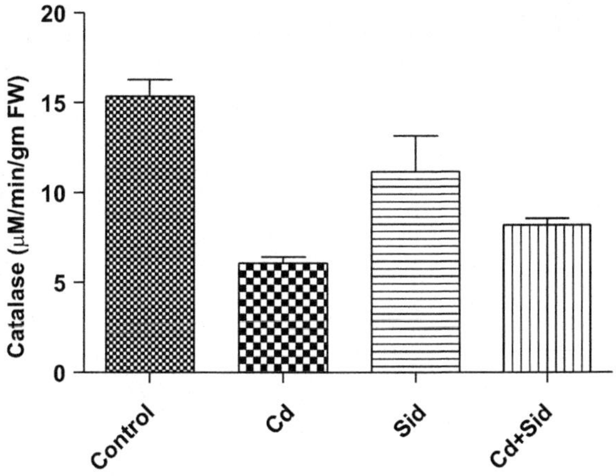

Fig. 19.9 CAT activity of Cd-stressed wheat seedlings treated with siderophore

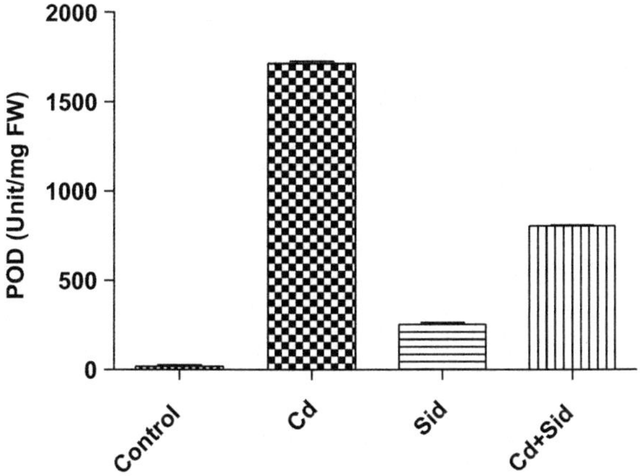

Fig. 19.10 POD activity of Cd-stressed wheat seedlings treated with siderophore

19.3.10 SOD Activity

SOD activity was maximum in case of cadmium-treated plants. It increased 64.56% as compared to control plants. There was great decrease in the SOD activity, about 51.23% in case of cadmium-treated plants with siderophore indicating that siderophore has shown the reduction in SOD activity (Fig. 19.11).

19.4 Conclusion

Siderophore biology is a rising and exciting field of research, and various new siderophores as bioremedial potential need to be discovered regularly. In this study, the toxic effect of cadmium has been reported to be reduced and lessened by utilizing the chelation capacity of hydroxamate siderophore of *Aspergillus*

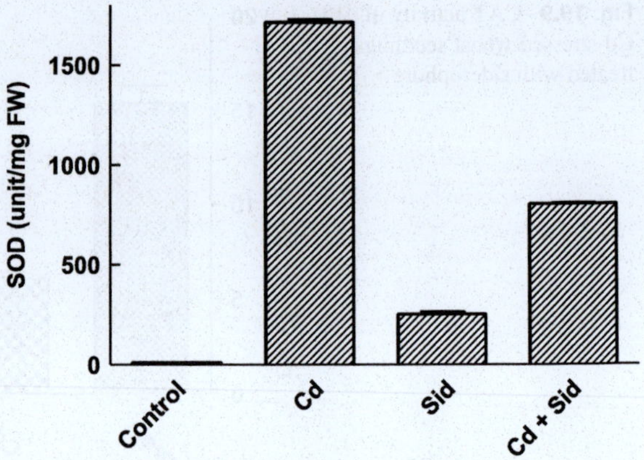

Fig. 19.11 SOD activity of Cd-stressed wheat seedlings treated with siderophore

nidulans origin. The toxic effect was reduced when the stressed plants were treated with siderophore concentration of 25SU/ml which without further affecting plant physiology restored the normal condition of plants. Thus, siderophores can be further investigated in order to use in high Cd-polluted soil to cultivate cereal crops like wheat which will be useful in decreasing stress caused by cadmium mitigation.

References

Aebi H (1984) Catalase in vitro. In: Methods in enzymology, vol 105. Academic Press, Florida, pp 114–121

Ali H, Khan E (2018) What are heavy metals? Long-standing controversy over the scientific use of the term 'heavy metals' – proposal of a comprehensive definition. Toxicol Environ Chem 100(1):6–19

Arnon DI (1949) Copper enzymes in isolated chloroplast, polyphenol- oxidase in *Beta vulgaris*. Plant Physiol 24:1–15

Baysse C, De Vos D, Naudet Y, Vandermonde A, Ochsner U, Meyer JM, Budzikiewicz H, Schäfer M, Fuchs R, Cornelis P (2000) Vanadium interferes with siderophore-mediated iron uptake in *Pseudomonas aeruginosa*. Microbiol (Reading) 146(Pt 10):2425–2434

Brien SO, David JH, Angus B (2014) Social evolution of toxic metal bioremediation in *Pseudomonas aeruginosa*. Proc R Soc B 281:20140858

Dhindsa RS, Dhindsa P, Thorpe T (1981) Leaf senescence correlated with increased levels of membrane permeability and lipid peroxidation and decrease levels of superoxide dismutase and catalase. J Exp Bot 32:93–101

Eisendle M, Oberegger H, Zadra I, Haas H (2003) The siderophore system is essential for viability of *aspergillus nidulans*: functional analysis of two genes encoding l-ornithine N 5-monooxygenase (sidA) and a non-ribosomal peptide synthetase (sidC). Mol Microbiol 49(2):359–375

Food and Agriculture Organisation of the United Nations. (2013) FAOSTAT. URL http://faostat.fao.org/site/567/DesktopDefault.aspx?PageID=567#ancor Accessed on 1 October 2013

Jeonga S, Moonb HS, Kyoungphile N (2014) Enhanced uptake and translocation of arsenic in Cretan brake fern (*Pteris cretica* L.) through siderophore arsenic complex formation with an aid of rhizospheric bacterial activity. J Hazard Mater 280:536–543

Johnstone T, Nolan E (2015) Beyond iron: non-classical biological functions of bacterial siderophores. Dalton Trans 44:6320–6339

Khan A, Gupta A, Singh P, Mishra AK, Ranjan RK, Srivastava A (2020) Siderophore-assisted cadmium hyperaccumulation in *Bacillus subtilis*. Int Microbiol 23(2):277–286

Khan A, Singh P, Kumar R, Das S, Singh RK, Mina U, Agrswal GK, Rakwal R, Sarkar A, Srivastava A (2021) Antifungal activity of siderophore isolated from *Escherichia coli* against *aspergillus nidulans* via iron-mediated oxidative stress. Front Microbiol 12:729032

Khan A, Singh P, Srivastava A (2018) Synthesis, nature and utility of universal iron chelator - Siderophore: a review. Microbiol Res 212-213:103–111

Kubier A, Wilkin RT, Pichler T (2019) Cadmium in soils and groundwater: a review. Appl Geochem 108:1–16

Kumari S, Khan A, Singh P, Dwivedi SK, Ojha KK, Srivastava A (2019) Mitigation of as toxicity in wheat by exogenous application of hydroxamate siderophore of *aspergillus* origin. Acta Physiol Plant 41:107

Lichtenthaler HK, Wellburn AR (1983) Determinations of total carotenoids and chlorophylls a and b of leaf extracts in different solvents. Biochem Soc Trans 11:591–592

Mehri I, Khessairi A, Turki Y (2012) Effect of dose-response of zinc and manganese on siderophore production. Am J Environ Sci 8:143–151

Nagajyoti PC, Lee KD, Sreekanth TVM (2010) Heavy metals, occurrence and toxicity for plants: a review. Environ Chem Lett 8:199–216

Nair A, Juwarkar A, Singh S (2007) Production and characterization of siderophores and its applications in arsenic removal from contaminated soil. Water Air Soil Pollut 180:199–212

Pakchung AA, Soe CZ, Lifa T, Codd R (2011) Complexes formed in solution between vanadium (IV)/(V) and the cyclic dihydroxamic acid putrebactin or linear suberodihydroxamic acid. Inorg Chem 50:5978–5989

Polle A, Junkermann W (1994) Inhibition of apoplastic and symplastic peroxidase activity from Norway spruce by the photooxidant hydroxymethyl hydroperoxide. Plant Physiol 104:617–621

Rizwan M, Ali S, Abbas T, Zia-Ur-Rehman M, Hannan F, Keller C, Al-Wabel MI, Ok YS (2016) Cadmium minimization in wheat: a critical review. Ecotoxicol Environ Saf 130:43–53

Ruggiero CE, Neu MP, Vanderberg LA (1999) Plutonium and uranium interactions with siderophores of aerobic soil microbes. J Inorg Biochem 74:282

Schalk IJ, Hannauer M, Braud A (2011) New roles for bacterial siderophores in metal transport and tolerance. Environ Microbiol 13(11):2844–2854

Singh P, Khan A, Kumar R, Kumar R, Singh VK, Srivastava A (2020) Recent developments in siderotyping: procedure and application. World J Microbiol Biotechnol 36:178

Singh P, Khan A, Kumar R, Ojha KK, Singh VK, Srivastava A (2022) *In silico* analysis of comparative affinity of phytosiderophore and bacillibactin for iron uptake by YSL15 and YSL18 receptors of *Oryza sativa*. J Biomol Struct Dyn 9:1–14

Yu S, Chunying T, Xin B, Jinsong L, Tao S, Liying D, Yu J, Juanjuan Q (2017) Optimization of siderophore production by *bacillus* sp. PZ-1 and its potential enhancement of phytoextration of Pb from soil. J Microbiol Biotechnol 27(8):1500–1512

Zou G, Gregory LB (2005) Synthesis and properties of different metal complexes of the siderophore desferriferricrocin. Biometals 18:63–74

Microbial Remediation of Heavy Metals 20

Norli Ismail, Amir Talebi, and Harlina Ahmad

Abstract

Heavy metals presence in wastewater are usually associated with the usage of metal-based chemical substances such as inorganic coagulant, metal salt, and metal-based processing. Heavy metal is defined as an essential element that can be found in Earth. Few amounts of heavy metals would be beneficial for human beings, but excessive amounts of heavy metals would become a catastrophic disaster for the environment. Therefore, it is crucial to verify the water quality, mostly when even just 1.0 mg/L of the concentration may contribute a higher impact on the environment. The existing or the remaining low concentration of heavy metals of post-chemical treatment of polluted water of wastewater remains troublesome due to incompliance effluent discharge. This chapter will highlight the removal of heavy metals through bioremediation and biotransformation. Concomitantly the mechanism and the roles of extracellular polymeric substances (EPS) will be discussed.

N. Ismail (✉)
Environmental Technology Division, School of Industrial Technology, Universiti Sains Malaysia, Penang, Malaysia

Renewable Biomass Transformation Cluster, School of Industrial Technology, Universiti Sains Malaysia, Penang, Malaysia
e-mail: norlii@usm.my

A. Talebi
Schahverdy Entwicklung und Konstruktion Maschinenbau, Baden-Württemberg, Germany

H. Ahmad
Environmental Technology Division, School of Industrial Technology, Universiti Sains Malaysia, Penang, Malaysia

R. P. Singh et al. (eds.), *Heavy Metal Toxicity: Environmental Concerns, Remediation and Opportunities*, https://doi.org/10.1007/978-981-99-0397-9_20

Keywords

Bioremediation · Biotransformation · Bioreduction · Bioextraction · Biosorption

20.1 Introduction

Microbial remediation is insinuated to applying microorganisms in reducing organic contaminants. In general, the microorganisms (endogenous or exogenous) are capable to dissociate the chemical structure of the contaminants by consuming them as a victual and food resource. Decomplexation and transformation of heavy metals by bacteria is a mechanism of heavy metals removal in water or wastewater. Heavy metals exist either in metalloids or in metallic elements and have relatively high atomic weight. Some heavy metals pose severe hazard even when present at low concentration. Pushkar et al. (2015) readdress the matters related to non-degradable heavy metals and thus remains as a persistent pollutant with most of the existing treatment both physical and chemical treatments.

Heavy metal pollution has a negative impact on the biogeochemical cycles and ocean productivity (Bong et al. 2010). Heavy metals usually distract the biogeochemical cycles of the environmental ecosystem and life cycle of the microorganisms in soils, water, and air, especially related to the conversion of organic carbon to biocarbon. Another negative effect of heavy metals in nature is physiological failure and undernourishment in plants and consequently, causing serious and permanent damage to human health. Heavy metals can accumulate in living organisms and are responsible for many metabolic and physiological disorders (Jin et al. 2018).

Previous research reported physicochemical and biological methods, such as precipitation, electrolytic processes, solvents extraction, ultrafiltration, activated carbon, ion exchange, coagulation-flocculation, and liquid membrane to name a few have been widely studied and applied in heavy metals removal of wastewater. Those treatment approaches works well for initial high concentrations of heavy metals. Anyhow the consequences pose a hazard to the environment associated with hazardous and toxic secondary pollution generation.

Biological methods, such as bioadsorption, bioremediation, or using plants as well as microorganisms such as bacteria, fungi, and microalgae, demonstrate a crucial advantage compared with the conventional physicochemical methods: being cost-effective and economically beneficial, green, sustainable, and in addition less wastes and by-products generation. Apart from above discussed advantages of bioremediation techniques, there is a very interesting supremacy of this method, in which heavy metals dissolved in the wastewater can be effectively recovered and precipitated in different beneficial and reusable forms, such as metal sulphides (Pavithra et al. 2020; Yin et al. 2019).

20.2 Physicochemical of Heavy Metals Removal and Recovery from Wastewater

Heavy metals play a role either as essential or as non-essential (toxic) elements. The essential elements refer to macro and micro whilst toxic elements refer to non-essential. Essential elements include Mn, Cu, Zn, Ni, Co, and Fe, and non-essential elements include Cd, Pb, and As. For many years, researchers and scientists have investigated numerous techniques and methods for heavy metals (HM) removal and recovery from aqueous solutions. The objective of some research is to achieve the highest possible efficiency, whilst others try to apply a method in which both efficiency and economic aspects are met. In this section we briefly review some most common and conventional methods of HM removal with special reference to advantages and disadvantages of each one of them. However, the focus of the current chapter is HM removal by bioremediation, which will be discussed rather long-windedly and comprehensively from common processes.

20.2.1 Precipitation

Precipitation is perhaps the most conventional HM removal technique amongst the physicochemical treatment methods. Precipitation is highly effective with noticeable low-cost competence, due to the cheap used chemicals. The method is easy, convenient, and straightforward, with high selectivity. Precipitation can be easily synthesized with further treatment methods and can be designed for large scale and volume (Azimi et al. 2017; Balladares et al. 2018; Carvajal-Fl'orez and Santiago-Alonso 2019; Pohl 2020).

Despite the many benefits of precipitation, the technique has shown some serious disadvantages. It is reported that the method shows a decline in efficiency at low HM concentrations and extra volume of chemicals/precipitating agents are required to be added. The treatment method is highly dependent on pH value, and finally, hazardous sludge generation which brings a disposal issue along, consequently.

20.2.2 Coagulation-Flocculation

Coagulation-flocculation is the formation of colloidal particles in aqueous solution and perception of certain species. This process has a similarity to precipitation and is considered as one of the most conventional and common treatment processes, particularly for industrial wastewater. Its simplicity, high efficiency, and high dewatering capacity makes coagulation and flocculation an ever-interesting heavy metals removal technique. However, high chemicals consumption, relatively high operational costs, and particularly high amount of sludge generation, can be counted as some disadvantages of coagulation and flocculation treatment (Carolin et al. 2017).

20.2.3 Membrane Separation

Membrane separation technologies with ease of operation and high efficiency are widely used in wastewater treatment. The separation process is generally based on pressure-driven filtration in which the membrane acts as a semipermeable feature which inhibits the contaminants crossing. The species crossing through the membrane is a function of different factors such as particles size, membrane pore size, contaminants concentration, and pressure (Zhu et al. 2019; Hampu et al. 2020; Abdullah et al. 2019; Saravanan et al. 2021).

Researchers have investigated different types of membrane separation such as reverse osmosis, nanofiltration, microfiltration, and ultrafiltration for heavy metals removal from wastewater. Despite the various advantages of membrane separation method, such as simplicity and functionality, high favourite selectivity, and, low chemicals requirement, has shown some serious difficulties and disadvantages: The system is very low in permeation rate, flux, and stability, along with membrane fouling and high maintenance costs (Hasanpour and Hatami 2020; Waldman et al. 2021).

20.2.4 Solvent Extraction and Adsorption

Solvent extraction is a method in which chemical varieties are separated based on their solubilities. Solvent extraction consists of a solvent that can dissolve another substance and is used to separate hazardous contaminants from sludge and sediments. Solvent extraction is based on liquid–liquid extraction (LLE) system, including two compartments: aqueous phase and organic phase. Aqueous phase is loaded (contaminated) by one particular or various solutes whilst the organic phase is an immiscible liquid (diluent). The organic phase according to its liquid extraction classification, could be with or without an extraction facilitator (carrier, extractant). The principle of LLE is based on the distribution of a solute from aqueous phase to the organic phase, due to the solute (Chang et al. 2010; Rajadurai and Anguraj 2020).

Solvent extraction method has a simple configuration and possesses high metal removal efficiency with direct metal recovery (Rangabhashiyam and Balasubramanian (2019). The method is not very effective at low concentrations of HMs as it causes secondary pollution, requires highly toxic and hazardous solvents and has high operational cost.

Adsorption is a conventional metal removal method which depicted advantages such as high efficiency and relatively cost-effective. The drawback is the low selectivity and a significant volume of chemical or toxic sludge production.

20.3 Removal of Heavy Metals in Contaminated Media by Biological Process

Bio-based treatment methods use plants, microbial species, plant enzymes, and living organisms to remove organic and inorganic pollutants from contaminated water and soil media. The most conventional species in this method can be mentioned are algae, fungi, yeasts, bacteria, and cyanobacteria and the removal process can be described as adsorption, degradation, deactivation, and transformation.

20.3.1 Bioremediation

Bioremediation techniques are talented substitutes for conventional wastewater treatment methods with an environmentally friendly approach. Bioremediation methods are a feasible, highly efficient, and cost-effective method for removing and/or recovering several contaminants from polluted effluents (Carvajal-Fl'orez and Santiago-Alonso 2019; Arora 2020; Iravani and Varma 2020).

Figure 20.1 illustrated the bioremediation of heavy metals consists of removal or extraction from aqueous media using plants and micro−/macroorganisms (Kumar and Bharadvaja 2020). The major role of HMs ions extraction by living microorganisms is mainly performed by cell wall components of rich electronegative groups (Rangabhashiyam and Balasubramanian 2019). Electronegativity is a chemical characteristic representing the disposition of an atom or a functional group to attract electrons towards itself. The electronegativity of an atom is affected by both its atomic number and the distance that its valence electrons reside from the charged nuclei (Pavithra et al. 2020). From this point of view, some functional groups with

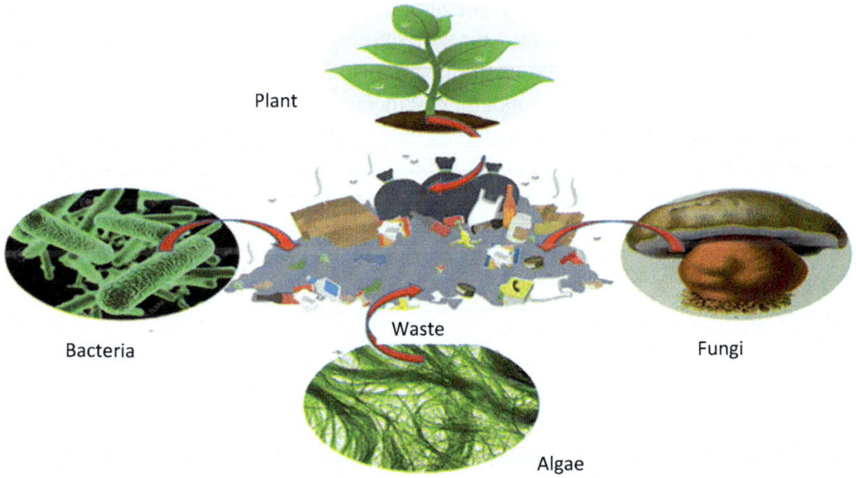

Fig. 20.1 Illustration of bioremediation principle (Kumar and Bharadvaja 2020)

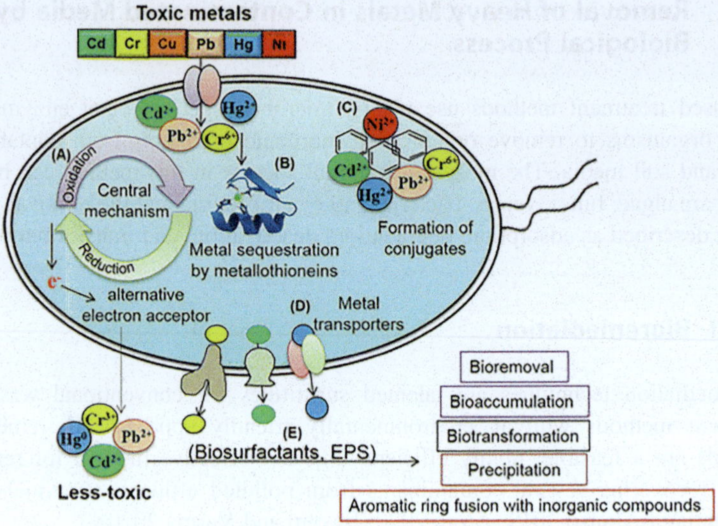

Fig. 20.2 Microbial bioremediation by: (**a**) oxidation–reduction of metals, (**b**) metal sequestration by metallothioneins, (**c**) conjugate formation with organic compounds/precipitation, (**d**) metal efflux by metal transporters followed by bioremoval by microbial products, and (**e**) bioremoval of metals by microbial products (biosurfactants or EPS) (Das et al. 2016)

powerful interactions with metal ions are ideal for HMs microbial remediation, such as carboxylic, phosphoryl, sulphonic, amino, and hydroxyl groups exhibited in Fig. 20.2.

It should be mentioned that microbial remediation is a general term for various treatment methods in which microorganisms play a major role in it. In fact, depending on the targeted heavy metals, the nature of the microorganism, the complexation between the heavy metal ion, microbe, and solubility, the micro remediation is divided into various types, including cell membrane transport, physical adsorption, and ion exchange complexation and biosorption (Abidli et al. 2022). Metal biotransformation is the fundamental mechanism of microbial remediation for decontaminating environments by diverse microbial groups through metal–microbe interaction (bioleaching, bioaccumulation, biotransformation, biomineralization, and biosorption). Exopolymer binding or extracellular polymeric substances (EPS) expedite bacterial cell-mediated immobilization and solubilization.

20.3.2 Phytoremediation

Phytoremediation is effective and cost-effective method relatively to remove heavy metals from the contaminated mediums. Various aquatic and terrestrial plants act as hyperaccumulators of heavy metals and are used for phytoremediation. The most conventional phytoremediation technique is called Rhizofiltration, in which using plants roots play the key role in absorption, concentration, and/or precipitation of the heavy metals (El Liethy et al. 2022). At present phytoremediation is extended into various scopes or branches addressing a specific application associated with plants application as follows: phytoextraction, phytostabilization, phytodegradation, phytostimulation, phytovolatilization, phytodesalination. Akansha et al. (2020) and others research quoted somewhere recommended phytoremediation to couple with either pre or post treatment (physical, chemical or biological) to enhance the phytoremediation efficiency.

The diminishing cause of phytoremediation is due to the slow growth rates of plants versus time taking biomass, plus the time consumer removal progression (Mustafa and Hayder 2021). One solution to rectify this problem is by using microbes or microbial remediation. Microbial remediation consists of biosorption, bioflocculation, bioaccumulation, biomethylation, biotransformation (oxidation–reduction), bioleaching, bioprecipitation, biomineralization, biosurfactant production etc., as summarized in Fig. 20.3 (Sharma 2021).

20.3.3 Biosorption, Bioreduction, and Biooxidation

Biosorption process consists of two major compartments: a solid phase which is also named as sorbent or biosorbent or adsorbent or biological material and a liquid phase (solvent, in most cases water) loaded or contaminated by a certain species to be treated (adsorbate, metals, or elements for instance). Biosorption of heavy metals lasts till the equilibrium is achieved between the heavy metals bound and less portion remaining in the wastewater. The degree of adsorbent affinity for the heavy metal determines its distribution between the solid and liquid phases. Another advantage of biosorption is that the process is metabolically independent, in which they consumed and non-living microbial biomass is still efficient in removing HM ions (Velkova et al. 2018). Table 20.1 presents a brief review of heavy metals removal using different types of algae as biosorbent.

Bioremediation by redox is one of the most advantageous pathways in heavy metals removal by microbial remediation. The process is recognized as bioreduction. One of the advantages of bioreduction of heavy metal ions is less harmful metallic compounds generation and by-products compared with conventional chemical reduction methods. It is reported that bioreduction shows a higher capability to remove heavy metals from aqueous solutions when the initial metal concentrations in the raw influent are less than 20 mg/L. Alongside the heavy metals' removal, bioreduction is also an attractive alternative specifically for the recovery of precious metals such as Au from wastewater for a reasonable economic value. Microbial remedial of heavy metals relies on different pathways from a simple adsorption or absorption process to a more complex mechanism such as reduction process which takes place via microbial surface enhancement (Iravani and Varma 2022; Birungi et al. 2020).

The biooxidation consists of two major agents, namely oxidizer and absorbent. Plant-based material is the most conventional natural absorbents or oxidizers adopted in metal biosorption. Bernard et al. (2018) summarized application of

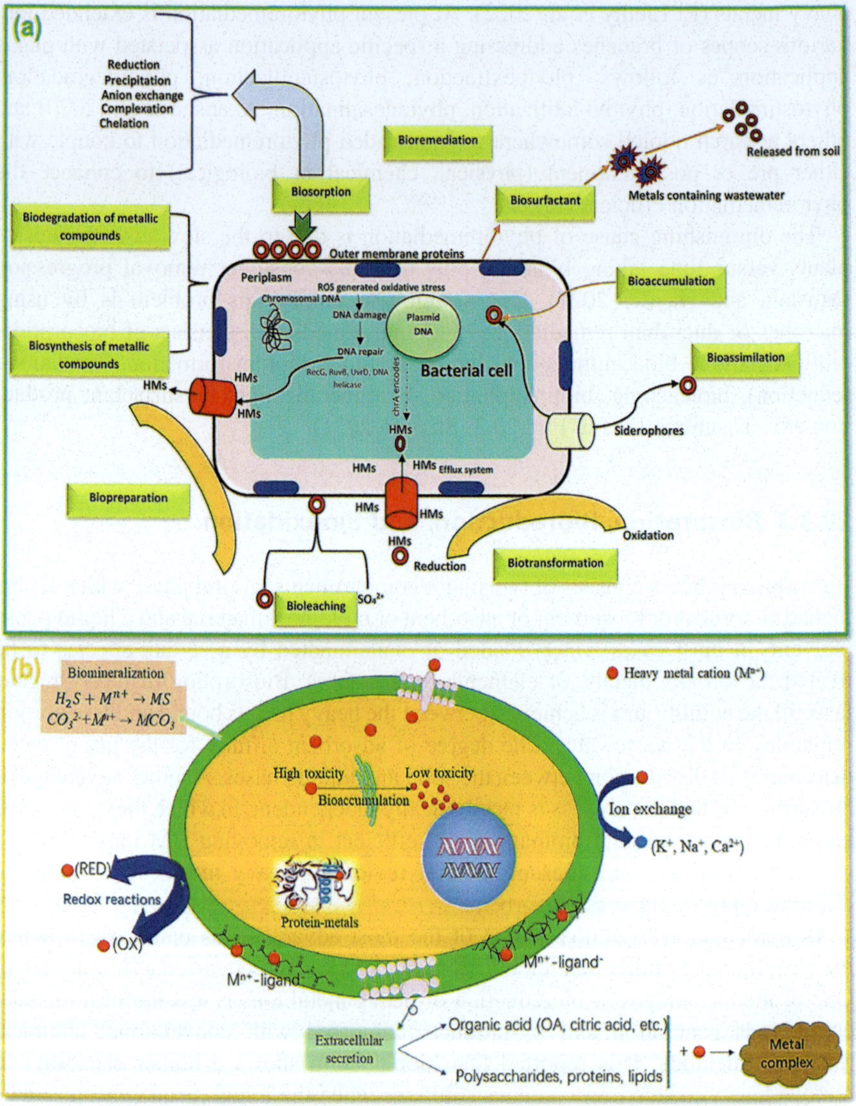

Fig. 20.3 Schematic illustration of (**a**) various types of bacterial interactions with HMs in contaminated WW, Reprinted and adapted from reference Sharma (2021). Copyright (2021), with permission from Elsevier; and (**b**) various mechanisms involved in bioremediation, Reprinted and adapted from reference Xing et al. (2020), Copyright (2020), with permission from Elsevier

microorganisms in HM removal via oxidation. The heavy metals taken up via biosorption are a metabolism-independent mechanism. The biosorption mostly ensues on the exterior cell of the microbial. The biosorption mechanism includes bioaccumulation, bio-conversion via redox reaction, metals sequestration and

Table 20.1 Biosorption data of biosorbent algae with metal sorbent at operating conditions

Biomass type	Metal	H	T (°C)	[mg/L]	Wt (g/L)	Ns (rpm)	Time (h)	Q (mg/g)	Reference
Calotropis procera (Ait)	Pb Cu	4.0 5.0	25	25-100	2.0	150	6.0	22.8 14.5	(Lodeiro et al. 2002)
Bifurcaria bifurcata	Cd	4.5	–	10-350	2.5	175	3.0	95	(Gina et al. 2002)
Oocystis	Cd Cu Pb Zn	7.5 5.0.5 5.5 5.5	–	28	0.5-35 0.5-25 10-30 2-12	28,51 4.4,60 16-80 0.5-157.5	60-80	72	(Sheikha et al. 2008)
Green algae	Zn	5.0	25	20-350	3.0		1.0	7.62	Kleinübinga et al. 2010)
Sargassum filipendula	Cu Ni	4.5	25	250	5.0	175	6.0	NA	
Filamentous alga of *Pithophora spswas*	Cd Cr(VI) Pb		25	2-30 2-30 2-30		0.17-14 0.13-4.9 0.12-10.3		9 days	Ramsenthil and Meyyappan 2010)
Sargassum sp.	Cr(III)	4.0	30	92,162	2-5	6	68.9	NA	Yavuz and Denizli 2006)
Microalgae	Cu Zn		30	50-250	5.0	150		0.66 0.72 mmol/g	Saravanan et al. 2021)
Brown Algae (*sargassum* sp.)	Cr Zn Ni Cu Cd Hg Fe	3.0 3.0 5.0 4.0 5.0 4.0 3.0	30 30 30 30 30 30 30	NA	NA	150 150 150 150 150 100 150	NA	20.2 15.4 26.1 18.6 22.2 14.8 14.6	Saravanan et al. 2021)

species-transformation. It is also referred to passive metals' uptake process either by motile or by dead biomass cells between the cell wall and surface layers through surface complexation.

20.3.4 Bioleaching

Microbial leaching or so-termed bioleaching is a process used for the extraction of precious metals from aqueous solutions or low-grade ores as a promising in promoting green and environmentally friendly leaching methods. Compared to conventional leaching based on acidification, bioleaching shows the advantage of eliminating large quantities of acid usage. For example, acid thiobacillus ferroxidase or iron-oxidizing bacteria can be used as a bioleaching agent to reduce the pH of the metal-leaching solution. Effectiveness of bioleaching is determined by inherent effeciency of the microorganism along with bioreactor's design and operational conditions (Barkusaraey et al. 2021).

The most used bioleaching microorganisms are *At. thiooxidans* and *At. ferrooxidans*. These microorganisms can oxidize the metal sulphide (insoluble) into corresponding sulphate, influenced by various factors such as temperature, pH, sludge concentration, and sludge properties (Gu et al. 2018; Yang et al. 2020). A broad variety of metals have been recovered using the bioleaching technique, such as Au, Ag, In, Co, Pb, Cr, Cd, Ni, Zn, As, and REEs (Eltarahony et al. 2021; Martínez and Argumedo-Delira 2019; Rizki and Tanaka 2019; Giebner et al. 2019). Table 20.2 presented a brief review of the most conventional bioleached metals.

20.3.5 Bioprecipitation

Bioprecipitation is a cutting-edge method for metal recovery aided by biologically produced metallic compounds, such as oxalates, phosphates, and sulphates, in suitable bioreactor configurations. Bioprecipitation is generally followed by metal recovery for further industrial applications (Wong and Rene 2017). In this method, precipitation of the metal ions (soluble in the aquatic phase) is a function of the bacteria's intention and characteristics in the self-defence mechanism. Non-biological metal precipitation occurs mainly due to reduction and formation of sulphate and phosphates. In contrast, interaction of bacterial cell surface with metal ions leads to metal precipitation independent of cellular metabolism (Eltarahony et al. 2021).

Sulphate-reducing bacteria are heterotrophic and require anaerobes which oxidize organic compounds or hydrogen for energy metabolism using sulphate as a terminal electron acceptor. It reduces sulphate to sulphide which consequently complexes with metal ions existing in the microorganisms' cells to metal sulphides formation (Sreedevia et al. 2022). For example, in a system for bioprecipitation of metal phosphates, the enzyme phosphatase releases inorganic phosphate from cellular organic phosphate, leading to metals/ radionuclides precipitation on the cell. The

Table 20.2 Metal ions bioleaching from industrial wastewater sludge, under different conditions

Type	Metal solubilization	Remediation condition	Reduction in sludge solids	N, P, and K loss rate	Pathogen reduction	References
Bioleaching	Zn(91%) Ni(81%) Cu(79%) Pb(52%) Cr(42%)	Initial pH 7; Temp. 30 °C; S^0 5 g/L; Inoculation 5% (v/v)		N (6–10%) P (15%–30%)	Below detection limit (4×10^3 cfu/100 ml^{-1})	Villar (2003)
Simultaneous sewage sludge digestion and metal leaching	Cd(47.03%) Cr(18.00%) Cu(80.72%) Ni(56.75%) Pb(17.6%) Zn(92.98%)	Initial pH 7; Temp. 25 °C; S^0 3 g/L; Inoculation 20% (v/v)	VSS (50%)	K (8.8–14.8%)		Meknasi et al. 2000)
Thermophilic bioleaching	Mn (73–100%) Zn(51–60%) Ni(38–52%) Cu(17–43%) Cr(1–38%)	Initial pH 7; Temp. 55 °C; S^0 2 g/L; Initial solid conc. 15 g/L	SS (33–48%) VSS (47%–67%)			Chen and Cheng (2019)
Autoheated thermophilic aerobic sludge digestion and metal bioleaching	Cu(76%) Mn(78.2%) Ni(79.5%) Zn(84.2%)	Initial pH 7.2; Temp. 55–60 °C; S^0 5 g/L; Initial solid conc. 40 g/L Inoculation 10% (v/v)	TS (21%) VS (27%) SS (27%) VSS (39%)			Jain et al. (2010)

process is facilitated by certain microorganisms capable to produce phosphatase, which can then hydrolyze organophosphate sources, phosphate ions (PO_4^{-3}) that react with metal ions towards the formation of metal phosphates precipitates (Yu et al. 2021), such as $Pb_3(PO_4)_2$ from Pb(II)). Similarly, Zhan et al. (2021) have reported that immobilized cells of *Citrobacter* sp. have successfully precipitated Cu, Cd, Pb, and U from glycerol-2-phosphate enriched solutions, and phosphatase catalyzed glycerol-2-phosphate cleavage released hydrogen phosphates, which precipitated metals extracellularly as insoluble metal phosphates.

References

Abdullah N, Yusof N, Lau WJ, Jaafar J, Ismail AF (2019) Recent trends of heavy metal removal from water/wastewater by membrane technologies. J Ind Eng Chem 76:17–38

Abidli A, Huang Y, Rejeb ZB, Zaoui A, Park CB (2022) Sustainable and efficient technologies for removal and recovery of toxic and valuable metals from wastewater: recent progress, challenges, and future perspectives. Chemosphere 292:133102

Akansha J, Nidheesh PV, Gopinath A, Anupama KV, Suresh Kumar M (2020) Treatment of dairy industry wastewater by combined aerated electrocoagulation and phytoremediation process. Chemosphere 253:126652

Arora PK (2020) Bacilli-mediated degradation of xenobiotic compounds and heavy metals. Front Bioeng Biotechnol 8:570307

Azimi A, Azari A, Rezakazemi M, Ansarpour M (2017) Removal of heavy metals from industrial wastewaters: a review. Chem Bio Eng Rev 4(1):37–59

Balladares E, Jerez O, Parada F, Baltierra L, Hernández C, Araneda E, Parra V (2018) Neutralization and co-precipitation of heavy metals by lime addition to effluent from acid plant in a copper smelter. Miner Eng 122(15):122–129

Barkusaraey FH, Mafigholami R, Faezi Ghasemi M, Gholam K (2021) Optimization of zinc bioleaching from paint sludge using Acidithiobacillus thiooxidans based on response surface methodology. J Environ Sci Health 56(11):1243–1252

Bernard EI, Stanley IRO, Grace OI, Ebere PA, Abraham OA, Ibe KE (2018) Toxicity and bioremediation of heavy metals contaminated ecosystem from tannery wastewater: a review. J Toxicol 2018:2568038

Birungi ZS, Shen N, Evans MNC, Roestorff M (2020) Recovery of rare earths, precious metals and bioreduction of toxic metals from wastewater using algae. In: Bharagava R (ed) Emerging eco-friendly green Technologies for Wastewater Treatment, *Microorganisms for sustainability*, vol 18. Springer, Singapore, pp 267–297

Bong C, Malfatti CF, Azam F, Obayashi Y, Suzuki S (2010) The effect of zinc exposure on the bacteria abundance and proteolytic activity in seawater. In: Hamamura N, Suzuki S, Mendo S, Barroso CM, Iwata H, Tanabe S (eds) Interdisciplinary studies on environmental chemistry - biological responses to contaminants. Terrapub, pp 57–63

Carolin CF, Kumar PS, Saravanan A, Joshiba GJ, Naushad M (2017) Efficient techniques for the removal of toxic heavy metals from aquatic environment: a review. J Environ Chem Eng 5(3): 2782–2799

Carvajal-Fl'orez E, Santiago-Alonso CG (2019) Technologies applicable to the removal of heavy metals from landfill leachate. Environ Sci Pollut Res 26(16):15725–15753

Chang SH, Teng TT, Norli I (2010) Extraction of Cu(II) from aqueous solutions by vegetable oil-based organic solvents. J Hazard Mater 181:868–872

Chen SY, Cheng YK (2019) Effects of sulfur dosage and inoculum size on pilot-scale thermophilic bioleaching of heavy metals from sewage sludge. Chemosphere 234:346–355

Das S, Hirak RD, Jaya C (2016) Genetic basis and importance of metal resistant genes in bacteria for bioremediation of contaminated environments with toxic metal pollutants. Appl Microbiol Biotechnol 100(7):2967–2984

El Liethy MA, Mohammed AD, El-K A, Abdelaal M, Marwa Waseem AH, Elgarhy AH, Kamika I, El-S GA, Mwaheb MA (2022) Temporal phytoremediation potential for heavy metals and bacterial abundance in drainage water. Sci Rep 12:8223

Eltarahony M, Kamal A, Zaki S, Abd-El-Haleem D (2021) Heavy metals bioremediation and water softening using ureolytic strains Metschnikowia pulcherrima and Raoultella planticola. J Chem Technol Biotechnol 96(11):3152–3165

Giebner F, Kaden L, Wiche O, Tischler J, Schopf S, Schlömann M (2019) Bioleaching of cobalt from an arsenidic ore. Miner Eng 131:73–78

Gina KY, Tanga YZ, Aziz MA (2002) Derivation and application of a new model for heavy metal biosorption by algae. Water Res 36:1313–1323

Gu T, Rastegar SO, Mousavi SM, Li M, Zhou M (2018) Advances in bioleaching for recovery of metals and bioremediation of fuel ash and sewage sludge. Bioresour Technol 261:428–440

Hampu N, Werber JR, Chan WY, Feinberg EC, Hillmyer MA (2020) Next generation ultrafiltration membranes enabled by block polymers. ACS Nano 14(12):16446–16471

Hasanpour M, Hatami M (2020) Application of three-dimensional porous aerogels as adsorbent for removal of heavy metal ions from water/wastewater: a review study. Adv Colloid Interf Sci 284: 102247

Iravani S, Varma RS (2020) Bacteria in heavy metal remediation and nanoparticle biosynthesis. ACS Sustain Chem Eng 8(14):5395–5409

Iravani S, Varma RS (2022) Genetically engineered organisms: possibilities and challenges of heavy metal removal and nanoparticle synthesis. Clean Technol 4:502–511

Jain R, Ashish P, Sreekrishnan TR, Dastidar MG (2010) Autoheated thermophilic aerobic sludge digestion and metal bioleaching in a two-stage reactor system. J Environ Sci 22(2):230–236

Jin Y, Luan Y, Ning Y, Wang L (2018) Effects and mechanisms of microbial remediation of heavy metals in soil: a critical review. Appl Sci 8:1336–1353

Kleinübinga SJ, Vieiraa RS, Beppua MM, Guibalb E, Carlos da Silvaa MG (2010) Characterization and evaluation of copper and nickel biosorption on acidic algae sargassum Filipendula. Mater Res 13(4):541–550

Kumar L, Bharadvaja N (2020) Microbial remediation of heavy metals. In: Shah MP (ed) Microbial bioremediation & biodegradation. Springer Singapore, Singapore, pp 49–72

Lodeiro P, Cordero B, Barriada JL, Herrero R, Sastre de Vicente ME (2002) Biosorption of cadmium by biomass of brown marine macroalgae. Bioresour Technol 96:1796–1803

Martínez EDM, Argumedo-Delira R, S'anchez-Viveros G, Alarc'on A, Mendoza-L'opez MR (2019) Microbial bioleaching of ag, au and Cu from printed circuit boards of mobile phones. Curr Microbiol 76(5):536–544

Meknasi YF, Tyagi RD, Narasiah KS (2000) Simultaneous sewage sludge digestion and metal leaching: effect of aeration. Process Biochem 36(3):263–273

Mustafa HM, Hayder G (2021) Recent studies on applications of aquatic weed plants. In phytoremediation of wastewater: a review article. Ain Shams Eng J 12(1):355–365

Pavithra KG, Kumar PS, Jaikumar V, Vardhan KH, SundarRajan P (2020) Microalgae for biofuel production and removal of heavy metals: a review. Environ Chem 18(6):1905–1923

Pohl A (2020) Removal of heavy metal ions from water and wastewaters by sulfur containing precipitation agents. Water Air Soil Pollut 231(10):503

Pushkar BK, Sevak PI, Singh A (2015) Isolation and characterization of potential microbe for bio-remediating heavy metal from Mithi river. Ann Appl BioSci 2(2):20–27

Rajadurai V, Anguraj BL (2020) Ionic liquids to remove toxic metal pollution. Environ Chem Lett 19(2):1173–1203

Ramsenthil R, Meyyappan R (2010) Single and multi-component biosorption of copper and zinc ions using microalgal resin. Int J Environ Sci Develop 1(4):298–301

Rangabhashiyam S, Balasubramanian P (2019) Characteristics, performances, equilibrium, and kinetic modeling aspects of heavy metal removal using algae. Bioresour Technol 5:261–279

Rizki IN, Tanaka ON (2019) Thiourea bioleaching for gold recycling from ewaste. Waste Manage 84:158–165

Saravanan A, Senthil Kumar P, Jeevanantham S, Karishma S, Tajsabreen B, Yaashikaa PR, Reshma B (2021) Effective water/wastewater treatment methodologies for toxic pollutants removal: processes and applications towards sustainable development. Chemosphere 280:130595

Sharma P (2021) Efficiency of bacteria and bacterial assisted phytoremediation of heavy metals: an update. Bioresour Technol 328:124835

Sheikha D, Ashour I, Abu Al-Rub FA (2008) Biosorption of zinc on immobilized green algae: equilibrium and dynamics studies. J Eng Res 5(1):20–29

Sreedevia SD, Suresh K, Guangming J (2022) Bacterial bioremediation of heavy metals in wastewater: a review of processes and applications. J Water Process Eng 48:102884

Velkova Z, Kirova G, Stoytcheva M, Kostadinova S, Todorova K, Gochev V (2018) Immobilized microbial biosorbents for heavy metals removal. Eng Life Sci 18(12):871–881

Villar LD, Garcia O Jr (2003) Assessment of anaerobic sewage sludge quality for agricultural application after metal bioleaching. Environ Technol 24(12):1553–1559

Waldman RZ, Gao F, Phillip WA, Darling SB (2021) Maximizing selectivity: an analysis of isoporous membranes. J Memb Sci 633:119389

Wong SJ, Rene ER (2017) Bioprecipitation-a promising technique for heavy metal removal and recovery from contaminated wastewater streams. MOJ Civil Engineering 2(6):191–193

Xing D, Magdouli S, Zhang J, Koubaa A (2020) Microbial remediation for the removal of inorganic contaminants from treated wood: recent trends and challenges. Chemosphere 258:127429

Yang W, Wei S, Ji L, Xiaolei Z (2020) Bioleaching of heavy metals from wastewater sludge with the aim of land application. Chemosphere 249:126134

Yavuz H, Denizli A, Güngünes H, Safarikova M, Safarik I (2006) Biosorption of mercury on magnetically modified yeast cells. Sep Purif Technol 52:253–260

Yin K, Wang Q, Lv M, Chen L (2019) Microorganism remediation strategies towards heavy metals. Chem Eng J 360:1553–1563

Yu W, Hu J, Yu Y, Ma D, Gong W, Qiu H, Hu Z, Gao HW (2021) Facile preparation of sulfonated biochar for highly efficient removal of toxic pb(II) and Cd1(II) from wastewater. Sci Total Environ 750:141545

Zhan J, Yang X, Zhang X, Wang Y, Cai X, Chen H (2021) Bioprecipitation facilitates the green synthesis of sulfidated nanoscale zero-valent iron particles for highly selective dechlorination of trichloroethene. J Environ Chem Eng 9(5):106050

Zhu Y, Fan W, Zhou T, Li X (2019) Removal of chelated heavy metals from aqueous solution: a review of current methods and mechanisms. Sci Total Environ 678:253–266

The manufacturer's authorised representative in the EU is Springer Nature Customer Service Centre GmbH, Europaplatz 3, 69115 Heidelberg, Germany. If you have any concerns regarding our products, please contact ProductSafety@springernature.com

Printed and bound by CPI Group (UK) Ltd, Croydon, CR0 4YY
01/10/2026
02225900-0016